Régis Gras, Einoshin Suzuki, Fabrice Guillet and Filippo Spagnolo (Eds.)

Statistical Implicative Analysis

Studies in Computational Intelligence, Volume 127

Editor-in-chief
Prof. Janusz Kacprzyk
Systems Research Institute
Polish Academy of Sciences
ul. Newelska 6
01-447 Warsaw
Poland
E-mail: kacprzyk@ibspan.waw.pl

Further volumes of this series can be found on our homepage: springer.com

Vol. 107. Margarita Sordo, Sachin Vaidya and Lakhmi C. Jain (Eds.)
Advanced Computational Intelligence Paradigms in Healthcare - 3, 2008
ISBN 978-3-540-77661-1

Vol. 108. Vito Trianni
Evolutionary Swarm Robotics, 2008
ISBN 978-3-540-77611-6

Vol. 109. Panagiotis Chountas, Ilias Petrounias and Janusz Kacprzyk (Eds.)
Intelligent Techniques and Tools for Novel System Architectures, 2008
ISBN 978-3-540-77621-5

Vol. 110. Makoto Yokoo, Takayuki Ito, Minjie Zhang, Juhnyoung Lee and Tokuro Matsuo (Eds.)
Electronic Commerce, 2008
ISBN 978-3-540-77808-0

Vol. 111. David Elmakias (Ed.)
New Computational Methods in Power System Reliability, 2008
ISBN 978-3-540-77810-3

Vol. 112. Edgar N. Sanchez, Alma Y. Alanís and Alexander G. Loukianov
Discrete-Time High Order Neural Control: Trained with Kalman Filtering, 2008
ISBN 978-3-540-78288-9

Vol. 113. Gemma Bel-Enguix, M. Dolores Jiménez-López and Carlos Martín-Vide (Eds.)
New Developments in Formal Languages and Applications, 2008
ISBN 978-3-540-78290-2

Vol. 114. Christian Blum, Maria José Blesa Aguilera, Andrea Roli and Michael Sampels (Eds.)
Hybrid Metaheuristics, 2008
ISBN 978-3-540-78294-0

Vol. 115. John Fulcher and Lakhmi C. Jain (Eds.)
Computational Intelligence: A Compendium, 2008
ISBN 978-3-540-78292-6

Vol. 116. Ying Liu, Aixin Sun, Han Tong Loh, Wen Feng Lu and Ee-Peng Lim (Eds.)
Advances of Computational Intelligence in Industrial Systems, 2008
ISBN 978-3-540-78296-4

Vol. 117. Da Ruan, Frank Hardeman and Klaas van der Meer (Eds.)
Intelligent Decision and Policy Making Support Systems, 2008
ISBN 978-3-540-78306-0

Vol. 118. Tsau Young Lin, Ying Xie, Anita Wasilewska and Churn-Jung Liau (Eds.)
Data Mining: Foundations and Practice, 2008
ISBN 978-3-540-78487-6

Vol. 119. Slawomir Wiak, Andrzej Krawczyk and Ivo Dolezel (Eds.)
Intelligent Computer Techniques in Applied Electromagnetics, 2008
ISBN 978-3-540-78489-0

Vol. 120. George A. Tsihrintzis and Lakhmi C. Jain (Eds.)
Multimedia Interactive Services in Intelligent Environments, 2008
ISBN 978-3-540-78491-3

Vol. 121. Nadia Nedjah, Leandro dos Santos Coelho and Luiza de Macedo Mourelle (Eds.)
Quantum Inspired Intelligent Systems, 2008
ISBN 978-3-540-78531-6

Vol. 122. Tomasz G. Smolinski, Mariofanna G. Milanova and Aboul-Ella Hassanien (Eds.)
Applications of Computational Intelligence in Biology, 2008
ISBN 978-3-540-78533-0

Vol. 123. Shuichi Iwata, Yukio Ohsawa, Shusaku Tsumoto, Ning Zhong, Yong Shi and Lorenzo Magnani (Eds.)
Communications and Discoveries from Multidisciplinary Data, 2008
ISBN 978-3-540-78732-7

Vol. 124. Ricardo Zavala Yoe
Modelling and Control of Dynamical Systems: Numerical Implementation in a Behavioral Framework, 2008
ISBN 978-3-540-78734-1

Vol. 125. Larry Bull, Ester Bernadó-Mansilla and John Holmes (Eds.)
Learning Classifier Systems in Data Mining, 2008
ISBN 978-3-540-78978-9

Vol. 126. Oleg Okun and Giorgio Valentini (Eds.)
Supervised and Unsupervised Ensemble Methods and their Applications, 2008
ISBN 978-3-540-78980-2

Vol. 127. Régis Gras, Einoshin Suzuki, Fabrice Guillet and Filippo Spagnolo (Eds.)
Statistical Implicative Analysis, 2008
ISBN 978-3-540-78982-6

Régis Gras
Einoshin Suzuki
Fabrice Guillet
Filippo Spagnolo
(Eds.)

Statistical Implicative Analysis

Theory and Applications

With 147 Figures and 74 Tables

Springer

Régis Gras
LINA, FRE 2729 CNRS
14 avenue de la Chaise
35170 Bruz, France
regisgra@club-internet.fr

Fabrice Guillet
LINA, FRE 2729 CNRS
Polytech'Nantes
rue C. pauc, BP 50609
44306 Nantes, cedex 3, France
Fabrice.Guillet@univ-nantes.fr

Einoshin Suzuki
Department of Informatics
Kyushu University
744 Motooka, Nishi, Fukuoka
819-0395, Japan
suzuki@i.kyushu-u.ac.jp

Filippo Spagnolo
Dipartimento di Matematica
Univesrità di Palermo
Via Archirafi n.34
90123 Palermo, Italy
spagnolo@math.unipa.it

ISBN 978-3-540-78982-6 e-ISBN 978-3-540-78983-3

Studies in Computational Intelligence ISSN 1860-949X

Library of Congress Control Number: 2008924359

Cover design: Deblik, Berlin, Germany

Printed on acid-free paper

9 8 7 6 5 4 3 2 1

springer.com

Preface

Statistical implicative analysis is a data analysis method created by Régis Gras almost thirty years ago which has a significant impact on a variety of areas ranging from pedagogical and psychological research to data mining. This new concept has developed into a unifying methodology, and has generated a powerful convergence of thought between mathematicians, statisticians, psychologists, specialists in pedagogy and last, but not least, computer scientists specialized in data mining.

Statistical implicative analysis (SIA) provides a framework for evaluating the strength of implications; such implications are formed through common knowledge acquisition techniques in any learning process, human or artificial. Therefore, the epistemological interest of SIA is, in my opinion, of universal interest for researchers. In many applications implications appear as "rules" and, as it is often the case, rules have exceptions. SIA provides a powerful instrument for quantifying the quality of a rule taking into account the reality of these exceptions. Many applications, especially in data mining, extract large sets of rules that are impossible to assimilate by humans and used efficiently in decision processes. Therefore, it is important to develop measures of interestingness for these rules and the success of SIA-based techniques in this direction is indisputable.

This volume collects significant research contributions of several rather distinct disciplines that benefit from SIA. Contributions range from psychological and pedagogical research, bioinformatics, knowledge management, and data mining.

The first applications of SIA were in the realm of didactics and this field is richly represented here by several contributions that focus on such diverse problems as didactics of algebra and geometry, the teaching of functions representations and graphing, Bayesian inference, and student representations of physical activities.

Interesting data mining applications authored by leading researchers in the field range from applying SIA in the study of rules produced by decision trees, association rules generated by the analysis of transactional data, temporal rules, measures of interestingness for various types of rules, and hierarchical organization of rules. A novel method for analyzing DNA microarrays is formulated using SIA concepts. Furthermore, applications of SIA to the study of ontologies and textual taxonomies, as well as applications to fuzzy knowledge discovery are also included.

We have here a new volume that confirms the validity of a novel and powerful statistical methodology, though many convincing applications. The contributors have done a masterful job of exposition.

After reading this book, I have in mind a few applications of SIA in my own research. I am convinced that the readers will find this volume as stimulating as I did.

Boston,
September, 2007

Prof. Dan A. Simovici
Department of Computer Science
University of Massachusetts at Boston University

Review Committee

All published chapters have been reviewed by at least 2 referees.

- Saddo Ag Almouloud (University of Sao Paulo, Brazil)
- Carmen Batanero (University of Grenada)
- Hans Bock (Aachen University, Germany)
- Henri Briand (LINA, University of Nantes, France)
- Guy Brousseau (University of Bordeaux 3, France)
- Alex Freitas (University of Kent, UK)
- Athanasios Gagatsis (University of Chyprius)
- Robin Gras (University of Windsor, Canada)
- Howard Hamilton (University of Regina, Canada)
- Jiawei Han (University of Illinois, USA)
- David J. Hand (Imperial College, London, UK)
- André Hardy (University of Namur, Belgium)
- Robert Hilderman (University of Regina, Canada)
- Yves Kodratoff (LRI, University of Paris-Sud, France)
- Pascale Kuntz (LINA, University of Nantes, France)
- Ludovic Lebart (ENST, Paris, France)
- Amédéo Napoli (LORIA, University of Nancy, France)
- Maria-Gabriella Ottaviani (University of Roma, Italy)
- Balaji Padmanabhan (University of Pennsylvania, USA)
- Jean-Paul Rasson (University of Namur, Belgium)
- Jean-Claude Régnier (University of Lyon 2, France)
- Gilbert Ritschard (Geneve University, Switzerland)
- Lorenza Saitta (University of Piemont, Italy)
- Gilbert Saporta (CNAM, Paris, France)
- Dan Simovici (University of Massachusetts Boston, USA)
- Djamel Zighed (ERIC, University of Lyon 2, France)

Associated Reviewers

Nadja Maria Acioly-Régnier, Angela Alibrandi, Jérôme Azé, Maurice Bernadet, Julien Blanchard, Catherine-Marie Chiocca, Raphaël Couturier, Stéphane Daviet, Jérôme David, Carmen Diaz, Pablo Gregori, Alain Kuzniak, Eduardo Lacasta, Dominique Lahanier-Reuter, Stéphane Lallich, Letitzia La Tona, Patrick Leconte, Rémi Lehn, Philippe Lenca, Elsa Malisani, Rajesh Natajaran, Pilar Orús, Gérard Ramstein, Ansaf Salleb, Aldo Scimone, Benoît Vaillant, Ingrid Verscheure

Manuscript coordinator

Bruno Pinaud (LINA, University of Nantes, France)

Acknowledgments

The editors would like to thank the chapter authors for their insights and contributions to this book.

The editors would also like to acknowledge the members of the review committee and the associated referees for their involvement in the review process of the book, and without whose support the book would not have been satisfactorily completed.

A special thank goes to H. Briand for his encouragements.

Thanks also to J. Blanchard who has managed the cyberchair web site.

Finally, we thank Springer and the publishing team, and especially T. Ditzinger and J. Kacprzyk, for their confidence in our project.

Nantes,
December 2007

Régis Gras
Einoshin Suzuki
Fabrice Guillet
Filippo Spagnolo

Contents

List of Contributors

Nadja Maria Acioly-Régnier
EA 3729, University of Lyon, France
acioly.regnier@wanadoo.fr

Carmen Batanero
Facultad de Educación, University of Granada, Spain
batanero@ugr.es

Serge Baquedano
PerformanSe SAS, Carquefou (Nantes), France
serge.baquedano@performanse.fr

Maurice Bernadet
LINA CNRS 2729, Polytechnic School of Nantes University, France
maurice.bernadet@univ-nantes.fr

Julien Blanchard
LINA CNRS 2729, Polytechnic School of Nantes University, France
Julien.Blanchard@univ-nantes.fr

Henri Briand
LINA CNRS 2729, Polytechnic School of Nantes University, France
henri.briand@univ-nantes.fr

Catherine-Marie Chiocca
ENFA, Castanet-Tolosan (Toulouse), France
catherine-marie.chiocca@educagri.fr

Raphaël Couturier
LIFC, University of Franche-Comte, France
raphael.couturier@univ-fcomte.fr

Marie-Caroline Croset
LIG CNRS 5217, University of Grenoble I, France
Marie-Caroline.Croset@imag.fr

Carmen Díaz
Facultad de Psicología, University of Huelva, Spain
carmen.diaz@dpsi.uhu.es

Jérôme David
LINA CNRS 2729, Polytechnic School of Nantes University, France
jerome.david@univ-nantes.fr

Stéphane Daviet
LINA CNRS 2729, Polytechnic School of Nantes University, France
stephane.daviet@univ-nantes.fr

Iliada Elia
Department of Education, University of Cyprus, Cyprus
ilada@ucy.ac.cy

Inmaculada de la Fuente
Facultad de Psicología, University of Granada, Spain
edfuente@ugr.es

Athanasios Gagatsis
Department of Education, University of Cyprus, Cyprus
gagatsis@ucy.ac.cy

Régis Gras
LINA CNRS 2729, Polytechnic School of Nantes University, France
regisgra@club-internet.fr

Fabrice Guillet
LINA CNRS 2729, Polytechnic School of Nantes University, France
fabrice.guillet@univ-nantes.fr

Pablo Gregori
Universitat Jaume I, Castellón, Spain
gregori@mat.uji.es

Pascale Kuntz
LINA CNRS 2729, Polytechnic School of Nantes University, France
pascale.kuntz@univ-nantes.fr

Alain Kuzniak
Didirem team, University of Paris 7, France
alain.kuzniak@orleans-tours.iufm.fr

Eduardo Lacasta
Departamento de Matemáticas, Public Universidy of Navarra, Spain
elacasta@unavarra.es

Dominique Lahanier-Reuter
THEODILE TEAM (E.A. 1764), University of Lille 3, France
dominique.reuter@univ-lille3.fr

Stéphane Lallich
ERIC, University of Lyon 2, France
stephane.lallich@univ-lyon2.fr

Rémi Lehn
LINA CNRS 2729, Polytechnic School of Nantes University, France
remi.lehn@univ-nantes.fr

Philippe Lenca
TAMCIC CNRS 2872, GET/ENST Bretagne, France
philippe.lenca@enst-bretagne.fr

Elsa Malisani
GRIM, University of Palermo, Italy
e.malisani@alice.it

Alain Morineau
Modulad, Rocquencourt, France
alain.morineau@modulad.fr

Jean-François Nicaud
LIG CNRS 5217, University of Grenoble I, France
Jean-Francois.Nicaud@imag.fr

Pilar Orús
Universitat Jaume I, Castellón, Spain
orus@mat.uji.es

Vincent Philippé
PerformanSe SAS, Carquefou (Nantes), France
vincent.philippe@performanse.fr

Vincent Pisetta
ERIC, University of Lyon 2, France
v-pisett@univ-lyon2.fr

Ricco Rakotomalala
ERIC, University of Lyon 2, France
ricco.rakotomalala@univ-lyon2.fr

Gérard Ramstein
LINA CNRS 2729, Polytechnic School of Nantes University, France
gerard.ramstein@univ-nantes.fr

Jean-Claude Régnier
University of Lyon 2, France
jean-claude.regnier@univ-lyon2.fr

Gilbert Ritschard
Dept of Econometrics, University of Geneva, Geneva, Switzerland
gilbert.ritschard@unige.ch

Aldo Scimone
GRIM, University of Palermo, Italy
aldo.scimone@libero.it

Filippo Spagnolo
GRIM, University of Palermo, Italy
spagnolo@math.unipa.it

Einoshin Suzuki
Graduate School of Information Science and Electrical Engineering, Kyushu University, Japan
suzuki@i.kyushu-u.ac.jp

Jana Trgalova
LIG CNRS 5217, University of Grenoble I, France
Jana.Trgalova@imag.fr

Benoît Vaillant
South Britany University, VALORIA, France
benoit.vaillant@univ-ubs.fr

Ingrid Verscheure
LEMME, University of Toulouse 1, France
gri.verscheure@wanadoo.fr

Miguel R. Wilhelmi
Departamento de Matemáticas, Universidad Pública de Navarra, Spain
miguelr.wilhelmi@unavarra.es

Djamel A. Zighed
ERIC, University of Lyon 2, France
abdelkader.zighed@univ-lyon2.fr

Introduction

Régis Gras[1], Einoshin Suzuki[2], Fabrice Guillet[1], and Filippo Spagnolo[3]

[1] LINA, CNRS UMR 6241,
Polytechnic Graduate School of Nantes University, France
regisgra@club-internet.fr, fabrice.guillet@univ-nantes.fr

[2] Graduate School of Information Science and Electrical Engineering,
Kyushu University, Japan
suzuki@i.kyushu-u.ac.jp

[3] Department of Mathematics,
University of Palermo, Italy
spagnolo@math.unipa.it

In the framework of data mining, which has been recognized as one of the ten emergent technologies for computer sciences, association rule discovery aims at mining potentially useful implicative patterns from data. Initially stimulated by researches in didactics of mathematics, the Statistical Implicative Analysis (SIA) offers an original statistical approach based on Implication Intensity measure, which is dedicated to rule extraction and analysis. Implication Intensity, the first method in SIA in its initial form, evaluates the interestingness of a rule $x \rightarrow y$ by the rarity of its number of counter-examples $(x\overline{y})$, according to its probability distribution under an independence hypothesis of x and y. This interestingness measure has been involving a large number of research works and applications due to its theoretical ethics and practical merits. Through a graphical interface, CHIC (Cohesive Hierarchical Implicative Classification) software allows easy use of various techniques in SIA for a wide range of users from experts in data analysis to practitioners with little background in computer science.

This book includes two complementary topics: on the one hand, theoretical works related to SIA, or linking SIA with other data analysis methods; and on the other hand, applied works illustrating the use of SIA in applicative domains such as: psychology, social sciences, bioinformatics and didactics. It should be of interest to developers of data mining systems as well as researchers, students and practitioners devoted to data mining and statistical data analysis.

R. Gras et al.: *Introduction*, Studies in Computational Intelligence (SCI) **127**, 1–7 (2008)
www.springerlink.com

Structure of the book

The book is structured in four parts. The first one gathers three general chapters defining the methodology and the concepts for the Statistical Implicative Analysis approach. The second part contains six chapters dealing with the use of SIA as a decision aid tool for the analysis of concept learning in the framework of education, teaching and didactics. In the third part, seven chapters illustrate the use of SIA as a methodological answer in various application fields. Lastly, the fourth part includes six chapters describing the extension of SIA and its application to capture rule interestingness in data mining.

PART I: METHODOLOGY AND CONCEPTS FOR SIA

- **Chapter 1: An overview of the Statistical Implicative Analysis (SIA) development**, by Gras and Kuntz, gives a broad overview of the Statistical Implicative Analysis which is a data analysis method devoted to the extraction and the structuration of quasi-implications. It offers a synthesis which both presents the basic statistical framework of the approach and details recent developments.

- **Chapter 2: CHIC: cohesive hierarchical implicative classification**, by Couturier, is concerned with a data analysis tool based on SIA, named CHIC. Its aim is to discover relevant implications between variables. It proposes two different ways to organize these implications into systems: i) under the form of an oriented hierarchical tree and ii) as an implication graph. Also, it produces a (non oriented) similarity tree based on the likelihood of the links.

- **Chapter 3: Assessing the interestingness of temporal rules with Sequential Implication Intensity**, by Blanchard *et al.*, discusses the interestingness of sequential rules which is a key problem in sequence analysis since the frequent pattern mining algorithms can produce huge amounts of rules. It defines an original statistical measure named Sequential Implication Intensity (SII) that evaluates the statistical significance of the rules according to a probabilistic model. Numerical simulations show that SII has unique features.

PART II: APPLICATION TO CONCEPT LEARNING IN EDUCATION, TEACHING AND DIDACTICS

- **Chapter 4: Student's Algebraic Knowledge Modelling: Algebraic Context as Cause of Student's Actions**, by Croset *et al.*. This chapter describes the construction of a student model in the field of algebra in the framework of Aplusix learning environment. Patterns of student behaviours are discovered by using SIA. This makes building implicative connections between algebraic contexts and student's actions possible.

- **Chapter 5: The graphic illusion of high school students**, by Lacasta and Wilhelmi. This chapter deals with the analysis of the relationship between the mathematical background on linear and quadratic functions, and the representation of functions (graphics, figures and so on). Factorial analysis shows a contradiction in the usual assumption of the existence of a "graphical conceptualization" of functions different from a "non-graphical" one. Nevertheless, the authors of textbooks and teachers show a trend to use the graphical representation of functions. In the context of proportionality, the SIA reveals the existence of a graphical illusion shared by high school students.

- **Chapter 6: Implicative networks of student's representations of Physical Activities**, by Chiocca and Verscheure, proposes to discuss the results of a questionnaire-based study of young people's attitudes/representations to team games and volleyball. Processing by CHIC software shows several networks of variables which make profiling kinds of students possible. The study of contributions of two additional variables, sex and gender, enables to improve choices of representative networks for later interviews. Interestingly and somewhat unexpectedly, while sex is a strong predictor of attitudes and dispositions to team sports and volleyball, gender is not.

- **Chapter 7: The structure of conversions among representations of functions: A comparison between the hierarchical clustering of variables, implicative statistical analysis and confirmatory factor analysis**, by Elia and Gagatsis, focuses on a comparative study of three statistical methods, namely the hierarchical clustering of variables, the implicative method, and the Confirmatory Factor Analysis, applied to experimental data describing the understanding of functions. The investigation concentrates on the structure of students' abilities to carry out conversions of functions from one mode of representation to another.

- **Chapter 8: Implications Between Learning Outcomes in Elementary Bayesian Inference**, by Diaz *et al.*, deals with the use of SIA in order to study some hypotheses about the interrelationships in students' understanding of different concepts and procedures after 12 hours of teaching elementary Bayesian inference. A questionnaire made up of 20 multiple choice items was used to assess learning of 78 psychology students. The results obtained suggest four groups of interrelated concepts: conditional probability, logic of statistical inference, probability models and random variables.

- **Chapter 9: Personal Geometrical Working Space: a Didactic and Statistical Approach**, by Kuzniak, studies the answers that pre-service teachers gave in a geometry exercise . The purpose is to improve the understanding of what we call the geometrical working space. A first study, based on the notion of geometrical paradigms, leads to a classification of students' answers. Then, statistical tools are used in order to fine-tune the previous analysis and to explain student evolution during their training.

PART III: A METHODOLOGICAL ANSWER IN VARIOUS APPLICATION FRAMEWORKS

- **Chapter 10: Statistical Implicative Analysis of DNA microarrays**, by Ramstein, focuses on the application of SIA to microarray gene expression data. The specificity of these data requires an adaptation of the concept of intensity of implication. More specifically, it introduces the concept of rank interval and shows that the integration of the implicative method in this framework is more efficient than correlation techniques. The method is applied to the most challenging problems encountered in gene expression analysis, namely the "discovery of gene coregulation, gene selection and tumour classification".

- **Chapter 11: On the use of Implication Intensity for matching ontologies and textual taxonomies**, by David *et al.*, is concerned with the validation of ontology matching. It is based on an extensional and asymmetric matching approach designed to find implicative tendencies between two textual taxonomies or ontologies. More precisely, the chapter focuses on experimental evaluations of a set of interestingness measures, selected according to their properties and semantics. The experiments performed on a benchmark show that the implication intensity delivers the best results.

- **Chapter 12: Modelling by Statistics in Research of Mathematics Education**, by Malisani *et al.*, deals with the theoretical and experimental relationships between factorial and implicative analyses for modeling in the framework of didactics of mathematics. A first experiment introduces the supplementary variables for studying some reasoning schemes on the solution of Goldbach's conjecture. A second one studies the aspects of unknown variables and the functional relation in problem-solving in the contexts of algebra and analytical geometry.

- **Chapter 13: Didactics of Mathematics and Implicative Statistical Analysis**, by Lahanier-Reuter, evaluates the assumption that "The

Didactics of mathematics has constantly regarded SIA as a profitable and heuristic method of data analysis". It shows some explanations such that: implicative links may be interpreted as rules and regulations connecting actions, discourses, etc., or as a group's characteristics. Some examples show how SIA can be used and what special research results it can provide. The chapter concludes on some recommendations.

- **Chapter 14: Using the Statistical Implicative Analysis for Elaborating Behavioral Referentials**, by Daviet *et al.*, is concerned with "PerformanSe Echo" assessment tool that helps human resources managers in evaluating the behavioral profile of a person along 10 bipolar dimensions. This chapter is interested in building a set of psychological indicators based on a population of 613 experienced executives who are seeking a job. The goal is twofold: first to confirm the previous validation study, then to build a relevant behavioral referential on this population.

- **Chapter 15: Fictitious Pupils and Implicative Analysis: a Case Study**, by Orus and Gregori, details a case study, in the context of Didactics of Mathematics, in which they adopt the methodology of using fictitious data in SIA. Unlike supplementary variables, the fact of adding fictitious data to the sample does modify the results, so caution is needed. Nevertheless, fictitious students are a tool for better understanding the data structure.

- **Chapter 16: Identifying didactic and sociocultural obstacles to conceptualization through Statistical Implicative Analysis**, by Acioly-Reignier and Régnier, aims at understanding the relationship between culture and cognition. The authors focus on both the roles of written culture and the teaching and learning strategies involved. The data were gathered through short interviews and questionnaire based surveys. SIA enabled them to determine the implicative rules between the responses and thus the pre-ordering of the responses. The results showed that some specific symbolic representations constitute didactical and/or socio-cultural obstacles.

PART IV: EXTENSIONS TO RULE INTERESTINGNESS IN DATA MINING

- **Chapter 17: Pitfalls for Categorizations of Objective Interestingness Measures for Rule Discovery**, by Suzuki, points out four pitfalls for the categorizations of objective interestingness measures for rule discovery: data bias, rule bias, expert bias, and search bias. The main objective of this chapter is to issue an alert for the pitfalls which are harmful to one of the most important research topics in data mining. The author also lists desiderata in categorizing objective interestingness measures.

- **Chapter 18: Inducing and Evaluating Classication Trees with Statistical Implicative Criteria**, by Ritschard *et al.*, highlights the interest of SIA for classification trees. It shows how Gras' implication index may be defined for rules induced from a decision tree, and that this index looks like a standardized residual of contingency tables. The first use concerns the a posteriori individual evaluation of the classification rules. The second use relies on assigning the most appropriate conclusion to each leaf of the tree. The practical usefulness of this statistical implicative view on decision trees is demonstrated through a full scale real world application.

- **Chapter 19: On the behaviour of the generalisations of the intensity of implication: a data-driven comparative study**, by Vaillant *et al.*, proposes a generalisation of interestingness measure for association rules, taking into account a reference point, chosen by an expert, in order to apprehend the confidence of a rule. This generalisation introduces new connections between measures, leads to the enhancement of some of them, and new parameterised possibilities. The behaviour of the parameterised measures is illustrated and discussed on classical datasets. This study highlights the different properties of each of them and discusses the advantages of the proposal.

- **Chapter 20: The TVpercent principle for the counterexamples statistic**, by Rakotomalala and Morineau, puts into practice the principle of test value percent criterion for the counterexamples statistic, which is the basis of SIA approach. It shows how to compute the test value and what the connection with the measures used by SIA is. The behavior of these measures is evaluated on a large dataset comprising several hundreds of thousands of transactions.

- **Chapter 21: User-System Interaction for Redundancy-Free Knowledge Discovery in Data**, by Lehn *et al.*, deals with applying techniques initially designed for redundancy reduction in functional dependencies to association rule reduction. Although the two kinds of relations have different properties, this method allows very concise representations that are easily understood by the decider and can be further exploited for automatic reasoning. This method is compared to other approaches and tested on synthetic datasets. The information loss resulting from the reduction is also discussed.

- **Chapter 22: Fuzzy Knowledge Discovery Based on Statistical Implication Indexes**, by Bernadet, is concerned with the application of SIA to fuzzy knowledge discovery. It explains how to adapt the statistical indexes to fuzzy knowledge. Yet, these indexes do not evaluate the associated fuzzy rules which depend on the chosen fuzzy operators. The best fuzzy operators are selected by applying the generalized modus ponens on the items of several databases and by comparing its results to the effective conclusions. By studying methods to aggregate fuzzy rules, this chapter shows that in order to keep classical reduction schemes, fuzzy operators must be chosen differently. However, one of these possible operator sets is also one of the best for processing the generalized modus ponens.

Part I

Methodology and concepts for SIA

An overview of the Statistical Implicative Analysis (SIA) development

Régis Gras and Pascale Kuntz

Laboratoire d'Informatique de Nantes Atlantique
Equipe COnnaissances & Décision
Site Ecole Polytechnique de l'Université de Nantes
La Chantrerie — BP 50609 — 44306 Nantes cedex 3
regisgra@club-internet.fr, pascale.kuntz@univ-nantes.fr

Summary. This paper presents an overview of the Statistical Implicative Analysis which is a data analysis method devoted to the extraction and the structuration of quasi-implications. Originally developed by Gras [11] for applications in the didactics of mathematics, it has considerably evolved and has been applied to a wide range of data, in particular in data mining. This paper is a synthesis which both briefly presents the basic statistical framework of the approach and details recent developments.

Key words: quasi-implication, implication intensity, implicative graph, implicative hierarchy, typicality

1 Introduction

Two important components are involved in the operational human processes of knowledge acquisition: facts and rules between facts or between rules themselves. Through one's own culture and one's own personal experience, the learning process integrates a progressive elaboration of these knowledge forms. It can be faced with regressions, questions or changes which arise from decisive quashing, but the knowledge forms contribute to maintain a certain equilibrium. The rules formed inductively become quite stable when their success number -which depends on their explicative or inferential quality- reaches a certain level of confidence. At first, it is often difficult to replace an initial rule by another when few counter-examples appear. If they increase, the confidence in the rule can decrease and the rule can be reajusted or even rejected. However, when confirmations are numerous and counter-examples are rare, the rule is robust and can stay in our minds. For instance, let us consider the acceptable rule "All Ferraris are red". Even if one or two counter-examples happen this rule is maintained, and it will be even confirmed again by new examples.

R. Gras and P. Kuntz: *An overview of the Statistical Implicative Analysis (SIA) development*, Studies in Computational Intelligence (SCI) **127**, 11–40 (2008)
www.springerlink.com

Hence, contrary to what happens in mathematics where rules do not allow for any exception, the rules considered in human sciences are considered to be acceptable when the number of counter-examples remains "tolerable" in view of the number of situations where they are positive and efficient. In data analysis, the problem is to determine a consensus criterion which quantifies the confidence quality level of the rule according to the user's requirements. Our approach rests on three epistemological assumptions. The criterion is statistical. It is non linearly, robust to noise (i.e. not very influenced by the first counter-examples), and it becomes very low if the counter-examples often reappear. Our choice can be questioned, however it has been confirmed in various situations.

1.1 From didactics to data mining

"If a question is more complex than another, then each pupil who succeeds in the first one should also succeed in the second one". Every teacher knows that this situation shows exceptions whatever the complexity degree between questions. The evaluation and the structuration of such implicative relationships between didactic situations are the generic problems at the origin of the development of the Statistical Implicative Analysis (SIA) [11]. These problems, which have also drawn attention from psychologists interested in ability tests [5, 27], have known a significant renewed interest in the last decade in data mining.

Indeed, quasi-implications, also called association rules in this field, have become the major concept in data mining to represent implicative trends between itemset patterns. In data mining, the paradigmatic framework is the so-called basket analysis where a quasi-implication $T_i \to T_j$ means that if a transaction contains a set of items T_i then it is likely to contain a set of items T_j too. For simplicity's sake, let us now on call "rule" a quasi-implication. In data mining, rules are computed on large size databases. From the seminal work of Agrawal *et al.* [1], numerous algorithms have been proposed to mine such rules. Most of them attempt to extract a restricted set of relevant rules, easy to interpret for decision-making. Yet, comparative experiments have shown that results may vary with the choice of rule quality measures (e.g. [13, 25]). In the rich literature devoted to this problem, interestingness measures are often classified into two categories: the subjective (user-driven) ones and the objective (data-driven) ones. Subjective measures aim at taking into account unexpectedness and actionability relatively to prior knowledge, while objective measures give priority to statistical criteria. Among the latter, the most commonly used criterion for quantifiying the quality of a rule $a \to b$ is the combination of the support (the frequency $f(a \wedge b)$) which indicates whether the items a and b occur reasonably often in the database, with the confidence (the conditional frequency). However, it is well-known that the confidence presents a major default: it is insensitive to the dilatation of $f(a)$, $f(b)$ and the database size. Other functions measure a link or an absence of link between the items but, like χ^2, they do not clearly specify the direction of the relationship. Moreover, in addition to rule filtering, rule structuring is necessary to highlight relationships and makes rule interpretation both easier and more accurate.

The SIA provides a complete framework to evaluate the interestingness of the rules and to structure them in order to discover relationships at different granularity levels. The underlying objective is to highlight the emerging properties of the whole system which can not be deduced from a simple decomposition into sub-parts (e.g. [30]). All these properties, which emerge from complex interactions -probably non linear-, contribute to the interpretation of the global nature of the system.

1.2 Contents of the paper

Section 2 presents the statistical framework to measure the rule quality: we first remind the reader the definition of the implication intensity for binary variables and propose different properties. Section 3 presents the extensions of the basic definition for different types of variables (modal, frequential, interval), and an entropic version adapted to large datasets. The following sections are concerned with rule structuration. Section 4 defines the implicative graph. Section 5 generalizes the notion of rule to the notion of R-rule (rule of rule), and section 6 describes the combinatorial structure of an implicative hierarchy whose elements are R-rules. Aids for analyzing these complex structures are developed in section 7 (significative levels of the implicative hierarchy) and section 8 (supplementary individuals and variables). An illustration from a real data corpus coming from a survey on teacher's perception of training in mathematics is presented in section 9.

2 The implicative intensity for the binary case

2.1 The basic situation

Let us consider a population E of n objects or individuals described by a finite set V of binary variables (attributes, criteria, scores, ...). We are here interested in the following question: "To what extent the variable b is true when the variable a is true" ? In other words, "do the subjects have a tendency for having b when we know that they have a? " In real-life situations — e.g. in human sciences— deductive theorems of the logical form $a \Rightarrow b$ are often difficult to establish because of the exceptions. Consequently, it is necessary to "mine" the dataset to extract rules reliable enough to conjecture causal relationships which structure the population. At the descriptive level, they allow to detect a certain stability in the structuration. And, at the predictive level, they allow to make assumptions. However, the rule mining processes require rigorous approaches which prevent a too flimsy empiricism.

2.2 The statistical framework

Our approach, based on the non-parametric test reasoning, is close to the Likelihood Linkage Analysis (LLA) developed by I.C. Lerman [26]. The quality measurement of an implicative relationship $a \rightarrow b$ is based on the unlikelihood of the counter-example number in the dataset i.e. cases where b is false when a is true [11, 12, 14]. To quantify this unlikelihood, we compare the deviation between the contingency and a theoretical model associated with a random drawing. In exploratory data analysis, we consider the deviation value and not just the H_0 acceptation/reject. This measure quantifies the "surprisingness" of the expert faced with a number of counter-examples improbably small for an independence assumed between the variables and for the cardinalities of the considered data.

More precisely, let us denote by $A \subset E$ the subset of individuals for which a is true, by $\overline{A}$ its complementary set and by $n_a = card(A)$ (resp. $n_{\overline{a}}$) the cardinal of A (resp. $\overline{A}$). The logical rule $A \Rightarrow B$ is true when $A \subset B$. However, this strict inclusion is exceptionally observed in real-life situations; in practice, it is quite common to observe a few subjects where a is true and b is false without having the general trend to have b when a is true contested. Consequently, we consider in the following quasi-rules —called rules for simplicity's sake— of the form $a \rightarrow b$.

2.3 Definitions

To accept or reject $a \rightarrow b$ it is quite common to consider the number $n_{a \wedge \overline{b}} = card\left(A \cap \overline{B}\right)$ of counter-examples. However, to quantify the surprisingness of the rule, this number must be relativized according to n, n_a and n_b. Intuitively, it is all the more surprising to discover that a rule has a small number of counter-examples as the data set is large. The objective of the *implicative intensity* is precisely to express the unlikelihood of $n_{a \wedge \overline{b}}$ in E.

We compare the observed number of counter-examples $n_{a \wedge \overline{b}}$ with the number of expected counter-examples for an independence hypothesis. Like I.C. Lerman with the similarity in LLA [26], we randomly draw two subsets X and Y of, respectively, n_a and n_b elements.

Definition 1. *The rule $a \rightarrow b$ is said to be admissible for a given threshold α if the probability of having the observed number of counter-examples $card\left(A \cap \overline{B}\right)$ greater than the expected number $card\left(X \cap \overline{Y}\right)$ is smaller than α:*

$$\Pr\left(card\left(X \cap \overline{Y}\right) \leq card\left(A \cap \overline{B}\right)\right) \leq \alpha$$

The distribution of $card\left(X \cap \overline{Y}\right)$ depends on the drawing pattern. When X and Y are draw with throw-in the distribution is Binomial, otherwise it is Hypergeometric.

Remark 1. For a certain process of drawing, the random variable $card\left(X\cap\overline{Y}\right)$ follows a Poissonian distribution $P\left(\lambda\right)$ with $\lambda=\frac{n_a n_{\overline{b}}}{n}$.

Let us consider a process where the individuals dynamically arrive e.g. a flow of transactions which fill up a database. We stop the process when there are n_a individuals with a true and n_b individuals with b true. Let $card\left(X\cap\overline{Y}\right)$ be the random variable associated with the counter-example number during the process. We suppose that the process checks three hypotheses : *(i)* the waiting times for the events (a and $\overline{b}$) are independent random variables, *(ii)* the distribution of the number of events which happen in the interval $[t, t+T]$ only depends on T, *(iii)* two events can not simultaneously happen.

Consequently, the number of events which happen during a fixed period follows a Poissonian distribution $P\left(\lambda\right)$ where λ is the cadence of the event arrival.

The probability of the event $(a=1)$ (resp. $(b=0)$) is estimated by the frequency n_a/n (resp. $n_{\overline{b}}/n$). Then, the probability of the joint event ($a=1$ and $b=0$) is estimated by

$$\frac{n_a}{n}.\frac{n_{\overline{b}}}{n}$$

Hence, for a flow of n individual, the arrivals of the event ($a=1$ and $b=0$) follow a Poissonian distribution with parameter $\lambda=\frac{n_a n_{\overline{b}}}{n}$.

Consequently,

$$\Pr\left(card\left(X\cap\overline{Y}\right)=s\right)=e^{-\lambda}\frac{\lambda^s}{s!}$$

and the probability that the chance leads to a greater number of counter-examples than those observed is defined by

$$\Pr\left(card\left(X\cap\overline{Y}\right)\leq card\left(A\cap\overline{B}\right)\right)=\sum_{s=0}^{card\left(A\cap\overline{B}\right)}\frac{\lambda^s}{s!}e^{-\lambda}$$

In the following, we consider the Poissonian distribution. In the classical approximation conditions, the other distributions converge on the Poissonian type.

Let us consider, for $n_{\overline{b}}\neq 0$, the standardized random variable $Q\left(a,\overline{b}\right)$:

$$Q\left(a,\overline{b}\right)=\frac{card\left(X\cap\overline{Y}\right)-\frac{n_a n_{\overline{b}}}{n}}{\sqrt{\frac{n_a n_{\overline{b}}}{n}}}$$

We denote by $q\left(a,\overline{b}\right)$ the observed value of $Q\left(a,\overline{b}\right)$ in the experimental realization. It is defined by

$$q\left(a,\overline{b}\right)=\frac{n_{a\wedge\overline{b}}-\frac{n_a n_{\overline{b}}}{n}}{\sqrt{\frac{n_a n_{\overline{b}}}{n}}}$$

This value measures a deviation between the contingency and the expected value when a and b are independent.

When the approximation is justified (e.g. $\lambda > 4$) the random variable $Q\left(a,\bar{b}\right)$ is approximatively $N\left(0,1\right)$-distributed.

Definition 2. *The implication intensity $\varphi\left(a,b\right)$ of the rule $a \rightarrow b$ is defined by*

$$\varphi\left(a,b\right) = 1 - \Pr\left(Q\left(a,\bar{b}\right) \leq q\left(a,\bar{b}\right)\right) = \frac{1}{\sqrt{2\pi}} \int_{q\left(a,\bar{b}\right)}^{\infty} e^{-\frac{t^2}{2}} dt$$

if $n_b \neq n$, and $\varphi\left(a,b\right) = 0$ otherwise.

Definition 3. *The implication intensity $\varphi\left(a,b\right)$ is admissible for a given threshold α if $\varphi\left(a,b\right) \geq 1 - \alpha$.*

The implication intensity measures the surprisingness to observe a small number of counter-examples. It is an inductive and informative quality measure. Consequently, if the rule is trivial —i.e. when B is small or equal to E— this surprisingness is small.

Proposition 1. *[12] Let us suppose that n_a is fixed and $A \subset B$. If n_b tends towards n, then $\varphi\left(a,b\right)$ tends towards 0.*

We set $\varphi\left(a,b\right) = 0$ if $n_b = n$ by continuity (consequence of the property 1). If $A \subset B$ then $\varphi\left(a,b\right)$ can be smaller than 1 when the surprisingness is not sufficient.

2.4 Comparison with some classical measures

The observed quasi-implication $q\left(a,\bar{b}\right)$ is not symmetrical. It is different from the Pearson's correlation coefficient $\rho\left(a,b\right)$ which measures the linkage between a and b.

Proposition 2. *[12] Let $\rho\left(a,b\right)$ be the value of the Pearson's correlation between the binary variables a and b. If $q\left(a,\bar{b}\right) \neq 0$ then*

$$\frac{\rho\left(a,b\right)}{q\left(a,\bar{b}\right)} = -\sqrt{\frac{n}{n_b n_{\bar{a}}}}$$

The variation of the implication intensity is different from the Loevinger's coefficient [27] and from the confidence $conf\left(a,b\right) = n_{a \wedge b}/n_a$. It increases non linearly with the increasing of E, A and B, and it decreases with the trivial situations. Moreover, the maximal intensity is not necessarily reached for the inclusion $A \subset B$; indeed, the inductive quality may be quite low, whereas $conf\left(a,b\right) = 1$ [13].

2.5 Stability of the implication intensity

The implication intensity is noise-resistant in the neighbourhood of $n_{a\wedge\bar{b}} = 0$ [13].

In the following, we study the sensitivity of $\varphi(a,b)$ for small variations of the parameters n, n_a, n_b and $n_{a\wedge\bar{b}}$. Previous numerical experiments have confirmed the influence of the parameter variations on φ [10, 13]. Here, we study the differentiation of q.

Let us consider the parameters n, n_a, n_b and $n_{a\wedge\bar{b}}$ as real numbers which satisfy the following inequalities: $n_{a\wedge\bar{b}} \leq \inf(n_a, n_b)$ et $\sup(n_a, n_b) \leq n$. In this case, q can be considered as a continuous differentiable function:

$$dq = \frac{\partial q}{\partial n}dn + \frac{\partial q}{\partial n_a}dn_a + \frac{\partial q}{\partial n_b}dn_b + \frac{\partial q}{\partial n_{a\wedge\bar{b}}}dn_{a\wedge\bar{b}}$$

To study the variability of q depending on $n_{\bar{b}}$, we replace $n_{\bar{b}}$ by $n - n_{\bar{b}}$, and consequently the sign in the partial derivative.

Example 2. Let us suppose that n_a is constant, and that n_b and $n_{a\wedge\bar{b}}$ may vary. Then,

$$\frac{\partial q}{\partial n_b} = \frac{1}{2}n_{a\wedge\bar{b}}\left(\frac{n_a}{n}\right)^{1/2}(n-n_b)^{-3/2} + \frac{1}{2}\left(\frac{n_a}{n}\right)^{1/2}(n-n_b)^{-1/2}$$
$$\frac{\partial q}{\partial n_{a\wedge\bar{b}}} = \frac{1}{\sqrt{\frac{n_{a\wedge\bar{b}}}{n}}}$$
$$\frac{\partial q}{\partial n_a} = 0$$

Consequently, if Δn_b and $\Delta n_{a\wedge\bar{b}}$ are positive, then $\Delta q(a,\bar{b})$ is positive. This property can be interpreted as follows: for fixed n and n_a, the implication intensity decreases when the numbers of the b examples and the $a \Rightarrow b$ counter-examples increase. The implication intensity is maximal for the observed values n_b and $n_{a\wedge\bar{b}}$, and minimal for $n_b + \Delta n_b$ and $n_{a\wedge\bar{b}} + \Delta n_{a\wedge\bar{b}}$.

To examine the sensibility of the implication intensity, we consider φ as a function of q:

$$\varphi(q) = \frac{1}{\sqrt{2\pi}}\int_q^{\infty} e^{-t^2}dt$$

By differentiation, we obtain

$$\frac{d\varphi}{dq} = -\frac{1}{\sqrt{2\pi}}e^{-q^2} < 0$$

This result confirms that the implication intensity decreases with q, and it gives the speed of the variation.

With a similar approach, let us compare the stability of φ with the stability of the confidence $conf(a,b)$. The sensibility of $conf$ to the variation of the counter-examples is defined by

$$\frac{\partial c}{\partial n_{a\wedge\overline{b}}} = -\frac{1}{n_a}$$

Consequently, as expected, the confidence increases when $n_{a\wedge\overline{b}}$ decreases. However, the variation of the decreasing speed is constant whatever n and n_b. This situation highlights the limits of the parameter role in the sensitivity of the measure.

3 Extensions to different types of variables

3.1 Modal and frequential variables

The basic situation

The first applicative framework of this research was concerned with the representation that the teachers have of their own practice [3]. In a survey, a set of teachers has been asked to order a list of significative words depending on their importance. The resulting implications were: "if I select a word x with the importance i_x then I select the word y with the importance $i_y \geq i_x$".

In this case, we consider modal variables $a \in [0, 1]$ which describe satisfaction degrees. A similar case appears in situations where the variable frequency can be interpreted as a pre-order on the set of the values given by the subjects. Such situation appears in didactics when we study the success frequency for a test composed of questions coming from different domains.

Formalization

Let us denote by $a(i)$ and $\overline{b}(i)$ the values of $i \in E$ for the modal variables a and $\overline{b}$, and by s_a and s_b their empirical standard-deviations.

Definition 4. *[24]. For a pair (a, b) of modal variables, the implication intensity, called the propension index, is defined by*

$$q_p\left(a, \overline{b}\right) = \frac{\sum_{i\in E} a(i)\,\overline{b}(i) - \frac{n_a n_{\overline{b}}}{n}}{\sqrt{\frac{\left(n^2 s_a^2 + n_a^2\right)\left(n^2 s_{\overline{b}}^2 + n_{\overline{b}}^2\right)}{n^3}}}$$

Proposition 3. *When a and b are binary variables then $q_p\left(a, \overline{b}\right) = q\left(a, \overline{b}\right)$.*

In this case, it is easy to prove that $n^2 s_a^2 + n_a^2 = n n_a$, $n^2 s_{\overline{b}}^2 + n_{\overline{b}}^2 = n n_{\overline{b}}$ and $\sum_{i\in E} a(i)\,\overline{b}(i) = n_{a\wedge\overline{b}}$.

This extension remains valid for the frequential variables and the positive numerical variables when they are normalized: $\widetilde{a}(i) = a(i) / \max_{i\in E} a(i)$.

A similar measure has been recently introduced by Régnier and Gras [29] for ranking variables associated with a total order on a set of choices presented to a judge population. In this case, the considered implication is "if an object i is ordered by the judges at a place p_i then an object j is ordered by the same judges at a place $p_j > p_i$".

3.2 Variables on intervals

The basic situation

Let us consider a given set of biometric data. The considered implication is "if the weight of a male is between 65 and 70 kgs then his height is between 1.70 and 1.76 m".

More generally, let us consider two real variables a and b with a finite number of values in the respective intervals $A = [a_1, a_2]$ and $B = [b_1, b_2]$. Roughly speaking, the problem consists in finding implicative trends between representative unions of sub-intervals of A and B.

Main steps of the heuristic

The problem is decomposed in two steps. First, we partition the intervals A and B in a finite number of sub-intervals $\{A_1, A_2, \ldots, A_p\}$ and $\{B_1, B_2, \ldots, B_q\}$ which depend on the structure of the a and b distributions: there is an internal statistical homogeneity in each A_i (resp. B_i) and a high dispersion between each pair A_i, A_j (resp. B_i, B_j). Second, we compute the most significative implicative trends between unions of A_i and unions of B_j.

We have adapted the k-means algorithm for the interval partitioning problem [16]. The quality criteria of the partition are the intra-class and the inter-class inertia. Let $\pi(A)$ and $\pi(B)$ be two partitions obtained by this approach which respectively contain n_A and n_B elements. We denote by $\Omega(\pi(A))$ (resp. $\Omega(\pi(B))$) the set of the $2^{n_{A-1}}$ (resp. 2^{n_B-1}) partitions of A (resp. B) composed of the unions of elements of $\pi(A)$ (resp. $\pi(B)$) associated with adjacent intervals in A (resp. in B). For instance, if $\pi(A) = \{A_1, A_2, A_3, A_4\}$ s.t. $A = A_1 \cup A_2 \cup A_3 \cup A_4$ then

$$\Omega(\pi(A)) = \begin{array}{c}\{\{\{A_1\}, \{A_2\}, \{A_3\}, \{A_4\}\}, \{\{A_1A_2\}, \{A_3\}, \{A_4\}\}, \ldots, \\ \{A_1A_2A_3A_4\}\}\end{array}$$

For each pair $(P_i, P_j) \in \Omega(\pi(A)) \times \Omega(\pi(B))$ (resp. $(P_j, P_i) \in \Omega(\pi(B)) \times \Omega(\pi(A))$) we compute the geometric mean of the implication intensities between each sub-interval of P_i (resp. P_j) and each sub-interval of P_j (resp. P_j). Let us denote by $\max_{AB}$ and $\max_{BA}$ the respective maximal values between $\Omega(\pi(A))$ and $\Omega(\pi(B))$ and between $\Omega(\pi(B))$ and $\Omega(\pi(A))$. The implication is optimal if there is a partitioning of A which corresponds to $\max_{AB}$ and a partitioning of interval of B which corresponds to $\max_{BA}$.

3.3 Interval variables

The basic situation

Let us consider a score distribution of a class for different subjects. The considered implication is "the sub-interval $[2; 5.5]$ in mathematics generally implies the sub-interval $[4.25; 7.5]$ in physics". These two sub-intervals belong to an "optimal" partition -according to the inertia- of the definition domains $[1; 18]$ and $[3; 20]$ of the scores in mathematics and in physics.

Main steps of the heuristic

The previous approach can be adapted to the interval variables, which are symbolic data. Let us consider two variables a and b which are associated with a series of intervals due to the measure imprecision: I_i^a (resp. I_i^b) is the interval of a (resp. b) for the individual $i \in E$. Let I^a (resp. I^b) be the interval which contains all the a (resp. b) values. We can define on I^a and I^b a partition which optimizes a given criterium. The intersections between I_i^a and I^a and between I_i^b and I^b follow a distribution that takes into account the common parts. Consequently, the problem is similar to the computation of the rules between the on-interval variables (we refer to [16] for details).

3.4 The entropic version of the implication intensity

The limits of the basic implication intensity for large datasets

Pertinent results have been obtained with the implicative intensity φ for various applications where the data corpuses are relatively small ($n < 300$). However, in data mining, numerical experiments have highlighted two limits of φ for large datasets. First, it tends to be not discriminant enough when the size of E dramatically increases (e.g. [8]); its values are close to 1 even though the inclusion $A \subset B$ is far from being perfect. Second, like numerous measures proposed in the literature, it does not take into account the contrapositive $\overline{b} \Rightarrow \overline{a}$ which could allow to reinforce the affirmation of the good quality of the implicative relationship between a and b, and the capacity to estimate the causality between the variables.

The entropic implication intensity

To overcome these difficulties, we have proposed to modulate the value of the surprise quantified by the implication intensity by taking into account both the imbalance between $card\,(A \cap B)$ and $card\,(A \cap \overline{B})$ associated with $a \Rightarrow b$ and the imbalance between $card\,(A \cap \overline{B})$ and $card\,(\overline{A} \cap \overline{B})$ associated with the contrapositive $\overline{b} \Rightarrow \overline{a}$ [6, 7, 18]. We have introduced a new measure, called the entropic implication intensity, based on the Shannon's entropy to non-linearly quantify these differences.

More precisely, let us first consider a weighted version of the implication intensity $\phi\,(a, b) = (\varphi\,(a, b)\,.\tau\,(a, b))^{1/2}$ where $\tau\,(a, b)$ measures the imbalance between $n_{a \wedge b}$ and $n_{a \wedge \overline{b}}$ and the imbalance between $n_{a \wedge \overline{b}}$ and $n_{\overline{a} \wedge \overline{b}}$. Intuitively, the surprise measured by ϕ must be softened (resp. confirmed) when the number of counter-examples $n_{a\wedge\overline{b}}$ is high (resp. small) for the rule and its contrapositive considering the observed numbers n_a and $n_{\overline{b}}$.

A well-known index for taking the imbalances into account non-linearly is the Shannon's conditional entropy. The conditional entropy $H_{b/a}$ of cases (a and b) and (a and $\overline{b}$) given a is defined by

$$H_{b/a} = -\frac{n_{a\wedge b}}{n_a}\log_2\frac{n_{a\wedge b}}{n_a} - \frac{n_{a\wedge\overline{b}}}{n_a}\log_2\frac{n_{a\wedge\overline{b}}}{n_a}$$

and, similarly, the conditional entropy $H_{\overline{b}/\overline{a}}$ of cases ($\overline{a}$ and $\overline{b}$) and (a and $\overline{b}$) given $\overline{b}$ is defined by

$$H_{\overline{a}/\overline{b}} = -\frac{n_{a\wedge\overline{b}}}{n_{\overline{b}}}\log_2\frac{n_{a\wedge\overline{b}}}{n_{\overline{b}}} - \frac{n_{\overline{a}\wedge\overline{b}}}{n_{\overline{b}}}\log_2\frac{n_{\overline{a}\wedge\overline{b}}}{n_{\overline{b}}}$$

We can here consider that these entropies measure the average uncertainty on the random experiments in which we check whether b (resp. $\overline{a}$) is realized when a (resp. $\overline{b}$) is observed. The complements of 1 for these uncertainties $I_{b/a} = 1 - H_{b/a}$ and $I_{\overline{b}/\overline{a}} = 1 - H_{\overline{b}/\overline{a}}$ can be interpreted as the average information collected by the realization of these experiments; the higher this information is, the stronger the guarantee of the quality of the implication and its contrapositive will be.

Intuitively, the expected behavior of the measure ϕ is determined by three phases:

1. a slow reaction to the first counter-examples (robustness to noise).
2. an acceleration of the reject in the neighborhood of the balance.
3. an increasing rejection beyond the balance -which was not guaranteed by the basic implication intensity φ.

Hence, in order to have the expected significance, our model must satisfy the following constraints:

1. Integrating both the information relative to $a \to b$ and that relative to $\overline{b} \to \overline{a}$ respectively measured by $I_{b/a}$ and $I_{\overline{a}/\overline{b}}$. A product $I_{b/a}.I_{\overline{a}/\overline{b}}$ is well-adapted to simultaneously highlight the quality of these two values.
2. Raising the conditional entropies to the power for a fixed number $\alpha > 1$ in the information definitions to reinforce the contrast between the different phases described below: $\left(\left(1 - H^{\alpha}_{b/a}\right).\left(1 - H^{\alpha}_{\overline{a}/\overline{b}}\right)\right)^{1/\beta}$ with $\beta = 2\alpha$ to remain of the same dimension as φ.
3. The need to consider that the implications have lost their inclusive meaning when the number of counter-examples is greater than half of the observation of a and $\overline{b}$. Beyond these values we consider that the terms $\left(1 - H^{\alpha}_{b/a}\right)$ and $\left(1 - H^{\alpha}_{\overline{b}/\overline{a}}\right)$ are equal to 0.

Let $f_a = \frac{n_a}{n}$ (resp. $f_{\bar{b}} = \frac{n_{\bar{b}}}{n}$) be the frequency of a (resp. $\bar{b}$) on E and $f_{a \wedge \bar{b}}$ be the frequency of the counter-examples. The proposed adjustment of the previous informations $I_{b/a}$ and $I_{\bar{a}/\bar{b}}$ can be defined by-

$$\widehat{I^{\alpha}}_{b/a} = 1 - H^{\alpha}_{b/a} = 1 + \left(\left(1 - \frac{f_{a \wedge \bar{b}}}{f_a}\right) \log_2 \left(1 - \frac{f_{a \wedge \bar{b}}}{f_a}\right) + \frac{f_{a \wedge \bar{b}}}{f_a} \log_2 \left(\frac{f_{a \wedge \bar{b}}}{f_a}\right) \right)^{\alpha}$$

if $f_{a \wedge \bar{b}} \in \left[0, \frac{f_a}{2}\right[$; otherwise, $\widehat{I^{\alpha}}_{\bar{a}/\bar{b}} = 0$
and

$$\widehat{I^{\alpha}}_{\bar{a}/\bar{b}} = 1 - H^{\alpha}_{\bar{b}/\bar{a}} = 1 + \left(\left(1 - \frac{f_{a \wedge \bar{b}}}{f_{\bar{b}}}\right) \log_2 \left(1 - \frac{f_{a \wedge \bar{b}}}{f_{\bar{b}}}\right) + \frac{f_{a \wedge \bar{b}}}{f_{\bar{b}}} \log_2 \left(\frac{f_{a \wedge \bar{b}}}{f_{\bar{b}}}\right) \right)^{\alpha}$$

if $f_{a \wedge \bar{b}} \in \left[0, \frac{f_a}{2}\right[$; otherwise, $\widehat{I^{\alpha}}_{\bar{a}/\bar{b}} = 0$.

Definition 5. *The imbalances are measured by $\tau(a,b)$ —called the inclusion index— defined by*

$$\tau(a,b) = \left(\widehat{I^{\alpha}}_{b/a} . \widehat{I^{\alpha}}_{\bar{a}/\bar{b}}\right)^{1/2\alpha}$$

and, the weighted version of the implication intensity —called the entropic implication intensity— is given by

$$\phi(a,b) = (\varphi(a,b) . \tau(a,b))^{1/2}$$

Example 3.

	b	$\bar{b}$	$\sum$
a	200	400	600
$\bar{a}$	600	2800	3400
$\sum$	800	3200	4000

	b	$\bar{b}$	$\sum$
a	400	200	600
$\bar{a}$	1000	2400	3400
$\sum$	1400	2600	4000

	b	$\bar{b}$	$\sum$
a	40	20	60
$\bar{a}$	60	280	340
$\sum$	100	300	400

Table 1. Distribution examples (a, b and c).

For the table 1.a, the implicative intensity is $\varphi(a,b) = 0.9999$. The entropic functions are $H_{a/\bar{b}} = 0 = H_{\bar{b}/\bar{a}}$. The weighting coefficient is $\tau(a,b) = 0$. And,

$\phi(a,b) = 0$ whereas the confidence $c(a,b)$ is equal to 0.333. The entropic functions moderate the implication intensity when the inclusion is bad.

For the table 1.b, the implication intensity $\varphi(a,b) = 1$. The entropic functions are $H_{a/\overline{b}} = 0.918$ and $H_{\overline{b}/\overline{a}} = 0.391$. The weighting coefficient is $\tau(a,b) = 0.6035$. And, $\phi(a,b) = 0.777$ and the confidence $c(a,b) = 0.666$.

The table 1.c proves that the correspondance between φ and ϕ is not monotonous. The intensity implication is lower for the table 1.c than for the table 1.b. And, it is the contrary for $\phi(a,b)$. Let us remark that the confidence is the same for the two tables.

4 The implicative graph

When computing the implication intensities between all pairs of variables of V, we obtain a square matrix M of numbers in $[0,1]$. The global structure of the relationships between the variables does not clearly appear. To highlight this structure we have associated a directed graph with M, called the *implicative graph* [2, 11].

Let Φ_α be the relationship defined on $V \times V$ by φ for a given threshold $\alpha \in [0,1]$: $a\Phi_\alpha b$ if and only if $\varphi(a,b) \geq \alpha$. The threshold α, which controls the implicative quality of the rules, is chosen by the user. The relationship Φ_α is reflexive, not symmetric and not transitive. However, it is interesting to consider the partial order relationships between the subsets of V. Consequently, we extend the relationship Φ_α: if $a\Phi_\alpha b$ and $b\Phi_\alpha c$ then we accept the transitive closure $a\Phi_\alpha c$ if and only if $\varphi(a,c) \geq 0.5$ i.e. when the implicative trend of a on c is better than the neutrality.

Hence, for a given threshold α, the graph $G_{M,\alpha}$ is defined as follows: its vertices are the variables of V, and there is an arc between a pair of variables (a,b) if and only if $a\Phi_\alpha b$.

Different options of the software CHIC allows to easily interact with the drawing of the graph.

5 From rules to R-rules

5.1 The basic situation

In the didactics of mathematics, one of the fundamental question is to identify the source of the problems -both didactical and epistemological- the pupil is faced with during his learning processes. These obstacles are based on the conceptions the pupil is building up. These conceptions are structured by simple or complex rules which together allow to elaborate the basis of a cognitive model.

This structuration is neither a simple union of rules nor a classical hierarchical structure where the variable classes are fit into partitions which are partially ordered by the relation "thiner than" which reflects the similarity between the class elements. To complete the information provided by the previous models, we have proposed the concept of R-rules (rules of rules) which are an extension of the quasi-implications: their premisses and their conclusions can be rules themselves [15, 17, 20, 23]. To guide the intuition a parallel can be drawn from the proof theory with the logical implication: $(X \Rightarrow Y) \Rightarrow (Z \Rightarrow W)$ describes an implication between the two theorems $X \Rightarrow Y$ and $Z \Rightarrow W$ previously established.

5.2 The R-rules and their interpretation

In the following, we consider binary variables. The R-rules are an extension of the classical binary rules $a \rightarrow b$ to rules of rules $R' \rightarrow R''$, which may be complex themselves. For instance $a \rightarrow (b \rightarrow c)$ is a R-rule between a variable a and a rule $(b \rightarrow c)$, and $(a \rightarrow b) \rightarrow (c \rightarrow d)$ is a R-rule between two rules $(a \rightarrow b)$ and $(c \rightarrow d)$. To indicate the complexity of the implication composition, we associate a complexity degree with each R-rule.

Definition 6. *The R-rules of degree 0 are variables of V. The R-rules of degree 1 are the simple quasi-implications of the form $a \rightarrow b$. A R-rule of degree i, $1 < i \leq p$, is a rule $R' \rightarrow R''$ between two R-rules R' and R'' whose respective degrees j and k satisfy $j + k = i - 1$.*

For instance, $a \rightarrow b$ is a R-rule of degree 1, $a \rightarrow (b \rightarrow c)$ a R-rule of degree 2 and $(a \rightarrow b) \rightarrow (c \rightarrow d)$ a R-rule of degree 3. When there is no ambiguity we denote by R a R-rule of degree greater or equal than 1.

The R-rules allow to express different levels of abstraction: (1) situation or object descriptions (conjunction of R-rules of degree 0), (2) implications between variables (R-rules of degree 1), and (3) implications between implications (some R-rules of degree greater than 1). Consequently, their interpretation may vary according to three typical cases:

1. when $R \rightarrow a$ then a may be interpreted as a quasi-consequence of R;
2. the R-rule $a \rightarrow R$ means that a R-rule R may be partially deduced from the observation of a. Moreover, although we here consider quasi-implications only, the intuition can be supported by Heyting algebra where an implication $a \Rightarrow (b \Rightarrow c)$ is equivalent to $(a\ AND\ b) \Rightarrow c$;
3. the R-rule $R' \rightarrow R''$ means that the property R'' is the quasi-corollary of a previous property R'

5.3 A measure of cohesion of the R-rules

The objective is to discover R-rules with a good implicative quality —called *cohesion* in the following— i.e. R-rules $R' \rightarrow R''$ with a strong implicative relationship between the components of R' and those of R''. For instance, it seems natural to form a R-rule $(a \rightarrow b) \rightarrow (c \rightarrow d)$ if the implicative relationships $a \rightarrow c$, $a \rightarrow d$, $b \rightarrow c$ and $b \rightarrow d$ are significant enough. Intuitively, this means that they must contrast with the disorder of a random experience.

The entropy is well-suited to measure this disorder. Let us first consider a R-rule $a \rightarrow b$ of degree 1. And, let Y be the random indicator variable of the event $\left(Q\left(a,\overline{b}\right) \geq q\left(a,\overline{b}\right)\right)$. The distribution of Y is defined by $\Pr(Y=1) = \varphi(a,b)$ and $\Pr(Y=0) = 1-\varphi(a,b)$. The entropy of this experience is $-p\log_2 p - (1-p)\log_2(1-p)$ where $p = \varphi(a,b)$.

The extreme values are 0 if $\varphi(a,b) = 0$ (by setting $0\log_2 0 = 0$) and 1 if $\varphi(a,b) = 0.5$. This last value is reached when $n_{a\wedge\overline{b}} = n_a n_{\overline{b}}/n$ i.e. when $n_{a\wedge\overline{b}}$ is equal to the expected mean. In this case, when $\varphi(a,b) < 0.5$, the meaning of the implication is lost and it seems natural to set te cohesion equal to 0.

Definition 7. *The cohesion $c(a,b)$ of a R-rule $a \rightarrow b$ of degree 1 is defined by*

$$c(a,b) = \left(1 - \left(-p\log_2 p - (1-p)\log_2(1-p)\right)^2\right)^{1/2}$$

if $p = \varphi(a,b) > 0.5$ and $c(a,b) = 0$ otherwise.

We square the entropy to reinforce the contrast between values in $[0,1]$ and the square root to the complement to 1 allows to measure the cohesion on a same scale as the entropy.

The generalization of this definition to R-rules of higher degree is guided by the following requirement: the cohesion of $R' \rightarrow R''$ must take into account both the cohesion of R' and R'' as well as the implicative relationships between the attributes of R' and those of R''. Let $\prec_R$ be the left right reading order on the variables which composed a R-rule. For instance, for $(a \rightarrow b) \rightarrow (c \rightarrow d)$ the order on $\{a,b,c,d\}$ is defined by $a \prec_R b \prec_R c \prec_R d$.

Then, a simple way to satisfy the previous requirements is to take the mean of the cohesions of R' and R'' and of the cohesions if each ordered pairs composed of one attribute of R' and one attribute of R'' in accordance with the permutation orders. Here we favour the geometric mean as it is equal to 0 as soon as the cohesion of one ordered pair is equal to 0 (i.e. when an implication is low or without surprise) and it is close to 1 when the cohesions of all ordered pairs are high.

Definition 8. *Let R be a R-rule of the form $R' \rightarrow R''$ where R' and R'' are respectively associated with the orders $a'_1 \prec_{R'} a'_2 \prec_{R'} \ldots a'_k$ and $a''_1 \prec_{R''} a''_2 \prec_{R''} \ldots a''_h$. The cohesion of R is defined by*

$$c(R) = \left(\prod_{i=1,k-1;j=2,k} c\left(a'_i, a'_j\right) \cdot \prod_{i=1,h-1;j=2,h} c\left(a_i, a''_j\right) \cdot \prod_{i=1,k;j=1,h} c\left(a'_i, a''_j\right) \right)^{2/r(r-1)}$$

where $r = k + h$.

6 The implicative hierarchy

6.1 The basic situation

Generally speaking, R-rules contribute to increasing the analysis richness. We do not solely extract facts or isolated behaviors, but more general conducts, revealing more global, less singular phenomena i.e. in didactics profound psychological representations. The different complexity degrees of the R-rules can be associated with a hierarchical structure which reflects the genesis of the "operating knowledge" developed by Piaget [28]. We go from one level to another by a process of reflecting abstraction: from object representation to representation of operations on the objects, then to representation of operations on the operations. This process involves a dynamical hierarchical point of view in contrast with the static point of view associated with a taxonomy. Hence, the individual description of the R-rules by aggregating simple rules is not sufficient. It is necessary to develop a global structure which reflects the emerging properties of the whole. Consequently, we have developed the concept of "implicative hierarchy" to structure the significant R-rules.

Let us introduce this notion by an example. A graphical representation of an implicative hierarchy on the variable set $V = \{a, b, c, d, e\}$ is given on figure 1. The elements of the implicative hierarchy $\overrightarrow{H}_V$ are R-rules:

$$\overrightarrow{H_V} = \{a, b, c, d, e, b \rightarrow c, e \rightarrow d, a \rightarrow (e \rightarrow d)\}$$

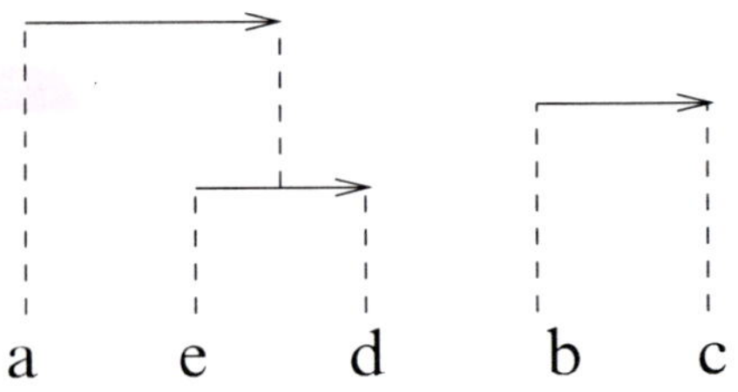

Fig. 1. Graphical representation of the implicative hierarchy $\overrightarrow{H}_V = \{a, b, c, d, e, b \rightarrow c, e \rightarrow d, a \rightarrow (e \rightarrow d)\}$

Note that contrary to hierarchies in classical hierarchical classification (HC) the tree associated with the implicative hierarchy is not necessarily connected. Intuitively, this means that it contains only significant R-rules according to the cohesion measure.

6.2 Definitions

The R-rules which composed an implicative hierarchy can be associated with k-permutations —called *classes* by analogy with the HC— that satisfy special interlocking conditions. For instance, in the example given below, the R-rule $a \rightarrow (e \rightarrow d)$ is associated with the permutation aed. And, this is the only possible association as the R-rule $(a \rightarrow e)$ associated with the permutation ae is not in $\overrightarrow{H}_V$. The class set H_V associated with $\overrightarrow{H}_V$ is

$$H_V = \{a, b, c, d, e, bc, ed, aed\}$$

The R-rules are deduced by a recursive decomposition of the non elementary classes of H_V. The class aed is the unique amalgamation of $a \in H_V$ and $ed \in H_V$. Since the class ed is associated with $e \rightarrow d$, the class aed is associated with $a \rightarrow (e \rightarrow d)$.

More formally, let Ω_V be the set of all k-permutations on the variable set V, $k = 1, p$. The elements C of Ω_V are strings with distinct characters. Let $\prec$ be the left-right reading order on the variables of a permutation of Ω_V as we defined it previously. In order to compare and combine the elements of Ω_V to form an implicative hierarchy, we define three operators on Ω_V, whose appelations are inspired by the set theory:

- *Intersection*. The intersection $C'\widehat{\cap}C''$ of two strings of Ω_V is the largest sub-string of contigous variables common to C' and C''. In case of equality we keep the first sub-string of C' according to $\prec$. If $C' = acdb$ and $C'' = cdab$ then $C'\widehat{\cap}C'' = cd$, and if $C' = abcd$ and $C'' = cdab$ then $C'\widehat{\cap}C'' = ab$.
- *Union*. The union $C'\widehat{\cup}C''$ of two distinct strings C' and C'' s.t. $C'\widehat{\cap}C'' = \emptyset$ is the concatenation of C' and C'' with C' first according to $\prec$. If $C' = aceb$ and $C'' = fgh$ than $C'\widehat{\cup}C'' = acebfgh$.
- *Difference*. For three strings C, C' and C'' of Ω_V s.t. $C = C'\widehat{\cup}C''$, the difference $C\widehat{-}C'$ between C and C' is C'' and the difference $C\widehat{-}C''$ between C and C'' is C'. If $C = abc$, $C' = ab$ and $C'' = c$ then $C\widehat{-}C' = c$ and $C\widehat{-}C'' = ab$.

Definition 9. *An implicative hierarchy H_V is a subset of permutations of Ω_V satisfying the three following requirements:*

1. *H_V contains the variables of V, called elementary classes*
2. *for each pair C', $C'' \in H_V$, $C'\widehat{\cap}C'' = \{\emptyset, C', C''\}$*
3. *for each non elementary class $C \in H_V$, there is a single pair C', $C'' \in H_V$ s.t. $C = C'\widehat{\cup}C''$*

From the condition 2, a hierarchy is a partially ordered set with the inclusion relation $\widehat{\subset}$ defined on Ω_V by: $C'\widehat{\subset}C''$ if and only if $C'\widehat{\cap}C'' = C'$. The condition 3 is required to recover all the classes of the hierarchy.

The isolated interpretation of a class of the hierarchy is tricky since it is a k-permutation which does not state the implication composition. For instance, if we analyse the class $aed \in H_V$ all alone, we do not know the exact meaning of aed: it could be either $a \rightarrow (e \rightarrow d)$ or $(a \rightarrow e) \rightarrow d$. However, the whole H_V class set allows to dispel ambiguity: $a \to (e \to d)$ is chosen as ed is a class of H_V.

Proposition 4. *[15] Each non elementary class C of an implicative hierarchy H_V can be associated with a unique R-rule.*

The R-rule set $\overrightarrow{H}_V$ associated with H_V can be graphically represented by a valuated binary directed tree:

- each of the elementary classes are located at a terminal node;
- each of the internal node is represented by an arrow which describles the R-rule subtended by the associated class;
- the height $h(C) \in R_+$ of each node C satisfies the following condition: for each node $C' \in H_V$ s.t. $C' \widehat{\subset} C$ then $h(C) > h(C')$.

6.3 Construction of an implicative hierarchy

The significant R-rules which form an implicative hierarchy are calculated by a incremental algorithm similar to the basic process of the classical HC. The amalgamation criterium is here the maximization of the cohesion.

At each level h_i of H_V, a new R-rule is built. It results from the amalgamation of two R-rules built at a previous level h_j, $0 < j < i$. More precisely,

- the initial level h_0 of H_V is composed of the variable set V;
- at h_1, two variables of V with the maximal cohesion are "grouped" together to form a R-rule of degree 1;
- at h_2, the R-rule is composed either of two variables not yet aggregated, called separate variables, or of the R-rule of degree 1 built at h_1 and a separate variable. The selected R-rule is the one with the maximal cohesion;
- at h_3, the R-rule may be of three types: a R-rule of degree 1 composed of two separate variables, a R-rule of degree 2 composed of a R-rule of degree 1 built at h_1 or h_2 and a separate variable, or a R-rule of degree 3 composed of the two R-rules of degree 1 built at h_1 and h_2.
- and so on. The process stops as soon each cohesion of the new potential R-rules is null.

For instance, for the implicative hierarchy of the figure 1, the process stops at h_3 if the cohesion is null for the R-rules $(a \to (d \to e)) \to (b \to c)$ and $(b \to c) \to (a \to (e \to d))$.

We refer to [15] for an algorithmic description of this algorithm and the analysis of its complexity.

The directed hierarchy H_V can be associated with a valuation which satisfies the ultrametric inequality.

Proposition 5. *[15] For any class C of H_V, let us define the height h of C by $h(C) = 1 - c(C)$ if C is non elementary and $h(C) = 0$ otherwise, where $c(C)$ is the cohesion of the R-rule associated with C. Let u be a dissimilarity on $V \times V$ defined by*

- $u(a,b) = 1$ *if a and b are not amalgameted in H_V,*
- $u(a,b) = h(C_{ab})$ *otherwise*

where C_{ab} is the smallest class of H_V which contains both a and b. The dissimilarity u is symmetric, positive and satisfies the ultrametric inequality:

$$u(a,b) \leq Max\{u(a,c), u(b,c)\}$$

for any $a, b, c \in V$.

From the Benzécri-Johnson theorem [4,22] this property a posteriory justifies our choice of the word "hierarchy".

7 The significative levels of the implicative hierarchy

7.1 The basic situation

Due to the multiplicity of the levels in the implicative hierarchy, it is necessary to highlight those which are the more relevant for the structuration process. In psycho-didactical or sociological applications, these levels seem to correspond to consistent and stable conceptions. Hence, they contribute to a finest interpretation of the set of the computed R-rules.

We have investigated two different approaches for this problem. The first one is based on a rank analysis used in HC by Lerman [26]: it compares the quality of the partitions obtained at each level of the hierarchy. The second one is more local [19]: it focusses on the quality of the R-rules built at each level. In the following, we present the first approach which is the only one to be implemented in the CHIC software [9].

7.2 A criterium to determine the significative levels

Let us note that the cohesion coefficient defined in the section 5.3 can be associated with a pre-ordering $\preceq_c$ on $P = V \times V - \{(a,a), (b,b), \ldots\}$:

$$(a,b) \preceq_c (c,d) \Leftrightarrow c(a,b) \leq c(c,d)$$

The idea consists in determining the levels of H_V which "better express" this pre-ordering. At each level h_k, two sets of variable pairs can be distinguished: the set A_k of the amalgameted variable pairs at h_k, and the set S_k of the separate variable pairs (not yet amalgameted to form a R-rule of degree ≥ 1). By construction, $A_k \cup S_k = P$.

Let $G_{\preceq_c}$ be the graph of $\preceq_c$. The set $G_{\preceq_c} \cap (A_k \times S_k)$ is composed of pairs of pairs which respect $\preceq_c$ at the level k. For instance, let us consider the variable set $V = \{a, b, e, f\}$ such that $c(a, b) < c(e, f)$. Let us suppose that at the level h_k the variables e, f and k are separate whereas the variables a and b are amalgameted in a class. Then, the pair $((e, f), (a, b)) \in G_{\preceq_c} \cap (A_k \times S_k)$.

The objective is now to measure the adequation between $G_{\preceq_c}$ and $A_k \times S_k$. Let us denote by Θ the set of all the pre-orderings on $P = V \times V - \{(a, a), (b, b), \ldots\}$ with the same cardinality as $\preceq_c$. We consider the random preordering G^*on Θ -with a uniform distribution-. From the theorem of Wald and Wolfowitz [31] we can deduce that the theoretical mean of $G^* \cap (A_k \times S_k)$ is $\mu = 1/2 \text{ card}(A_k \times S_k)$ and its standard deviation is

$$\sigma = \frac{1}{12}(\text{card}(A_k \times S_k)(\text{card}G^* + 1))$$

The adequation between $G_{\preceq_c}$ and $A_k \times S_k$ at the level h_k is measured by

$$s(\preceq_c, k) = \frac{\text{card}(G_{\preceq_c} \cap (A_k \times S_k)) - \mu}{\sigma}$$

Definition 10. *A level h_k of the implicative hierarchy H_V is significative if it is a local maximum of $s(\preceq_c k)$: $s(\preceq_c, k-1) < s(\preceq_c, k) < s(\preceq_c, k+1)$*

If $G_{\preceq_c} \cap (A_k \times S_k) = A_k \times S_k$ then the partition $A_k \times S_k$ on $V \times V$ associated with the structuration at h_k is in total accordance with the preordering induced by the cohesion.

8 Typicality and contributions

8.1 The basic situation

Like in factorial analysis, we introduce the notion of "additional variable": it does not contribute to the computation of the relationships involved in the implicative hierarchy, but it brings an additional information for its interpretation(e.g. age, sex, social-professional category).

Our objective is to identify individuals, or individual groups, and additional variables which contribute to class forming at each level of the implicative hierarchy.

8.2 A representation space

Let C be the class built at the level h_k of the hierarchy H_V. This class results from the amalgamation of two classes $C' \in H_V$ and $C'' \in H_V$ not amalgameted at the previous level h_{k-1}.

The variable pair (a, b) is a *generic pair* at h_k if $\varphi(a,b) \geq \varphi(i,j)$ for any $i \in C'$ and $j \in C''$. The *generic intensity* at h_k is denoted by $\varphi_k = \varphi(a,b)$. This pair characterizes the most noticeable implicative effect for a given class.

Moreover, the classes C' and C'' are themselves the results of an amalgamation at a lower level. Hence, at each level h_g, $g \leq k$, of H_V, we can determine a generic pair: the resulting vector $(\varphi_1, \varphi_2, \ldots, \varphi_k) \in [0,1]^k$ is called the *implicative vector* of the class C built at h_k.

A similar representation can be used for evaluating the impact of an individual on the formation of a path on the implicative graph $G_{M,\alpha}$. Let us consider a path P of length k on $G_{M,\alpha}$ with a transitive closure (i.e. each arc is associated with a rule with an implication intensity greater than 0.5). Then, P contains $k(k-1)/2$ transitive arcs. A pair (a,b) of P is generic if $\varphi(a,b) \geq \varphi(i,j)$ for any $i, j \in P$.

The vectors $(\varphi_1, \varphi_2, \ldots, \varphi_k) \in [0,1]^k$ form a representation space where the individuals can be projected. In the following, we precise the properties of this space for an implicative hierarchy. They could be similarly defined for an implicative graph.

8.3 Implicative power of an individual on a class

In this subsection, we define a dissimilarity on $E \times H_V$ to measure the "proximity" between an individual $i \in E$ and a class $C \in H_V$.

We first check if the individual i is in accordance with the implication of the generic pair (a,b) of C at the level h_k. Let us denote by $a(i)$ (resp. $b(i)$) the binary variable which characterizes the presence/absence of a (resp. b) for i. The contribution of i to the pair (a,b) is defined by

- $\varphi_{i,k} = 1$ if $a(i) = 1$ or 0 and $b(i) = 1$
- $\varphi_{i,k} = 0$ if $a(i) = 1$ and $b(i) = 0$
- $\varphi_{i,k} = p \in]0,1[$ if $a(i) = b(i) = 0$

In practice, p is set to the neutral value 0.5.

Any individual i is associated with a k-dimensional vector$(\varphi_{i,1}, \varphi_{i,2}, \ldots, \varphi_{i,k})$ which characterizes its contribution to the k generic pairs of the class C built at h_k. An individual whose components are equal to the implicative vector $(\varphi_1, \varphi_2, \ldots, \varphi_k)$ is called the *optimal typical individual.*

We measure the typicality of i in C by the χ^2 distance between the distributions $(1 - \varphi_g)$ and $(1 - \varphi_{i,g})$, for $g = \overline{1,k}$. In contrast with the usual Euclidean distance, it allows to compare $\varphi_g - \varphi_{i,g}$ to φ_g and to normalize the distance effect for large φ_g.

Definition 11. *The implicative distance $d_2(i, C)$ between an individual $i \in E$ and a class $C \in H_V$ built at the level h_k is defined by*

$$d_1(i,C) = \left(\frac{1}{k} \sum_{g=1}^{k} \frac{(\varphi_g - \varphi_{i,g})^2}{1-\varphi_g} \right)^{1/2}$$

If it exists g s.t. $\varphi_g = 0$ we set $(\varphi_g - \varphi_{i,g}) / (1 - \varphi_g) = 0$. In this case, the generic implication is maximal and thus it exists an excellent implicative relationship for all the individuals $i \in E$ $(\varphi_{i,g} = 1)$.

Remark 2. Let us consider a class $C \in H_V$ at the level h_k. We can define a metric space structure on E with

$$d_C(i,j) = \left(\frac{1}{k} \sum_{g=1}^{k} \frac{(\varphi_{i,g} - \varphi_{j,g})^2}{1-\varphi_g} \right)^{1/2}$$

for any $(i,j) \in E^2$.

The distance $d_C(i,j)$ measures the behavior difference between i and j considering C. It defines a discrete topological C-structure on E. Let us consider the vectors $(\varphi_{i,1}, \varphi_{i,2}, \ldots, \varphi_{i,k})$ and the norm $\left\| \overrightarrow{i} - \overrightarrow{j} \right\| = d_C(i,j)$. This topology is equivalent to the previous one (similarly to the duality in correspondence analysis). The elements of the diagonal matrix of the symmetrical operator associated with the quadratic form which defines d_C are $(k(1-\varphi_i))^{-1}$ for $i = 1, k$. Let us remark that the semantic of the vector sum is not precised in the SIA. Nevertheless, it could be interesting to characterize the individuals which belong to a ball of a given diameter with a given center (e.g. the optimal individual).

8.4 Individual and group typicalities

Definition 12. *The typicality $\gamma(i,C)$ of an individual $i \in E$ for a class $C \in H_V$ is defined by the ratio between the distance $d_1(i,C)$ and the maximal value of the distance on the individual set:*

$$\gamma(i,C) = 1 - \frac{d_1(i,C)}{Max_{j \in E} d_1(j,C)}$$

The maximal distance $d_1(j,C)$ is reached by the individuals with null or very low $\varphi_{i,k}$. They are contrasting with the generic rules. And, the typicality of i is large when i is different from these individuals.

A straightforward extension of the previous definition allows to define the typicality $\gamma(G,C)$ of a individual group $G \subset E$:

$$\gamma(G,C) = \frac{1}{card(G)} \sum_{i \in G} \gamma(i,C)$$

In practice, an operational tool is required to evaluate the statistical significancy of a group typicality. The basic idea consists in partitioning E in two opposite groups E_1 and E_2 with regards to their typicalities $\gamma(E_1, C)$ and $\gamma(E_2, C)$ in C. This dispersion can be measured by the inter-class inertia. The barycenter $\overline{\gamma}$ of the typicalities $\gamma(E_1, C)$ and $\gamma(E_2, C)$ is defined by

$$\overline{\gamma} = \frac{1}{n}\left(card\left(E_1\right)\gamma\left(E_1, C\right) + card\left(E_2\right)\gamma\left(E_2, C\right)\right)$$

By construction, $\overline{\gamma}$ is also the barycenter of all the individual typicalities in E. Consequently, the inter-class inertia is

$$V_E = \frac{card\left(E_1\right)}{n}\left(\gamma\left(E_1, C\right) - \overline{\gamma}\right)^2 + \frac{card\left(E_2\right)}{n}\left(\gamma\left(E_2, C\right) - \overline{\gamma}\right)^2$$

Definition 13. *An individual group $G_C^* \subset E$ is optimal for a class $C \in H_V$ if its typicality is greater than the typicality of its complementary set in E, and if it constitutes with this later a bi-partitioning which maximizes V_E. This partition is said to be significant.*

It is interesting to detect the group or the additional variable associated with the greatest typicality for the optimal group. We measure the surprisingness of the proportion of concerned individuals. Let $\{E_i\}_i$ be a given partition of E. It can be defined by an additional variable. For each class E_i, we consider the random variable X_i which is a random subset of E of cardinality $card\left(E_i\right)$, and the random variable Z_i defined by $Z_i = card\left(E_i \cap G_C^*\right)$. The variable Z_i follows a Binomial distribution with parameters $card\left(E_i\right)$ and $card\left(G_C^*\right)/n$ [21].

Definition 14. *The most typical group of the class C is the subset $E_i \subset E$ which minimizes the probability p_i on the set $\{p_i = \Pr\left(Z_i > card\left(X_i \cap G_C^*\right)\right)\}_i$.*

The probability $p_i$is an error of the first kind: the risk of making a mistake when considering that the group is not typical.

8.5 Contribution

The contribution is different from the typicality: it measures the individual and additional variable responsabilities for the existence of a rule or a R-rule between the variables of V.

Let us consider two variables $a \in V$ and $b \in V$ linked by a rule $a \rightarrow b$ at the first level h_1 of the implicative hierarchy H_v. The contribution of an individual i to (a, b) is defined by $\varphi_{i,1}$. This notion can be extended to the formation of a class C at the level h_k.

Definition 15. *The distance $d_2(i, C)$ between an individual $i \in E$ and a class C at the level h_k of the implicative hierarchy H_V is defined by*

$$d_2(i, C) = \frac{1}{k} \sum_{g=1}^{k} (1 - \varphi_{i,g})^2$$

The contribution $\theta(i, C)$ of $i \in E$ to $C \in H_V$ is defined by $\theta(i, C) = 1 - d_2(i, C)$.

The maximal value of $\theta(i, C)$ is equal to 1; it is reached for an individual i whose components $\varphi_{i,g}$ are all equal to 1.

The concepts defined in the previous sections can be easily adapted to the distance d_2. In practice, the contribution is often easier to interpret than the typicality.

9 Illustration

We illustrate the applicative interest of the different concepts presented below on a data set stemming from a survey of the French Public Education Mathematical Teacher Society on the level in mathematics of pupils in the final year of secondary education and the perception of this subject [8]. In parallel with evaluation tests for students, a set of 311 teachers have been asked on the objectives of the training in mathematics (table 2 presents some items used in the following) and their opinions about commonly shared ideas on this subject (table 3). For each proposition, the teacher could answer "I agree with this idea" (positive opinion), "I disagree" (negative opinion) or "I partially agree".

The figure 3 presents a part of the directed hierarchy obtained on the set composed of the objectives and the different modalities for the opinions (51 items). The interpretation of the whole set of rules is far beyond the scope of this paper. Nevertheless, we have selected some of them, easy to interpret for a non specialist in education theory, to show the use of a directed hierarchy on a real-life corpus. As for the complementarity of this structure with a more classical approach based on the relationship representation by a graph, it is highlighted in figure 2. The vertex set V of this graph contains the same items as those selected for figure, and there is an arc between two vertices a_i and a_j of V if and only if $\varphi(a_i, a_j) \geq 0.5$ and for any $a_k \in V$, $\varphi(a_i, a_k) < 0.5$ and $\varphi(a_j, a_k) < 0.5$ (e.g. [2]). The choice of the threshold comes from the fact that beyond 0.5 the implicative tendency (e.g. $a_i \to a_j$) is better than neutrality. It is important to note that, due to the non transitivity of the relationship on A induced by φ, the existence of two arcs of the form (a_i, a_j) and (a_j, a_k) does not entail the existence of the arc (a_i, a_k). For instance, in figure, we can not deduce a relationship between the items E and $OP7$.

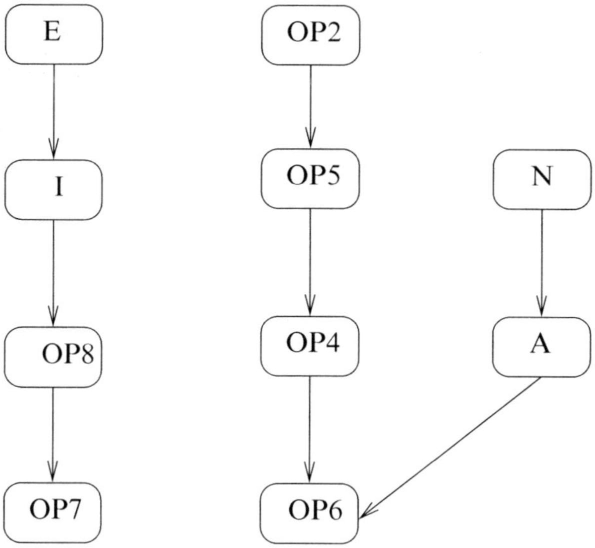

Fig. 2. A part of the implicative graph on the items of the survey on the training in mathematics

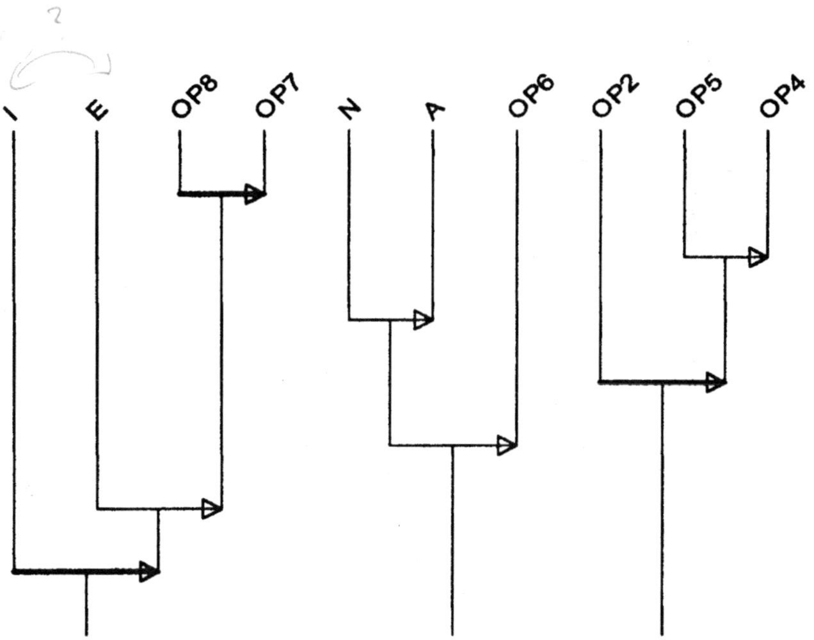

Fig. 3. A part of the directed hierarchy on the items of the survey on the training in mathematics

On the other hand, beside the binary rules, most of the R-rules of the total directed hierarchy involve three or four items. The interpretation of rules with more attributes are generally more difficult to interpret. Nevertheless, they provide more information than the set of the implied binary rules.

The R-rule $(N \to A) \to OP6$ has the following meaning: if know-how acquisition must be accompanied by knowledge acquisition, then the teacher ask for well-defined programs. In this case, focussing on knowledge requires a predefined charter from the institution. The R-rule allows to give a more synthetical interpretation than the binary rules: these are concerned with the behaviour, as seen within the behavioural framework, whereas the R-rule here describes a conduct of a higher order which determines the behaviour. Teachers who consider that the objective C (Preparation to civic and social life) is not relevant are mostly responsible for this R-rule. They have a very restrictive representation of the teaching of maths, focussed on the subject, and their teaching conforms to national standard without any questioning.

The R-rule $(OP2 \to (OP5 \to OP4))$ can be interpreted as follows: if I wish to keep up the complete problem for the A-level exam and if the importance given to the demonstration in maths is subordinated to a fixed scale of grading, then I conform to the national syllabus instructions. This rule corresponds to a class of teachers subjected to the institution and conservative in their educational choices. They consider that, in France, the land of Descartes, the demonstration is the foundation of the mathematical activity and that the complete problem at the exam is the evaluation criterion. For them, the syllabuses and the grading scales defined by the institution are essential to teaching and assessment. We find again a very classical teaching conception based on an explicit and unconditional support to the institution.

Contrary to the previous ones, the R-rule $(I \to (E \to (OP8 \to OP7)))$ can be interpreted as a sign of an openminded didactic conception. Indeed, it means that if a teacher lays the emphasis on the critical mind development and the imagination and creativity, then he considers that a personal training of the pupils in the search of examples and counter-examples is sufficient for discovering divisibility features by themselves. This R-rule reveals a relationship between the non-dogmatic behaviours of the teacher and the wish to place the pupil in a situation of personal research.

A	Knowledge acquisition
B	Preparation to professional life
C	Preparation to civic and social life
D	Preparation to examinations
E	Development of imagination and creativity
I	Development of critical mind
N	Know-how acquisition

Table 2. Some items from the list of the objectives of training in mathematics

OP1	It's true that maths are an element of selection
OP2	For the A-level exam, I prefer a complete problem with different parts rather than independent questions
OP3	In my grading system, I give more importance to the reasoning than to the result
OP3	When I correct, I prefer a very detailed grading system
OP5	The demonstration is the only rigourous way to do maths
OP6	I prefer well-defined programs precising what I must do and not do
OP7	In the last form of secondary education, a pupil should be able to recognize whether a number written in the base 10 is divisible by 4
OP8	In the last form of secondary education, a pupil should be able to give an example or counter-example of the following statement: if two applications f and g are strictly increasing on a given interval, then the product $f \times g$ is also increasing.
OPX	Individual estimation of a size (e.g. width, length)

Table 3. Some items from the list of the commonly shared ideas in the teaching of maths

We study now the additional information brought by the supplementary variable which defines the main option of the cursus: Scientific (S), Economic and Social (ES), Arts (A) and Technology (T). The observed distribution of the variable is: $S = 155$, $ES = 68$, $A = 22$ and $T = 66$.

Let us consider the class $C = (E \rightarrow (OP8 \rightarrow OP7)) \rightarrow OPX$. This rule corresponds to a class of teachers which give importance to imagination and personnal research. The most typical modality for this variable is S (scientific). Indeed, 116 teachers of the option S on 155 are in the optimal group $G^*_C$of cardinality 201. Let X be a random subset of same cardinality as S (155) and Z be random variable defined by the intersection of X and the optimal group G^*. Then, Z follows a binomial distribution of parameters 155 and $201/311 = 0.656$. The probability for Z to be greater than 116 is the risk 0.00393. The analysis of the series of the risks associated with the different options S, A and T shows that the most typical modality of the class C is S. The pair (S, C) is said to be mutually specific. Similarly, the most typical modality of the rule $B \rightarrow K$ is T; consequently, the pair $(T, (B, K))$ is mutually specific. It confirms that the teachers in technical cursus consider that the mathematics should be useful for the professional life (B) and consequently for the other disciplines.

The computation of the contribution of S to C shows that 111 teachers on 311 participate to the optimal group. The number of teachers of S has decreased (from 116 to 67), and its proportion in the optimal group is significantly lower than for the typicality computation. The teachers of S are the most typical, i.e. in accordance with the general behavior of the population. However, their contribution to the four involved variables is lower than the contribution of the teachers of the other cursus. The risk is equal to 0.0251: it is more than 6 times greater than the typicality. This remark illustrates the nuances brought by the two concepts: typicality and contribution.

10 Conclusion

In this paper we have proposed an overview of the Statistical Implicative Analysis. Beyond the results, we have related the genesis of the considered problems which arise from questions of experts in different fields. The theoretical basis is quite simple, but the numerous questions on the original assumptions, which do not appear here, have lead to modifications and sometimes to deeper revisions. Fortunately, the proposed answers go beyond the original framework, and SIA is now a data analysis method, based on a non symmetrical approach, which has been shown to be relevant for various applications.

In the next future, we are planning to consider new problems: *(i)* the extension of SIA to vectorial data, *(ii)* and to fuzzy variables, *(iii)* the integration of missing data, *(iv)* the redundant rule reduction. We are also interested in the complementarity of SIA with other approaches, in particular with decision trees (see Ritschard's paper in this book). And, we will obviously carry on exploring real-life data sets and confronting our theoretical tools to experimental analysis to make them evolve.

References

1. R. Agrawal, T. Imielinsky, and A. Swami. Mining association rules between sets of items in large databases. In *Proc. of the ACM SIGMOD'93*, pages 679–696. AAAI Press, 1993.
2. M. Bailleul. Des réseaux implicatifs pour mettre en évidence des relations. *Mathématiques, Informatique et Sciences Humaines*, 154:31–46, 2001.
3. M. Bailleul and R. Gras. L'implication statistique entre variables modales. *Mathématiques et Sciences Humaines*, 128:41–57, 1995.
4. J.P Benzécri. *L'analyse des données (vol. 1): Taxonomie*. Dunod, Paris, 1973.
5. J.M. Bernard and S. Poitrenaud. L'analyse implicative bayesienne d'un questionnaire binaire : quasi-implications et treillis de galois simplifié. *Mathématiques, Informatique et Sciences Humaines*, 147:25–46, 1999.
6. J. Blanchard, P. Kuntz, F. Guillet, and R. Gras. Mesure de la qualité des règles d'association par l'intensité entropique. *Revue des Nouvelles Technologies de l'Information-Numéro spécial Mesures de qualité pour la fouille de données*, RNTI-E-1:33–44, 2004.
7. J. Blanchard, P. Kuntz, G. Guillet, and R. Gras. Implication intensity: From the basic definition to the entropic version - chapter 28. In *Statistical Data Mining and Knowledge Discovery*, pages 475–493. CRC Press - Chapman et al., 2003.
8. A. Bodin and R. Gras. Analyse du préquestionnaire enseignants. *Bulletin de l'Association des Professeurs de Mathématiques de l'Enseignement PUblic*, 425:772–786, 1999.

9. R. Couturier and R. Gras. C.h.i.c. : Traitement de données avec l'analyse implicative. *Revue des Nouvelles Technologies de l'Information*, RNTI-II:679–684, 2005.
10. L. Fleury. *Extraction de connaissances dans une base de données pour la gestion de ressources humaines.* PhD thesis, Université de Nantes, 1996.
11. R. Gras. *Contribution à l'étude expérimentale et à l'analyse de certaines acquisitions cognitives et de certains objectifs en didactique des mathématiques.* PhD thesis, Université de Rennes 1, 1979.
12. R. Gras, S. Ag Almouloud, M. Bailleul, A. Larher, M. Polo, H. Ratsimba-Rajohn, and A. Totohasina. *L'implication statistique - Nouvelle méthode exploratoire de données.* La Pensee Sauvage editions, France, 1996.
13. R. Gras, R. Couturier, J. Blanchard, H. Briand, P. Kuntz, and P. Peter. Quelques critères pour une mesure de qualité de règles d'association. *Revue des Nouvelles Technologies de l'Information*, RNTI-E-1:197–202, 2004.
14. R. Gras, E. Diday, P. Kuntz, and R. Couturier. Variables sur intervalles et variables-intervalles en analyse statistique implicative. In *Proc. of Société Francophone de Classification*, pages 166–173. Université des Antilles-Guyane, 2001.
15. R. Gras and P. Kuntz. Discovering r-rules with a directed hierarchy. *Soft computing*, 1:46–58, 2005.
16. R. Gras, P. Kuntz, and H. Briand. Les fondements de l'analyse statistique implicative et quelques prolongements pour la fouille de données. *Mathématiques et Sciences Humaines*, 154:9–29, 2001.
17. R. Gras, P. Kuntz, and H. Briand. Hiérarchie orientée de régles généralisées en analyse implicative. *Extraction des Connaissances et Apprentissage*, 17-3:145–157, 2003.
18. R. Gras, P. Kuntz, R. Couturier, and F. Guillet. Une version entropique de l'intensité d'implication pour les corpus volumineux. *Extraction des Connaissances et Apprentissage*, 1-2:69–80, 2001.
19. R. Gras, P. Kuntz, and J.-C. Régnier. Significativité des niveaux d'une hiérarchie orientée en analyse statistique implicative. *Revue des Nouvelles Technologies de l'Information*, RNTI-C-1:39–50, 2004.
20. R. Gras and A. Larher. L'implication statistique, une nouvelle méthode d'analyse de données. *Mathématiques, Informatique et Sciences Humaines*, 120:5–31, 1992.
21. R. Gras and H. Ratsimba-Rajohn. Analyse non symétrique de données par l'implication statistique. *RAIRO-Recherche opérationnelle*, 30-3:217–232, 1996.
22. S.C. Johnson. Hierarchical clustering scheme. *Psychometrika*, 32:241–254, 1967.
23. P. Kuntz, R. Gras, and J. Blanchard. Discovering extended rules with implicative hierarchies. In *Proc. of the new frontiers of statistical data mining and knowledge discovery*, pages 166–173. Knoxville, Tennesee, 2001.
24. J.B. Lagrange. Analyse implicative d'un ensemble de variables numériques: application au traitement d'un questionnaire aux réponses modales ordonnées. *Revue de statistique appliquée*, 46(1):71–93, 1998.
25. P. Lenca, P. Meyer, B. Vaillant, P. Picouet, and S. Lallich. Evaluation et analyse multi-critères des mesures de qualité des règles d'association. *Revue des Nouvelles Technologies de l'Information*, RNTI-E-1:219–246, 2004.
26. I.C. Lerman. *Classification et analyse ordinale des données.* Dunod, Paris, 1981.
27. J. Loevinger. A systemic approach to the construcion and evalation of tests of ability. *Psychological Monographs*, 61, 1947.

28. J. Piaget. *Le jugement et le raisonnement chez l'enfant.* Delachaux et Niestlé, 1967.
29. J.-C. Régnier and R. Gras. Statistique de rangs et analyse statistique implicative. *Revue de Statistique Appliquée*, LIII:5–38, 2005.
30. L. Seve. *Emergence, complexité et dialectique.* Odile Jacob, Paris, 2005.
31. A. Wald and J. Wolfowitz. Statistical tests based on permutations of the observations. *Ann. Math. Stat.*, 15, 1944.

CHIC: Cohesive Hierarchical Implicative Classification

Raphaël Couturier

Computer Science Laboratory of University of Franche-Comte (LIFC),
IUT de Belfort-Montbeliard, BP 527, 90016 Belfort, France
raphael.couturier@univ-fcomte.fr

Summary. CHIC is a data analysis tool based on SIA. Its aim is to discover the more relevant implications between states of different variables. It proposes two different ways to organize these implications into systems: i) In the form of an oriented hierarchical tree and ii) as an implication graph. Besides, it also produces a (non oriented) similarity tree based on the likelihood of the links between states. The paper describes its main features and its usage.

Key words: data mining tool, oriented hierarchical tree, implication graph, similarity tree, CHIC.

1 Introduction

Statistical Implicative Analysis was initiated by Gras [7, 8]. The first goal of this method was to define a way of answering the question: "If an object has a property, does it also have another one? ". Of course the answer is rarely true. Nevertheless it is possible to notice that a trend is appearing. SIA aims at highlighting such tendencies in a set of properties. SIA can be considered as a method to produce association rules. Compared to other association rule methods, SIA distinguishes itself by providing a non linear measure that satisfies some important criteria. First of all, the method is based on implication intensity that measures the degree of astonishment inherent in a rule. Hence, some trivial rules that are potentially well known to an expert are discarded. In fact, a rule of the form $A \Rightarrow B$ is considered trivial if almost all objects of the population have property B. In this case, the implication intensity is close to 0 and this is not the case when rules can be considered as surprising. This implication intensity may be reinforced by the degree of validity that is based on Shannon's entropy, if the user chooses this computation mode. This measure does not only take into account the validity of a rule itself, but its counterpart too. Indeed, when an association rule is estimated as valid, i.e. the set of items A is strongly associated with the set of items B, then

R. Couturier: *CHIC: Cohesive Hierarchical Implicative Classification*, Studies in Computational Intelligence (SCI) **127**, 41–53 (2008)
www.springerlink.com

it is legitimate and intuitive to expect that its counterpart is also valid, i.e. the set of non-B items is strongly associated with the set of non-A items. Both the implication intensity and the degree of validity can be completed by a classical utility measure based on the size of the support rule and are combined to define a final relevance measure that inherits the qualities of the three measures (with the entropic theory), i.e. it is noise-resistant as the rule counterpart is taken into account and it only selects non trivial rules. For further information the reader is invited to consult [9]. Based on that original measure, CHIC, given a set of data, enables one to extract association rules. CHIC and SIA have been used in wide domain areas, for example [3, 4, 6, 14].

Based on the implication intensity and the similarity intensity CHIC allows to build two trees and one graph. The most classical tree is a similarity tree (usually known as dendogram). It is based on the similarity index defined by Lerman [13]. In a similar way, the implication intensity can be used to build an oriented hierarchy tree. The implication intensity can also be used to define an implication graph, which lets the user select the association rules and the variables he or she wants.

In opposition to most of other multidimensional data analysis methods, the SIA establishes the following properties between the variables it handles:

- relationship between variables are dissymmetrical
- the association measures are non linear and are based on probabilities
- the user can use graphical representations which follow the semantic of the relationship

For example, most of the following methods: the factor analysis, the discriminant analysis or the preference analysis are based on metric space distances. Most of hierarchical classification methods use proximity or similar indexes. So, relationships between variables are essentially symmetric. Moreover, most of the times those relationships vary linearly with observation parameters. Some methods are built on measures based on probabilities which simplifies the results interpretation. Some papers present comparison between different measures. Interested readers can consult [12] or the chapter entitled "On the behaviour of the generalisation of the intensity of implication : a data-driven comparative study" in this book.

Section 2 addresses the variables that can be handled in CHIC, their format and the options that may help users. In Section 3 some details are explained on the way to compute the association rules. Section 4 presents the similarity and the hierarchy tree. Section 5 describes the implication graph. In Section 6 some other features of CHIC are presented. Section 7 gives an illustration with interval variables and computation of typicality and contribution. Finally Section 8 concludes this paper.

2 Variables

Initially CHIC as the SIA was designed to handle binary variables. Later, SIA was enhanced by other kinds of variables and so was CHIC. Currently, CHIC allows the user to handle binary variables, frequency variables, variables over intervals and interval-variables. The case of binary variables is obviously the simplest one. Ordinal variables (also called nominal ones) can be coded using as many binary variables as number of categories. Frequency variables take a real value between 0 and 1. This kind of variables allows the user to include the case of discrete variables which only takes a fixed number of values (or modalities) ranging between 0 and 1. Of course, the way of defining modalities is very important, because it strongly affects the results of CHIC whether the values of modalities are close to 0 or 1. This remark is also true concerning the frequency variables. It should be noticed that ordinal variables are also coded using frequency ones. The user must pay attention to the way real variables are transformed into frequency ones. Several strategies are available depending on the values. If the values are positive, they can be divided by the maximum value. Another possibility resides in considering that the minimum value represents 0 and the maximum represents 1, all the other variables are proportionally distributed between the minimum and the maximum values. If a real variable has both positive and negative values, it is possible to split the variables into two variables, one for positive values and another one for negative values. In this case, previous remarks are true for both new variables. However, it is possible to consider that the minimum value (even if it is negative) represents 0 and the maximum represents 1. In this case, all other values are transformed into the interval $[0, 1]$.

Variables over intervals and interval-variables are used to model more complex situations. Both these kinds of variables are explained in the following section. Variables over intervals allow to stage the following problem. In fact, the conversion of a real variable into a frequency one may imply difficult choices from the user's point of view, as explained previously. Using the same real values, a variable over intervals proceeds differently. It consists in decomposing values of a variable into a given number of intervals. The number of intervals is chosen by the user and then the algorithm of dynamic clouds [5] automatically constitutes the intervals which have distinct bounds. This algorithm has the particularity of building intervals by minimizing inertia in each interval. Then each interval is represented by a binary variable and an individual has value 1 if it belongs to the interval and 0 otherwise. Using such a decomposition, an individual belongs only to one interval. Hence, the number of variables increases with this method. Let us take an example. Assume that we have a set of individuals and that for each of them we have their weight and height. Then, assume that everybody weighs between $40kg$ and $140kg$ and the height ranges between $140cm$ and $200cm$. Figure 1 shows an example with few individuals, values have been chosen arbitrarily. Supposing that we are interested in decomposing each variable into 4 intervals, according

	weight p	height p	woman s	man s
i1	95	200	0	1
i2	43	140	1	0
i3	75	186	0	1
i4	60	174	1	0
i5	110	183	0	1
i6	140	180	0	1
i7	100	176	0	1
i8	79	172	1	0
i9	69	170	1	0
i10	45	165	1	0
i11	65	180	0	1
i12	60	175	0	1
i13	120	175	0	1
i14	80	160	1	0
i15	100	180	0	1
i16	121	175	0	1
i17	97	162	1	0
i18	72	176	0	1
i19	40	160	1	0
i20	57	175	1	0

Fig. 1. A simple example of data with interval variables and supplementary variables

to the distribution of both variables, it is possible that we obtain the following intervals $[40, 60[$, $[60, 95[$, $[95, 110[$, $[110, 140]$ for the weights which are respectively called *weight*1, *weight*2, *weight*3 and *weight*4 and the following intervals $[140, 165[$, $[165, 174[$, $[174, 186[$, $[186, 200]$ for the heights which are respectively called *height*1, *height*2, *height*3 and *height*4. In the follow-up to this computation, all the unions of the intervals of a variable are considered. So with variable *height*, there are also intervals *height*12, *height*23, *height*34, $height1-3$, $height2-4$, $height1-4$. Intervals of the form *nameAB* correspond to the union of two consecutive intervals, for example *height*23 corresponds to the union of *height*2 and *height*3. Intervals of the form $nameA-B$ correspond to the union of all the consecutive interval between *nameA* and *nameB*, for example $height1-3$ corresponds to the union of *height*1, *height*2 and *height*3. Of course the most interesting feature of the interval variables consists in trying to make smaller partitions with these intervals, i.e. merging some intervals together in order to know which intervals are naturally close to each other. CHIC implements such an algorithm which is mathematically described, for example, in [7,8]. With the previous example, if other variables inform on the habits of individuals, it is then possible to obtain information about possible relationships between these other variables and the weights and the heights of people. For example, it is possible to know that people measuring between $140cm$ and $180cm$ are best suited to doing some particular things or that other people with such and such habits weigh principally between $90kg$ and

$150kg$. Of course the number of intervals may have a great influence on the result.

Whereas for a variable over intervals each individual takes a value 1 for only one interval, the particularity of an interval-variable is that an individual takes some values on different intervals. Moreover the intervals can be contiguous and represent a discrete decomposition, as it is the case using an automatic decomposition method like the dynamic clouds one, but they can also be defined by the user according to appropriate criteria. Taking the previous example with the weight and height, a user may prefer to state that people may be thin, normal or healthy and that they can be small, normal or tall. Nonetheless, the particularity of an interval-variable is that an individual may take values between several intervals but the sum of all its values must be less or equal to 1. In most cases, the sum will be equal to 1, but this is not mandatory. Roughly speaking, it is far from being easy to classify objects and individuals because opinions may frequently diverge on the fact that somebody or something should be described as "small" or "normal" for example. Consequently, saying that someone is rather slim may be expressed by assigning this individual with 0.75 thin and 0.25 normal. It should also be noted that this allows the user to handle fuzzy variables which are very useful in several problems [2]. The fuzziness characteristic comes from either a human appreciation, which by definition is subjective, or by an inaccurate measurement process which for some reason introduces a bias. In any case, CHIC uses the standard methods presented in the next section.

As to the data format, CHIC uses the CSV (comma separated values) format, a standard in spreadsheet tools. Labels for individuals are recorded on the first row and labels for variables are recorded on the first column. Values of individuals are represented into a 2-dimensional array. The values for each variable of an individual are stored into a line in this array (the first element is the name of the individual). The values for each individual of a variable are stored into a column in this array (the first element is the name of the variable). Of course, the nature of the values in the array differs according to the kind of variables (binary, frequency variable, ...).

As explained in the article of Gras and Kuntz in this book, supplementary variables can be used in CHIC in order to explain some important facts. This kind of variables do not intervene in the computation but it is used to give sense for the computation of typicality and contribution. Let us take an example. Assuming that we want to study the impact of a new tramway in a part of a town and that in this aim a survey has been performed. This survey gathers several information concerning the needs and the hopes of this project's potential users. Of course the gender of the people questioned is given. For example, some rules such as: working people living far from their work are generally very interested in the project, or family with young children are also very favorable to it, may be extracted. Using the gender of people as a supplementary variable, it is then possible to know if people that are responsible for the construction of the previous rules are rather men or women or if there is no distinction.

Before starting any computation with an appropriate presentation of the results, the user must choose some options. The most important one is choosing the computation method: either the classical one or the entropic one. This criterion will produce different results. The entropic version of the implication does not only take into account the validity of a rule itself, but its counterpart too. Usually, the entropic version is best suited for large data set. It is also more severe than the classical one which produces more intensive rules but is totally inappropriate with a large data set.

3 How to efficiently compute association rules

At the start of a computation, CHIC first computes rules by choosing a similarity analysis or an implicative one. The computation of association rules (for implicative analysis) is based on the algorithm described by Agrawal [1]. This algorithm can efficiently compute conjunction rules. Roughly speaking, in order to produce rules with n variables, it consists in computing all the occurrences of all the possible tuples of variables of size 1 up to n. For example, assuming that we have 5 variables labeled from A to E and that we are interested in finding rules composed of 3 variables, i.e. rules of the form $A \wedge B \Rightarrow C$, then the algorithm seeks frequent co-occurrences of all the variables, of all the pairs of variables and of the triplets of variables. With the example, the triplets are: ABC, ABD, ABE, ACD, ACE, ADE, BCD, BCE and CDE. If the user chooses a given threshold α and assuming that $|B| < \alpha$ (where $|B|$ represents the cardinality of B) and all other variables have a larger cardinality, then only 4 triplets may have a cardinality greater than α (ACD, ACE, ADE and CDE), but this is not necessarily the case. Then for all the tuples whose occurrences are greater than the threshold, it is possible to produce the rules. For example with the following tuples ABC, AB and BC, AC, it is possible to compute the intensity of the two rules $A \wedge B \Rightarrow C$, $B \wedge C \Rightarrow A$ and $A \wedge C \Rightarrow B$.

Even if the user specifies a high threshold, the number of rules produced by the algorithm with large data sets may be quite huge. Moreover, it is possible that some rules are very similar, that is why we have introduced an originality criterion. It allows one to select only conjunction rules that are original, i.e. rules for which sub-rules are not so trivial. Consider, for example, the following rule $A \wedge B \Rightarrow C$, this rule is original if its intensity is high and if the rules $A \Rightarrow C$ and $B \Rightarrow C$ have a small intensity measure.

As the definition of the originality has never been described formally we will explain it in details. Let $A_m \Rightarrow B$ be an association rule such that the antecedent A_m is a set of m properties, a set of properties is also viewed as a conjunction. As soon as a rule $A_i \Rightarrow B$, where A_i is a subset of A_m, has a high implication intensity then, the originality of $A_m \Rightarrow B$ decreases.

Therefore, the originality of $A_m \Rightarrow B$ depends on the implication intensity associated with each rule $A_i \Rightarrow B$, $i \in \{1, \ldots, m-1\}$, such that A_i is a subset of A_m that contains i properties. For each $i \in \{1, \ldots, m-1\}$ there is $\binom{m}{i}$ subsets of A_m that contains i properties. We note $A_m^{i,j}$ the $j-th$ subset with $i \in \{1, \ldots, m-1\}$ and $j \in \{1, \ldots, \binom{m}{i}\}$.

Thus, the originality of $A_m \Rightarrow B$ depends on the $2^m - 2$ parts of A_m (empty set and A_m are not considered). That is to say, the implication intensity of $A_m \Rightarrow B$ depends on the set of rules $R_m = \cup_{i \in \{1,\ldots,m-1\}} \{A_m^{i,j} \Rightarrow B \mid j \in \{1, \ldots, \binom{m}{i}\}\}$.

We propose to measure the originality of the rule $A_m \Rightarrow B$ as the geometrical average of the difference of the implication intensity associated with $A_m \Rightarrow B$ and with each rule of R_m. Finally, the coefficient of originality of $A_m \Rightarrow B$ is defined as follow.

The originality of an association rule $A_m \Rightarrow B$, where A_m is a conjunction of m properties, is measured by:

$$\texttt{Originality}(A_m \Rightarrow B) =$$

$$\left[\left(\prod_{i=1}^{m-1} \left(\prod_{j=1}^{\binom{m}{i}} \texttt{ImpInt}(A_m \Rightarrow B) - \texttt{ImpInt}(A_m^{i,j} \Rightarrow B) \right) \right) \right]^{\dfrac{1}{2^m - 2}}$$

Where $\texttt{ImpInt}(A \Rightarrow B)$ is the implication intensity of the rule $A \Rightarrow B$.

With this definition, the number of "original" rules decreases often drastically. In general, when the number of conjunctions increases, the number of trivial rules also increases. That is why the criterion of originality provides an efficient way of only having interesting rules. We plan to compare our originality measure with other measures that could produce similar results in future works.

4 Similarity and hierarchy tree

Once CHIC has computed the whole set of rules according to the parameters that the user has chosen, it can build a tree with some of the rules. This tree may be seen as a classification oriented or not in function of the kind of computation (similarity or implication). There are some common principles in building both trees. In the following a rule is called a class and it is composed of two variables in their simplest form. At each level of the classification, CHIC selects the class with the highest intensity (of similarity or implication). In order to know how to compute the intensity of a class and to have an explanation of the variables used in figures 2, 3 and 4, one should refer to the chapter of R. Gras and P. Kuntz in this book.

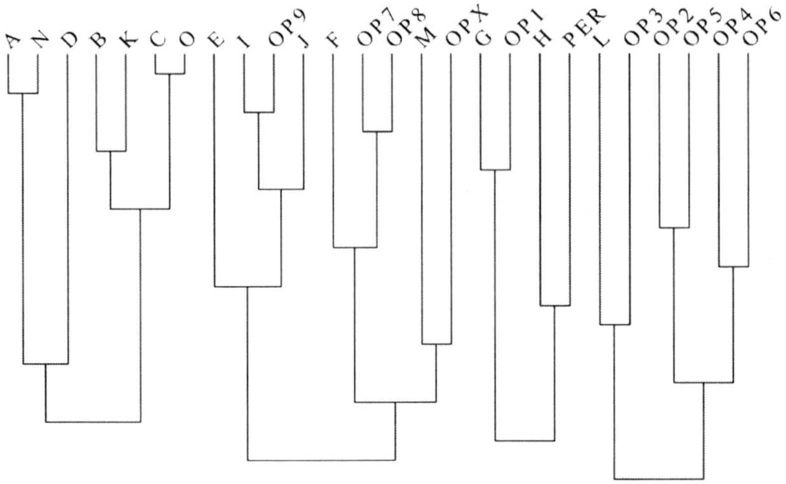

Fig. 2. An example of a similarity tree

Then, at each step, CHIC computes a set of new classes with all the existing ones. In order to build a new class, CHIC either aggregates an existing class with a variable which has not been aggregated in another class yet or aggregates two existing non aggregated classes. Nonetheless each couple of variables between the two classes must have a valid intensity, i.e. greater than 0.5. For example the formation of a class $((a, b), c)$ entails that classes (a, c) and (b, c) are meaningful from the analysis point of view (similarity or implication). The class $((a, b), c)$ represents the rule $(a \Rightarrow b) \Rightarrow c$ with the implicative analysis and represents the fact that a and b are similar and that this class is similar to c from the similarity point of view. For more details on the class formation, interested readers are invited to read the article of Gras and Kuntz in this book and in [10].

If the user is interested to know how the tree without one or more variables would look like, he can simply deselect them in the item toolbox. It should be noticed that this toolbox is available for all kinds of representation provided by CHIC (trees of graph). Unfortunately, a modification of the variables involved in the computation (even a small one) implies a complete rebuilding of the tree. This step of class construction is strongly dependent on the number of variables (the algorithm has a complexity which depends on the factorial of the number of variables).

Before running an analysis the user can choose in the computation options to highlight the significant level in the tree.

Figure 2 shows a similarity tree and Figure 3 shows a hierarchy tree. For the latter one, significant levels are pointed out. They are represented by a red line (in CHIC). Each significant level means that the current level

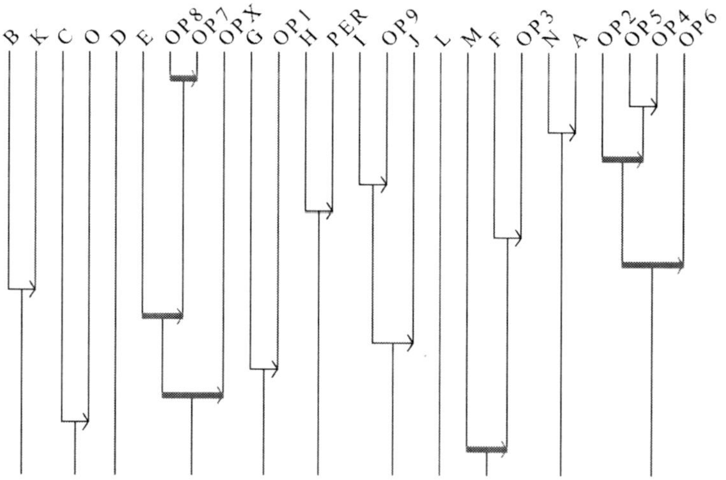

Fig. 3. An example of a hierarchy tree

is more significant than the previous one and than the next one, which is not significant by definition. For more details on its construction, interested readers should refer to the definition given in this book and in [11].

The similarity index is computed by the classical theory or by the entropic one. The last one should be preferred with a large number of individuals. Moreover, the construction of the similarity tree with the classical index leads to only one class that gather all the others. On the contrary, with the entropic version of the similarity index, the algorithm very frequently builds more than one class. In fact, according to similarities of data, the number of classes varies.

5 Implication graph

As explained previously, both classifications (based on the similarity and implication) in CHIC only select some of the rules and ignore some other rules by constructing the tree. If all the rules are required to point out an interesting feature, the implication graph may be preferred since in this graph, the user can see the rules that have a greater intensity than a given threshold. In fact, four thresholds are available and CHIC uses different colors to quickly show which rules are the most important. In Figure 4 some rules are represented in an implication graph. An arrow is used to show the implication between two variables (the rule $A \Rightarrow B$ is represented by an arrow from A to B). As the number of rules may be large, the user has the possibility to select only some variables. Hence, only rules with present variables are represented. This consequently reduces the number of rules. Moreover, in order to make the graph more readable, CHIC uses an automatic graph drawing algorithm which tries to minimize the number of crossings between rules. By default, transitive closures are not displayed on the implication graph. A simple click with the mouse in the toolbox displays them. CHIC computes them once and for all for each new graph. Afterwards, even if the user selects or deselects some variables, changes the thresholds of the rules, chooses or not to display transitive closures, then CHIC only displays the graph without any computation. This allows users to stress the important features of their data. Besides, the graph drawing procedure may be time consuming with large graph, so it is not used automatically.

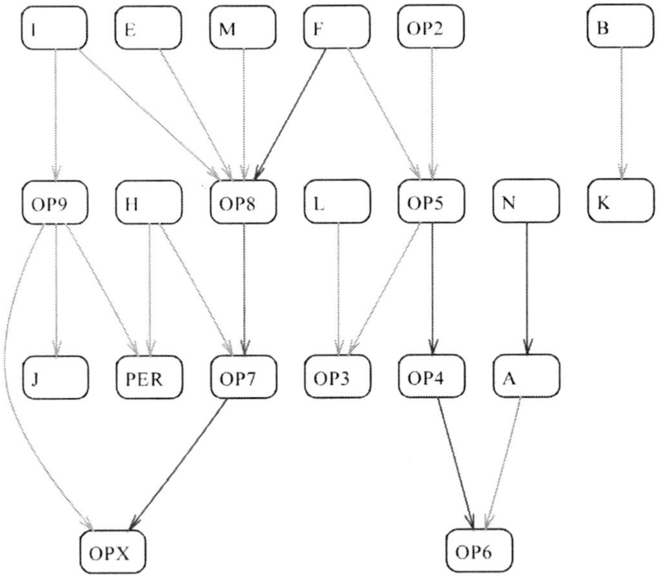

Fig. 4. An example of an implication graph

6 Other possibilities

In addition to the traditional representation modes previously described, CHIC provides some practical features. For each of the graphic representation it is possible to compute the contribution and the typicality of an individual to a given rule. In the same way, it is possible to compute the contribution and the typicality of a set of individuals to a given rule.

The notion of typicality is defined by the fact that some individuals are "typical" of the behavior of the population. They contribute well to the creation of a rule, i.e. with a similar intensity to the rule. For example if a rule

$A \Rightarrow B$ has an implicative intensity equals to 0.7, then the most typical individuals respectively have values close to 0.5 and 1 for A and B (those values depend on how the rule was created, i.e. what computation mode was chosen, and especially the cardinality of the set A and B). By opposition, the notion of contribution is defined to measure if individuals are more responsible for the creation of the rule than the other ones. With the previous example, the most contributive individuals are those who have 1 for the variables A and B. So, the notions of typicality and contribution are different. In the same way, the notion of typicality (resp. contribution) of a set of individuals (or of a category of individuals) is defined in order to know if a considered set of individuals is typical (resp. contributive) to a rule. In order to have formal definitions of those notions, one should refer to the chapter of R. Gras and P. Kuntz in this book.

7 An illustration with interval variables and computation of typicality and contribution

This section is intended to give a simple and concrete example with the two interval variables of Section 2. Figure 5 shows an implicative graph issued from data of Figure 1. The two interval variables *weight* and *height* are automatically split into 4 intervals by CHIC as described in Section 2.

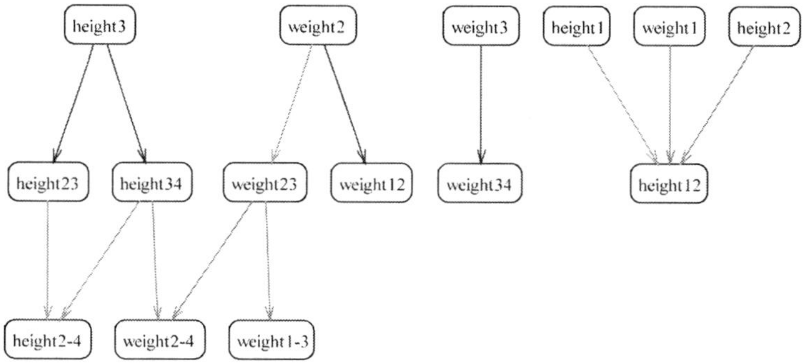

Fig. 5. An example of a implication graph

In the graph some interesting rules are visible. For example, we can see the rules: $weight1 \Rightarrow height12$ and $height34 \Rightarrow height2\text{-}4$. Because the number of individuals is small, and consequently not significant, and because the values of this set have been arbitrarily generated, nothing else than: "light individuals have generally a rather small height and that tall individuals are not light ones" can be concluded. Nevertheless these rules show an implication between the partitions of the two variables. Considering that these data can have a sense for the expert, then we could have computed the typicality and the contribution of the group of individuals. For example, concerning the rule *height*34 ⇒ *weight*2-4, CHIC determines that the variable *man* contributes the most to it (the error is 0.00638 for man, so close to 0; the error is equal to 1 for woman, so it does not contribute at all). In the opposite way the most typical variable to rule *weight*1 ⇒ *height*12 is the variable *woman* (the error is 0.00499 for woman, so this is a very good typicality; the error equals 1 for man, so it does not contribute at all). Both those results are not surprising analyzing the data.

8 Conclusion

CHIC implements almost all methods and techniques described in SIA. In this chapter we have described the main features of CHIC. First the variables that can be handled are described. These different kinds of variables enable to model several particular cases frequently encountered. Some options of CHIC are briefly given in order to allow users to know what they can do with CHIC. Then, the three main representations are presented. The similarity tree and the hierarchy tree respectively provide a non-oriented and an oriented classification. The implication graph, which is the most interactive, allows users to mine their data and highlights the important rules that may interest the expert. For those three representations some functionalities are given. Although the theory of SIA is far from being simple for a novice user, CHIC allows to benefit from SIA results and distinguishes itself from other data mining tools by its particular features.

References

1. R. Agrawal, T. Imielinski, and A. Swami. Mining association rules between sets of items in large databases. In *Proceedings of the 1993 ACM SIGMOD International Conference on Management of Data*, pages 207–216, 1993.
2. G. Bojadziev and M. Bojadziev. *Fuzzy sets, fuzzy logic, applications.* World scientific, 1996.
3. R. Couturier. Un système de recommandation basé sur l'ASI. In *Troisième rencontre internationale de l'Analyse Statistique Implicative (ASI3)*, pages 157–162, 2005.
4. R. Couturier, R. Gras, and F. Guillet. Reducing the number of variables using implicative analysis. In *International Federation of Classification Societies, IFCS 2004*, pages 277–285. Springer Verlag: Classification, Clustering, and Data Mining Applications, 2004.
5. E. Diday. La méthode des nuées dynamiques. *Revue de statistique appliquée*, 19(2):19–34, 1971.
6. G. Froissard. CHIC et les études docimologiques. In *Troisième rencontre internationale de l'Analyse Statistique Implicative (ASI3)*, pages 187–197, 2005.

7. R. Gras. Panorama du développement de l'A.S.I. à travers des situations fondatrices. In *Actes de la 3ème Rencontre Internationale A.S.I.*, pages 9–33. Université de Palerme, 2005.
8. R. Gras, S. Ag Almouloud, M. Bailleul, A. Lahrer, M. Polo, H. Ratsimba-Rajohn, and A. Totohasina. *L'implication Statistique.* La Pensée Sauvage, 1996.
9. R. Gras, R. Couturier, J. Blanchard, H. Briand, P. Kuntz, and P. Peter. *Quelques critères pour une mesure de qualité de règles d'association. Un exemple : l'implication statistique*, chapter Mesures de qualité pour la fouille de données, pages 3–32. RNTI-E-1, Cepaduès Editions, 2004.
10. R. Gras and P. Kuntz. Discovering R-rules with a directed hierarchy. *Soft Computing, A Fusion of Foundations, Methodologies and Applications*, 1:46–58, 2005.
11. R. Gras, P. Kuntz, and J.C. Régnier. Significativité des niveaux d'une hiérarchie orientée. *Classification et fouille de données, RNTI-C-1, Cépaduès-Editions*, pages 39–50, 2004.
12. P. Lenca, P. Meyer, P. Vaillant, P. Picouet, and S. Lallich. *Evaluation et analyse multi-critères de qualité des règles d'association*, chapter Mesures de qualité pour la fouille de données, pages 219–246. RNTI-E-1, Cépaduès, 2004.
13. I. C. Lerman. *Classification et analyse ordinale des données.* Dunod, 1981.
14. P. Orus and P. Gregori. Des variables supplémentaires et des élèves "fictifs", dans la fouille didactique de données avec CHIC. In *Troisième rencontre internationale de l'Analyse Statistique Implicative (ASI3)*, pages 279–291, 2005.

Assessing the interestingness of temporal rules with Sequential Implication Intensity

Julien Blanchard, Fabrice Guillet, and Régis Gras

Knowledge & Decision (KOD) research team
LINA — FRE CNRS 2729
Polytechnic School of Nantes University, France
julien.blanchard@polytech.univ-nantes.fr

Summary. In this article, we study the assessment of the interestingness of sequential rules (generally temporal rules). This is a crucial problem in sequence analysis since the frequent pattern mining algorithms are unsupervised and can produce huge amounts of rules. While association rule interestingness has been widely studied in the literature, there are few measures dedicated to sequential rules. Continuing with our work on the adaptation of implication intensity to sequential rules, we propose an original statistical measure for assessing sequential rule interestingness. More precisely, this measure named Sequential Implication Intensity (SII) evaluates the statistical significance of the rules in comparison with a probabilistic model. Numerical simulations show that SII has unique features for a sequential rule interestingness measure.

Key words: Temporal Data Mining, Event Sequences, Interestingness Measures for Sequential Rules, Rule Significance.

1 Introduction

Frequent pattern discovery in sequences of events[1] (generally temporal sequences) is a major task in data mining. Research work in this domain consists of two approaches:

- discovery of frequent *episodes* in a long sequence of events (approach initiated by Mannila, Toivonen, and Verkamo [12, 13]),
- discovery of frequent *sequential patterns* in a set of sequences of events (approach initiated by Agrawal and Srikant [1, 17]).

The similarity between *episodes* and *sequential patterns* is that they are sequential structures, i.e., a structure defined with an order (partial or total). Such a structure can be, for example:

[1] Here we speak about sequences of qualitative variables. Such sequences are generally not called time series.

J. Blanchard et al.: *Assessing the interestingness of temporal rules with Sequential Implication Intensity*, Studies in Computational Intelligence (SCI) **127**, 55–71 (2008)
www.springerlink.com

breakfast then *lunch* then *dinner*

The structure is described by its frequency (or support) and generally by constraints on the event position, like a maximal time window "less than 12 hours stand between *breakfast* and *dinner*" [5, 10, 14, 17, 18].

The difference between *episodes* and *sequential patterns* lies in the measure of their frequency: frequency of *episodes* is an intra-sequence notion [5, 10, 14, 18–20], while frequency of *sequential patterns* is an inter-sequence notion [1, 8, 16, 17, 21] (see [11] for a synthesis on the different ways of assessing frequency). Thus, the frequent *episode* mining algorithms search for structures which often recur inside a single sequence. On the other hand, the frequent *sequential pattern* mining algorithms search for structures which recur in numerous sequences (independently of the repetitions in each sequence). These last algorithms are actually an extension to sequential data of the frequent itemset mining algorithms, used among other things to generate association rules [2, 9].

Just as the discovery of frequent itemsets leads to the generation of association rules, the discovery of *episodes*/*sequential patterns* is often followed by a sequential rule generation stage which enables predictions to be made within the limits of a time window [5, 10, 14, 16–19, 21]. Such rules have been used to predict, for example, stock market prices [5] or events in a telecommunication network [14, 18]. A sequential rule can be for instance:

$$breakfast \xrightarrow{6\text{h}} lunch$$

This rule means "if one observe *breakfast* then one will certainly observe *lunch* less than 6 hours later".

In this article, we study the assessment of the interestingness of sequential rules. This is a crucial problem in sequence analysis since the frequent pattern mining algorithms are unsupervised and can produce a huge number of rules. While association rule interestingness has been widely studied in the literature (see [3] for a survey), there are few measures dedicated to sequential rules. In addition to frequency, one mainly finds an index of confidence (or precision) that can be interpreted as an estimation of the conditional probability of the conclusion given the condition [5, 10, 14, 16–19, 21]. A measure of recall is sometimes used too; it can be interpreted as an estimation of the conditional probability of the condition given the conclusion [18, 19]. In [5] and [10], the authors have proposed an adaptation to sequential rules of the J-measure of Smyth and Goodman, an index coming from mutual information[2]. Finally, an entropic measure is presented in [20] to quantify the information brought by an episode in a sequence, but this approach only deals with episodes and not with prediction rules.

These measures have several limits. First of all, the J-measure is not very intelligible since it gives the same value to a rule $a \xrightarrow{\omega} b$ and to its opposite

[2] The J-measure is the part of the average mutual information relative to the truth of the condition.

$a \xrightarrow{\omega} \bar{b}$, whereas these two rules make conflicting predictions. Confidence and recall, vary linearly, which makes them rather sensitive to noise. Above all, these measures increase with the size of the time window chosen. This behavior is absolutely counter-intuitive since a rule with a too large time window does not contribute to making good quality predictions. Indeed, the larger the time window, the greater the probability of observing the conclusion which follows the condition in data, and the less significant the rule. Another major problem, which concerns confidence, recall, and J-measure, is that these indexes are all frequency-based: the phenomena studied in data are considered only in a relative way (by means of frequencies) and not in an absolute way (by means of cardinalities). Thus, if a sequence is made longer by repeating it x times one after the other, the indexes do not vary[3]. Statistically, the rules are all the more reliable since they are assessed on long sequences yet. In the end, a good interestingness measure for sequential rules should therefore decrease when the size of the time window is too large, and increase with sequence enlargement. These essential properties have never been highlighted in the literature.

Continuing with our work begun in [4], on the adaptation of implication intensity to sequential rules [6,7], we propose in this article an original statistical measure for assessing sequential rule interestingness. More precisely, this measure evaluates the statistical significance of the rules in comparison with a probabilistic model. The next section is dedicated to the formalization of the notions of *sequential rule*, *example of a rule*, and *counter-example of a rule*, and to the presentation of the new measure, named *Sequential Implication Intensity (SII)*. In section 3, we study *SII* in several numerical simulations and compare it to other measures.

2 Measuring the statistical significance of sequential rules

2.1 Context

Our measure, *SII*, evaluates sequential rules extracted from **one unique sequence**. This approach can be easily generalized to several sequences, for example by computing an average or minimal *SII* on the set of sequences. Rules are of the form $a \xrightarrow{\omega} b$, where a and b are episodes (these ones can even be structured by intra-episode time constraints). However, in this article, we restrict our study to sequential rules where the episodes a and b are two single events.

---

[3] We consider here that the size of the time window is negligible compared to the size of the sequence, and we leave aside the possible side effects which could make new patterns appear overlapping the end of a sequence and the beginning of the following repeated sequence.

The studied sequence is a continuous sequence of instantaneous events (adaptation to discrete sequences is trivial). It is possible that two different events occur at the same time. This amounts to using the same framework as the one introduced by Mannila, Toivonen, and Verkamo [14]. To extract the appropriate cardinalities from the sequence and compute SII, one only needs to apply their episode mining algorithm named Winepi [13, 14] (or one of its variants). In the following, we stand at the post-processing stage by considering that Winepi has already been applied on the sequence, and we directly work on the episode cardinalities that have been discovered. Here again, our approach could be generalized to other kinds of sequences, for which other episode mining algorithms have been proposed. For example, Höppner has studied sequences with time-interval events that have a non-zero duration and can overlap [10].

2.2 Notations

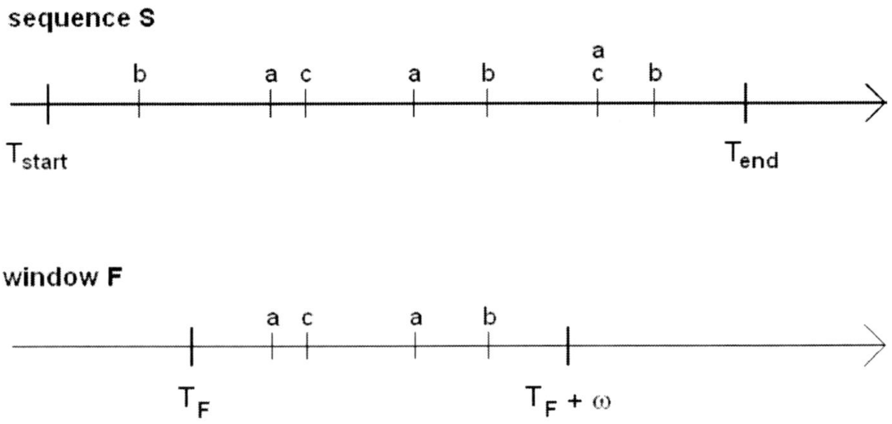

Fig. 1. A sequence S of events from $E = \{a, b, c\}$ and its window F of size ω beginning at T_F.

Let $E = \{a, b, c, \ldots\}$ be a finite set of *event types*. An *event* is a couple (e, t) where $e \in E$ is the type of the event and $t \in \mathbb{R}_+$ is the time the event occurred. It must be noted that the term event is often used to refer the event type without reducing intelligibility.

An *event sequence* S observed between the instants T_{start} and T_{end} is a finite series of events

$$S = \Big((e_1, t_1), (e_2, t_2), (e_3, t_3), \ldots, (e_n, t_n)\Big)$$

such that:

$$\forall i \in \{1..n\}, (e_i \in E \wedge t_i \in [T_{start}, T_{end}])$$
$$\forall i \in \{1..n-1\}, t_i \leq t_{i+1}$$
$$\forall (i,j) \in \{1..n\}^2, t_i = t_j \Rightarrow e_i \neq e_j$$

The size of the sequence is $L = T_{end} - T_{start}$.

A *window* on a sequence S is a subsequence of S. For instance, a window F of size $\omega \leq L$ beginning at the instant $t_F \in [T_{start}, T_{end} - \omega]$ contains all the events (e_i, t_i) from S such as $t_F \leq t_i \leq t_F + \omega$.

In the following, we consider a sequence S of events from E.

2.3 Sequential rules

We establish a formal framework for sequence analysis by defining the notions of *sequential rule*, *example of a rule*, and *counter-example of a rule*. The examples and counter-examples of a sequential rule have never been defined in the literature about sequences.

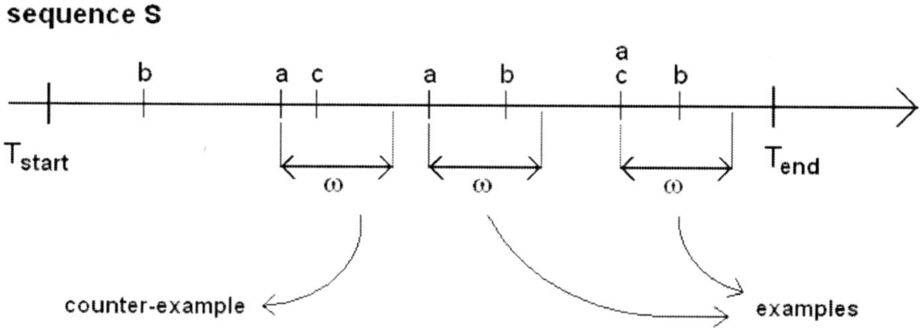

Fig. 2. Among the 3 windows of size ω beginning on events a, one can find 2 examples and 1 counter-example of the rule $a \xrightarrow{\omega} b$.

Definition 1 A **sequential rule** is a triple (a, b, ω) noted $a \xrightarrow{\omega} b$ where a and b are events of different types and ω is a strictly positive real number. It means: "if an event a appears in the sequence then an event b certainly appears within the next ω time units".

Definition 2 The **examples** of a sequential rule $a \xrightarrow{\omega} b$ are the events a which are followed by at least one event b within the next ω time units. Therefore the number of examples of the rule is the cardinality noted $n_{ab}(\omega)$:

$$n_{ab}(\omega) = \left| (a,t) \in S \mid \exists (b,t') \in S, 0 \leq t' - t \leq \omega \right|$$

Definition 3 The **counter-examples** of a sequential rule $a \xrightarrow{\omega} b$ are the events a which are not followed by any event b during the next ω time units. Therefore the number of counter-examples of the rule is the cardinality noted $n_{a\bar{b}}(\omega)$:

$$n_{a\bar{b}}(\omega) = \left|(a,t) \in S \mid \forall (b,t') \in S, (t' < t \vee t' > t + \omega)\right|$$

Contrary to association rules, n_{ab} and $n_{a\bar{b}}$ are not data constants but depend on the parameter ω.

The originality of our approach is that it treats condition and conclusion in very different ways: the events a are used as references for searching the events b, i.e. only the windows which begin by an event a are taken into account. On the contrary, in the literature about sequences, the algorithms like Winepi move a window forward (with a fixed step) over the whole sequence [14]. This method amounts to considering as examples of the sequential rule any window that has an event a followed by b, even if it does not start by en event a. In comparison, our approach is algorithmically less complex.

Let us note n_a the number of events a in the sequence. We have the usual equality $n_a = n_{ab} + n_{a\bar{b}}$. A sequential rule $a \xrightarrow{\omega} b$ is completely described by the quintuple $(n_{ab}(\omega), n_a, n_b, \omega, L)$. The examples of a sequential rule now being defined, we can specify our measure for the frequency of the rules:

Definition 4 The **frequency** of a sequential rule $a \xrightarrow{\omega} b$ is the proportion of examples compared to the size of the sequence:

$$frequency(a \xrightarrow{\omega} b) = \frac{n_{ab}(\omega)}{L}$$

With these notations, the confidence, recall, and J-measure are given by the following formula:

$$confidence(a \xrightarrow{\omega} b) = \frac{n_{ab}(\omega)}{n_a}$$

$$recall(a \xrightarrow{\omega} b) = \frac{n_{ab}(\omega)}{n_b}$$

$$J\text{−}measure(a \xrightarrow{\omega} b) = \frac{n_{ab}(\omega)}{L} \log_2 \frac{n_{ab}(\omega)L}{n_a n_b} + \frac{n_{a\bar{b}}(\omega)}{L} \log_2 \frac{n_{a\bar{b}}(\omega)L}{n_a(L - n_b)}$$

2.4 Random model

Following the implication intensity for association rules [6], the sequential implication intensity SII measures the statistical significance of the rules $a \xrightarrow{\omega} b$. To do so, it quantifies the unlikelihood of the smallness of the number of counter-examples $n_{a\bar{b}}(\omega)$ with respect to the independence hypothesis between the types of events a and b. Therefore, in a search for a random model, we suppose that the types of events a and b are independent. Our goal is to determine the distribution of the random variable $\mathcal{N}_{a\bar{b}}$ (number of counter-examples of the rule) given the size L of the sequence, the numbers n_a and n_b of events of types a and b, and the size ω of the time window which is used.

We suppose that the arrival process of the events of type b satisfies the following hypotheses:

- the times between two successive occurrences of b are independent random variables,
- the probability that a b appears during $[t, t+\omega]$ only depends on ω.

Moreover, two events of the same type cannot occur simultaneously in the sequence S (see section 2.2). In these conditions, the arrival process of the events of type b is a Poisson process of intensity $\lambda = \frac{n_b}{L}$. So, the number of b appearing in a window of size ω follows Poisson's Law with parameter $\frac{\omega.n_b}{L}$. In particular, the probability that no event of type b appears during ω time units is:

$$p = \mathrm{P}(Poisson(\frac{\omega.n_b}{L}) = 0) = e^{-\frac{\omega}{L}n_b}$$

Therefore, wherever it appears in the sequence, an event a has the fixed probability p of being a counter-example, and $1-p$ of being an example. Let us repeat n_a times this random experiment to determine the theoretical number of counter-examples $\mathcal{N}_{a\bar{b}}$. If ω is negligible compared to L, then two randomly chosen windows of size ω are not likely to overlap, and we can consider that the n_a repetitions of the experiment are independent. In these conditions, the random variable $\mathcal{N}_{a\bar{b}}$ is Binomial with parameters n_a and p:

$$\mathcal{N}_{a\bar{b}} = Binomial(n_a, e^{-\frac{\omega}{L}n_b})$$

When permitted, this Binomial distribution can be approximated by another Poisson distribution (even in the case of "weakly dependent" repetitions — see [15]).

Definition 5 The **sequential implication intensity (*SII*)** of a rule $a \xrightarrow{\omega} b$ is defined by:

$$SII(a \xrightarrow{\omega} b) = \mathrm{P}(\mathcal{N}_{a\bar{b}} > n_{a\bar{b}}(\omega))$$

Numerically, we have:

$$SII(a \xrightarrow{\omega} b) = 1 - \mathrm{P}(\mathcal{N}_{a\bar{b}} \leq n_{a\bar{b}}(\omega)) = 1 - \sum_{k=0}^{n_{a\bar{b}}(\omega)} \mathrm{C}_{n_a}^{k} (e^{-\frac{\omega}{L} n_b})^k (1 - e^{-\frac{\omega}{L} n_b})^{n_a - k}$$

3 Properties and comparisons

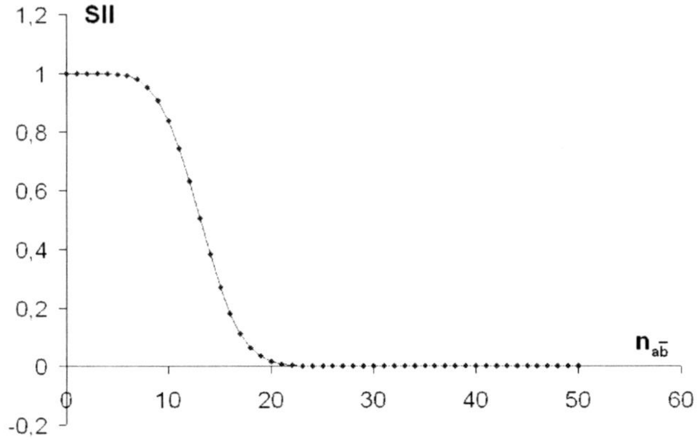

Fig. 3. *SII* w.r.t. the number of counter-examples.

SII quantifies the unlikelihood of the smallness of the number of counter-examples $n_{a\bar{b}}(\omega)$ with respect to the independence hypothesis between the types of events a and b. In particular, if $SII(a \xrightarrow{\omega} b)$ is worth 1 or 0, then it is unlikely that the types of event a and b are independent (deviation from independence is significant and oriented in favor of the examples or of the counter-examples). This new index can be seen as the complement to 1 of the p-value of a hypothesis test. However, following the implication intensity, the aim here is not testing a hypothesis but actually using it as a reference to evaluate and sort the rules.

In the following, we study *SII* in several numerical simulations and compare it to confidence, recall, and J-measure. These simulations point out the intuitive properties of a good interestingness measure for sequential rules.

3.1 Counter-example increase

In this section, we study the measures when the number $n_{a\bar{b}}$ of counter-examples increases (with all other parameters constant). For a rule $a \xrightarrow{\omega} b$,

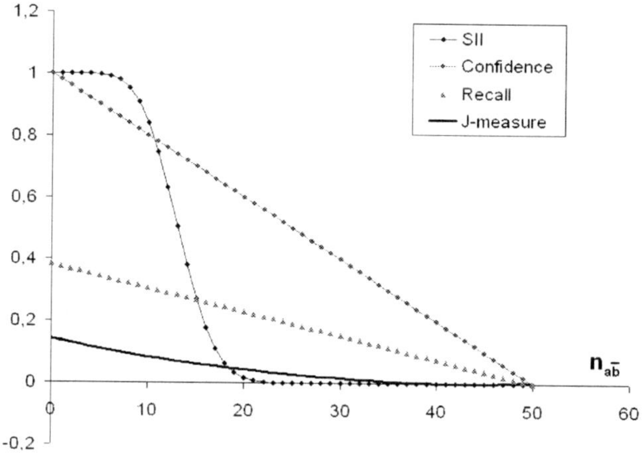

Fig. 4. SII, confidence, recall, and J-mesure w.r.t. the number of counter-examples. $n_a = 50$, $n_b = 130$, $\omega = 10$, $L = 1000$

this can be seen as making the events a and b more distant in the sequence while keeping the same numbers of a and b. This operation transforms events a from examples to counter-examples.

Fig. 4 shows that SII clearly distinguishes between acceptable numbers of counter-examples (assigned to values close to 1) and non-acceptable numbers of counter-examples (assigned to values close to 0) with respect to the other parameters n_a, n_b, ω, and L. On the contrary, confidence and recall vary linearly, while J-measure provides very little discriminative power. Due to its entropic nature, the J-measure could even increase when the number of counter-examples increases, which is disturbing for a rule interestingness measure.

3.2 Sequence enlargement

We call sequence enlargement the operation which makes the sequence longer by adding new events (of new types) at the beginning or at the end. For a rule $a \xrightarrow{\omega} b$, such an operation does not change the cardinalities $n_{ab}(\omega)$ and $n_{a\bar{b}}(\omega)$ since the layout of the events a and b remain the same. Only the size L of the sequence increase.

Fig. 5 shows that SII increases with sequence enlargement. Indeed, for a given number of counter-examples, a rule is more surprising in a long sequence rather than in a short one since the a and b are less likely to be close in a long sequence. On the contrary, measures like confidence and recall remain unchanged since they do not take L into account (see Fig. 6). The J-measure varies with L but only slightly. It can even decrease with L, which is counter-intuitive.

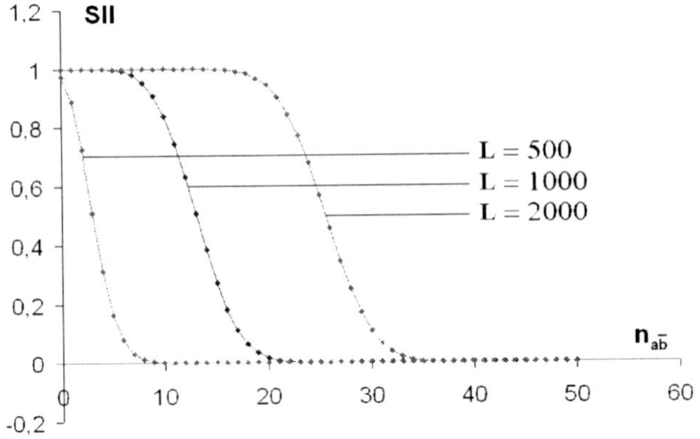

Fig. 5. SII with sequence enlargement.
$n_a = 50$, $n_b = 130$, $\omega = 10$

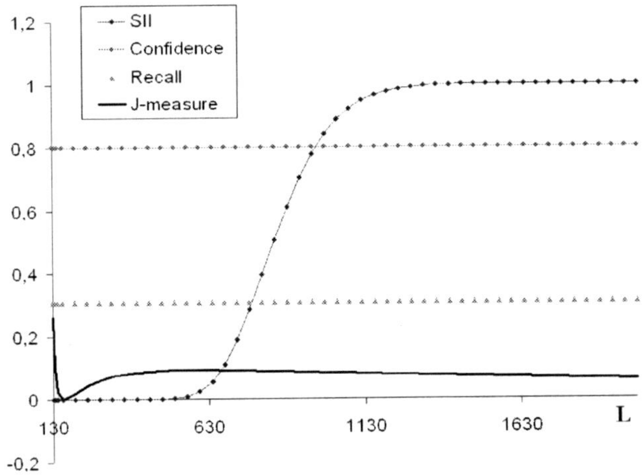

Fig. 6. SII, confidence, recall, and J-mesure with sequence enlargement.
$n_a = 50$, $n_b = 130$, $n_{a\bar{b}} = 10$, $\omega = 10$

3.3 Sequence repetition

We call sequence repetition the operation which makes the sequence longer by repeating it γ times one after the other (we leave aside the possible side effects which could make new patterns appear by overlapping the end of a sequence and the beginning of the following repeated sequence). With this operation, the frequencies of the events a and b and the frequencies of the examples and counter-examples remain unchanged.

Fig. 7 shows that the values of SII are more extreme (close to 0 or 1) with sequence repetition. This is due to the statistical nature of the measure.

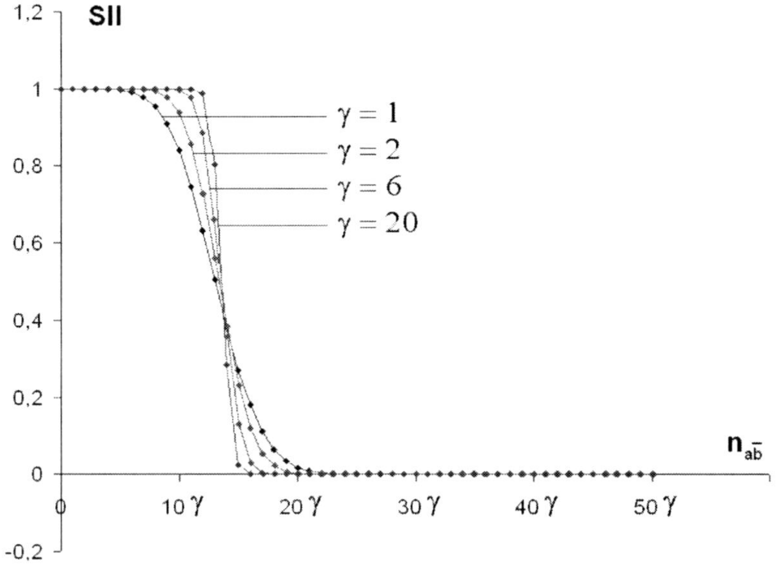

Fig. 7. *SII* with sequence repetition.
$n_a = 50 \times \gamma$, $n_b = 130 \times \gamma$, $\omega = 10$, $L = 1000 \times \gamma$

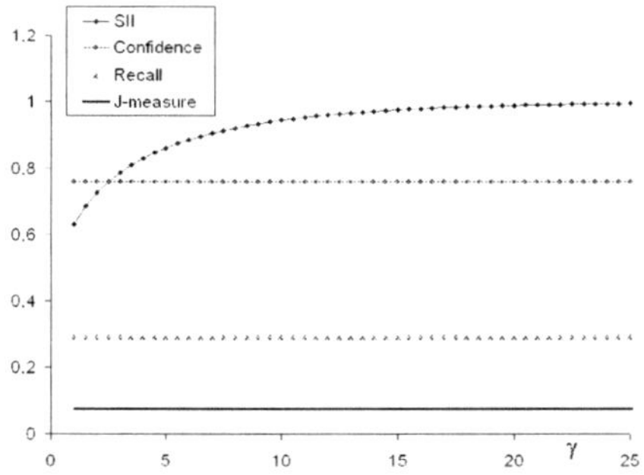

(a) $n_{a\bar{b}} = 12 \times \gamma$

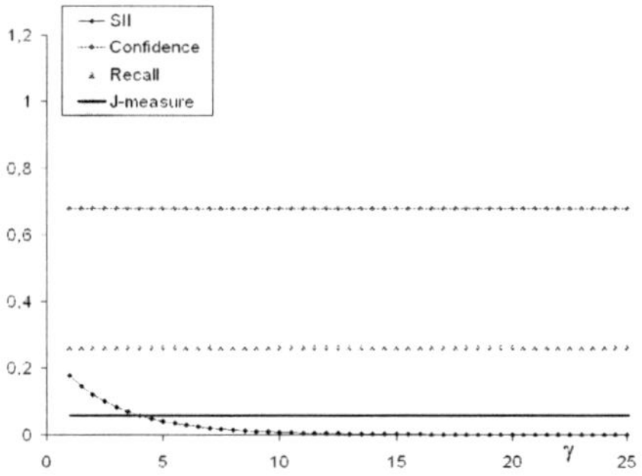

(b) $n_{a\bar{b}} = 16 \times \gamma$

Fig. 8. *SII*, confidence, recall, and J-mesure with sequence repetition.
$n_a = 50 \times \gamma$, $n_b = 130 \times \gamma$, $\omega = 10$, $L = 1000 \times \gamma$

Statistically, a rule is all the more significant when it is assessed on a long sequence with lots of events: the longer the sequence, the more one can trust the imbalance between examples and counter-examples observed in the sequence, and the more one can confirm the good or bad quality of the rule. On the contrary, the frequency-based measures like confidence, recall, and J-measure do not vary with sequence repetition (see Fig. 8).

3.4 Window enlargement

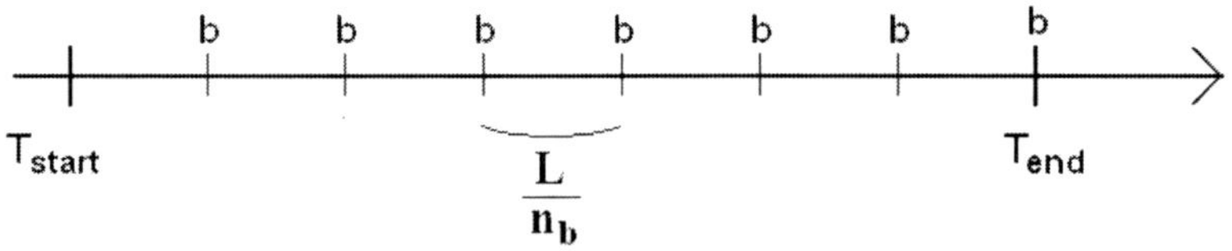

Fig. 9. A sequence where the events b are regularly spread.

Window enlargement consists of increasing the size ω of the time window. As the function $n_{a\bar{b}}(\omega)$ is unknown ($n_{a\bar{b}}$ is given by a data mining algorithm, it depends on the data), we model it in the following way:

$$n_{a\bar{b}}(\omega) = n_a - \frac{n_a n_b}{L}\omega \ , \qquad \text{if } \omega \leq \frac{L}{n_b}$$

$$n_{a\bar{b}}(\omega) = 0 \ , \qquad \text{otherwise.}$$

This is a simple model, considering that the number of examples observed in the sequence is proportional to ω: $n_{ab}(\omega) = \frac{n_a n_b}{L}\omega$. The formula is based on the following postulates:

- According to definitions 2 and 3, n_{ab} must increase with ω and $n_{a\bar{b}}$ must decrease with ω.
- If $\omega = 0$ then there is no time window, and the data mining algorithm cannot find any example[4]. So we have $n_{ab} = 0$ and $n_{a\bar{b}} = n_a$.
- Let us consider that the events b are regularly spread over the sequence (Fig. 9). If $\omega \geq \frac{L}{n_b}$, then any event a can capture at least one event b within the next ω time units. So we are sure that all the events a are examples, i.e. $n_a = n_{ab}$ and $n_{a\bar{b}} = 0$.

In practice, since the events b are not regularly spread over the sequence, the maximal gap between two consecutive events b can be greater than $\frac{L}{n_b}$. So the

[4] We consider that two events a and b occurring at the same time do not make an example.

threshold $\omega \geq \frac{L}{n_b}$ is not enough to be sure that $n_a = n_{ab}$. This is the reason why we introduce a coefficient k into the function $n_{a\bar{b}}(\omega)$:

$$n_{a\bar{b}}(\omega) = n_a - \frac{n_a n_b}{L} \frac{\omega}{k} , \qquad \text{if } \omega \leq \frac{kL}{n_b}$$

$$n_{a\bar{b}}(\omega) = 0 , \qquad \text{otherwise.}$$

The coefficient k can be seen as a non-uniformity index for the events b in the sequence. We have $k = 1$ only if the events b are regularly spread over the sequence (Fig. 9).

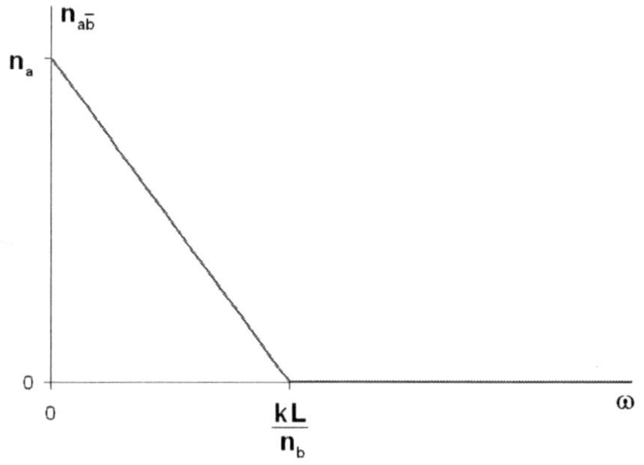

Fig. 10. Model for $n_{a\bar{b}}(\omega)$.

With this model for $n_{a\bar{b}}(\omega)$, we can now study the interestingness measures with regard to ω and k. Several interesting behaviors can be pointed out for SII (see illustration in Fig. 11):

- There exists a range of values for ω which allows SII to be maximized. This is intuitively satisfying[5]. The higher the coefficient k, the smaller the range of values.
- If ω is too large, then $SII = 0$. Indeed, the larger the time window, the greater the probability of observing a given series of events in the sequence, and the less significant the rule.
- As for the small values of ω (before the range of values which maximizes SII):
  - If $k \approx 1$, then n_{ab} increases fast enough with ω to have SII increase (Fig. 11 at the top).

[5] When using a sequence mining algorithm to discover a specific phenomenon in data, lots of time is spent to find the "right" value for the time window ω.

– If k is larger, then n_{ab} does not increase fast enough with ω. *SII* decreases until n_{ab} becomes more adequate (Fig. 11 at the bottom).

On the other hand, confidence (idem for recall) increases linearly with ω (see Fig. 12 with a logarithmic scale). Above all, the three measures confidence, recall, and J-measure do not tend to 0 when ω is large[6]. Indeed, these measures depend on ω only through $n_{a\bar{b}}$, i.e. the parameter ω does not explicitly appear in the formulas of the measures. If ω is large enough to capture all the examples, then $n_{a\bar{b}} = 0$ is fixed and the three measures become constant functions (with a good value since there is no counter-example). This behavior is absolutely counter-intuitive. Only *SII* takes ω explicitly into account and allows rules with too large time window to be discarded.

4 Conclusion

In this article, we have studied the assessment of the interestingness of sequential rules. First, we have formalized the notions of *sequential rule*, *example of a rule*, and *counter-example of a rule*. We have then presented the *Sequential Implication Intensity (SII)*, an original statistical measure for assessing sequential rule interestingness. *SII* evaluates the statistical significance of the rules in comparison with a probabilistic model. Numerical simulations show that *SII* has interesting features. In particular, *SII* is the only measure that takes sequence enlargement, sequence repetition, and window enlargement into account in an appropriate way.

To continue this research work, we are developing a rule mining platform for sequence analysis. Experimental studies of *SII* on real data (Yahoo Finance Stock Exchange data) will be available soon.

References

1. R. Agrawal and R. Srikant. Mining sequential patterns. In *Proceedings of the international conference on data engineering (ICDE)*, pages 3–14. IEEE Computer Society, 1995.
2. R. Agrawal and R. Srikant. Fast algorithms for mining association rules. In Jorge B. Bocca, Matthias Jarke, and Carlo Zaniolo, editors, *Proceedings of the twentieth international conference on very large data bases (VLDB 1994)*, pages 487–499. Morgan Kaufmann, 1994.
3. J. Blanchard. *Un système de visualisation pour l'extraction, l'évaluation, et l'exploration interactives des règles d'association.* PhD thesis, Université de Nantes, 2005.
4. J. Blanchard, F. Guillet, and H. Briand. L'intensité d'implication entropique pour la recherche de règles de prédiction intéressantes dans les séquences de

[6] This does **not** depend on any model chosen for $n_{a\bar{b}}(\omega)$.

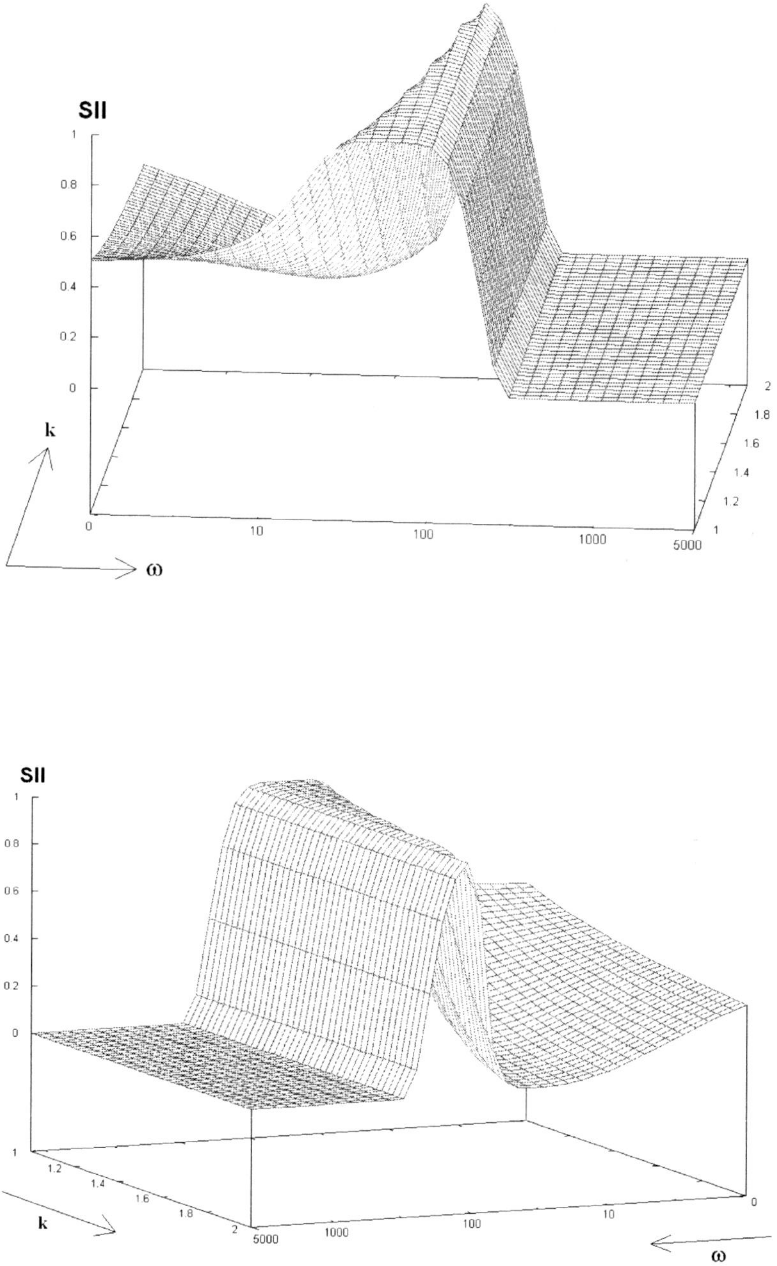

Fig. 11. *SII* with window enlargement.
$n_a = 50$, $n_b = 100$, $L = 5000$

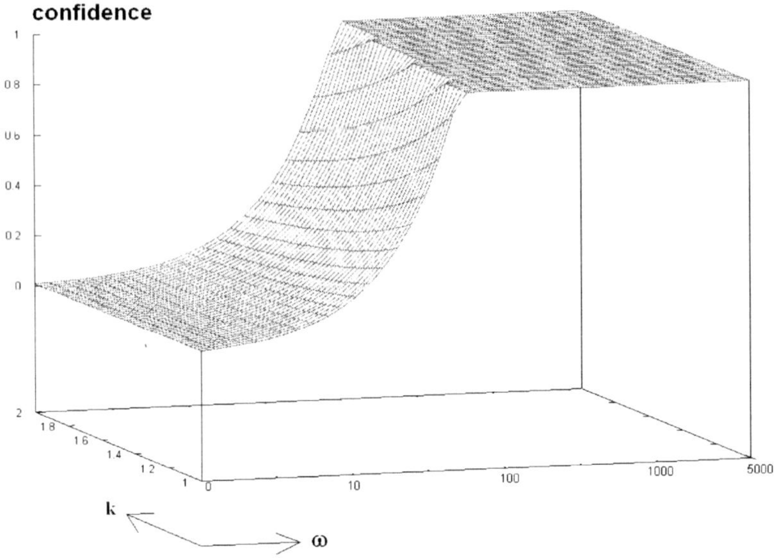

Fig. 12. Confidence with window enlargement.
$n_a = 50$, $n_b = 100$, $L = 5000$

pannes d'ascenseurs. *Extraction des Connaissances et Apprentissage*, 1(4):77–88, 2002. Actes des journées Extraction et Gestion des Connaissances (EGC) 2002.

5. G. Das, K.-I. Lin, H. Mannila, G. Renganathan, and P. Smyth. Rule discovery from time series. In R. Agrawal, P. E. Stolorz, and G. Piatetsky-Shapiro, editors, *Proceedings of the fourth ACM SIGKDD international conference on knowledge discovery and data mining*, pages 16–22. AAAI Press, 1998.
6. R. Gras. *L'implication statistique : nouvelle méthode exploratoire de données.* La Pensée Sauvage Editions, 1996.
7. R. Gras, P. Kuntz, R. Couturier, and F. Guillet. Une version entropique de l'intensité d'implication pour les corpus volumineux. *Extraction des Connaissances et Apprentissage*, 1(1-2):69–80, 2001. Actes des journées Extraction et Gestion des Connaissances (EGC) 2001.
8. J. Han, J. Pei, and X. Yan. Sequential pattern mining by pattern-growth: Principles and extensions. In W. W. Chu and T. Y. Lin, editors, *Recent Advances in Data Mining and Granular Computing (Mathematical Aspects of Knowledge Discovery)*, pages 183–220. Springer-Verlag, 2005.
9. J. Hipp, U. Güntzer, and G. Nakhaeizadeh. Algorithms for association rule mining a general survey and comparison. *SIGKDD Explorations*, 2(1):58–64, 2000.
10. F. Höppner. Learning dependencies in multivariate time series. In *Proceedings of the ECAI'02 workshop on knowledge discovery in spatio-temporal data*, pages 25–31, 2002.
11. M. Joshi, G. Karypis, and V. Kumar. A universal formulation of sequential patterns. Technical report, University of Minnesota, 1999. TR 99-021.

12. H. Mannila and H. Toivonen. Discovering generalized episodes using minimal occurrences. In *Proceedings of the second ACM SIGKDD international conference on knowledge discovery and data mining*, pages 146–151. AAAI Press, 1996.
13. H. Mannila, H. Toivonen, and A. I. Verkamo. Discovering frequent episodes in sequences. In *Proceedings of the first ACM SIGKDD international conference on knowledge discovery and data mining*, pages 210–215. AAAI Press, 1995.
14. H. Mannila, H. Toivonen, and A I. Verkamo. Discovery of frequent episodes in event sequences. *Data Mining and Knowledge Discovery*, 1(3):259–289, 1997.
15. Sheldon M. Ross. *Introduction to Probability Models*. 2006. 9th edition.
16. M. Spiliopoulou. Managing interesting rules in sequence mining. In *PKDD'99: Proceedings of the third European conference on principles of data mining and knowledge discovery*, pages 554–560. Springer-Verlag, 1999.
17. R. Srikant and R. Agrawal. Mining sequential patterns: generalizations and performance improvements. In *EDBT'96: Proceedings of the fifth International Conference on Extending Database Technology*, pages 3–17. Springer-Verlag, 1996.
18. X. Sun, M. E. Orlowska, and X. Zhou. Finding event-oriented patterns in long temporal sequences. In Kyu-Young Whang, Jongwoo Jeon, Kyuseok Shim, and Jaideep Srivastava, editors, *Proceedings of the seventh Pacific-Asia conference on knowledge discovery and data mining (PAKDD2003)*, volume 2637 of *Lecture Notes in Computer Science*, pages 15–26. Springer-Verlag, 2003.
19. G. M. Weiss. Predicting telecommunication equipment failures from sequences of network alarms. In *Handbook of knowledge discovery and data mining*, pages 891–896. Oxford University Press, Inc., 2002.
20. J. Yang, W. Wang, and P. S. Yu. Stamp: On discovery of statistically important pattern repeats in long sequential data. In Daniel Barbará and Chandrika Kamath, editors, *Proceedings of the third SIAM international conference on data mining*. SIAM, 2003.
21. M. J. Zaki. SPADE: an efficient algorithm for mining frequent sequences. *Machine Learning*, 42(1-2):31–60, 2001.

Part II

Application to concept learning in education, teaching, and didactics

Student's Algebraic Knowledge Modelling: Algebraic Context as Cause of Student's Actions

Marie-Caroline Croset, Jana Trgalova, and Jean-François Nicaud

LIG Laboratory, MeTAH Team
46, Av. Felix Viallet
38031 Grenoble Cedex, France
{Marie-Caroline.Croset, Jana.Trgalova, Jean-Francois.Nicaud}@imag.fr

Summary. In this chapter, we describe a construction of a student model in the field of algebra. For gathering the data, we have used the Aplusix learning environment, which allows students to make freely calculation steps and records all the students' actions. One way to build and update the student model is to precisely follow what the student is doing, by means of a detailed representation of cognitive skills. We are interested in persistent and reproducible actions, i.e., the same action done by a student in different algebraic contexts, rather than in a local student action. For discovering patterns of student behaviours, we use a statistical implicative analysis which makes possible seeking for stability of the actions and determining the contexts where they appear. This theory allows us to build implicative connections between algebraic contexts and actions.

Key words: Student model, Interactive Learning Environments, algrebraic transformations, errors

1 Introduction

Teachers need information about students learning of reasoning processes in order to be able to take appropriate didactical decisions: knowing where the student has difficulties, what he/she masters, how his/her knowledge evolves... They pay particular attention to errors: "errors are not only the effect of ignorance, of uncertainty, of chance [...], but the effect of a previous piece of knowledge which was interesting and successful, but which now is revealed as false or simply not adapted" [1]. Usually established by hand assignment, the collection and analysis of precise and individual information about student's knowledge is a slow and bothersome work, especially if there is a large student body.

Interactive Learning Environments (ILEs) make it possible to overcome these difficulties: automatic data collections are carried out and constitute what is called a "student model". Sison and Shimura define a student model

M.-C. Croset et al.: *Student's Algebraic Knowledge Modelling: Algebraic Context as Cause of Student's Actions*, Studies in Computational Intelligence (SCI) **127**, 75–98 (2008)
www.springerlink.com

as "an approximate, possibly partial, primarily qualitative representation of student knowledge about a particular domain, or a particular topic or skill in that domain, that can fully or partially account for specific aspects of student behaviour" [2]. Student models not only inform teachers and researchers about the students' knowledge state, they also guide artificial tutors in the choice of exercises to be presented to the students. They need first an accurate behaviour diagnosis and second a technique to analyze the resulting data, in order to select which errors should be corrected, or mentioned. Indeed, a systematic and repeated remediation cannot be considered: Sleeman has pointed out that error-specific and automatic feedback is no more useful to the student than generic remediation [3], unless to detect patterns in behaviours and cognitive reasons for these behaviours. Brown and Van Lehn have developed a "Repair theory" to provide procedures that will account for bugs [4]. The authors see a bug appearance as an attempt at a repair when the student's knowledge leads to an impasse. They are concerned with systematic bugs that outline regularities, which constitutes an interesting point of view.

Following the same idea, in order to avoid a systematic ineffective remediation we are looking for organized rather than isolated errors, in the particular field of algebra. Our data are collected thanks to the Aplusix learning environment (cf. section 3). We have chosen the statistical implicative analysis (SIA) [5] and the CHIC software where this theory is implemented to outline stabilities. The reasons for this choice are described in section 4. We assume that the SIA allows us:

- to determine individually which tasks present patterns of stable errors (section 7),
- to outline student stable behaviours in terms of implicative links between algebraic contexts and actions (section 6),
- to model behaviour groups for a population of students (section 8).

2 Interactive Learning Environments for Algebra

Since 2003, our team has been engaged in research on student modelling in the field of algebra; especially in the area of transformational (rule-based) activities [6] such as expanding, collecting like terms, factoring, and solving equations. Expression transformation is concerned with changing the form of the expression or equation in order to maintain equivalence.

The choice of an ILE to gather data is crucial. Indeed, transformation steps in most ILEs do not depend only on the student's decisions: they also depend on the degree of initiative left to the student by the environment. The existing systems for algebra are quite different in terms of possible actions, ranging from Computer Algebra Systems (CASs)[1], such as Maple or Derive, where

[1] In a CAS, the transformations are made by the system and often the computer itself selects the sub-expressions to which the rules are applied. In addition, and

the system solves transformational tasks, to ILEs such as Aplusix where the student has to perform all the transformations with the given expressions.

To avoid combinatorial explosion some ILEs limit the student action to one calculation step. The expected step is often very simple. Such is for example the case of Cognitive Tutors [7]: each student action is required to be on an interpretable path. At each step, the student action is compared to applicable rules in the model and immediate feedback is conventionally provided. If the student action matches one of the applicable rules, the tutor accepts the action and applies the rule to update the internal representation of the problem state. If the student action does not match the action of any applicable rule in the model, the action does not register and the tutor provides a brief message in the hint window. In case of ambiguity about the interpretation of the student action, the student is presented with a disambiguation menu to identify the appropriate interpretation of the action. The feedback is immediate and immediate error correction is required. As a consequence, Cognitive Tutors make it possible to follow the student very closely but, as Mc Arthur points out, "because each incorrect rule is paired with a particular tutorial action [...], every student who takes a given incorrect step gets the same message, regardless of how many times the same error has been made or how many other errors have been made" [8].

Other authors ask the students to mark the rule they wish to apply. In MathXpert [9] for example, the student selects a sub-expression, then chooses a rule in a menu providing the rules that are applicable to this sub-expression, see Fig. 1. The chosen rule is then automatically applied. The opportunity to make mistakes in such an environment is strongly restricted. The T-algebra environment [10] differs from MathXpert in the fact that it is the student who has to write the result of the rule application (when the system is in 'free mode'). However, the rules menu presented by the system is also contextual: it depends on the selected sub-expression.

A completely different approach is used in the Aplusix environment [11]: this microworld leaves the students entirely free to produce the expressions they wish, without specifying which rules they should apply. This environment allows students to apply several rules in one single step, as they do in the usual paper and pencil environment. Therefore, with such environments, we are closer to the real mental processes of students. For this reason, we have chosen Aplusix environment for our study of students' actions.

---

contrary to the ILEs, the commands of a CAS are very powerful: simplify, factorise, expand, solve, differentiate, integrate, and so on.

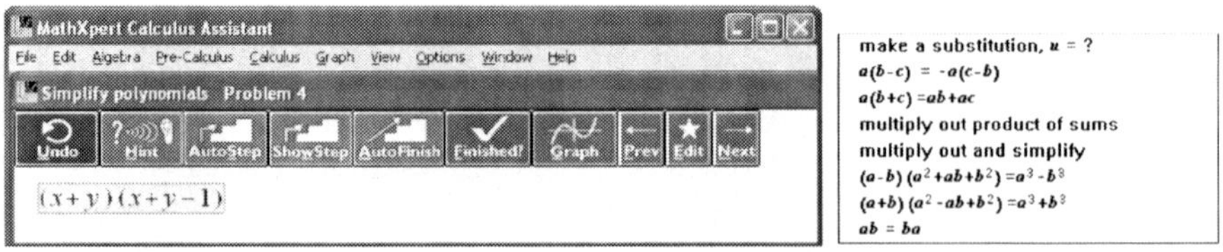

Fig. 1. Screenshot of MathXpert: rules menu is contextual to the selected expression

3 Aplusix Learning Environment

3.1 Presentation

Four different types of exercises, built by teachers or researchers, can be proposed to students: calculate numerical expressions, expand and reduce polynomial expressions, factor polynomial expressions, and solve equations, inequalities or systems of linear equations. In each situation, an expression and an instruction are given to the students. To solve the exercise, the students can duplicate the expression and modify it, see Fig. 2. The transformation of an expression into another one is called a student's step. In the test mode no feedback is given to students, while in the training mode epistemic feedback is generated in terms of indicators giving the state of expressions and the correctness of the student's calculations. Aplusix records in text files the following information about the student's actions: time, keyboard or mouse actions, and expression obtained. The files can be viewed by the student, the teacher or the researcher thanks to a 'replay system' included in the software.

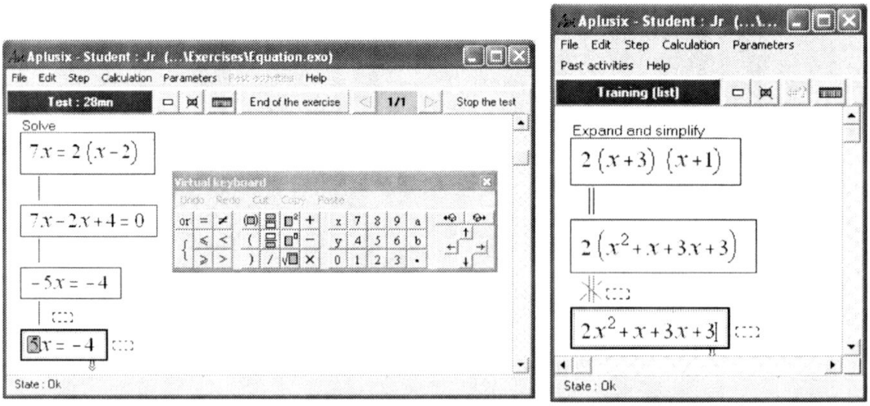

Fig. 2. Screenshots of Aplusix. On the left, the student is in test mode, he/she has done several transformations in each step. On the right, the student is in training mode, with feedback about correctness

Aplusix permits the occurrence of complex errors and actions in one student's step, as shown in Fig. 2. As a result, the understanding of the student reasoning is complicated and the difficulty of providing a diagnosis of mistakes increases. It is then necessary to subdivide a student's step into elementary steps. This is done by an automatic process: the rules diagnosis provided by Anaïs, presented in the following section.

3.2 Rules Diagnosis

The files used for an automatic analysis consist only of the calculation steps validated by the student (i.e., the student's steps): corrections, hesitations and time are not taken into account. Software, called Anaïs, has been developed to analyse students' productions. It is based on rules established from a didactical analysis and gathered in a library. The analysis consists of searching for the best sequence of the rules (correct or incorrect) that can explain a given student's step. The process of the analysis is as follows: from the expression that is the source of the student's step, Anaïs develops a tree by applying all the rules applicable to this expression.

- The application of a rule produces a new node. Anaïs thus gradually builds a research tree, at each level choosing the node to be developed by using a heuristic that takes into account the goal (the expression resulting from the student's step).
- When the process is successful, the goal may be reached by several paths, each of them being a diagnosis. The selection of the best diagnosis is based on a cost of paths expressed in terms of the number and kind of rules applied.

The Anaïs software provides a diagnosis in the form of a sequence of intermediate elementary stages (rewriting rules) to explain the steps produced by the student, as shown in Tab. 1.

We call *elementary step* an automatic intermediate stage provided by Anaïs. An elementary step has an initial expression and a final (intermediate) expression.

A single rule explains each elementary step. A step can belong to one of four different tasks, whatever the type of the exercise: expansion, factoring, collecting like terms, and movement, cf. Tab. 1. An exercise of the equation-solving type may involve factoring, collecting like terms, and movement steps.

Remark 1. One can question the interest of leaving such a large freedom to the student's answers and thus having to reconstruct the intermediate steps. Indeed, the ILEs presented in section 2 do not need to do this complex work. However, when a student solves an algebra exercise, he/she does not always see an expression transformation as the application of a rule, with an initial and a final expression. Requiring that students cite the rules applied may be

Student's step	(Automatic) elementary steps	Initial expression	Final expression	Associated automatic rules	Associated Task
$7x - 2x + 4 = 0$	$7x - 2x \mapsto 9x$	$7x - 2x$	$9x$	$ax + bx \mapsto (a - b)x$ (incorrect)	Reduction
$\mapsto 9x = -4$	$9x + 4 = 0 \mapsto 9x = -4$	$9x + 4 = 0$	$9x = -4$	$a + b = 0 \mapsto a = -b$ (correct)	Movement

Table 1. Example of a student's step decomposition into automatic elementary steps

of didactical interest, but it may lead us away from the student's real way of thinking. The freedom the students have while working with Aplusix puts us as close as possible to their real mental processes.

4 The Choice of Statistical Implicative Analysis

One of the difficulties in catching stable errors is to define what stability is: what is a good threshold to decide when a rule can be considered as having been regularly used? The most common and spontaneous technique for catching stability is to count the number of opportunities for a rule application and divide it by the number of effective student's applications [12,13]. Authors do not define very precisely what is called opportunity for a rule application. They recognize themselves that "different mal-rules have widely different 'opportunities' of occurring, and in some cases the number of opportunities is impossible to quantify without looking closely at individual students protocols" [13].

4.1 Difficulties in Defining what Stable Behaviours Are

Let us take an example to show the difficulty of stability definition. Two students, A and B, are asked to collect like terms in the five expressions given in Tab. 2.

$3x - 5 + 3x$	$E1$
$3x + 3x - 5$	$E2$
$-5 + 3x + 3x$	$E3$
$7x - 8 + 3 + 7x$	$E4$
$5x \times 2x$	$E5$

Table 2. Expressions proposed to student A and B

Let us call R, the rule: $ax + bx \mapsto (a - b)x$.

The four expressions $E1$, $E2$, $E3$, and $E4$ present an opportunity to apply R. Clearly, $E5$ does not present an opportunity for application of R because the initial expression is a product instead of a sum. Therefore, it seems that there are four opportunities for application of R. Let us see what the answers of students for the four expressions are:

Let us suppose that student A's answer is -5 for each of the first four expressions (and, for example, $10x^2$, for $E5$). This means that student A has used R for these four expressions: the frequency of application of R associated to student A is then $4/4$. It seems that the behaviour of student A is very stable with respect to this rule.

Let us suppose now that student B's answers are respectively: -5, $6x - 5$, $-5 + 6x$ and -5 for the first four expressions (and, for example, $10x^2$, for $E5$). This would mean that the student B has used R only for $E1$ and $E4$. The frequency of application of R for the student B would then be $2/4$. Does it mean that the application of R for the student B is unstable? We do not believe that.

Balacheff explains that two different behaviours (using two different rules here) "can appear as conflicting but this inconsistency can be explained either by the time evolution or by the situation/context" [14]. The algebraic context of the four expressions is not the same: in $E1$ and $E4$, the minus sign is between the two monomials. Moreover, the minus sign is placed side by side with $3x$, respectively $7x$. In the expressions $E2$ and $E3$, there is also a minus sign, but it is not between the two equal monomials. The didactical variable, which is the position of the minus sign in the expressions, does not take the same values for each expression, and it influences the behaviour of student B. We can say that the behaviour of student B is stable, with respect to the minus sign position: for him/her, the expressions $E2$ and $E3$ do not present an opportunity to apply R. Each student has his own conception of the opportunity for applying the rule R. For this reason, it does not seem possible to objectively count the number of opportunities for a rule application. The application opportunity is a subjective notion: a "transfer from one situation to another one is not an obvious process, even if in the eyes of an observer these situations are isomorphic" [15]. We see that stability definition depends on what is called opportunity.

4.2 Algebraic Context as Source of Behaviour

In our work, we decided not to have a priori ideas of what constitute opportunities for a given rule application. We suppose that the sources of errors in incorrect transformations are principally in the characteristics of the expression, such as the degree of the initial expression, the nature of its coefficients, the presence of a minus sign, and so on. We need to describe precisely the algebraic characteristics of the initial expression to which a rule is applied and we are looking for the algebraic context that can 'better' explain the rule utilisation for each student.

4.3 Statistical Implicative Analysis

Since we are looking for causes of students' errors and not just correlations, the choice of SIA seems worthwhile. Indeed, the SIA approach makes it possible to find implicative links between attributes, in our case between algebraic characteristics and rules. In addition, this technique takes into account the number of times that a context appears relative to the other contexts. For example, if a student S_1 uses a rule R_1 ten times in a context C_0, and he/she uses a rule R_2 five times in the same context C_0, then the quasi-implication $C_0 \rightarrow R_1$ can be evaluated.[2] Let us consider another student, S_2, who uses the rule R_1 100 times in the context C_0 and the rule R_2 50 times in the same context. The frequency of R_1 application by both students is the same (2/3), but the SIA approach makes a distinction between 10/15 and 100/150.

5 The Conditions of the Expected Quasi-Implications

5.1 Algebraic Context Variables

Let us consider a rule set, $\{R_k\}_{1 \leq k \leq p}$.[3] To this set, we associate *algebraic context variables*, $\{V_i\}_{1 \leq i \leq n}$ that are the main characteristics of initial expressions on which each rule R_k can be applied. Each algebraic context variable, V_i, (just called variable in what follows) can be assigned values noted $\{\overline{V}_{ij}\}_{1 \leq j \leq m_i}$.

We call *contexts* the vectors $(\overline{V}_{1j_1}, \overline{V}_{2j_2}, \ldots, \overline{V}_{nj_n})$, associated to an expression. The number of such vectors is $\prod_{i=1}^{i=n} m_i$. We denote a context by C_l, where $l \in \{1, \ldots, \prod_{i=1}^{i=n} m_i\}$.

Remark 2. Since we will use the CHIC software, we prefer to have variables with binary values. Therefore, we consider the values $\{\overline{V}_{ij}\}_{1 \leq j \leq m_i}$ as new variables, called *binary context variables*, which take 0 or 1 as values. When necessary, the distinction between binary context variables and algebraic context variables will be made.

We illustrate this with an example. Let us consider the expression $3x - 5$. The operator and the presence of a minus sign are two algebraic context variables of the expression. The operator can be one of the five binary context variables which are times, plus, minus (e.g. the expression $-(2x + 7)$), bracket and exponent. These binary context variables can be assigned binary values:

[2] Quasi-implication is called a "rule" by the authors of the SIA approach. In order not to confuse it with what we call algebraic rules, we will call it implication or quasi-implication, noted $\rightarrow$, while the transformation of an algebraic expression is noted $\mapsto$.

[3] In general, rule set is associated to a task or a part of task.

the expression has or does not have the particular operator. The second variable, presence of a minus sign, has one binary variable: itself. A context extract of the expression $3x - 5$ is then (Plus, Presence of minus sign).

Obviously, variables depend on tasks: variables for factoring or for movement are not the same. We assume that the variables have impact on the use of a rule by a student. This means that the rule used depends on the value of a variable.

We will consider a behaviour as *stable* if the student uses the same rule each time (or almost each time) that he/she is in the same algebraic context. Description and choice of variables are determining factors for catching stability and this is a difficult task. The values depend on didactical decisions, as we will show in the next subsection.

5.2 Creation of the Algebraic Context Variables List

The choice of the main characteristics of an expression is based on a two-steps procedure:

- a praxeological analysis [16] of textbooks in the field of transformational activities,
- a didactical analysis, together with the construction of the rules library.

Chevallard, in his Anthropological Theory of Didactics (ATD), describes mathematical knowledge both as a means and a product of activity, which form the "praxeological organization", a union of practice and discourse about practice. The basic elements of the anthropological model (types of problems, techniques, technologies and theories) make it possible to analyze mathematical textbooks and organize problems by types. What makes a difference between two types of problems constitutes our first list of variables. For example, two types of problems related to the task of expansion are *expand* $a(b+c)$ and *expand* $(a+b)(c+d)$. One of the variables can thus be the number of the terms in the sum. For example, in the first case, this variable will be assigned the value $(1, 2)$ while in the second case, the value $(2, 2)$.

This work is not sufficient. Indeed, if we limit ourselves to textbook analysis, we remain within the institutional contract. One thing that can cause errors is a rupture of this contract. Therefore, we need to complete the first list with what we think may be missing in the textbooks. For example, in the task of expansion very few textbooks propose problems with three factors. We have then considered the number of factors as a new variable associated to the task of expansion.

In addition, we have to describe what the values of the variables are and which are relevant for a change of the student's behaviour. For example, the degree of an expression is obviously a variable. But what values are pertinent for this variable? When interviewed, some students pointed out that they knew how to deal with expressions of degree 2 but not with expressions of degree 3 and more. This leads us to suppose that the fact that an expression is of degree 3 or 4 or 5 makes no difference for a student. The values of the expression degree will then be $0, 1, 2$ and greater than or equal to 3.

Not all these variables are as yet what Brousseau calls *didactical variables* [17]: we are not sure that they provoke a change of strategy or a change of rules use. However, we think that they can have an impact on strategies for some students. The SIA analysis will say which of the variables are didactical and which are not. Indeed, the SIA analysis will give results in terms of implications between variables and actions. The variables that emerge in these implications will be considered to be didactical variables.

5.3 Presentation of the Files Analysed

Experimentation with Aplusix has been carried out for different purposes by teachers and researchers [18]. However, experimentation presented in this chapter was all conducted in the test mode (i.e., without information about correctness of students' steps, see section 3). Log files were gathered in a database and analysed by Anaïs to provide a sequence of elementary steps for each student's step.

We can query (with the Structured Query Language — SQL) the database and get result sets. These sets are tables which can be recorded in the CSV format and directly used by the CHIC software. We will call these tables *CHIC tables*.

The lines of the CHIC tables used in section 6 and 7 consist of the elementary steps from the automatic diagnosis for *individual* students. At least one CHIC table is associated to one student. The columns, called *attributes*, consist of the binary context variables and the actions. The actions can be either the rules diagnosed by Anaïs (section 6) or a collection of rules (section 7), according to what we want to model: a precise task or a set of tasks. The values assigned to the rules are binary: either the rule is used or not. If an elementary step in line i is explained by the rule which is in column j, there is a 1 in the cell (i, j). The binary context variables are, as their name indicates, binary. There is 1 in the cell (i, k) each time the initial expression of the step i can be described by the (binary) variables of column k. An example of line is shown in Tab. 3.

The files used in section 8 are not exactly the same and will be explained in time.

Remark 3. On a line, it is not possible to have 1 in the columns of two binary variables that depend on the same algebraic context variable while there can be as many 1's as there are context variables.

5.4 Implications

Four kinds of implications between attributes, binary context variables, $\overline{V}_{ij}$ and rules, R_k, are envisaged. Let us explain what they mean:

Elementary Steps	R_1: $ax + bx \mapsto (a - b)x$	R_2: $ax + bx \mapsto (a + b)x$	Decimal coefficients	Integer coefficients	Plus operator
$7x - 2x \mapsto 9x$	1	0	0	1	1

Table 3. Extract of CHIC table analysed by the software CHIC

- $\overline{V}_{ij} \to R_k$. This implication means that when expression variables V_i take the value $\overline{V}_{ij}$, the rule R_k is almost always used by the student. This is the most important implication for us.
- $R_k \to \overline{V}_{ij}$. This implication means that when the rule R_k is used,$\overline{V}_{ij}$ is often the context in question. The contrapositive is easier to understand: when the context is not $\overline{V}_{ij}$, R_k is not used.
- $\overline{V}_{ij} \to \overline{V}_{rs}$, with $r \neq i$[4]. This implication means that the algebraic variable $\overline{V}_{ij}$ mathematically implies the variable $\overline{V}_{rs}$. This occurs when an expression in the data can be described both by $\overline{V}_{ij}$ and $\overline{V}_{rs}$. For example, in grade 8, students do not know how to solve equations of degree 2 in the canonical form, i.e. $ax^2 + bx + c = 0$. However, they can solve such equations if the left member is a product of two linear factors and the right member is 0. The variables 'degree 2' and 'member 0' would be then strongly correlated.
- $R_k \to R_s$. This implication does not appear in our work. Each line is a single elementary step: it means that, in the line, there is one and only one rule in the whole set of rules that can be used.

An option in the CHIC software makes it possible to consider only the first two implications: we select the rules, as the 'principal vertex' of the graph (see Couturier chapter).

6 An Accurate Student's Model: the Case of Factoring

Experiments were conducted in the field of factoring. They were built in order to collect data on this domain. Of course, only the elementary steps of the factoring task are collected in the CHIC table for analysis: steps concerning collecting like terms or expanding are not taken into account. Here, the actions are the rules associated to the elementary steps from the automatic diagnoses. We have modelled students individually: the CHIC table represents here the data collected about *one* student.

[4] The index r cannot be equal to i: a value of a variable cannot imply another value of the same variable.

6.1 The Attributes

The Actions.

In the rules library, there are 20 rules concerning factoring. We present here the results for one student. Five rules have been diagnosed for this student:

- *Correct*: the factoring is correct.
- *ErMinus*: the factoring is erroneous and the mistake is about a minus sign. For example, $(5x+1)x-(1+5x)y \mapsto (5x+1)(x+y)$.
- *ErNothing*: when the cofactor of the common factor is 1, some students think that there remains "nothing" when the common factor is withdrawn. For example, $(x+3)(x+2)+(x+3) \mapsto (x+3)(x+2)$. This transformation can be explained by the loss of a term but interviews with students show that the concept of "nothing" was behind this transformation.
- *ErOther*: other kind of factoring errors.
- *NoInf*, meaning NoInformation: the student has not answered this task. Note that it is not a step but an expression: there is no final expression because the student has stopped solving the task.

The Context Variables.

There are 36 binary context variables associated to the factoring task. The six context variables are the nature of the common factor, its visibility, its degree, its position, the nature of its cofactors and the presence and position of a minus sign. Each variable is decomposed into binary variables. In what follows, we explain only those that appear in the implicative graph. Each of them is illustrated by an example.

- The nature of the common factor can be numeric (e.g., $6x+3$), monomial (e.g., $3x+15x^2$), a sum of two terms (e.g., $(5x+1)x-(1+5x)y$), a sum of three terms (e.g., $x(x+2+y)-(x+y+2)(1+5x)$), or a product (e.g., $x(x-4)+(x^2-4x)(x+1)$).
- The visibility of the common factor depends on its nature. Let us take the example of a sum of two terms as a common factor. Its visibility can be obvious (e.g., $(x+3)(x+2)+(x+3)$), commuted (e.g., $(5x+1)x-(1+5x)y$), opposite (e.g., $(x+3)(x+2)+(-x-3)$), commuted-opposite (e.g., $(x+3)(x+2)+(-3-x)$), disconnected (e.g., $x+(x+2)\times x+2$), multiple (e.g., $-6-3x+(1+x)(-2-x)$) or bi-multiple (e.g., $-6-3x+(1+x)(-4-2x)$).
- The nature of the cofactors can be numeric, unit, monomial, sum, product or identical. For example, in the expression $(x+3)(x+2)+(x+3)$, the nature of the cofactors is respectively *sum* $(x+2)$ and *unit* (1); while in the expression $x(x-4)(x+1)+(x-4)^2$, the cofactors of the common factor $(x-4)$ are on the one hand the *product* $x(x+1)$ and on the other hand, the sum $(x-4)$, which is *identical* to the common factor.

6.2 The Experimentation

Contrary to the cases presented in sections 7 and 8, this experimentation carried out with grade 8 students was organized especially to collect data about factoring. Therefore, the exercises were built in order to cover a maximum of types of the above mentioned variables. 10752 exercises could have been created by taking the direct product of values of the variables. Of course, it is not realistic to hope to find students who will work on so many exercises. Moreover, some combinations are not interesting from a didactical point of view: for example, it is not interesting to have two units for the two cofactors, as in the expression $(x + 5) + (x + 5)$. Moreover, if the common factor is a product, we have decided not to accept a product as cofactor in order not to generate excessively complicated expressions. For example, we have not considered expressions like $x(x + 1) + x(x + 1)(x + 2)(x + 4)$, where the common factor is a product $x(x+1)$ and one of the cofactors is also a product $(x + 2)(x + 4)$. For these reasons, only 41 expressions were kept (see Tab. 4 for an extract). The student whose results are presented here worked for 70 minutes on these expressions without any help or feedback. She tried to solve the whole range of the 41 exercises.

Examples	Common factor									Cofactors				
Expressions	FNumericObvious	FMonoObvious	FSum2	Fsum2Obvious	FSum2Opp	Fsum2Commut	Distribution	Distribution	Distribution	Distribution	Distribution	Expected value	Distribution	Distribution
$2y^2 + 2$	1	0	0	0	0	0	0	0	0	1	0	1	0	0
$-7x - 3x^2$	0	0	0	0	0	0	0	0	0	0	1	0	1	0
$3x^2 + yx$	0	1	0	0	0	0	0	0	0	0	0	1	1	0
$3(x + 8)^2 + x^2 + 8x$	0	0	1	0	0	0	0	1	0	0	0	1	1	0
$(x - 3)(1 - 4x) - 5(3 - x)$	0	0	1	0	0	0	1	0	0	0	1	0	0	1
$-6 - 3x + (2 - x)(-4 - 2x)$	0	0	1	0	0	0	0	0	1	0	1	0	0	0
$(x - 4) + (x - 4)x$	0	0	1	1	0	0	0	0	0	1	0	1	0	0
$-(x + 3 + y) + x(y + x + 3)$	0	0	0	0	0	0	0	0	0	1	0	1	0	0

Table 4. Extract of the Experimentation Exercises. The Case of Factoring

6.3 The Results

The implicative graph was then built from a file with 41 lines and 36 + 5 attributes. A first work with one premise is presented; a second one follows with two premises.

One premise.

Three sets of implications emerge from the analysis and are presented in Fig. 3:

- The student did not transform the expression, *NoInf*, when the common factor or one of the cofactors is a product (*FProduct* and *CofProduct*). For example, she did not treat exercises like $x(x+2)(x+1)+3(1+x)(2+x)$, where the common factor is a product. She also did not answer when the common factor is a disconnected sum (*FSum2TermDisc*), or an opposite one (*FSum2TermOpp*). For example, the expression $(x+(x+2)\times x+2)$ and $-5x-8+(5x+8)(x+2)$ contributed respectively to the implication ($NoInf \rightarrow FSum2TermDisc$) and ($NoInf \rightarrow FSum2TermOpp$). These results are interesting. Even when the student does not answer, it means something at a cognitive level: considering a product as a common factor seems too difficult for this student.
- The student performed a correct factoring when the common factor was numeric (*Fnum*), or monomial (*Fmono*). For example, the student correctly factorized the expression $-3y-12$ or $3x+9x(x+2)$.
- No interesting information about the *ErNothing* rule use emerges at this level. A necessary condition for the use of *ErNothing* is the presence of a unit cofactor in the expression (*CofUnit*). Indeed, this rule cannot be applied if the cofactor is not a unit. It is this information that appears in the implicative graph. We will see that with two premises we have more information about the use of this rule.

We also added a new action, called *Transformation* that is the opposite of the action *NoInf*. It means that the student transforms an expression either correctly or not. The following implications emerge at threshold 89 (see Couturier chapter): ($Fsum2TermObvious \rightarrow Transformation$) and ($Fsum2TermCommut \rightarrow Transformation$). When the common factor is a sum, either obvious or commuted, the student tries to transform the expression, the transformation being correct or not.

Fig. 3. Implicative graph with one premise. Threshold at 86. Case of factoring

Two premises.

The implications are more precise. One of the implications with two premises is interesting to describe. The student uses the rule ErNothing especially in the cases where the common factor is an obvious sum of two terms, *Fsum2TermObvious*, and of course, one of the cofactors is a unit *Cofunit*, cf. Fig. 4. The transformation $(x+3)(x+2)+(x+3) \mapsto (x+3)(x+2)$ contributes to this implication. When looking at the data, we see that indeed, the student did not use this rule when she was confronted with an expression like $5x+5$, even if the last cofactor is also a unit.

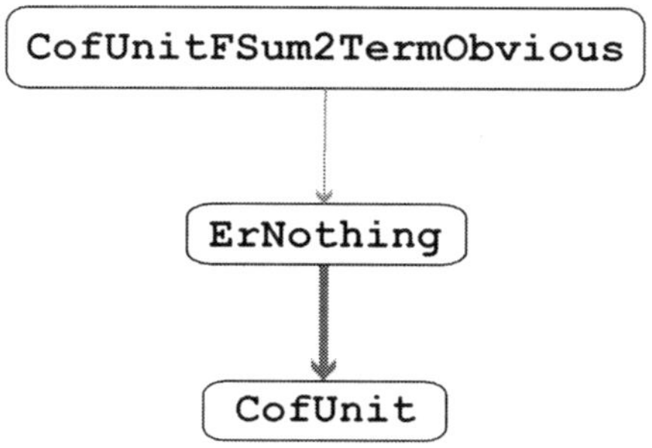

Fig. 4. Part of the implicative graph with two premises. Threshold at 90. Case of factoring

The study of the data of this student shows that she has stable behaviour in the field of factoring. She has correct and incorrect stable actions, well fixed in her behaviour. The SIA approach made it possible to detect stable behaviours according to contexts. It answered one of our questions.

7 Detection of a Task that is the Source of a Student's Errors

To model a student's behaviours, one needs to obtain enough data from this student. This is one of our difficulties. This difficulty was overcome in the previous analysis (section 6) by building special experiments about one precise task. But teachers rarely have the possibility of leaving their students at work on the same task for a long time. In addition, students learn while they work: their knowledge evolves. We cannot model students after too long a time of activities if we want to determine a state of their knowledge. Which task can we, then, model with a set of exercises? Do we have to model each task? And if so, do we have enough data? The SIA allows to answer these questions. It makes it possible to detect which task causes the most errors in a set of tasks, whatever the exercises are. For that, the process is quite similar to the previous one: the CHIC table concerns only one student but it includes many tasks done by her/him. The lines of the CHIC table are again elementary steps, the columns are binary variables and actions. However, in contrast to section 6, the actions are not the rules but sets of rules.

7.1 The Attributes

The Actions.

For this analysis, actions take only two values: either the elementary step is correct (called *Correct*) or it is not (called *Error*). No distinction between the kinds of errors is made.

The Context Variables.

Given an expression, there are 15 binary variables associated to these two actions, described by six variables: the degree of the expression, the operator, the nature of coefficients, the presence of minus sign, the presence of exponent applied to a number and the task associated to the elementary step. The six variables are explained in what follows:

- The degree of the expression, *InDeg*, can be assigned infinitely many values. We assume that there are only four values that have different impact on student's behaviors: the degree $0, 1, 2$ and greater than or equal to 3.
- The operator, *InOp*, can be: plus, times, exponent, parentheses, minus. The expression $x - 2$ has a plus operator while the expression $-(x + 3)$ has a minus as operator.
- The coefficients can be integer, *WIntCoef*, fractional, decimal or irrational.
- The fact that there is a minus sign in the expression, *WMinus*, as in the expression $x - 2$.
- The fact that there is exponent that is not applied to variables, as in the expression $x + 3^2$.
- The tasks associated to an elementary step are those described in section 3: expansion, factorisation, collecting like terms, *Collect*, and movement. The task information concerns the step much more than it does the source expression. However, we prefer to consider it as a context variable rather than an action.

7.2 Some Results

The treated CHIC table can contain as many elementary steps as the student has done. In the following example, 99 steps have been diagnosed. The table is then a 99×15 matrix. The student was chosen from an experiment that especially concerned movement tasks. The results of the implicative graph are presented in Fig. 5. This analysis shows that this student (grade 9) seems to master the movement task: the implication *Movement* → *Correct* appears. However, he does not master the task of collecting like terms, especially when the expression contains a minus sign and integer coefficients. This information appears at threshold 93.

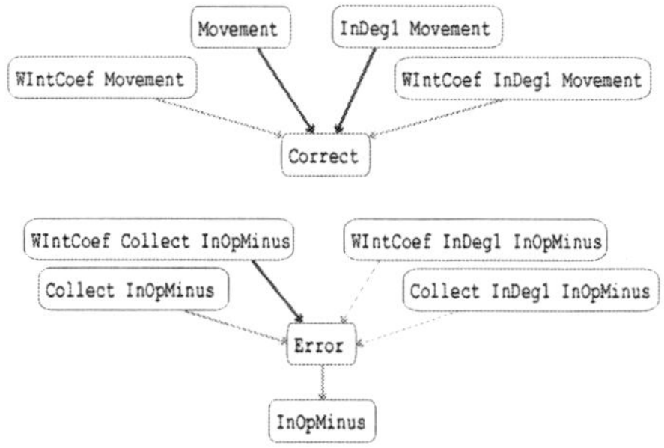

Fig. 5. Implicative graph with three premises, threshold 87. Example of a result for detecting a task that is the source of the student's errors. In this case, the task of collecting like terms, especially in the presence of a minus sign, seems to cause regular difficulties for this student

This treatment answers to our previous questions: the SIA makes it possible to detect tasks that do not present difficulties to the student and those that do. The student in question has difficulties with the specific task of collecting like terms. It is this task that can be interesting to model, as it was done in section 6: selection of the elementary steps concerning this task among the whole 99 steps, association of the context variables in this task and analysis of the resulting implicative graph. The last part can sometimes be complex. In the next section, we will try to overcome this problem.

8 Seeking for Main Behaviours

The analysis of the factoring implicative graph presented in section 6 (Fig. 3), is quite simple for those who have taken time to understand each attribute. On the other hand, it is not possible to give this kind of result to a teacher who would have as many graphs as students in his/her classroom. In addition, some results can be much more complex. An expert cannot be present all the time to analyze the implicative graph associated to each student. For this reason, we try to detect the main possible behaviour groups for a given task in order, in the near future, to automatically associate a student's work to one of these groups that would already be analysed.

8.1 Behaviour Groups

Previous CHIC tables have elementary steps as lines, actions and context variables as columns. For detecting main behaviours, we proceed in another way.

This time, we consider a population of n students. The attributes are the ordered pair (C_l, R_k), where C_l is a context (see section 5.1), and R_k a rule. For each student S, a first new CHIC table, T_S, is constructed in which the lines are again elementary steps, but the columns are the ordered pairs (C_l, R_k)[5]. Let (C_l, R_k) be the r-column. There is 1 in the cell (q, r) if the context C_l describes the elementary step of the line q and the rule R_k is the rule automatically associated to the step.

Let us consider a student S. The *frequency*, $F_S(C_l, R_k)$, of the occurrence of the ordered pair (C_l, R_k) for the student S, is defined by the number of times where R_k was used by the student S in the context C_l divided by the number of times where the student S was confronted with the context C_l.

Remark 4. If the student S has never met the context C_l, the frequency $F_S(C_l, R_k)$ is equal to the mean of the frequencies of (C_l, R_k) of the whole set of students.

A second new CHIC table is created: The columns are again the ordered pairs (C_l, R_k) while each line concerns a particular student. The value of the cell (q, r) is the frequency of the ordered pairs (C_l, R_k) for the student who is in the line q. If there are n students in the considered population, then there are n lines in the CHIC table. The number of columns is the number of possible ordered pairs, namely $p\prod_{i=1}^{i=n} m_i$.

We are looking for groups of ordered pairs (C_l, R_k). We call a *behaviour group* a set of (C_l, R_k) that is representative of a part of the population. For that, we use a similarity tree viewable by the CHIC software: the attributes are grouped according to the similarity of their use by the considered population. For example, a behaviour group consisting of (C_1, R_2) and (C_3, R_1) means that a group of students have used the rules R_2 and R_1 in the context C_1, or respectively C_3.

8.2 The Case of Collecting Like Terms

We are concerned by a part of the task of collecting like terms: the sum of two terms of the same degree, when the degree obtained is correct. In another words, we consider only elementary steps like $ax^m + bx^m \mapsto cx^m$, where m can be null, a and b can be positive or negative. We do not consider transformations like $ax^m + bx^m \mapsto cx^p$, where p is not equal to m. In addition, we consider only four possibilities for the coefficient c: c is obtained as a sum depending on a and b described below. We set aside the case where c is equal to a or b, or ab, and so on. We look for behaviour groups for this sub-task.

[5] If we take the notation used in the context definition (section 5.1), there are $p\prod_{i=1}^{i=n} m_i$ couples.

The Rules.

The actions correspond to the four rules concerning this sub-task:

- *Cor* is the correct calculation of the sum $a + b$ (e.g., $2x - 5x \mapsto -3x$ or $2x^3 + 5x^3 \mapsto 7x^3$).
- *PlusOp*, meaning PlusOpposite, is the sum of a and the opposite of b. The rule can be written as $a+b \mapsto a-b$ (e.g., $2x - 5x \mapsto 7x$ or $2x+5x \mapsto -3x$).
- *OpPlus*, meaning OppositePlus, is the sum of the opposite of a and b. The rule can be written as $a + b \mapsto -a + b$ (e.g., $2x - 5x \mapsto -7x$ or $2x + 5x \mapsto 3x$).
- *OpPlusOp*, meaning OppositePlusOpposite, is the sum of the opposite of a and the opposite of b. The rule can be written as $a + b \mapsto -a - b$ (e.g., $2x - 5x \mapsto -3x$ or $2x + 5x \mapsto -7x$).

The Contexts.

We decided to restrict the context variables associated to this task. We chose only two variables: sign of a and b, order of $|a|$ and $|b|$.[6]

The first variable can be decomposed into four binary variables: (*sign of a is plus, sign of b is plus*), (*sign of a is minus, sign of b is plus*), (*sign of a is plus, sign of b is minus*) or (*sign of a is minus, sign of b is minus*). The second variable can be decomposed into three binary variables: $|a|$ *is smaller than* $|b|$, $|a|$ *is greater than* $|b|$ or $|a|$ *is equal to* $|b|$.

For example, in the expression $2x - 5x$, the sign of a is plus and of b is minus. We denote this information by PM ("Plus, Minus"). In this same expression, $|a|$ is smaller than $|b|$. This information is denoted $C1$. The binary variable $|a|$ *is greater than* $|b|$ is denoted $C2$, and the variable a *is equal to* b is denoted Eg.

Therefore, there are 12 possible contexts for the task of collecting like terms denoted by juxtaposing the two binary variables. For example, the context $C1PM$ means that the context is (*sign of a is plus, sign of b is minus,* $|a|$ *is smaller than* $|b|$).

The Attributes:

Ordered pair (Context, Rule). Since there are twelve contexts and four rules, there are 48 attributes. Their name is obtained by the association of the rule name and the context name. For example, the attribute *CorC1PM* means the use of a correct rule, *Cor*, in the context $C1PM$, while the attribute *PlusOpC2MP* means the use of the rule *PlusOp* in the context $C2MP$.

---

[6] We are aware that we do not take into account some important variables such as the degree of the monomial, a presence of a minus sign in the expression of the student's step, or the nature of the coefficients. One of the reasons for this restriction is that the automation of the process can be long and the work is still in process: we want to be sure that we obtain results before continuing coding other variables.

The Experiment.

As was mentioned in the section 7, experiments were designed with the purpose of obtaining data about collecting like terms. However, this task appears in most of exercises that were proposed in these experiments. In this example, the students population consists of a group of three grade 8 classes of 87 students in total ($n = 87$). The table is then an 87×48 matrix.

Each student worked on more or less 28 exercises. In total, 2584 elementary steps concern the chosen sub-task. We obtain the average of 29.7 elementary steps per student.

Some Results.

In what follows, we describe five of the behaviour groups that are highlighted by the treatment of a similarity tree, see Fig. 6.

The behaviour class ($OpPlusOpC2MM$, $OpPlusOpC1MM$) can be interpreted as follow: when confronted with a sum of two negative terms, a and b, the students from this behaviour class sum $|a|$ and $|b|$. For example, they transform the expression $-2x-5x$ into $7x$ or $-5-2$ into 7. This can mean that they understand that they have to add $|a|$ and $|b|$, but they make a mistake for the sign of the result, making it always positive.

The second behaviour class ($CorC1PP$, $CorC1MP$, $OpPlusOpC2MP$, *OpPlusOpC1PM*, $OpPlusOpC2MM$, $OpPlusOpC1MM$) looks more precisely at the meaning of the previous behaviour class. Let us take the example of the six attributes of this class given in Tab. 5.

Attribute	Example	Example name
$CorC1PP$	$2+5 \mapsto 7$	$T1$
$CorC1MP$	$-2+5 \mapsto 3$	$T2$
$OpPlusOpC2MP$	$-5+2 \mapsto 3$	$T3$
$OpPlusOpC1PM$	$2-5 \mapsto 3$	$T4$
$OpPlusOpC2MM$	$-2-5 \mapsto 7$	$T5$
$OpPlusOpC1MM$	$-5-2 \mapsto 7$	$T6$

Table 5. Example of the six attributes of the class ($CorC1PP$, $CorC1MP$, $OpPlusOpC2MP$, $OpPlusOpC1PM$, $OpPlusOpC2MM$, $OpPlusOpC1MM$)

These students sum correctly when they are confronted with a sum of two positive terms, or with a sum of a negative term (-5), smaller (in absolute value) than a positive term (2), ($T1, T2$). When they are confronted with a sum of a negative term greater (in absolute value) than a positive term, the negative term being on the left ($T3$) or on the right ($T4$), the students calculate $|-5+2|$. The cases $T5$ and $T6$ have been already explained in the previous class.

We can interpret this group as follow: the students know that for a sum of one positive and one negative terms, they have to subtract the smaller term from the greater one; however, they do not know that the result must have the same sign as the greater term. For a sum of two terms with same sign, they know they have to add the two terms. However, they do not know that the sign of the result is the same as the sign of the terms. That is particularly visible when the two terms are negative. What they may be doing is that they multiply the signs of the two terms, as if it were a multiplication. They apply the rule: $\Delta|a| + \Delta|b| \mapsto |a| + |b|$, where Δ is the sign of the term a and b.[7] This interpretation was verified by replaying the work of the students who contributed the most to this class, via the 'replay system' in Aplusix. In particular, we observed that these students are never wrong about the sign in the case of a multiplication of two terms: they mastered the sign rules for multiplication. In order to verify this hypothesis it is possible to insert rules about multiplication of two terms in the CHIC table and see whether such groups emerge.

The third behaviour class (*CorC2MP*, *CorEgMM*, *CorC1MM*, *CorC2MM*) is a class of correct actions.

The fourth & fifth ones (*PlusOpC1PP*, *OpPlusC2PP*, *OpEgPP*, *OpPlus-C1PP*, *PlusOpC1PP*) seem strange: the students subtract one term from the another even if neither of them has minus sign: the contexts are all positive terms (*PP*). This is probably due to the choice of the variables: we have selected only two context variables. If we consider and add the variable "presence of a minus sign in the source of the student's step", we would perhaps have an explanation for this class. Behaviour classes such as those generalising behaviours of the students *A* and *B* presented in section 4 may appear.

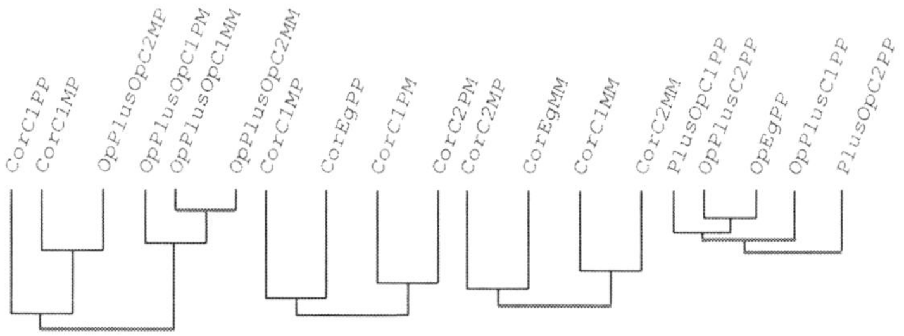

Fig. 6. Part of the similarity tree

We obtained a collection of ordered pairs (context, rules) that are representative of stable behaviours for a part of the students population. It was possible to explain these behaviour groups cognitively. We could thus reach a

[7] Application of an incorrect rule leads to a correct result, in the case of the sum of two positive terms.

higher level description of behaviours than in the previous sections. A direct application of these results would be modelling didactical decisions in terms of mapping didactical explanations to the identified behaviour groups, in order to address to each student an appropriate and easily understandable message.

9 Conclusion

Errors, and more globally behaviours in which we are the most interested, are those that are "persistent and reproducible". Catching stable errors and characterizing types of these errors could allow teachers, didacticians and artificial tutors to take adequate and personalized didactical decisions instead of using a systematic and repetitive feedback.

Data on which our stability research is based are collected in the learning environment for algebra, Aplusix. Each student's action is viewed as a rule application. Our purpose was to determine rules that are used regularly by the student and to link them to the algebraic context in which the student used them. We aimed at both a detailed description of the student's learning state and a high cognitive level of its description. However, mapping a single action to stable behaviours is not an obvious task.

Our research work proceeded in two phases. First, we have established a list of the sub-tasks that provoke regular errors. For that, we needed to determine a good granularity of the sub-tasks. We used the Statistical Implicative Approach (SIA) to determine, in a students population, the variables that cause a stable erroneous behaviour. Second, a systematic analysis of each sub-task allowed us to outline the main behaviour groups.

Thanks to the Statistical Implicative Analysis theory, a certain stability of behaviours has been observed by associating algebraic context variables to actions. It has been possible to determine for a student the possible tasks and sub-tasks that are interesting to model because they provoke stable behaviours in the student. Moreover, the SIA has permitted to point out the algebraic context variables that are source of actions: implicative links between variables and student's actions have been provided as a result of this analysis at a very fine grain size. A statistical implicative analysis of frequencies of ordered pairs (algebraic context, action) for a students' population has allowed behaviour groups to emerge: groups of ordered pairs the most used by a part of the population or those that show a very high level of stability of frequencies.

Using a statistical approach in a didactical research is an original method: most of the didactical research cannot afford to deal with a large student body. In addition, interactive learning environments providing models of students reach either a very fine level of description where steps are not linked together, or a coarse grain size with general information not linked to a precise domain (such as information about whether the student learned best with a directed or explanatory learning tasks [19]).

This research work is based on a description of the algebraic context variables that is a deep and hard work and that has important consequences on the modelling results. The obtained results seem to prove the relevance of our choices: students' actions are explained by the presence of some of these algebraic context variables. Our work also used elementary steps provided by an automatic diagnosis done within the learning environment. It is therefore dependent on the quality of this diagnosis and we need to study carefully the consequences of a bad diagnosis.

The results presented show a great didactical interest both for teachers and artificial tutors designers. Automatic diagnosis of students' knowledge state in terms of correct and erroneous rules applied in particular algebraic contexts provides the teacher with the opportunity to focus remediation to the very source of the incorrect behaviour. The presented students' knowledge modelling opens the possibility to build a model of didactical decisions allowing to produce an appropriate feedback as a response to students' action.

References

1. G. Brousseau. Les obstacles épistémologiques et les problèmes en mathématiques. *Recherches en Didactique des Mathématiques*, volume 4–2, pages 165–198, 1983.
2. R. Sison, M. Shimura. Student Modeling and Machine Learning. *International Journal of Artificial Intelligence in Education*, volume 9, pages 128–158, 1998.
3. D. H. Sleeman, A. E. Kelly, R. Martinak, R. D. Ward, J. L. Moore. Studies of Diagnosis and Remediation with High School Algebra Students. *Cognitive Science*, volume 13, pages 551–568, 1989.
4. J. S. Brown, K. Van Lehn. Repair theory: A generative theory of bugs in procedural skills. *Cognitive Science*, volume 4, pages 379–426, 1980.
5. R Gras. L'analyse implicative: ses bases, ses développements. *Educaçao Matematica Pesquisa*, volume 4:2, pages 11–48, 2004.
6. C Kieran. The core of algebra: Reflections on its Main Activities. *ICMI Algebra Conference*, Melbourne, Australia, pages 21–34, 2001.
7. J. R. Anderson, A. T. Corbett, K. R. Koedinger, R. Pelletier. Cognitive Tutors: lessons learned. *The journal of the learning sciences*, volume 4:2, pages 167–207, 1995.
8. D. McArthur, C. Stasz, M. Zmuidzinas. Tutoring Techniques in Algebra. *Cognition and Instruction*, volume 7:3, pages 197–244, 1990.
9. M. Beeson. Design Principles of Mathpert: Software to support education in algebra and calculus. In N. Kajler, editor, *Computer-Human Interaction in Symbolic Computation*, Springer-Verlag, Berlin, Heidelberg, New York, pages 89–115, 1998.
10. R. Prank, M. Issakova, D. Lepp, V. Vaiksaar. Using Action Object Input Scheme for Better Error Diagnosis and Assessment in Expression Manipulation Tasks. *Maths, Stats and OR Network*, Maths CAA Series, 2006.
11. J. F. Nicaud, D. Bouhineau, H. Chaachoua. Mixing microworld and CAS features in building computer systems that help students learn algebra. *International Journal of Computers for Mathematical Learning*, volume 9:2, 2004.

12. A. T. Corbett, J. R. Anderson. Knowledge Tracing: Modeling the Acquisition of procedural knowledge. *User modeling and user-adapted interaction*, volume 4, pages 253–278, 1995.
13. S. J. Payne, H. R. Squibb. Algebra Mal-Rules and Cognitive Accounts of Error. *Cognitive Science*, volume 14:3, pages 445–481, 1990.
14. N. Balacheff. Les connaissances, pluralité de conceptions. Le cas des mathématiques. In P. Tchounikine, editor, *Actes de la conférence Ingénierie de la connaissance*, Toulouse, pages 83–90, 2000.
15. S. Soury-Lavergne, N. Balacheff. Baghera Assessment project: Designing an Hybrid and Emergent Educational Society. *Cahier du laboratoire Leibniz*, volume 81, 2003.
16. Y. Chevallard. Concepts fondamentaux de la didactique: perspectives apportées par une approche anthropologique. *Recherches en didactique des mathématiques*, La Pensée Sauvage, volume 12:1, Grenoble, pages 73–111, 1992.
17. G. Brousseau. *Théorie des situations didactiques.* La Pensée Sauvage, Grenoble, 1998.
18. J. F. Nicaud. Modélisation cognitive d'élèves en algèbre et construction de stratégies d'enseignement dans un contexte technologique. Project report of the "Ecole et sciences cognitives" research programme, *Cahier du laboratoire Leibniz*, volume 123, 2005.
19. M. Quafafou, A. Mekaouche, H. S. Nwana. Multiviews learning and intelligent tutoring systems. *Proceedings of Seventh World Conference on Artificial Intelligence in Education*, volume 7, 1995.

The graphic illusion of high school students

Eduardo Lacasta and Miguel R. Wilhelmi

Departamento de Matemáticas, Universidad Pública de Navarra
31006 Pamplona (Navarra), Spain
{elacasta, miguelr.wilhelmi}@unavarra.es

Summary. The factorial analysis of the relationship between the mathematical background on linear and quadratic functions, on the one hand, and the representation of functions (graphics, figures and so on) on the other hand, stands in contradiction to the usual assumption of the existence of a "graphical conceptualization" of functions, different from the "non-graphical conceptualization". Nevertheless, both the authors of scholar texts and the teachers involved in this research tend to use the graphical representation of functions. In the context of proportionality, the Statistical Implicative Analysis of students' preferences regarding the kind of graphical representation reveals the existence of a graphical illusion shared by high school students.

Key words: Mathematics Education, function, graphics, Statistical Factorial Analysis, Statistical Implicative Analysis.

1 The function and its graphic: its basis and its form

In current teaching there is a trend that favors the presentation of mathematical notions (notably that of functions) in a visual way. This trend relies on curricular developments and in changes of teaching methods based on the growth and development of information and communication technologies (ICT), such as graphic, symbolic and programmable calculators or mathematical software as CABRI, Derive, Maple or Mathematica. These curricular developments and these changes in teaching are in many cases influenced by the Principles and Standards for School Mathematics [9].

> "[Technology Principle] Electronic technologies (calculators and computers) are essential tools for teaching, learning, and doing mathematics. They furnish visual images of mathematical ideas [...] Instructional programs from pre-kindergarten through grade 12 should enable all students to understand patterns, relations, and functions [Standard Algebra]. In grades 6–8 all students should:

E. Lacasta and M.R. Wilhelmi: *The graphic illusion of high school students*, Studies in Computational Intelligence (SCI) **127**, 99–117 (2008)
www.springerlink.com

- represent, analyze, and generalize a variety of patterns with tables, graphs, words, and, when possible, symbolic rules;
- relate and compare different forms of representation for a relationship;
- identify functions as linear or nonlinear and contrast their properties from tables, graphs, or equations [14]".

In school, the introduction and the development of the notion of function often observe an intuitive approach based nearly exclusively upon graphic language. This teaching practice may have the following risk: "to only teach properties of the functions that are specific to the graphic context" [5]. Therefore it is necessary to analyze the priority of graphic language in the teaching of Mathematics.

Numerous studies in Mathematics Education are theoretically based and experimentally supported by the importance of visualisation in teaching ([3, 6–8, 10, 16, 18], etc.). A naïve or hasty application of these studies may excessively emphasize this intuitive and graphic approach of the notion of function, with the illusion that the students will have the ability to identify and represent the same concept in different representations, and the flexibility in moving from one representation to another. Therefore, these applications ignore the phenomenon of compartmentalization [6].

All this has lead us to study the role of cartesian graph of functions (CGF) in Secondary teaching. We will examine students according to their competence in the resolution of presented problems: a) in an exclusively textual way, b) with a numeric table or c) by means of a graphic. Also, the relationship of those competences with the preferences declared by the students (for the way of presentation of the functions) is examined.

Does some mathematical competence include the other competences? Is a graphically competent student also competent in textually presented problem solving and in table presented problems? What role does the Cartesian graph of functions play in problem solving, especially in the case of relationships between linear functions and proportionality?

Lacasta identifies and describes five different functions of CGF like "material context (*milieu matériel*)" [12]. This complexity of CGF suggests that it is necessary to redefine the role attributed to GCF in Secondary school. It is also necessary to relate this material context to other forms of representing functions, such as the numeric table and the text.

In fact, the object-representation dichotomy is problematic. Font, Godino and D'Amore justify that each object-representation pairs (without segregation) permits a subset of practices of the whole set of practices that are considered the unique and holistic meaning of the object [7]. However, in each subset of practices, the object-representation pair (without segregation) is different, in that it makes different practices possible.

The objective of this work is to contrast the following hypotheses:

[H1] There is no relationship of statistical inclusion amongst the sets of students defined by the three competences (textual, numeric or graphic). In figure 1 a graphic representation of this hypothesis is shown.
This hypothesis expresses the idea that competent students exist especially in the resolution of problems given in a certain presentation way. But the field of the each competence —Ctx, Ctb, Cg— would be no more than an approach to the global mathematical competence, with appreciable areas of failure.

[H2] Students prefer a given form of presentation because it improves their competence in the outlined problems (figure 1).
This hypothesis expresses the idea that the perception about the difficulty of a given question (following the way of presentation) is an indicator of efficiency in problem solving: The students are aware of their real capacities.

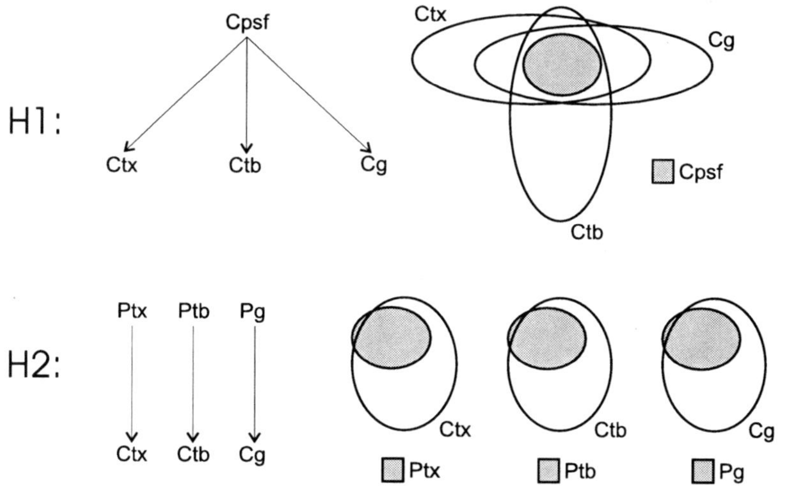

Caption
Cpsf: competence in problems solving involving linear and quadratic functions
Ctx: competence in textually presented problems solving
Ctb: competence in problems solving, when the function is represented by a numerical table
Cg: competence in problems solving, when the function is graphically represented
Ptx: student's preference for the textual presentation of the function
Ptb: student's preference for the tabular presentation of the function
Pg: student's preference for the graphical presentation of the function

Fig. 1. Graphic representation of hypotheses 1 and 2

The contrast of the hypotheses will give us information of the running of the didactic system, this is, information of the mechanisms that determine the construction and communication processes of knowledge relative to functions in high school. However, we have shown how the implication between two variables (competence or attitudinal) doesn't necessarily represent a cause-effect relationship and, therefore, it doesn't allow to establish teaching guidelines.

To contrast these hypotheses two questionnaires were given to a sample of 87 students. The description of the answers to the first questionnaire was carried out by means of a factorial analysis (Factorial Analysis of Correspondences and Main Components Analysis). The description of the answers to the second questionnaire was carried out by means of a Statistical Implicative Analysis (SIA).

A more general goal of this paper is to show how:

1. The SIA contributs to validation or refutation of hypotheses.
2. The SIA can be integrated with other statistical analyses (FAC and MCA, in particular).
3. The SIA yields conclusions that cannot be obtained with other forms of analysis by allowing the contrast between *a priori* and *a posteriori* analysis.

Shortly, in this paper we apply the SIA to a notable problem in mathematics didactic (the role of representations in the learning of mathematics) and we establish certain fundamental aspects of the SIA too.

2 *A priori* analysis and factorial analysis

To determine the existence of a graphic way of solving problems, we gave 87 Spanish Secondary school students a first questionnaire (see Appendix). The mathematical knowledge proposed was the following: the reading of intersections, the sign of functions for the intervals of variable x, the comparison of functions, the extrapolation, etc. We have limited our study to the use of polynomial functions of first and second degrees, which are known by all the students in the sample.

The functions in this questionnaire have been represented either by means of a Cartesian graph (G), a numerical table (T) or an algebraic formula (F). We have defined variables according to the knowledge proposed and its way of presentation. Each knowledge has been proposed following the three ways of presentation: G, T and F. In table 1 the set of defined variables is shown.

An *a priori* analysis establishes a model according to which the students' competences vary depending on the three kinds of presentation G, T and F. However, the results of the questionnaire show that the examined competences are not grouped following the *a priori* model.

	its	sig	com	reg	ext	max	Cp	Cd
G	1G	2G	3G	4G	5G	IpG	CpG	CdG
F	1F	2F	3F	4F	5F	IpF	CpF	CdF
T	1T	2T	3T	4T	5T	IpT	CpT	CdT

Caption
The way of presentation and the knowledge concerned for every question is determined respectively by margins of line and column
F: algebraic formula representation
T: numerical table representation
its: reading of the intersections
sig: sign of the functions for intervals of variable x
com: comparison of functions
rég: partition of the plane by functions.
ext: extrapolation
max: maxima and minima
The "max" column corresponds to questions on the identification of maxima on the quadratic function (identification on the parabola "Ip")
The "Cp" column corresponds to the calculation of increasing/decreasing intervals of the parabola
The "Cd" column corresponds to the determination of the increasing/decreasing character of the line
The numbers correspond to the groups of questions, from 1 to 5

Table 1. Variables of questionnaire 1

The empirical data show a gap between the *a priori* model and the procedures really observed. The used method was the factorial analysis (Factorial Analysis of Correspondence —FAC— and Main Components Analysis —MCA), since it was about "revealing symmetrical relationship and establishing discriminative factors in a population by the way of variables". In figures 2 and 3 the first two main factors are shown.

In the planes of the two main first factors (figures 2 and 3), the pupils' success to questions appear grouped according to the mathematical problem and not according to the kinds of presentation of the function (G, F or T).

The kind of presentation of the questions does not influence the global success of students in the questionnaire. Thus, there is no sign of the existence of the graphic *conception* [1] different from a non graphic conception of mathematical notions. Contrary wise, the student's responses vary according to mathematical notions and not according to the way it is presented (G, T or F).

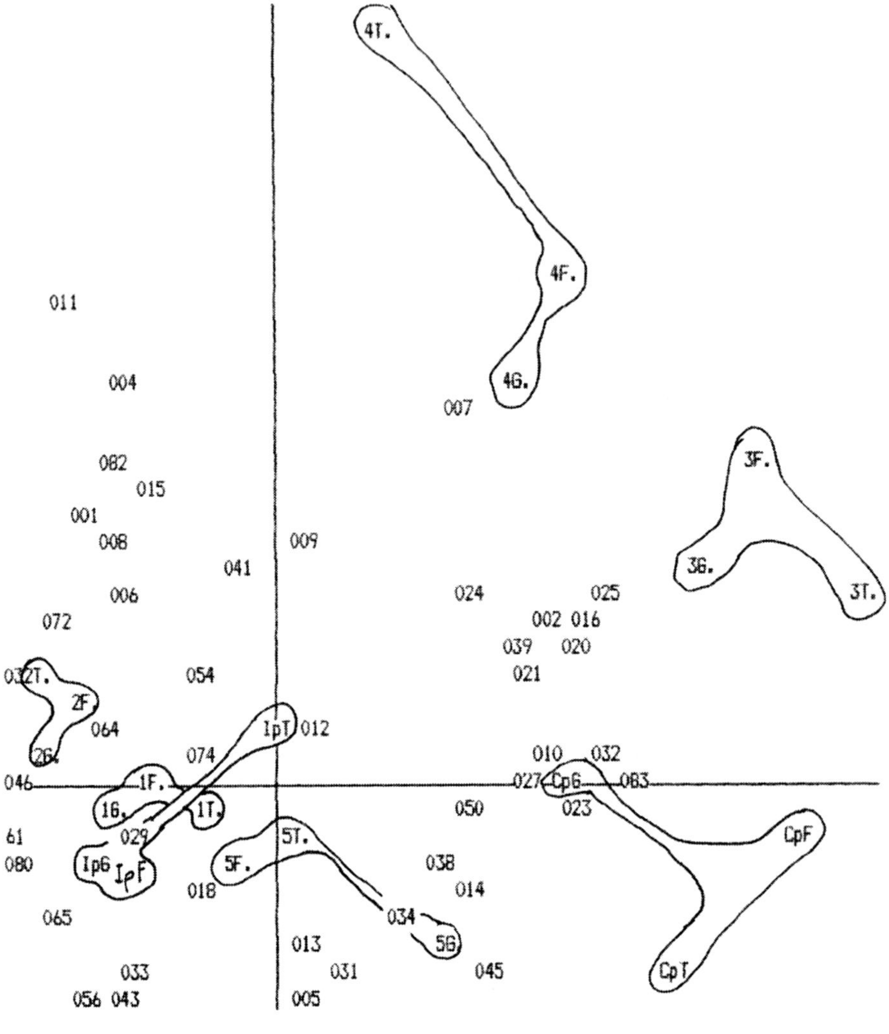

Fig. 2. First questionnai1.8 FAC plan of the success matrix

In the present conditions of math teaching in Secondary, the graphs[1] don't seem to play a crucial role in the learning of functions. Evidently, the graph on its own cannot support all mathematical knowledge that it represents, especially in the case of relationships between linear functions and proportionality.

[1] The concerned graph is always the Cartesian graph. The function is a characteristic notion of the Secondary education. In Primary education the influence of the graphic presentation of other mathematical notions can condition the ability of pupils in problem solving [6].

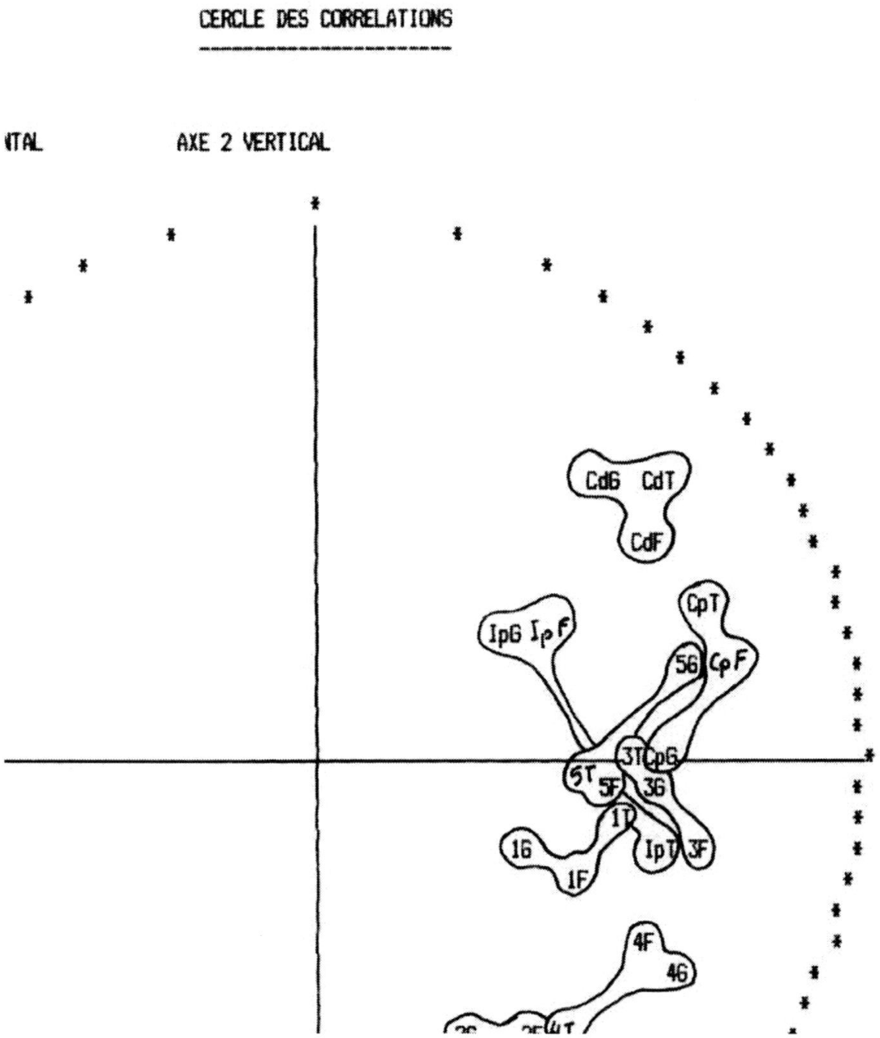

Fig. 3. First questionnai1.8 MCA plan of the success matrix

Pantziara, Gagatsis and Pitta-Pantazi provide an evidence of external validity of our conclusions through inferencial statistic (t-test) [15]. These authors reach a similar conclusion: "The results of the study suggest that the presence of the diagrams did not increase students' ability in solving the non routine problems [...] The results of the study show that the efficient use of a diagram did not imply the successful solution of a problem and reversely the successful solution of a problem did not imply the efficient use of the accompanied diagram" [15, p. 495].

3 Ostensive use of the graph in teaching

Teaching cannot do without the resource to show a part for the whole, this is, to identify by means of examples the generic mathematical objects. The *ostention* is the following didactic phenomenon: the teacher that has shown a copy of, for example, an increasing function, has the illusion that he has really communicated the mathematical notion of an increasing function.

> "We postulate the appearance of didactic phenomena [...] on the difficulties of generating [...] ostensive instruments, necessary for the progress of the mathematical activity, and frequently valued culturally by the school, but which cannot receive a clear mathematical status. This contradiction between, on one hand, the phenomenon of the ostensive reduction and, on the other hand, the cultural valuation of the ostensive instruments, considered indispensable for a 'significant' mathematical activity, doesn't seem to be able to be solved, in the usual didactical contracts, but by means of a certain mathematical activity" [2, p. 106].

Some time ago [11], we verified that most of high school teachers prefer, in the teaching of the functions, the conditions that better allow an ostensive *didactic contract* [4]; that is to say, high school teachers prefer the graphic representation of functions in the measure that it favours an ostensive contract.

The importance assigned by teachers to the Cartesian graph is based on a "false transparency" attributed to this graph. In other words, the represented function would be directly "visible" on the Cartesian graph that represents it. The didactical phenomenon of the "false transparency" can be described by the following computer science metaphor: Cartesian graph is for the teacher an editor WYTIWYG (What You 'Think' Is What You Get), whereas for the student in many cases it is an editor WYSIWYG (What You 'See' Is What You Get).

The *evidence illusion* [12] is a more general phenomenon: the professor observes the notion that he wants to teach in the representation of an object, while the student doesn't transcend to the representation; this is, the student sees the representation like a mere representation ("as such").

This vision of the teachers is based on the following belief (implicitly accepted): to achieve the learning of complex analytic notions, it is necessary to use a presentation that is easy for the students to learn; that would be the graph. Said otherwise, the graphical context facilitates the learning of complex notions. "[...] But many students cannot utilize their visual representations to advance in their problem solving" [17, p. 315].

4 Similarity Analysis and Implicative Analysis

The first questionnaire is a general and exploratory analysis. We need empirical data to analyze the observed facts in the case of relationships between linear functions and proportionality. For it, we need a new questionnaire and a specific statistical method to achieve non symmetrical relationships among curricular, attitudinal and competence variables (SIA). In fact, the application of the SIA to the first questionnaire doesn't determine additional information of the achieved by the precedent analyses (FAC and MCA).

A second questionnaire on the notion of proportionality was given to students. The students had to solve the same kind of mathematical problems: given four pairs of values, they had to determine whether the respective magnitudes were proportional. The enunciation of the problems was given in three different ways: G, T or F. The students had to say if the magnitudes were proportional or not and justify their answers.

Before carrying out these tasks, some students had been trained on the graphic representation of functions but others had not. All of them stated their preferences on the way of presentation of the tasks: textual, through a numeric table or by means of graphics.

The second questionnaire searches contrasting hypotheses. The empirical master plan must take into account several variables, susceptible of having a previous control of their impact on the contingent data. The types of variables are: *problem variables* (modes of presentation of the problems, the formulations, type of relation between problem variables) and an *individual variable* (type of classes to which the students belong). A Chi-square test shows that student's responses are similar in different types of observation according to problem variables. This fact allows us to define some dichotomic variables (RTx, RTb, RG). We assign the value 1 to a variable if an individual satisfactorily carries out at least 70% of the questions associated to this variable. Otherwise, we assign value 0.

Based on these facts (type of received training and stated preferences) and the answers of the students to the questionnaire we defined the following variables:

- A curricular variable (Erf): training to the representation of functions.
- Three attitudinal variables (PTx, PTb, PG): preferences to the different kinds of presentation.
- Five competence variables (Da, RG, REU, RTx, RTb): Students success to answer the questions based on type of presentation.

All the variables are binary (0, absence; 1, presence). Frequencies and percentages of variables appear in table 2.

The *Similarity Analysis* [13] shows that the preference of presentation with tables or text is related to the students' success in problem-solving proposed by means of numeric tables (figure 4). Nevertheless, the preference for graphic representation is only linked to the numerical resolution of graphically

Variable	RTx	RTb	RG	REU	Ngi	Erf	Da	PTx	PTb	PG
Frecuency	62	72	46	67	61	54	17	18	22	28
Percent (%)	71.27	82.76	52.88	77.02	70.12	62.07	19.55	20.69	25.29	32.19

Table 2. Frequencies and percentages of variables

proposed questions. But this preference is neither associated to graphic representation training nor to the students' success in the resolution of graphic problems.

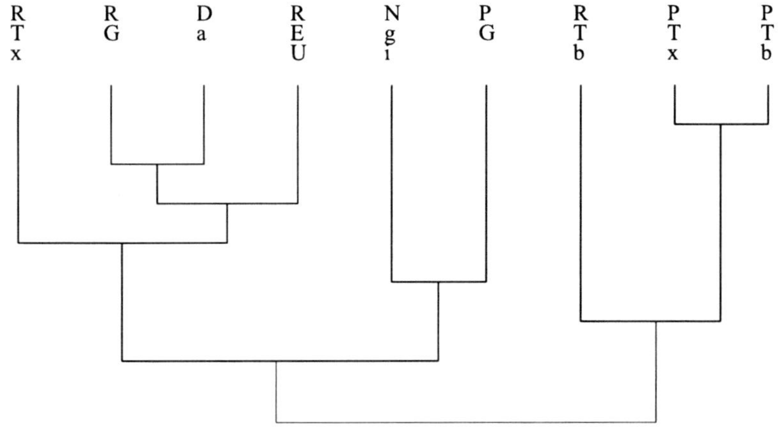

Caption
Ngi: numerical resolution when the problem is presented by an incomplete graph
RG: success in graphically presented problems
RTx: success in textually presented problems
RTb: success in numerical table presented problems
REU: global success in the questionnaire
PTx: preference for textual presentation
PTb: preference for tabular presentation
PG: preference for graphical presentation

Fig. 4. Similarity graph

Table 2 reveals that students encounter greater difficulties in graphically presented problems than in textually or in numerical table presented problems. In turn, the textual tasks are more difficult than the tasks involving tabular information. Moreover, in the similarity graph the connections of students' success at the graphic tasks, at the textual tasks and at the tabular tasks are relatively weak. These remarks suggest that some kind of compartmentalization in students' performance across the various representations exists.

The success in the resolution of graphic problems is linked to the technique of determination of the proportionality by means of the alignment of the points with the origin in the graphic. The training on representation of functions is linked to this success and this technique too.

REU comprehends competences RTx, RTb and RG. The similarity graph shows that the contribution of the competences RTx and RG to REU are more outstanding than the competence RTb.

The implicative graph (figure 5) shows that the set of students stating their preference for graphic language is not included (not even quasi-included) statistically in the set of successful students in the graphical tasks: that is, the graphic preference (PG) is implicatively isolated; the bigger implicative index involving PG is 53 (PG⟶Ngi, see table 3). Furthermore, there is no implication in any direction between graphical training (Erf) and global success in the questionnaire (REU).

Variables	RTx	RTb	RG	REU	Ngi	Erf	Da	PTx	PTb	PG
RTx	0	91	99	0	0	0	0	0	0	0
RTb	0	0	0	0	0	0	0	0	0	0
RG	99	97	0	100	75	97	0	0	0	0
REU	0	98	0	0	0	0	0	0	0	0
Ngi	68	59	0	90	0	0	0	0	0	0
Erf	60	72	0	88	75	0	0	0	0	0
Da	67	0	100	87	0	100	0	58	61	0
PTx	72	94	72	73	0	0	0	0	91	0
PTb	0	85	0	50	0	0	0	0	0	0
PG	0	0	0	0	53	0	0	0	0	0

Table 3. Implicative index

The results obtained through the Statistical Implicative Analysis (SIA) determine an evidence of the existence of a graphic illusion of secondary students.

The Implicative Analysis shows that the set of students defined by graphic success is included statistically in the sets defined by the textual and numeric success (figure 5). The success in the graphic resolution is only achieved by students able to also solve the problems presented by the means of a numerical table and by the means of text. In other words, only some competent students in the resolution on tabular and textual problems achieve the graphical competence.

These results refute hypothesis H1. Against what was affirmed in H1, the empirical data of the behaviours exhibited in the given sample establish a statistical inclusion relationship between the mathematical competences and the method of representation used. In fact, students that solved the problems

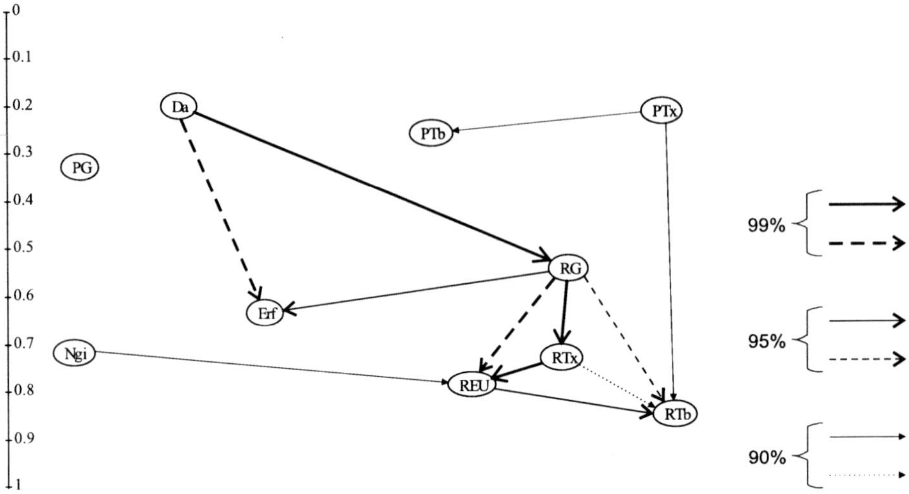

Caption
Dashed arrows: transitivity of the statistical implication
Continuous arrows: implication (99%, 95%, 90%)
Vertical scale: ratio of students giving evidence of the respective variables

Fig. 5. Implicative graph

presented graphically were in general able to solve these problems presented in a tabular and textual way, but not *vice versa*.

Hypothesis H1 suggests the existence of three independent competences to (numerical table, textual and graphical) that would be different approaches to a same mathematical knowledge. There would exist three different conceptions for these mathematical notions. However, the refutation of H1 comforts the thesis formulated in the *a priori* analysis (section 2): students' responses vary according to mathematical notions and not according to the way it is presented (G, T or F).

The refutation of H1 also questions the following extended belief among high school teachers: students learn more easily the notions of functions using graphic representation.

On the other hand, the implicative graph (figure 5) refutes hypothesis H2 partially. Indeed, the sets of students that prefer the tabular or textual presented problems are included in the set of the students that solve those problems satisfactorily. However, the set of students that prefer the graphic presentation is not included statistically in the set of students that solve the graphical represented problems.

5 Synthesis and conclusions

We used a sample of 87 Spanish High School students (13 years old). The functions in a first questionnaire have been presented either by means of a Cartesian graph (G), a numerical table (T) or an algebraic formula (F). The factorial analysis has allowed to explore the data and it has detected the statistical proximity (*symmetrical relationship*) of the behaviours, according to the treated notions. Indeed, the proximity doesn't obey the form of the tasks (graphic or non graphic presentation), but to the involved notions.

In a second questionnaire, on one hand, the proximity of some variables is detected in the Similarity Analysis. This way, the relationship between the variable "instruction in graphic representation" and the variable "detection of the proportion by the way of a graphical technique" is a very strong one. In addition, the relationship between the "preference for the tabular presentation" and the "success in the resolution of tabular problems" is notable too.

On the other hand, the SIA detects statistically significant relationships between the preference for textual presentation and the success in textually presented problems, that the Similarity Analysis does not show. We also observe that the preference for the graphic presentation doesn't imply the success in its resolution.

These facts take us to conclude that a *graphic illusion* exists amongst these students, since similarity relationships do not exist (neither implicative relationship) between the preference for the graphic presentation and the competence in the resolution of problems presented graphically.

The SIA contributes new elements, even though of delicate interpretation. The empirical data doesn't allow to justify the application of a specific teaching method. In fact, there is a radical rupture between explanatory didactics and normative (or technique) didactics based on two facts:

1. *The statistical implication is not transitive.* The competence in the determination of the proportionality by means of the alignment of the points in the graphic with the origin (Da) implies the success in graphically presented problems (RG). The competence RG implies the success in textually presented problems (RTx). However, it's not possible to deduce that "Da" implies "RTx". Consequently, it is not possible to establish the maxim: "If the students are competent in 'Da', then most of them solve the presented problems in a textual way (RTx)".
2. *The statistical implication admits an interpretation (mathematically isomorphic) according to the Set Theory*[2]. Given a set E, we defined P (Q respectively) like the set of elements x in E ($x \in E$) that verify the proposition p (q respectively):

$$P = \{x \in E \,|\, p(x)\} \; (Q = \{x \in E \,|\, q(x)\} \text{ respectively})$$

[2] The Boolean algebra rules can be written in both set and logic notation.

Then "p implies q" is equivalent to affirm that $P \subseteq Q$. This fact explains that the success in graphically presented problems (RG) implies graphical training (Erf). Only a subset of students who have received graphical training have succeeded in graphically presented problems (RG implies Erf).

3. *The statistical implication doesn't necessarily determine a cause-effect relationship.* The success in graphically presented problems implies the success in the textually or tabullary presented problems. This fact doesn't determine a preferential order in teaching for the different kinds of representation. There is no cause-effect relationship either between the resolution of graphic problems and that of other types of problems.

The results in SIA are static, that is, they represent a moment of the school reality. The fact "p implies q" doesn't mean that "p happens before q", neither that "p causes q". In this case, the conclusion is that only some competent students in tabular and textual contexts reach the graphic competence.

In Mathematics Education, the statistical analysis of empirical data doesn't explain the running of didactical systems; neither does it determine strategies for intervention and control. However, the statistical methods allow to confront the *a priori* analysis with the *a posteriori* analysis. Also, the SIA finds conclusions non detectable from other statistical methods; these conclusions allow to interpret more efficiently the didactical systems and to value the strategies designed for the intervention in these systems.

Some students perceive the graphic representation as an easy and intuitive instrument of reaching the mathematical knowledge, but this perception is contradicted. What role do the educational practices of high school teachers play in this perception? Studies are needed that would allow to establish normative approaches for teaching improvements of functions in high school and for training improvements of those teachers.

References

1. N. Balacheff. *Conception, connaissance et concept.* Séminaire DidaTech, Université Joseph Fourier (Paris), 1995.
2. M. Bosch and Y. Chevallard. La sensibilité de l'activité mathématique aux ostensifs. objet d'étude et problématique. *RDM*, pages 77–124, 1999.
3. E. G. Bremigan. An analysis of diagram modification and construction in students' solutions to applied calculus problems. *JRME*, pages 248–277, 2005.
4. G. Brousseau. *Theory of didactical situations in mathematics.* Kluwer Academic Publishers, 1997.
5. G. Chauvat. Courbes et fonctions au college. *Px*, pages 23–44, 1999.
6. I. Elia, A. Gagatsis, and R. Gras. Can we 'trace' the phenomenon of compartmentalization by using the implicative statistical method of analysis? an application for the concept of function. In R. Gras, F. Spagnolo, and J. David,

editors, *Troisième Rencontre Internationale A.S.I. Analyse Statistique Implicative. Quaderni di Ricerca In Didattica of G.R.I.M.*, pages 175–185 (Supplemento 2(15)), 2005.
7. V. Font, J. D. Godino, and B. D'Amore. An ontosemiotic approach to representations in mathematics education. *FLM*, pages 2–7, 2007.
8. G. A. Goldin. Representational systems, learning and problemen solving in mathematics. *JMB*, pages 137–165, 1998.
9. K.J. Graham and F. Fennell. Principles and standards for school mathematics and teacher education: Preparing and empowering teachers. *SSM*, pages 319–328, 2001.
10. T. Graham and J. Sharp. An investigation into able students' understanding of motion graphs. *TMA*, pages 128–135, 1999.
11. E. Lacasta. *Les graphiques cartésiens de fonctions dans l'enseignement secondaire des mathématiques: illusions et contrôles.* Université Bordeaux 1, 1995.
12. E. Lacasta. *Sur la théorie des situations didactiques*, chapter Les modes de fonctionnement du graphique cartésien de fonctions comme milieu, pages 249–262. La Pensée Sauvage, Gronoble, 2005.
13. I.C Lerman. *Classification et analyse ordinale des données.* Dunod, 1981.
14. NCTM. *Principles and Standards for School Mathematics.* E-version available in: [http://standards.nctm.org/]. Autor, 2000.
15. M. Pantziara, A. Gagatsis, and D. Pitta-Pantazi. The use of diagrams in solving non routine problems. In R. Gras, F. Spagnolo, and J. David, editors, *Proceedings of the 28th Conference of the International Group for the Psychology of Mathematics Education*, pages 489–496 (Vol. 3), 2004.
16. T. Romberg, E. Fennema, and T. Carpenter. *Integrating research on the graphical representation of functions.* Lawrence Erlbaum Associates, Inc., 1993.
17. D. A. Stylianou. On the interaction of visualization and analysis: the negotiation of a visual representation in expert problem solving. *JMB*, pages 303–317, 2002.
18. W. Zimmerman and S. (Eds.) Cunningham. *Visualization in teaching and learning mathematics.* Mathematical Association of America, 1993.

Appendix: first and second questionnaires

First questionnaire

The polynomial functions f, g, h, F_1, G_1, H_1, F_2, G_2 and H_2 are defined for any real number and are of degrees 1 or 2.

[A] The graphs for f, g and h are:

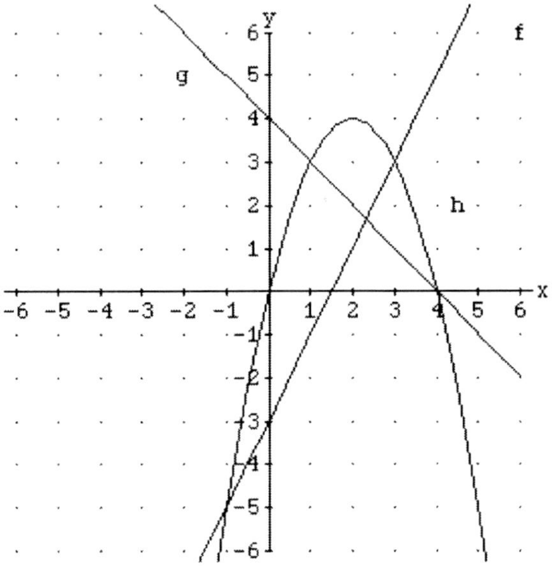

[B] The formulas for F_1, G_1 and H_1 are: $F_1(x) = x^2 + 2x - 3$; $G_1(x) = x + 3$; $H_1(x) = 1 - x$

[C] The numerical tables for F_2, G_2 and H_2 for some values of x are:

x	-6	-5	-4	-3	-2	-1	0	1	2	3	4	5	6	7
$F_2(x)$	-1	0	1	2	3	4	5	6	7	8	9	10	11	12
$G_2(x)$	-21	-12	-5	0	3	4	3	0	-5	-12	-21	-32	-45	-60
$H_2(x)$	-9	-7	-5	-3	-1	1	3	5	7	9	11	13	15	17

1. Find the values of x satisfying:
   a) $f(x) = h(x)$
   b) $F_1(x) = H_1(x)$
   c) $G_2(x) = H_2(x)$
2. Are the following statements true or false?
   a) If $x < 0$, then $f(x) < 0$, $g(x) < 0$ and $h(x) > 0$
   b) If $x > 1$, then $F_1(x) < 0$, $G_1(x) > 0$ and $H_1(x) < 0$
   c) If $x > 1$, then $F_2(x) > 0$, $G_2(x) < 0$ and $H_2(x) > 0$
3. For what intervals of x are the following inequalities true?
   a) $f(x) > h(x) > g(x)$

   b) $G_1(x) > H_1(x) > F_1(x)$
   c) $F_2(x) > H_2(x) > G_2(x)$
4. Find at least a pair of numbers (x, y) satisfying the following conditions:
   a) $y < f(x)$, $y > g(x)$, $y < h(x)$
   b) $y > F_1(x)$, $y < G_1(x)$, $y < H_1(x)$
   c) $y < F_2(x)$, $y > G_2(x)$, $y > H_2(x)$
5. Starting from $x = 6$ (that is to say, for $x > 6$), which of the following inequalities or statements are always true?
   a) $f(x) > g(x) > h(x)$
   b) $F_1(x) > G_1(x) > H_1(x)$
   c) $F_2(x) > G_2(x) > H_2(x)$
   d) It is not possible to assure none of them, because they can change for further values that 6.
6. Complete the following table, putting "yes" or "not" in each box:

Graph	∃ max or min	Curve	Straight	If:	Increasing	Decreasing
f				$x < 2$		
				$x > 2$		
g				$x < 1.5$		
				$x > 1.5$		
h				$x < 4$		
				$x > 4$		
F_1				$x < -2$		
				$x > -2$		
G_1				$x < -3$		
				$x > -3$		
H_1				$x < 1$		
				$x > 1$		
F_2				$x < -5$		
				$x > -5$		
G_2				$x < -1$		
				$x > -1$		
H_2				$x < -1.5$		
				$x > -1.5$		

Second questionnaire

1. In a stationary store 4 packages of notebooks are sold. Package A has 3 notebooks and it costs 225 euro cent (ec.); package B has 5 notebooks and it costs 375 ec.; package C has 10 notebooks and it costs 750 ec.; and package D has 15 notebooks and it cost 1125 ec. Is there a reduction for buying more notebooks? Why?
2. Tests of the French High Speed train are being made in a long, plain and straight line portion of railway. The time to travel several distances has been timed. The results are given in the following table:

Time (minutes):	3	5	9	12
Distance (Km.):	18	30	54	72

Based on the obtained results, does the train maintain its speed? Why?

3. In the Post Office, we are informed that to send packages to a same destination, the prices are according to the package's weight and are given by the following graph:

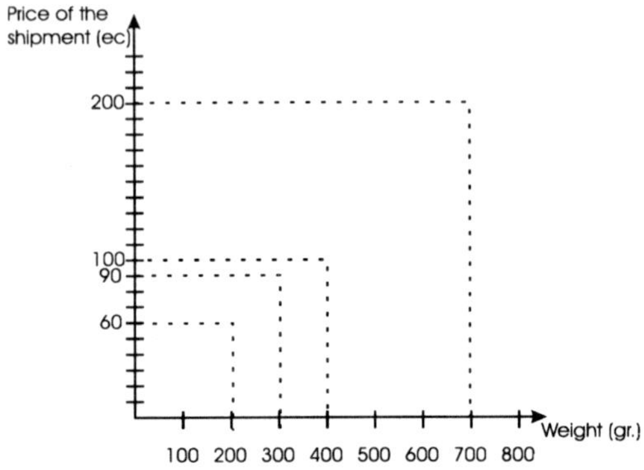

Is the shipment similarly expensive in all the cases or is there some variation according to the package's weight? Why?

4. In a warehouse there are some closed packages that contain cups of the same type. These packages have labels with the number of cups and the total weight. This data is represented in the following graph:

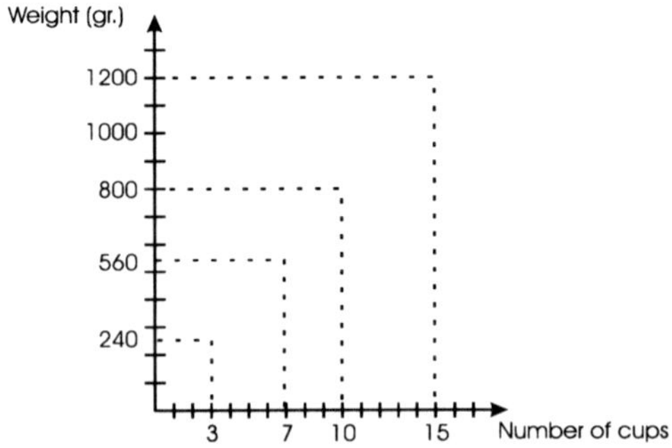

Are the labels correct or has there been some error? Why?

5. In another stationary store 4 packages of notebooks are also sold. The number of notebooks that each package contains and its corresponding price is given in the following table:

Time (minutes):	3	5	8	20
Distance (Km.):	225	375	675	1 350

Is there a reduction when buying more notebooks? Why?

6. In a freight forwarding agency we are informed that to send packages to a same destination, for a package of 200 grs. 60 ec is paid.; for a package of 300 grs., 90 ec.; for a package of 500 grs., 150 ec. and for a package of 600 grs., 180 ec. Represent this data graphically:

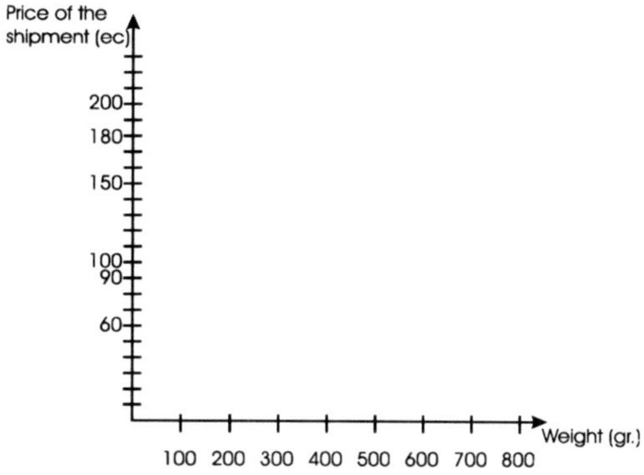

Is the shipment similarly expensive in all the cases or is there some variation according to the package's weight? Why?

7. In another warehouse there are also some closed packages that contain cups of the same type. These packages have some labels with the number of cups and the total weight. On package A is placed the following label: "3 cups, 240 grs.". On the package B: "6 cups, 480 grs.". On the package C: "12 cups, 1040 grs.". On the package D: "16 cups, 1280 grs.". Are the labels correct or has there been some error? Why?

8. Tests of the Spanish High Speed train are also being made in a long, plain and straight line portion of railway. The obtained results are the following: it takes 2 minutes in traveling 12 kilometers, 4 minutes in traveling 24 kilometers, 8 minutes in traveling 42 kilometers and 10 minutes in traveling 66 kilometers. Represent this data graphically. Does the train maintain its speed? Why?

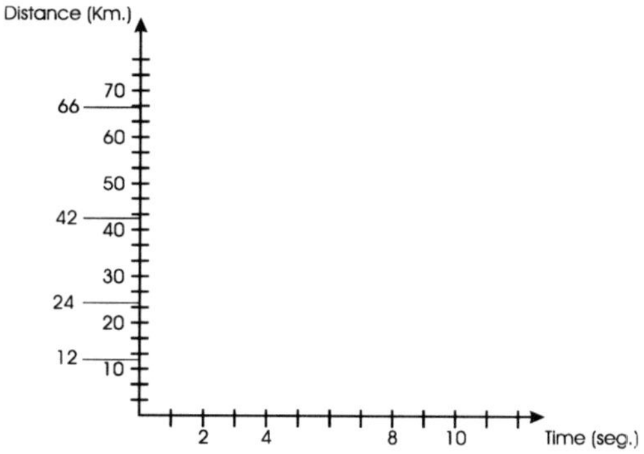

9. Which of these questions seemed easier to you? Why?

Implicative networks of student's *representations* of Physical Activities

Catherine-Marie Chiocca[1] and Ingrid Verscheure[2]

[1] E.N.F.A., dept CLEF, B.P. 22686, 31 326 Castanet-Tolosan, France
catherine-marie.chiocca@educagri.fr
[2] L.E.M.M.E., Paul Sabatier University, Bat. 3R1B2, 118 Rte de Narbonne, 31 062 Toulouse Cedex 4, France
gri.verscheure@wanadoo.fr

Summary. The proposal reports on and discusses results of a questionnaire-based study of young people's attitudes (*representations*) to team games and volleyball (in the context of physical education lessons). This questionnaire was given to students in French agricultural high school. Treatment use software CHIC. Questions approached attitudes, values and dispositions (*representations*) of students about physical education, and, more particularly about volleyball. Several networks of variables appear which make it possible to profile different kinds of students. Study of contributions of two additional variables, sex and gender, highlighted networks, makes it possible to improve choices of representatives networks students for later studies based on interviews. Interestingly and somewhat unexpectedly, while sex is a strong predictor of attitudes and dispositions to team sports and volleyball, gender is not.

Key words: gender, sex differences, representations, sport, network

1 Introduction

We are interested in studying teaching as it is actually done and not as it could be done. The descriptive approach to effective teaching practices is relatively important in our work [15].

However, an earlier approach [16] to the differential dynamics of didactic interactions according to gender in Physical Education (PE), particularly in the case of the volleyball attack, had demonstrated the need of questioning students about their *representations* of Physical and Sports Activities (PSA) and volleyball. We call *representations* attitudes, values and dispositions.
We began our research by classifying students according to their *representations*, then studied the influence of the sex and gender variables on the profiles that were revealed. The differences between the teachers' interactions

C.-M. Chiocca and I. Verscheure: *Implicative networks of student's representations of Physical Activities*, Studies in Computational Intelligence (SCI) **127**, 119–130 (2008)
www.springerlink.com

and certain representative student profiles was subsequently described and analysed.

In this paper, we have limited our discussion to the results of the questionnaire. We wanted to explore the relationships between the sex and gender variables — as revealed by the BSRI test (the Bem Sex-Role Inventory, [3]) which was validated for physical education by [7] — and *representations* of the volleyball attack.

We begin here by looking at the emergence of implicative networks that appeared to structure the data. Then we show that the sex variable had a greater influence than gender on students' *representations*. This result leads to a wider discussion of the validity of the BSRI test as a mean of determining how students' *representations* to PE are determined by gender.
The initial question resulted from didactic research into Physical Education relating to the volleyball attack. It concerns how the didactic contract is applied differently to males and females by PE teachers. Characteristic PE teaching content refers to actions: changes in motor control related to ability Bouthier and David [4] demonstrated that it is essential to take *representations* into account when teaching Physical and Sports Activities in school.

We have tried to understand how students' *representations* of sex and/or gender differ with respect to PE and team sports in general and their *representations* of the internal logic of volleyball in particular. In our research, *representations* were regarded as variables, that we wanted to relate to the subject variables (the students' sex and gender), with the aim of discussing the relationships between sex, gender and *representations* of volleyball, in order to clarify our didactic approach.

2 Processing data using CHIC software program

Data was collected in clusters. A questionnaire (an extract from the questionnaire can be found in the annexe) was put to students (in scientific, technological and professional streams) in agricultural high schools in the Midi-Pyrénées region, in order to determine their *representations* of PE, team sports and volleyball.
The questionnaire was divided into three main sections: an initial question to establish students' gender, based on items from the BSRI (For a discussion of the relevance of the BSRI test for determining gender attitudes, see [15]); a second section with questions about PE in general (its usefulness, students' preference for a particular discipline, views on mixed-sex sports education, teacher's sex, etc); and a third section consisting of questions about team sports and volleyball in particular (based on word association tests and on semantic differenciator test inspired from Osgood, [12]).

The volleyball word association test was adapted from free-association tests, which are verbal productions, used to study the social psychology of *representations* [1]. Beginning with a starter word — "volleyball" was used here — these tests consisted in asking subjects to say any words or expressions which came to mind (a maximum of seven here). The spontaneous nature and the projective dimension of this production reveal the semantic universe of the subject under study more quickly and easily than in an interview [13].

Semantic differenciator test, inspired from Osgood is a quantitative method for analysing connotations, consisting in associating a word with pairs of opposing representative adjectivesx [10]. Connotation is the intensive definition of a word. For example: the word "crow" evokes the colour black, a bad omen and a series of implicit or explicit meanings [11]. It is a general tool for finding meaning (words, figures, etc.) and the pairs of adjectives have to be adapted for each case: here, for volleyball. Our work was based on research by David [5] who used a semantic differentiator adapted for rugby to reveal the differential aspects of *representations* of rugby among a mixed-sex PE group. The pairs of bipolar scales are built using antonymous adjectives drawn from word lists. The semantic differentiator can be considered as an attitude scale for predicting behaviour. Moreover it helps build a fairly readable image of the *representations* of subject groups (particularly functional *representations*), an aspect which is particularly interesting to us as far as volleyball is concerned.

These two techniques were borrowed from methodologies used to explore students' *representations*. They were chosen because they were complementary. The word association test gave the *representations* a "fixed" aspect, whereas the semantic differentiator was based on tactical and dynamic ideas of volleyball.

According to Bailleul [2] "Statistical Implicative Analysis (SIA) is a particularly effective tool for studying *representations* and revealing their organisational structures". We used the CHIC software program (Cohesive Hierarchical Implications Classification) to process the answers in the questionnaire which related to *representations* of team sports and volleyball.

The CHIC software program [8,9] performs implicative analyses, the goal being to identify how much, statistically speaking, a particular answer to a particular item leads to another answer to another item, thus determining the reliability of the "quasi-implications" between variables. The software then produces an implication diagram of the variables, leading to the identification of networks of answers, themselves made up of "implicative chains". The implicative chains are graphic illustrations showing the implications between two or more variables. The networks are combinations of several chains all leading to the same variable. The various chains have similar or identical meaning and are used to interpret the observations [2].

3 Some results

3.1 Characteristics of the population under study

At the end of 2002, questionnaires were sent out to nine General Education and Agricultural Technology high schools in the Midi-Pyrénées region. After several reminders, 507 useable questionnaires were returned between December 2002 and February 2003. They were distributed after class by the PE teachers themselves, to all form four students, regardless of their teaching stream. The students were classified according to different forms of the gender variable, in accordance with the BSRI method, based on the median split method. Four forms of the gender variable were established from this: androgynous (A), non-differentiated (ND), feminine (F) and masculine (M). The gender and sex variables are independent: for instance, a female can have a masculine or androgynous gender. The division of students according to gender was quite well-balanced: 26 per cent were non-differentiated, 23 per cent were feminine, 25 per cent were masculine and 26 per cent were androgynous. In order to study the influence of the different forms of sex and gender variables on *representations* of volleyball, we initialised sex and gender as "additional variables". "Because of their multi-dimensional nature, the implicative analysis and the implication diagram enable us to move beyond simply acknowledging the existence of a relationship between two variables, in order to reveal meaningful oriented implicative networks" [2].

3.2 Three networks at the 0.70 threshold

At the 0.70 threshold implication level, very long chains (up to eight variables) may be found, as well as relatively distinct networks. We shall call these networks A, B and C. We shall begin by looking at the networks obtained at the 0.70 threshold. Then we shall discuss these three networks further, indicating where the different forms of the sex and gender variables had the greatest influence on the chains.

Network A

The first network consisted of different chains involving the "team spirit" and "I like team sports" variables. This network appeared to be characterised by "involved" variables (resulting from both the semantic differentiator and the word association test), all referring to the feminine characteristics of volleyball expressed in the literature [6, 14]: terms such as "to return it", "don't get hurt", "to train" or "to ensure".

A certain work dimension ("to train", technical moves ("to return it") and within certain limits ("don't get hurt") led to the development of knowing how to "play". This, combined with concerns that could almost be considered as hygienically oriented ("gentle", "to relax", "a fun activity") made me "feel good". "Feeling good", together with the "making progress" condition (the second work reference in this network), including the "control" elements which enabled me to be a "team player" and "feel good" in a team activity ("I like team sports"). This *representation* of volleyball (or even team sports in general) seemed well-balanced, as it had the work dimension on one side and the

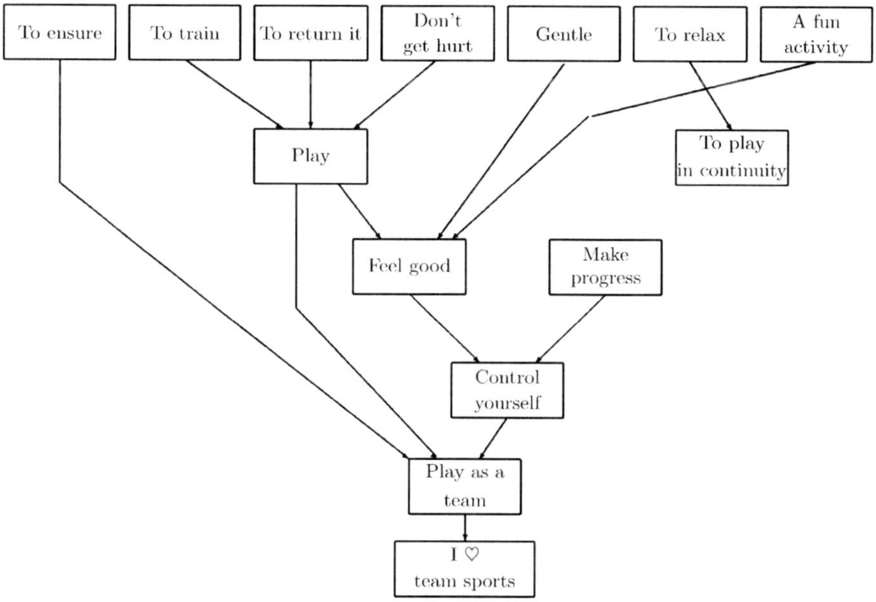

Fig. 1. Network A

enjoyment dimension on the other, while it excluded the competitive dimension.

In this network, we mainly found ideas relating to progress, team spirit and enjoying the activity. In our opinion these ideas embodied three of the essential dimensions of volleyball's internal logic: playing as a team, making progress and having fun, although they overlooked the idea of it also involving a test of strength.

Network B

With themes like "positive feelings" and "a fun activity" implying that "I like volleyball" and that it is "a team sport", we can hypothesise that the *representations* which formed this network's structure were positive attitudes towards this activity. Volleyball also seemed to be associated with a high regard for physical qualities (in particular this category included words such as "tall" and "jump high") as well as mobility. On the other hand, the game element seemed to be represented by "to play in continuity" (as confirmed by game-related words such as "upwards") and tactics.

Network C

In the third network, what predominated was the *representation* of the sport as a test of team strength ("match", "to win", "to be ready to fight", "play as

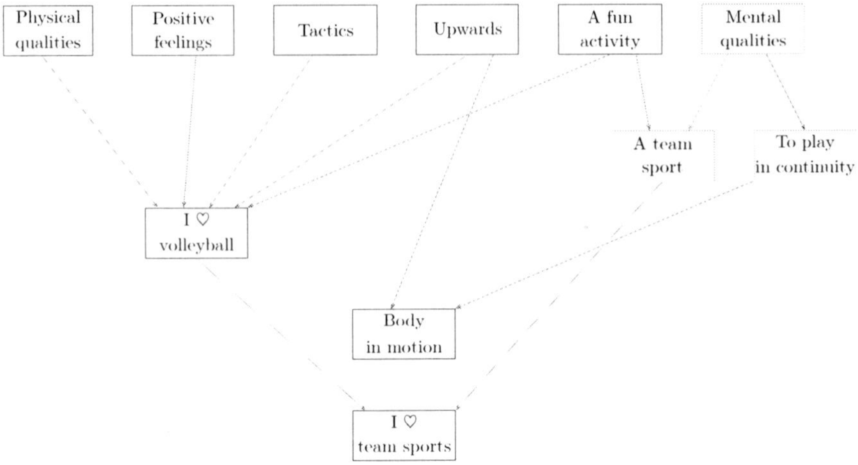

Fig. 2. Network B

a team"), characterised by a break in the exchange, either by force ("attack", "rough") or by cunning ("get clever", "to risk").

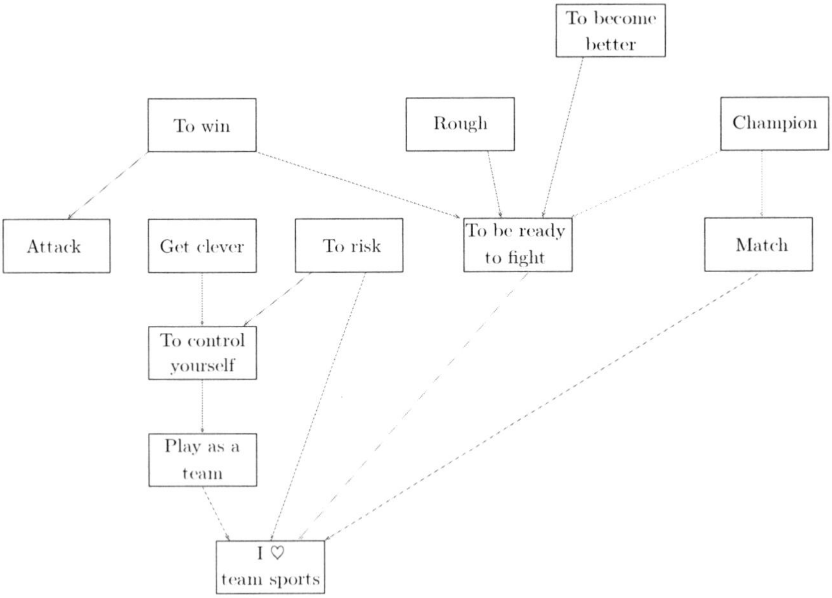

Fig. 3. Network C

Moreover, the representation of victory, of what needs to be done in order to win (as a team) was very significant in this network. Fundamentally, volleyball was represented in this network as a competitive sport.
In Network A, the students' *representation* seemed to be that in order to progress and enjoy playing volleyball, you need to train. The main ideas we found were progress, team spirit and enjoying the activity.

In Network B, the team dimension in volleyball was predominant and students seemed to think that playing volleyball in teams required particular tactics and qualities (mental and physical).

In Network C, the main *representation* that emerged was that of volleyball as a team activity, where the goal was to attack and win (either using force or dummy moves).

In our opinion, these three networks seemed to structure all the data that was collected, in particular the relationships between the different variables from the questions about team sports, the semantic differentiator and the volleyball word association test. We shall now discuss which forms of sex and gender variables had the greatest implications for these networks.

3.3 The influence of the additional "sex" and "gender" variables on the networks revealed by the 0.70 threshold

For each chain, we calculated the influence of the additional variables: two forms for sex (female and male) and four for gender (androgynous, non-differentiated, feminine and masculine). Each of these additional variables had a different influence on the formation of chains in the various networks. We consider that the influence of an additional variable may be included in our comments as an answer to our hypotheses, provided that its error rate is less than 0.10 (a standard criterion for deciding whether a variable significantly explains a particular phenomenon).

Influence of additional variables on Network A

When we looked at the chains that finished with: "play", "feel good", "control yourself", "play as a team" which were implied by the "I like team sports" variable; we noticed that regardless of what the "previous" variable was (to return it, to train, to relax, don't get hurt, gentle), the female form of the additional sex variable significantly influenced these chains (error rates: 0.0438, 0.0137, 0.042).

On the "a fun activity — feel good — control yourself — play as a team" chain, it was again the female form of the sex variable which had a significant influence, with an error rate of 0.0482.

The fact that the female form of the sex variable characterised all these chains led us to hypothesise that females consider volleyball to be a PSA, whose important aspects are team spirit, making progress and having fun.

It therefore seemed that the *representations* in Network A were formed around the idea that you first have to train and make progress before being able to enjoy volleyball. This would correspond more to the feminine characteristics encountered when playing volleyball as described in the literature [6, 14].

Influence of additional variables on Network B

The female form of the sex variable characterised several chains.

- On the "a fun activity — a team sport — I like team sports" chain, the female form of the sex variable had a significant influence, with an error rate of: 0.0891.
- On the "a fun activity — I like volleyball" chain, the female form of the sex variable had a significant influence, with an error rate of: 0.0764.
- On the "mental qualities — to play in continuity — I like team sports" chain, the female form of the sex variable had a significant influence, with an error rate of: 0.0863.
- On the "mental qualities — to play in continuity — body in motion" chain, the female form of the sex variable had a significant influence, with an error rate of: 0.0972.

However, two forms of the gender variable had a significant influence as additional variables on two other chains:

- On the "mental qualities — a team sport — I like team sports" chain, the feminine form of the gender variable had a significant influence, with an error rate of: 0.0621.
- On the "physical qualities — I like volleyball — I like team sports" chain, the masculine form of the gender variable had a significant influence, with an error rate of: 0.0983.

This network's representation, containing *representations* of volleyball as a fun team activity where certain physical and mental qualities are needed, seemed more comparable with the "female sex". However, the "feminine gender" had more influence where "mental qualities" were concerned; whereas with "physical qualities" the "masculine gender" was predominant. We thus suggest the idea of a network which is mainly mixed-sex in its formation or *representations*.

Influence of additional variables on Network C

This network suggested that volleyball was represented as a team opposing relationship, where the main interest came from playing matches and tackling the opponent. Volleyball is all about team spirit, attacking and playing matches.

The male form of the sex variable had a significant influence on several chains in this network:

- the "to win — to be ready to fight — I like team sports" chain, with an error rate of: 0.0184.
- the "rough — to be ready to fight — I like team sports" chain, with an error rate of: 0.0449.
- the "to win — attack — I like team sports" chain, with an error rate of: 0.041.
- the "to risk — I like team sports" chain, with an error rate of: 0.00572.

However, the androgynous form of the gender variable had a significant influence on the following chains:

- "champion — to be ready to fight — I like team sports", with an error rate of: 0.0189.
- "champion — match — I like team sports" with an error rate of: 0.0246.

This network seemed more comparable with *representations* of an opposing relationship in volleyball. The masculine form of the additional gender variable had a significant influence on several chains, while the androgynous form was predominant in one chain. It would rather be males, therefore, who would see volleyball as a sport of continual opposition. The representation of adversity (towards opponents as well as oneself) was revealed by this network in two forms: an aggressive manner ("attack", "rough") and a cunning manner ("get clever", "to control yourself", "to risk"). We therefore suggest that this network was more "male" than the previous two.

4 Conclusion

In view of the results, we can say that the sex variable appeared more often than the gender variable, as an additional variable with a significant influence on these chains. Only the "feminine", "androgynous" and "masculine" forms of the gender variable each had a significant influence on a single chain (the non-differentiated form of this variable did not predominate in any of the chains), whereas both forms of the sex variable ("female" and "male") played a greater role.

Furthermore, Network B highlighted volleyball's team dimension and the need for implementing tactics and have particular mental and physical qualities. This network, which could be summarized as: "playing volleyball as a team requires tactics and mental qualities", seemed closer to the "female" form; although the "feminine gender" was implied when the answers stated that mental qualities were required, while the "masculine gender" was implied when physical qualities were concerned. The "female" form's influence on this chain should thus be moderated.

Network C was more synonymous with *representations* of volleyball as an opposing relationship. There was the significance of the matches, as well as the desire to win, attack and play as a team. The additional variable having the

greatest influence on this network was the "male" form, with the androgynous gender also making a contribution to the "ready to fight" chain.
This level of analysis led us to believe that the students could have sexual *representations* of volleyball, although sometimes one should make adjustments according to gender.
In our other studies, not covered here, we drew parallels between classes that were revealed using ascendant hierarchical classification (AHC) and networks revealed using CHIC. We can therefore form student typologies according to sex, that sometimes need to be adjusted to take gender into account.

While sex is a strong predictor of attitudes and dispositions to team sports and volleyball, gender is not. We found that females and males prefer masculine forms of team games, and that both see volleyball as relatively gender-neutral. Both, females and males, appear to like volleyball, but for different reasons. While the females enjoy the cooperative effort of keeping the game going and so for continuity, the males show a strong preference for scoring points and so for discontinuity.

References

1. J.-C. Abric. *Pratiques sociales et représentations.* PUF, Paris, 1994.
2. M. Bailleul. Mise en évidence de réseaux orientés de représentations dans deux études concernant des enseignants stagiaires en iufm. In *Actes des journées sur la fouille dans les données par la méthode d'analyse statistique implicative*, 2000.
3. S.L. Bem. The measurement of psychological androgyny. *Journal of Consulting and Clinical Psychology*, vol. 42, n. 2:pp. 155–162, 1974.
4. D. Bouthier and B. David. Représentation et action: de la représentation initiale à la représentation fonctionnelle des aps en eps. *Méthodologie et didactique de l'éducation physique et sportive*, Ed. G. Bui-Xuan:pp. 233–249, 1989.
5. B. David. Rugby mixte en milieu scolaire. *Revue Française de Pédagogie*, n. 110:pp. 51–61, 1995.
6. Davisse. Sport, école et société: la part des femmes. *Paris Ed. Actio*, pages pp. 174–263, 1991.
7. Fontayne, Sarrazin, and Famose. The bem sex-role inventory: validation of a short-version for french teenagers. *European Review of Applied Psychology*, 50, n. 4:pp. 405–416, 2000.
8. R. Gras, S. Almouloud, M. Bailleul, and A. Larher. *L'implication statistique. Nouvelle méthode exploratoire de données.* La Pensée Sauvage, 1996.
9. R. Gras and P. Kuntz. The implicative statistical analysis — its theoretical foundations. Kluwer, 2007.
10. Jodelet. L'association verbale. *in P. Fraisse et J. Piaget, Traité de psychologie espérimentale*, fasc. VIII:p/97–153, 1972.
11. R. Menahem. Le différenciateur sémantique, le modèle de mesure. *L'année psychologique*, 68:pp. 451–465, 1968.
12. C.E. Osgood, G.J. Suci, and P.H. Tannenbaum. *The mesurement of meanings.* Chicago, University of Illinois Press, 1957.

13. M-L. Rouquette and P. Rateau. *Introduction à l'étude des représentations sociales.* Grenoble, PUG, 1998.
14. Tanguy. Le volley: un exemple de mise en oeuvre didactique. Echanges et controverses. n. 4:p. 7–20, 1992.
15. I. Verscheure, C. Amade-Escot, and C.-M. Chiocca. Représentations du volley-ball scolaire et genre des élèves: pertinence de l'inventaire des rôles de sexe de Bem? In *RFP 154*. 2006.
16. I. Verscheure and C. Amade-Escot. Gender difference in the learning of volley-ball attack. In *In Procedings AIESP Congress: Professionnal preparation and social needs*, 2002.

Appendix

The word association test was introduced in the following way: "What does the word "volleyball" make you think about? Can you give me some other words (between 5 and 7) that it brings to mind?"

The students gave us a total of 2386 words (an average of 4.7 words per pupil); including 521 different words, with some being quoted very often. For example: the word "net" was quoted 201 times, the word "ball" 198 times; the word "smash" 175 times, the word "pass" 102 times and the word "team" 97 times...

To help process this information, we grouped the words together into several categories. We began by grouping together words with the same root (ball, volleyball), then words which seemed similar in terms of the research questions, or in terms of their meaning. For example, we categorised the word "team" together with other word groups including it, such as "good team atmosphere", "team mates", "team", "solid team", "be in a team", "team game", "play as a team", etc. We combined the "team" words with others relating to the "collective" theme (e.g.: "learn to play together", "good group spirit", "communal", "colleagues", "closeness", "trust your partners", "understanding between players", "mutual support", "group"). This category was entitled "the communal aspect of volleyball". We continued in the same way for all 521 words, which were eventually grouped together into twenty categories.

To test the validity of grouping the words in this way, we called upon the "judging method". Two volleyball experts examined the words in our selected categories, to let us know whether or not they agreed with the categories the words had been put into. There was agreement on more than 80 per cent of the words, so we retained the classification and, after discussion, revised the word categorisation until a consensus was reached.

Osgood's semantic differentiator was introduced in the following way:

Here are a series of terms evoking volleyball (in a broad sense). According to which term at either end of the scale means the most to you, circle one figure only for each line.

Categories	Number of occurences
A fun activity	96
Team spirit aspect	287
Attack	274
Cooperation	214
Difficult	21
Sexual aspect	25
Equipment/material (the ball in particular)	258
Movement/energy	44
Fear/pain	55
Mental qualities	44
Physical qualities	78
Opposing relationship	103
Reference to beach games	87
Rules/limits (the net in particular)	306
Negative feelings	38
Positive feelings	49
Tactics	99
Technique	210
Upwards	42
Not initialised	24

Return	3	2	1	0	1	2	3	Attack
Make progress	3	2	1	0	1	2	3	Become athletic
Score a point	3	2	1	0	1	2	3	Play as a team
Prolong the exchange	3	2	1	0	1	2	3	Interrupt the exchange
Become strong	3	2	1	0	1	2	3	Get clever
Be the best	3	2	1	0	1	2	3	Learn to control yourself
Watch the ball	3	2	1	0	1	2	3	Watch the opponent
Rough	3	2	1	0	1	2	3	Gentle
to be ready to fight	3	2	1	0	1	2	3	Don't get hurt
Play	3	2	1	0	1	2	3	to win
Be a champion	3	2	1	0	1	2	3	Feel good
Static	3	2	1	0	1	2	3	Mobile
Make progress	3	2	1	0	1	2	3	Relax
to train	3	2	1	0	1	2	3	Match
Precision	3	2	1	0	1	2	3	Force
Be good enough	3	2	1	0	1	2	3	to risk
Rupture	3	2	1	0	1	2	3	to play in continuity

Table 1. Classification of words into twenty categories

A comparison between the hierarchical clustering of variables, implicative statistical analysis and confirmatory factor analysis

Iliada Elia and Athanasios Gagatsis

Department of Education, University of Cyprus
P.O. Box 20537, 1678 Nicosia Cyprus
{ilada, gagatsis}@ucy.ac.cy

Summary. This study aims to gain insight about the distinct features and advantages of three statistical methods, namely the hierarchical clustering of variables, the implicative method and the Confirmatory Factor Analysis, by comparing the outcomes of their application in exploring the understanding of function. The investigation concentrates on the structure of students' abilities to carry out conversions of functions from one mode of representation to others. Data were obtained from 587 students in grades 9 and 11. Using Confirmatory Factor Analysis, a model, that provides information about the significant role of the initial representations of conversions in students' processes, is developed and validated. Using the hierarchical clustering and implicative analysis, evidence is provided to students' compartmentalized thinking among representations. These findings remain stable across grades. The outcomes of the three methods were found to coincide and to complement each other.

Key words: implicative analysis, hierarchical clustering, CHIC, Confirmatory Factor Analysis, function, representation.

1 Introduction

Nowadays the centrality of representations in teaching, learning and doing mathematics seems to become widely acknowledged. A basic reason for this emphasis is that mathematical concepts are accessible only through their semiotic representations [1]. Kaput suggests that representations are "integrated" with mathematics. In certain cases, representations, such as graphs, are so closely connected with a mathematical concept, such as function, that it is difficult for the concept to be understood and acquired without the use of the corresponding representation [2].

A given representation, however, cannot describe thoroughly a mathematical concept, since it highlights only a part of its aspects [3]. This justifies

I. Elia and A. Gagatsis: *A comparison between the hierarchical clustering of variables, implicative statistical analysis and confirmatory factor analysis*, Studies in Computational Intelligence (SCI) **127**, 131–162 (2008)
www.springerlink.com

the strong support in the mathematics education community that students can grasp the meaning of mathematical concepts by experiencing multiple mathematical representations [3–6].

The concept of function is central to mathematics and its applications. It emerges from the general inclination of humans to connect two quantities, which is as ancient as mathematics itself. There is a large body of literature on the understanding of functions that focuses mainly on the role of different representations. For example, a number of studies have shown that students tend to have difficulties in transferring information related to functions gained in one representational context to another [1, 3].

Recent studies have used different methods of analysis to investigate students' abilities and their structure in using various representations of function. A few studies have employed the hierarchical clustering of variables [7,8], while other studies have used the hierarchical classification in combination with the implicative method [3, 9, 10]. There are also a number of studies that have attained their outcomes by using Confirmatory Factor Analysis (CFA) [11]. An important question that arises is whether each of these studies would have resulted in similar or congruent findings, if statistical methods other than the one applied, were conducted on its data. Another crucial issue concerns the aspects of a study that each statistical analysis serves better and helps more efficiently to make sense of. In an attempt to tackle these questions, the intent of this study, which focuses on students' abilities in transferring functions from one representation to another, is to apply all of the three aforementioned statistical methods of analysis on the same sample data and compare their outcomes.

2 Theoretical considerations: The role of representations on the understanding of functions

Students experience a wide range of representations from their early childhood years onward. A main reason for this is that most mathematics textbooks today make use of a variety of representations more extensively than ever before in order to promote understanding.

The use of multiple representations has been strongly connected with the complex process of learning in mathematics, and more particularly, with the seeking of students' better understanding of important mathematical concepts [12, 13], such as function. Given that a representation cannot describe fully a mathematical construct and that each representation has different advantages, using various representations for the same mathematical situation is at the core of mathematical understanding [1]. Ainsworth, Bibby and Wood [14] suggest that the use of multiple representations can help students develop different ideas and processes, constrain meanings and promote deeper understanding. By combining representations students are no longer limited by the strengths and weaknesses of one particular representation. Kaput [15] claims that the use of more than one representation or notation system helps students to obtain a better picture of a mathematical concept.

The ability to identify and represent the same concept through different representations is considered as a prerequisite for the understanding of the particular concept [1, 16]. Besides recognizing the same concept in multiple systems of representation, the ability to manipulate the concept with flexibility within these representations as well as the ability to "translate" the concept from one system of representation to another are necessary for the mastering of the concept [5] and allow students to see rich relationships [16].

Duval [1, 17] maintains that mathematical activity can be analyzed based on two types of transformations of semiotic representations, i.e. treatments and conversions. Treatments are transformations of representations, which take place within the same register that they have been formed in. Conversions are transformations of representations that involve the change of the register in which the totality or a part of the meaning of the initial representation is conserved, without changing the objects being denoted.

Some researchers interpret students' errors as either a product of a deficient handling of representations or a lack of coordination between representations [13, 18]. The standard representational forms of some mathematical concepts, such as the concept of function, are not enough for students to construct the whole meaning and grasp the whole range of their applications. Mathematics instructors, at the secondary level, traditionally have focused their teaching on the use of the algebraic representation of functions [19]. Sfard [20] showed that students were unable to bridge the algebraic and graphical representations of functions, while Markovits, Eylon and Bruckheimer [21] observed that the translation from graphical to algebraic form was more difficult than the reverse. Sierpinska [4] maintains that students have difficulties in making the connection between different representations of functions, in interpreting graphs and manipulating symbols related to functions. Gagatsis and Christou [11] developed a model involving some critical paths relating the conversions from one type of representation to another. These paths indicated that the conversion from one representation to another is not a straightforward task. For example, students' ability to translate a function from its graphical to the algebraic form was the result of students' understanding of three other conversions: (a) the conversion of a function from graphic to verbal form, (b) the conversion from verbal to graphic function, and (c) the conversion from verbal to algebraic form of a function. A possible reason for this kind of behaviour is that most instructional practices limit the representation of functions to the translation of the algebraic form to the graphic form and not the reverse. Furthermore, Aspinwall, Shaw and Presmeg [22] suggested that in some cases the visual representations create cognitive difficulties that limit students' ability to translate between graphical and algebraic representations.

Lack of competence in coordinating multiple representations of the same concept can be seen as an indication of the existence of compartmentalization, which may result in inconsistencies and delays in mathematics learning at school. This particular phenomenon reveals a cognitive difficulty that arises from the need to accomplish flexible and competent translation back and forth between different modes of mathematical representations [1]. Elia et al. [10] examined whether pupils of grade 9 (14 year olds) accomplished the conversions among different modes of representation of functions (i.e. graphic, symbolic, verbal). They found that different types of conversion among representations of the same mathematical content were approached in a completely distinct way. These results provided a strong indication of pupils' compartmentalized thinking and use of the various representations of functions and therefore their deficiencies in the understanding of the concept.

3 Aim and research questions

The aim of the study is to combine and compare the outcomes of CFA, hierarchical clustering of variables and implicative method on the same sample data concerning students' abilities in carrying out conversions among different representations of functions. A main concern is to gain insight about the distinct features, advantages and limitations of each of the three statistical methods in a central topic of mathematics education, namely the understanding of the concept of function, and to examine whether they coincide or even complement each other through students' observed performance on this particular subject.

In the light of the above, the following research questions are formulated:

- Which are the common features of the outcomes of the three statistical methods? To what extent there is consistency between the results derived from these processes?

- For which aspects of the study is the application of each statistical method more appropriate and open to complementary use?

4 Method

4.1 Participants, instrument and variables

The sample of the study consisted of 587 students of grades 9 and 11 in Greece. Specifically, 183 of the students were of grade 9 (14 years of age) and 404 of them were of grade 11 (16 years of age). Two tests were constructed and administered to the participants. The first test (A) consisted of six tasks in which students were given the graphic representation of an algebraic relation and were asked to translate it to the verbal and symbolic form respectively. The second test (B) consisted of six tasks (involving the same algebraic relations with test A) in which students were asked to translate each relation from its verbal representation to the graphical and to the symbolic representation respectively. Students had to carry out 12 conversions in each test, that is, 24 conversions in total. For each type of conversion the following types of algebraic relations were examined: $y < 0$, $xy > 0$, $y > x$, $y = -x$, $y = 3/2$, $y = x - 2$ based on a relevant research of Raymond Duval [23]. The former three tasks correspond to inequalities and thus regions of points, while the latter three tasks correspond to functions.

Each test included an example of an algebraic relation in a graphic, verbal and symbolic form to help students understand what they were asked to do. The example is illustrated in Table 1.

Graphic representation	Verbal representation	Symbolic representation
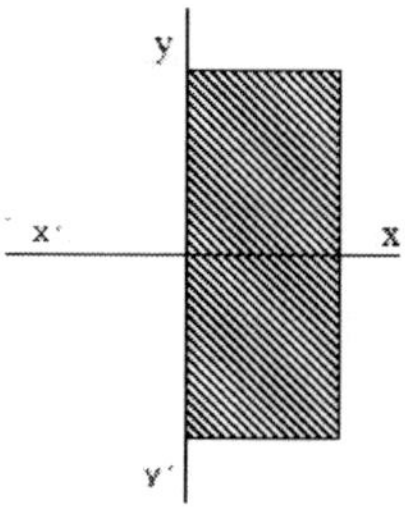	It represents the region of the points having positive abscissa.	$x > 0$

Table 1. An example of the tasks included in the test

Correct responses to the tasks were assigned a score of 1, while incorrect answers were given a score of 0. The variables used for the analyses of the data corresponded to students' responses to the tasks and were symbolized as follow: V11a, V12a, V21a, V22a, V31a, V32a, V41a, V42a, V51a, V52a, V61a, V62a, V11b, V12b, V21b, V22b, V31b, V32b, V41b, V42b, V51b, V52b, V61b and V62b. The symbolism used for the variables of the data is explained below:

(a) "a" stands for Test A, and "b" stands for Test B

(b) The first number after "v" stands for the number of the task in the test i.e. 1: $y < 0$, 2: $xy > 0$, 3: $y > x$, 4: $y = -x$, 5: $y = 3/2$, 6: $y = x - 2$

(c) The second number stands for the type of conversion for each test, i.e. for Test A, 1: graphic to verbal representation, 2: graphic to symbolic representation; for Test B, 1: verbal to graphic representation, 2: verbal to symbolic representation.

4.2 Data analysis

This section is distinguished into two parts. The first part involves a brief overview of the rationale, the components and basic concepts of structural equation modeling and CFA, while the second part concentrates on the underlying principles, elements and structure of the implicative statistical method and the hierarchical classification of variables.

Structural Equation Modeling and CFA

"Structural equation modeling (SEM) is a statistical methodology that takes a hypothesis testing (i.e. confirmatory) approach to the multivariate analysis of a structural theory bearing on some phenomenon" [24]. This theory concerns "causal" relations among multiple variables [25]. These relations are represented by structural, namely regression equations, which can be modeled in a pictorial way to allow a better conceptualization of the involved theory.

SEM differs from the more traditional multivariate statistical techniques in at least three dimensions: First, with the use of SEM the analysis of the data is approached in a confirmatory manner rather than in an exploratory way, making hypothesis testing more accessible and easier, compared with other multivariate procedures. Second, whereas SEM gives the estimates of measurement errors, the "conventional" multivariate methods cannot assess or correct for these parameters. Third, SEM involves not only observed but also latent (unobserved) variables, whereas the older techniques incorporate only observed measurements.

Latent and observed variables are two of the most basic concepts of SEM. Latent variables (i.e. factors) represent theoretical or abstract constructs that can not be observed or measured directly and are rather assumed to lie behind certain observed measures. Examples of latent variables in this study are the abilities of students to convert functions from one mode of representation to another. The measurement of latent variables is obtained indirectly by associating it with other variable/s that is/are observable. The latent variable is based on some behaviour supposed to represent it. This behaviour refers to scores on a particular instrument, like the test of this study, and in turn these measured scores are called observed variables.

Factor analysis is a well known statistical technique for examining associations between observed and latent variables. The covariation among a set of observed variables is investigated to get information on their underlying latent factors. Of primary interest is the strength of the regression paths from the factors to the observed variables. Exploratory Factor Analysis (EFA) and Confirmatory Factor Analysis (CFA) are two basic types of factor analysis. EFA is employed to determine how the observed variables are connected to their underlying constructs in situations where these links are unknown. By contrast, CFA, which is the type of analysis employed here, is used in situations where the researcher aims to test statistically whether a hypothesized linkage pattern between the observed variables and their underlying factors exists. This a priori hypothesis draws on knowledge of related theory and past empirical work in the area of the study. The basic steps that a researcher follows for carrying out CFA are described below: The model is specified based on knowledge of relevant theory and previous empirical research. Using a model-fitting program, such as EQS, the model is analyzed so that the estimates of the model's parameters with the data are derived. Then the tenability of the model is tested based on data that involve all the observed variables of the model. In other words, what is tested is how well the observed data fit the a priori structure. If the hypothesized model is not consistent with the data the model is respecified and the fit of the revised model with the same data is evaluated [24, 26].

The number of levels that the latent factors are away from the observed variables determines whether a factor model is called a first-order, a second-order or a higher order model. Correspondingly, factors one level removed from the observed variables are labeled first-order factors while higher-order factors which are hypothesized to account for the variance and co-variance related to the first-order factors are termed second-order factors. A second or a higher order factor does not have its own set of measured variables. In this study a second-order and a third-order model will be considered. A structural equation model involves two basic types of components: the variables and the processes or relations among the variables. A schematic representation of a model, which is termed path diagram, provides a visual interpretation of the relations that are hypothesized to hold among the variables under study. The basic notation and schematic presentation of the variables and the relations that are used in path diagrams are described below.

The observed or measured variables, which constitute the actual data of the study, are often designated as Vs and are shown in rectangles. The unmeasured variables, which are hypothetical and represent the structural organization of the phenomenon under study, are of three types: a) the latent factor which is designated as F and represented in the path diagram in ellipses or circles; b) a residual associated with the measurement of each observed variable which is referred to as error and designated as E; and c) a residual or the error associated with the prediction of each factor, which is termed disturbance and designated as D. Residual terms indicate the imperfect measurement of the observed variables and the imperfect prediction of the unobserved factor. Although both kinds of residuals represent errors, the former is termed error, and the latter is referred to as disturbance, to distinguish the error in measurement from error in prediction. For example, in the model of Figure 1, "V112a"-"V612a" and "V112b-V612b" represent the measured variables, "Gvs", "Vgs" and "Gvs" stand for the latent factors, "E1"–"E12" correspond to the residuals of the observed variables and "D1"–"D12" refer to the residuals of factors.

One type of the relations involved in a model is the structural regression coefficients indicating the impact of one variable on another. They are represented by one-way arrows. For example, the unidirectional arrows leading from the Factor "Gvs" (Figure 1) to the six observed variables (V112a-V612a) indicate that the scores on the latter variables are "caused" by the factor "Gvs". These relations are called "factor loadings". Similarly, unidirectional arrows from one factor to another imply that a factor causes or predicts another factor, e.g., in Figure 1 the arrows starting from "AbGV" and pointing toward "Gvs" and "Vgs" imply that "AbGV" predicts "Gvs" and "Vgs". A second type of processes used in a model is the impact of the errors in the measurement of the observed variables and in the prediction of the latent factors. The impact of random measurement errors on the observed variables and errors in the prediction of factors are represented as one-way arrows pointing from Es (e.g., E1-E12) and Ds (e.g., D1, D2) respectively, to the corresponding variables, as shown in Figure 1. A third type of processes involved in a model are the covariances or correlations between pairs of variables, which are represented as curved or (sometimes) straight two-way arrows (e.g., E1-E2 or E7-E8) as illustrated in Figure 1.

Bentler's [27] EQS program was used for testing the CFA models in this study. The estimation method that was used in EQS was maximum likelihood solution. The tenability of a model can be determined by using the following measures of goodness-of-fit: X^2, CFI (Comparative Fit Index) and RMSEA (Root Mean Square Error of Approximation) [28]. The following values of the three indices are needed to support model fit: The observed values for X^2/df should be less than 3, the values for CFI should be higher than .9 and the RMSEA values should be lower than .06.

Implicative Statistical Analysis and Hierarchical Clustering of Variables

For the analysis of the collected data, the hierarchical clustering of variables and Gras's implicative statistical method have been also conducted using a computer software called C.H.I.C. (Classification Hierarchique, Implicative et Cohesitive), Version 3.5. [29]. These methods of analysis determine the hierarchical similarity connections and the implicative relations of the variables respectively [30, 31]. For this study's needs, similarity and implicative diagrams have been produced from the application of the analyses on the whole sample and on each age group of students. The implications of the analyses were based on the classical theory.

The hierarchical clustering of variables [32] is a classification method which aims to identify in a set V of variables, sections of V, less and less subtle, established in an ascending manner. These sections are represented in a hierarchically constructed diagram using a similarity statistical criterion among the variables. The similarity stems from the intersection of the set V of variables with a set E of subjects (or objects). This kind of analysis allows the researcher to study and interpret clusters of variables in terms of typology and decreasing resemblance. The clusters are established in particular levels of the diagram and can be compared with others. This aggregation may be indebted to the conceptual character of every group of variables.

The construction of the hierarchical similarity diagram is based on the following process: Two of the variables that are most similar to each other with respect to the similarity indices of the method are joined together in a group at the highest (first) similarity level. Next, this group may be linked with one variable in a lower similarity level or two other variables that are combined together and establish another group at a lower level, etc. This grouping process goes on until the similarity or the cohesion between the variables or the groups of variables gets very weak. In this study the similarity diagrams allow for the arrangement of the variables, which correspond to students' responses in the tasks of the tests, into groups according to their homogeneity.

The implicative statistical analysis aims at giving a statistical meaning to expressions like: "if we observe the variable a in a subject, then in general we observe the variable b in the same subject" [30, 33]. Thus the underlying principle of the implicative analysis is based on the quasi-implication: "if a is true then b is more or less true". An implicative diagram represents graphically the network of the quasi-implicative relations among the variables of the set V. In this study the implicative diagrams contain implicative relations, which indicate whether success to a specific task implies success to another task related to the former one.

It should be noted that the present paper is related to the ones of Elia et al. [10] and Gagatsis and Christou [11], whose basic findings are included in the theoretical section (2).

5 Results

5.1 The outcomes of CFA

Before carrying out CFA, we examined the hypothesis that the data of our sample come from a normal population. The values of skewness and kurtosis, each divided by the corresponding standard errors, for the whole sample (0,6 and -3,2) and for each age group (Grade 9: 1,9 and -1,9; Grade 11: 0,0 and -2,1) indicated that the data were normally distributed.

Next, a series of models were tested and compared. Specifically, the first model was a third-order CFA model which was designed on the basis of the results of the study by Elia *et al.* [10]. It involved one third-order factor which was hypothesized as accounting for all variance and covariance related to the second-order factors. The second-order factors represented students' abilities to carry out conversions of algebraic relations with the graphic (Test A) and the verbal mode (Test B) respectively as the source representation. Each of the second-order factors were assumed to explain the variance and covariance related to three first-order factors measured by the observed variables corresponding to the 12 conversions of Test A and the 12 conversions of Test B respectively. The former three first-order factors were distinguished with respect to the conceptual characteristics of the tasks, that is, whether they involved a function or not or the kind of the function involved. The latter three first-order factors were differentiated with reference to their conceptual characteristics, but also to their type of conversion and more specifically their target representation (graphic or symbolic). The fit of this model was poor [X^2 (244) = 957.889; CFI = .831; RMSEA = .071, 90% confidence interval for RMSEA = 0.066–0.075], indicating that the particular structure was not appropriate to describe the structure of the abilities in performing conversions among the different modes of representation of the algebraic relations included in the two tests. The structure of this model was based on a combination of the results of the two age groups given by Elia *et al.* [10]. Consequently, some relationships in the model that applied for grade 9 did not apply for grade 11 and vice versa. This could be an explanation for the inconsistency of this model with the data of the whole sample.

Nevertheless, a main concern of this study was to validate a CFA model that could capture the structural organization underlying the processes of the students of both age groups in the conversions among different representations of functions. Therefore, a critical commonality that was observed between the two age groups in Elia *et al.*'s [10] study was the inconsistency in students' performance when dealing with conversions of a different source representation. This strong tendency among the students of both age groups, as well as, Duval's findings suggesting that a major difficulty in mathematics learning is the compartmentalization among registers of representations, formed a basis for the second model that we have examined [1]. The second model (Figure 1) involves two first-order factors and one second-order factor on which the first order-factors are regressed. The first-order factors stand for the two types of conversions of the six algebraic relations involved in the two tests with respect to their initial (source) representation, i.e. conversions from a graphic form to a verbal and to a symbolic form (Gvs) and conversions from a verbal form to a graphic and to a symbolic form (Vgs). Each factor was measured by six observed variables, which stand for the conversion tasks of the six mathematical relations having a common initial representation. In Figure 5.1 we have to note that:

1. "AbGV" stands for the ability to carry out conversions of functions and other algebraic relations having the graphic or the verbal form as the source representation.
2. "Gvs" stands for the ability to carry out conversions of functions and other algebraic relations from the Graphic form to the verbal and the symbolic form.
3. "Vgs" stands for the ability to carry out conversions of functions and other algebraic relations from the Verbal form to the graphic and the symbolic form.
4. E1–12 stand for the errors of the variables.
5. D1 and D2 stand for the errors of the latent factors.

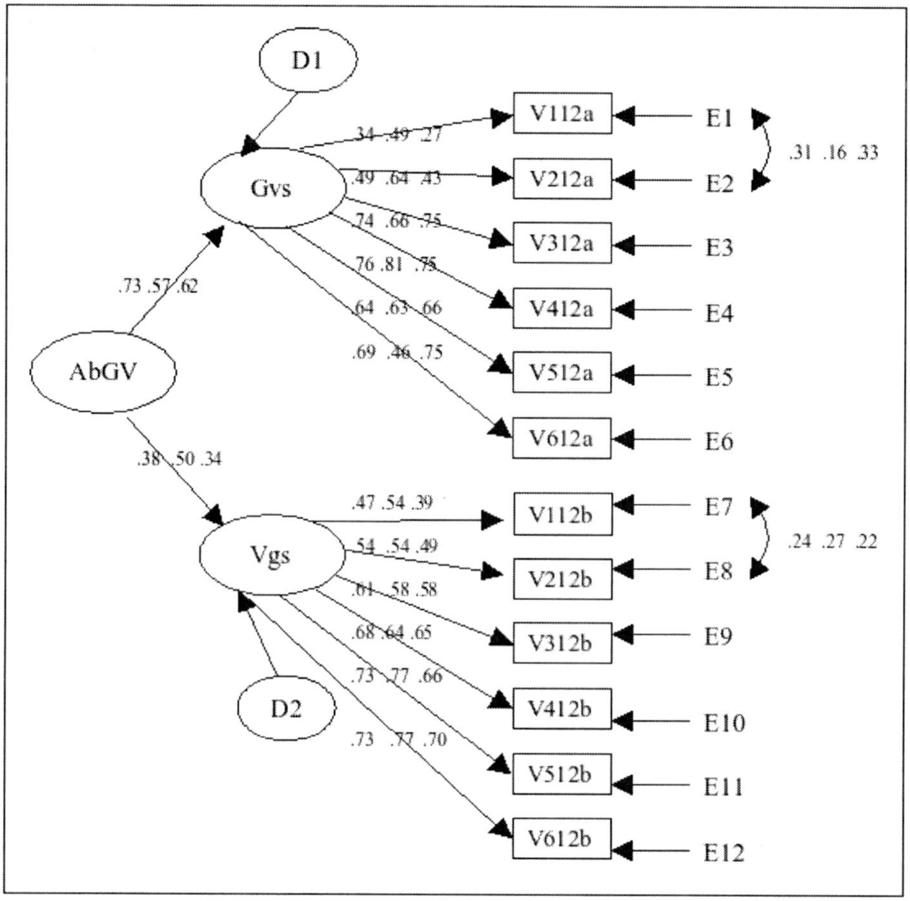

Fig. 1. The elaborated model for the conversions among different modes of representation of algebraic relations, with factor loadings for students of the whole sample and of grades 9 and 11, separately.

6. The first, second and third coefficient of each parameter stand for the application of the model on the performance of the students of the whole sample, of grades 9 and 11, respectively.

It is noteworthy that the number of the observed variables is half as the number of the observed variables of the first model. In the tests the conversion items of each algebraic relationship formed one task, having the same starting representation, and students normally treated the two conversion items of the same relationship "simultaneously". Therefore, we considered that it would be more meaningful to integrate the variables that corresponded to the conversions of the same mathematical relation and the same initial representation, despite their difference in the target representation. We combined each pair of variables (e.g., V11a and V12a) involving the same relation in the same test into one variable (e.g., V112a).

The second-order factor represents the general ability to perform conversions of algebraic relations starting from a graphic or a verbal form of representation (AbGV). The fit of this model was good [X^2 (50) = 165.730; CFI = .943; RMSEA = .063, 90% confidence interval for RMSEA = 0.052–0.073], verifying that the type of the conversion and more specifically the type of the starting representation of a conversion does have an effect on the process of carrying out conversions of functions and other algebraic relations among different representations.

As written above, combining the variables, the degrees of freedom of the models are decreased from 250 to 50. This is due to the fact that the number of data points, i.e. variances, covariances of the observed variables (how much information we have with respect to our data), is significantly decreased. In particular, while the data points of the model involving 24 observed variables were 300 ($24 \times 25/2$), the corresponding elements of the model involving half of the variables were 78 ($12 \times 13/2$). The number of estimable parameters in the former model was 50, while the number of parameters to be estimated in the latter model was 28. Given that the degrees of freedom stand for the difference between the data points and the structural parameters, the former model has 250 degrees of freedom, while the latter one has only 50.

Attention is drawn to the fact that the relation of the first-order factor standing for the conversion having the graphic mode as the starting representation to the second-order factor, corresponding to the general ability in representational conversions of functions, is considerably stronger than the relation of the other first-order factor. Thus, the conversions involving the verbal mode as the starting representation seem to have considerable autonomy from the conversions involving the graphic form as the source representation.

To test for possible differences between the two age groups in the structure described above, confirmatory factor analysis was applied to test the second-order model separately on each ge group. The model was found to be consistent with the data of both groups [grade 9: X^2 (50) = 108.354; CFI = .911; RMSEA = .082, 90% confidence interval for RMSEA = 0.061–0.102, grade 11: X^2 (50) = 141.846; CFI = .926; RMSEA = .068, 90% confidence interval for RMSEA = 0.054–0.080].

The regression coefficients of the scores in the conversion tasks of the algebraic relations standing for regions of points (V112a, b: $y < 0$, V212a, b: $xy > 0$) onto the corresponding first-order factors representing students' abilities in the conversions having as a source representation the graphic or the verbal form were lower than the regression coefficients of the other variables. Thus, the conversion of these algebraic relations seems to have considerable autonomy from the conversions of relations involving functions. Moreover, students' abilities to carry out the conversions of these inequalities are significantly correlated to each other in each test. The regression coefficient of the conversion of the constant function (V512a: $y = 3/2$) with the graphic form as the source representation was also lower compared to the respective coefficients of the variables standing for the other functions in Test A. This is an interesting finding to be discussed later in combination with the results of the other statistical methods.

5.2 The outcomes of the hierarchical clustering of variables and the implicative method of analysis

Each observed variable in the CFA model (Figure 1) represented students' "unified" score at the conversions of each algebraic relation from the graphic representation to the verbal and the symbolic representation (Test A) or from the verbal representation to the graphic and the symbolic representation (Test B). To make the comparison between the three statistical techniques under study feasible, we considered that it would be meaningful to include these variables also in the implicative analysis and the hierarchical classification.

Before elaborating on the results of the hierarchical clustering of variables and the implicative method, we will explicate our predictions concerned with the structure of the similarity and implicative diagrams involving the variables of the study. First, we expect that similarity and implicative relationships will be primarily established among the variables corresponding to the conversions of the same starting representation, namely the graphic and the verbal representations. This hypothesis is based on well documented findings suggesting the fragmentary way of students' thinking when dealing with different types of representations [1,6]. A second prediction is that close relationships will be formed among variables standing for the conversion tasks of functions because of their common perceptual and conceptual features. Third, we expect that success on the conversion of functions would entail success on the conversions of inequalities in the implicative diagrams. We consider the understanding of inequalities (graphically, regions of points) and principally of the inequality $y < 0$ and other similar ones as prerequisite for the understanding and use of more complex relations and functions such as $y = -x$ or $y = x - 2$.

The symbolism used for the variables of figure 2 (and the figures that follow) is the following:

1. "a" stands for Test A, and "b" stands for Test B.
2. The first number after "v" stands for the number of the task in the test. i.e. 1: $y < 0$, 2: $xy > 0$, 3: $y > x$, 4: $y = -x$, 5: $y = 3/2$, 6: $y = x - 2$
3. The second number stands for the source representation of the conversion corresponding to each test, i.e. for Test A, 1: graphic representation; for Test B: verbal representation.

Figure 2 illustrates the similarity relations among the "condensed" variables corresponding to grade 9 and 11 students' responses to the tasks of the two tests. Two distinct clusters of variables are established in the hierarchical similarity diagram. The first cluster involves students' responses to the tasks

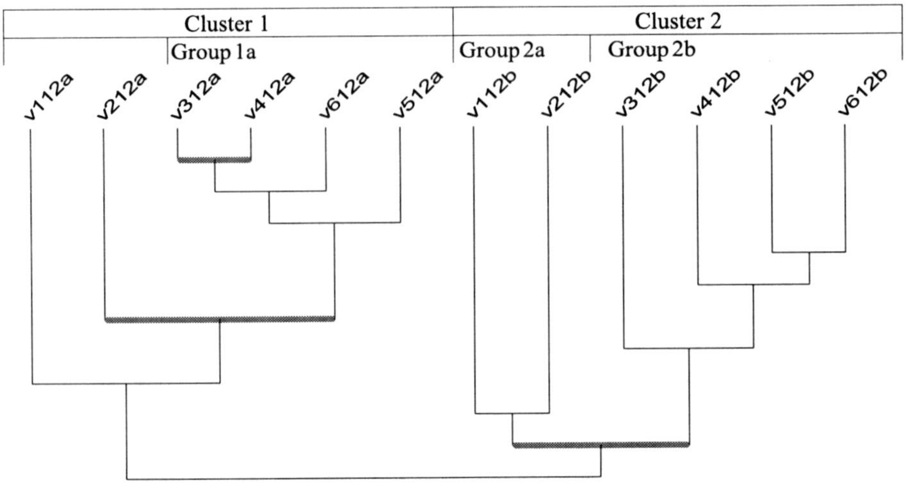

Fig. 2. The hierarchical similarity diagram among the responses of students of grades 9 and 11 to the conversion tasks of Test A and Test B (C.H.I.C. 3.5)

of Test A, while the second cluster is comprised of students' responses to the tasks of Test B. This suggests that students dealt with the conversions of the six algebraic relations which had the graphic mode as the initial representation consistently. Students exhibited consistency, in a lower level though, also in the conversions of the corresponding algebraic relations which had the verbal form as the initial representation. These findings verify the first prediction concerned with the variables' relationships, suggesting the establishment of close connections among students' responses to the conversions involving the same starting representation. The weak similarity relation between the two clusters suggests that students approached conversions of a different source representation in a distinct way, despite involving the same algebraic relations. This behaviour may be a consequence of students' compartmentalized way of thinking in different modes of representation.

The similarity relations within each cluster of variables are also of great interest since they can be seen as indications of students' way of understanding of the particular algebraic relations and of carrying out the particular types of conversion. Each cluster involves one similarity group, namely Groups 1a and 2b, in which the variables of the third task (v312a or v312b), the fourth task (v412a or v412b), the fifth task (v512a or v512b) and the sixth task (v612a or v612b) are linked together. On one hand, the similarity linkage among the variables referring to tasks 4, 5 and 6 which involve function provides evidence to the second prediction of our analysis, stating that close relationships will be established among variables standing for the conversion tasks of functions. On the other hand, the similarity connection of these variables with the variable referring to the conversion task 3 of the inequality $y > x$ extends our expectation. It is suggested that students carried out the conversions of functions, that is, tasks 4–6, as well as the algebraic relation which involved a function, that is, task 3, using similar processes. The conversions of the graphic or the verbal form of $y < 0$ (v112a or v112b) and of $xy > 0$ (v212a or v212b) were carried out differently from the conversions of the other relations probably due to their distinct properties as they did not entail functions. Whereas students tackled the conversions of the relations $y < 0$ (v112b) and $xy > 0$ (v212b) having the verbal form as a source representation similarly to each other thus forming a similarity pair (Group 2a), this was not the case in the conversions of the respective relations (v112a and v212a) with the graphic form as the initial representation. This finding provides further support to the assertion that the graphic form and the verbal form of the same mathematical content stimulated the use of distinct conversion processes by the students.

Figure 3 shows the implicative relations among the "condensed" variables corresponding to ninth and eleventh graders' responses to the tasks of the two tests. The diagram involves a dual implicative chain which indicates the hierarchical ordering of the conversion tasks with respect to their level of difficulty on the basis of students' performance. One branch of the chain, namely Branch A, involves mainly the responses to the conversions of the algebraic relations with the graphic representation as the initial one (Test A). The other branch, termed Branch B, is comprised mainly of the responses to the conversions of the algebraic relations with the verbal representation as the initial one (Test B). The establishment of these implicative branches gives further support to the first prediction for the analyses stating that implicative relationships will be primarily formed among the variables corresponding to the conversions of the same starting representation, namely the graphic and the verbal form. Both branches stem from the same variable which stands for the response to the conversion of the graphic representation of the function $y = x - 2$ (v612a). This is the most complex task of both tests and students who provided a correct solution at it, succeeded at all of the other conversion tasks of both tests. Students' great difficulty to the sixth conversion task of each test is verified also by their lowest success rates on the particular task in both grades as illustrated in Tables 5 and 6 of the Appendix. These tables present the means and frequency of occurrences of the two age groups, respectively, to all of the tasks of the two tests.

Considering "Branch B", students who accomplished the most difficult task of both tests succeeded at the corresponding task in Test B, meaning the conversion of the verbal representation of the same algebraic relation (v612b). Carrying out the latter task implied success at the task of the same type of conversion involving the function $y = -x$ (v412b), which in turn entailed correct performance in the conversions of the function $y = -3/2$ (v512b). These implicative relationships give further support to the second prediction for the analyses suggesting the formation of implicative relationships among variables standing for the conversion tasks of functions. The conversions of the constant function from a verbal to a symbolic or to a graphic representation

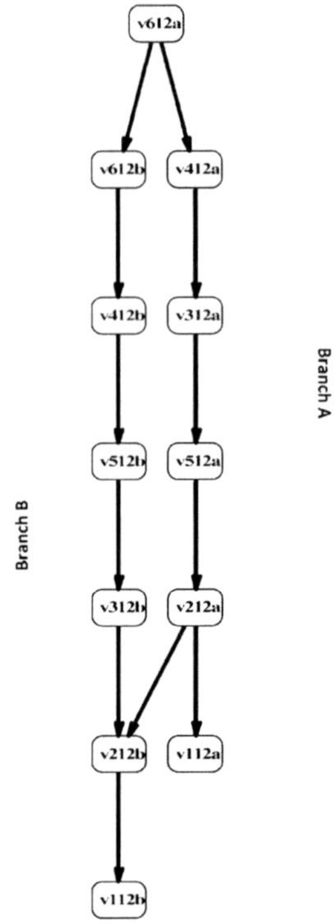

Fig. 3. The implicative diagram among the responses of students of grades 9 and 11 to the conversion tasks of Test A and Test B (C.H.I.C. 3.5)

seemed to be easier than the corresponding conversions of any other function involving the variable of x. The conversions of functions were more difficult, though, than the conversions of inequalities standing for regions of points in Test B (see Appendix). Moreover, success at the conversions of the verbal representation of functions implied success at the conversions of the same source representation of algebraic inequalities. Success at the conversions of the constant function, entailed success at the conversions of the algebraic relation $y > x$ (v312b), which sequentially implied success at the corresponding conversions of the relation $xy > 0$ (v212b). Students who performed correctly at the latter task, were able to carry out the simplest conversion task of Test B which involved the relation $y < 0$ (v112b). Therefore, our third prediction that success on the conversions of functions would entail success on the conversions of inequalities in the implicative diagrams was also verified.

Considering "Branch A", carrying out the most difficult task (v612a) entailed success at the conversions of the function $y = -x$ from a graphic to a verbal and to a symbolic representation (v412a). Success at the latter task implied success at the easier task incorporating the algebraic relation

$y > x$ (v312a). Consecutively, students who carried out the conversions of the graphic representation of $y > x$ were successful at the corresponding conversions of the constant function $y = 3/2$ (v512a). Success at the latter conversion task implied success at the conversions of the relation $xy > 0$ (v212a) corresponding to a region of points, which in turn entailed success at the conversion of the graphic form of the relation, $y < 0$ (v112a), representing a region of points as well. On one hand, these results indicate that in general the hierarchical ordering of the tasks based on students' performance to Test A is congruent with Test B, providing further evidence to the two latter predictions which suggest the establishment of close implicative relationships among the variables of the conversions of functions and of the implications of success at these conversions to success at the conversions of regions of points. On the other hand, unlike Branch B, the variable of the conversion of $y > x$ intervenes among the variables of the conversions of functions in Branch A, which is not completely in line with these predictions. The conversion of the graphic representation of the relation $y > x$, which despite being an inequality, it actually involves a function, was more complex than the corresponding conversion of the constant function and success at the task of the former relation implied success at the task of the latter relation.

It is worth noting that besides the implicative relation between the variables referring to the two conversion tasks that involved the same algebraic relation $y = x - 2$ (v612a-v612b), there is another implicative connection linking variables of the two tests. In particular, success at the conversions of the relation $xy > 0$ having the graphic mode as the initial representation (Test A: v212a) implied success at the conversion of the same relation with the verbal form as the initial representation (Test B: v212b). On one hand, the fact that only two implicative relations are formed between the variables of the two tests suggests that students' success at most of the tasks of Test A was independent from their success at the tasks of Test B. Thus, support is provided to the almost compartmentalized ways by which the students approached the conversions of a different source representation despite involving the same mathematical content. On the other hand, the fact that in both relations success at a task of Test A entailed success at a task of Test B provides evidence to the more difficult character of the conversion starting with a graph relatively to a conversion starting with a verbal description.

The next figures illustrate the results of the hierarchical clustering of variables and the implicative method for the students of Grade 9 and 11 separately. These results are generally congruent with the outcomes referring to the whole sample elaborated above, and therefore they are in line with the predictions concerned with the similarity and implicative relationships of the variables, with only a number of minor deviations.

Figure 4 illustrates the hierarchical similarity diagram of the "condensed" variables corresponding to grade 9 students' responses to the tasks of the two tests. In line with the general similarity diagram for the whole sample, two distinct clusters of variables are formed, namely Cluster 1 and 2, which correspond to Test A and B respectively, indicating lack of consistency in ninth graders' performance between conversions with a different source representation.

Within Cluster 1 two similarity groups are formed. Group 1a involves the variables of the conversions of a relation represented as a region of points, $xy > 0$ (v212a) and of a constant function $y = 3/2$, represented as a horizontal line parallel to the axis xx' (v512a). The transformations of these relations were approached differently from the transformations of the functions $y = x-2$ (V612a) and $y = -x$ (v412a) and the algebraic relation incorporating a function $y > x$ (V312a). The transformations of the three latter relations were tackled similarly due to their common functional character, thus establishing a group, namely Group 1b. Despite their common properties with the constant function, the conversion of this relation (v512a) was carried out differently, indicating students' difficulty to realize that the graph of a horizontal line represents a function and handle it as such. This finding deviates from the second prediction stating that close similarity relationships would be formed among the variables, which correspond to conversions of functions. Explaining verbally what the graph $y < 0$ (v112a) represented was carried out in a different way from the graphs of the other relations of the test probably because of its distinct perceptual characteristics, which made its interpretation easier. In Cluster 2 two similarity groups are also identified. Group 2a involves the

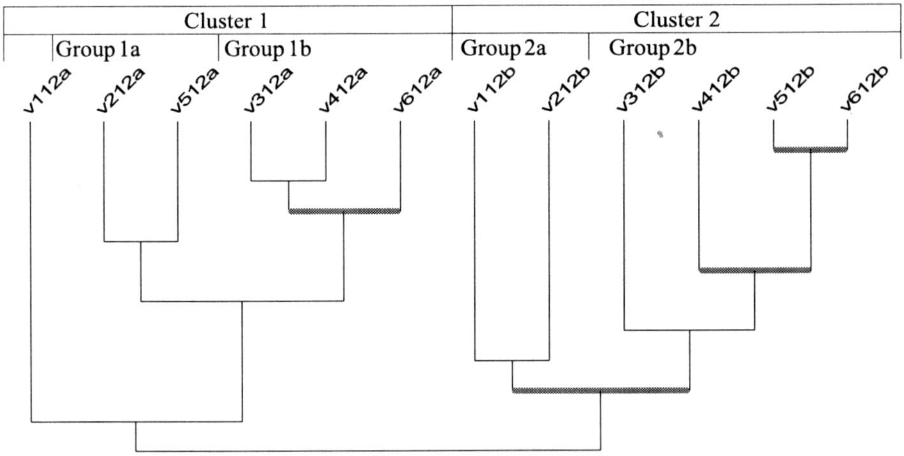

Fig. 4. The hierarchical similarity diagram among the responses of students of grade 9 to the conversion tasks of Test A and Test B (C.H.I.C. 3.5)

responses to the conversions of the verbal representation of two algebraic relations corresponding to regions of points (v112b and v212b). Group 2b, which presents a stronger similarity than Group 2a, is comprised of the responses to the conversions of the three functions of the test (v412b, v512b, v612b) and the algebraic relation which involves a function (v312b: $y > x$). Thus, the conceptual components of the algebraic relations, distinguished by whether they involve a function or not, seem to differentiate students' processes in the conversions of a verbal representation (Test B).

Figure 5 illustrates the implicative diagram of the "condensed" variables corresponding to grade 9 students' responses to the tasks of the two tests. The results of the implicative analysis are in line with the similarity relations explained above. Two separate "chains" of implicative relations among the variables are formed with respect to the test they refer to, namely Chain A and Chain B. This suggests that success at the conversions of the algebraic relations in the two tests depended primarily on their initial representation. The commonality of their content did not have a role. For instance, students who succeeded at the conversion of a function with the graphic form as the source representation did not necessarily succeed at the conversion of the same function with the verbal form as the initial representation. Students carried out the conversions by activating compartmentalized processes based on the source representation of the conversion.

The two implicative chains have a similar structure, which stems from the conceptual components of the tasks. In particular, the conversions of the functions $y = x - 2$ (v612a or v612b) and $y = -x$ (v412a or v412b) were the most difficult tasks, and success at them implied success at the conversions of almost all the other algebraic relations in each test. The conversions of the constant function (v512a or v512b) and the algebraic relation involving a function (v312a or v312b) were less complex. Students exhibited the greatest facility at the conversion tasks of the algebraic relations standing for regions of points ($y < 0$ or $xy > 0$). Figure 6 illustrates the hierarchical similarity diagram of the variables corresponding to grade 11 students' responses to the tasks of the two tests. The structure of the similarity relations in this diagram is analogous to the structure in the diagram concerning grade 9 students, as two similarity clusters are established with respect to the source representation of the conversions. A main difference though is that the first cluster (Cluster 1), which refers to the conversions of graphic representations, involves one similarity group (Group 1a), in which the variable of the fifth task (v512a) is linked to the variables of the third (v312a), fourth (v412a) and sixth (v612a) task. Thus, students carried out the conversion of the constant function using similar processes with the ones when performing the conversions of the other algebraic relations involving functions. This similarity is weaker though than the similarities among the conversions of the other relations (3, 4 and 6). Eleventh graders' increased consistency in comparison with the ninth graders' consistency indicates the older students' realization that the graph of the relation $y = 3/2$ represents a function despite its dissimilar perceptual form relatively to the graphs of the other relations involving functions. However, students performed the conversions of the other relations, i.e. $y < 0$ (v112a

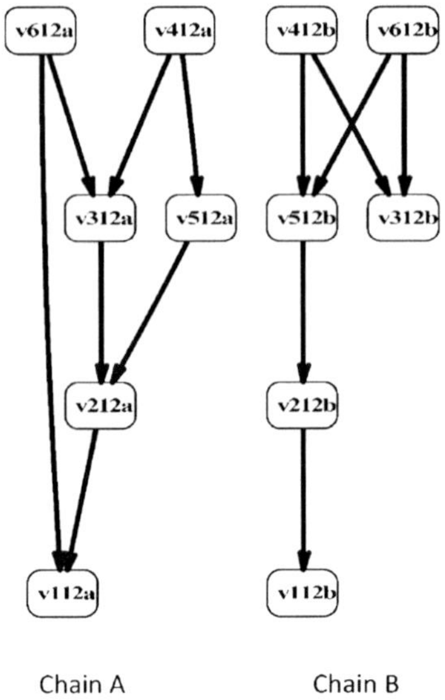

Fig. 5. The implicative diagram among the responses of students of grade 9 to the conversion tasks of Test A and Test B (C.H.I.C. 3.5)

or v112b) and $xy > 0$ (v212a or v212b), starting from a verbal or a graphic form in a different way, probably because they did not represent functions.

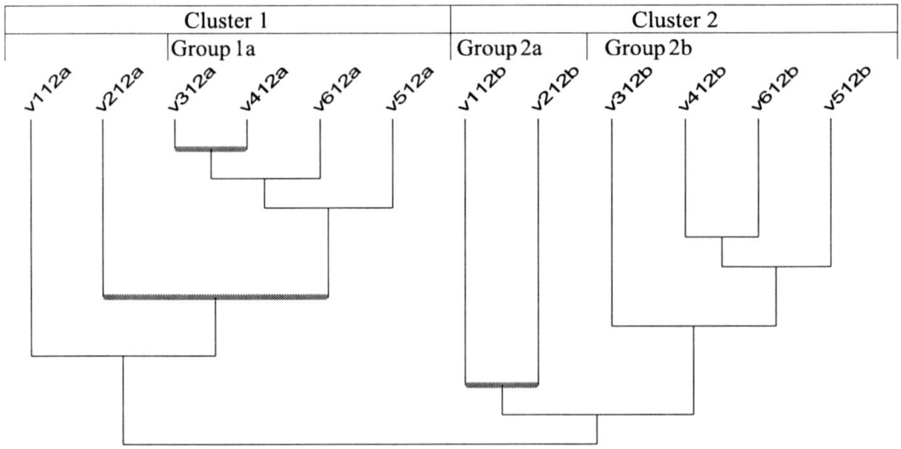

Fig. 6. The hierarchical similarity diagram among the responses of students of grade 11 to the conversion tasks of Test A and Test B (C.H.I.C. 3.5)

Figure 7 shows the implicative diagram of the variables corresponding to grade 11 students' responses to the tasks of the two tests. This implicative diagram is congruent with the implicative diagram concerning grade 9 students. Specifically, a basic commonality is the establishment of two distinct chains of implicative relations, among the variables with respect to the type of conversion or the test they refer to, namely Chain A and Chain B. Furthermore, like in the ninth graders' similarity diagram, success at the conversions of the algebraic relations involving functions implied success at the conversions of the relations corresponding to regions of points in each test.

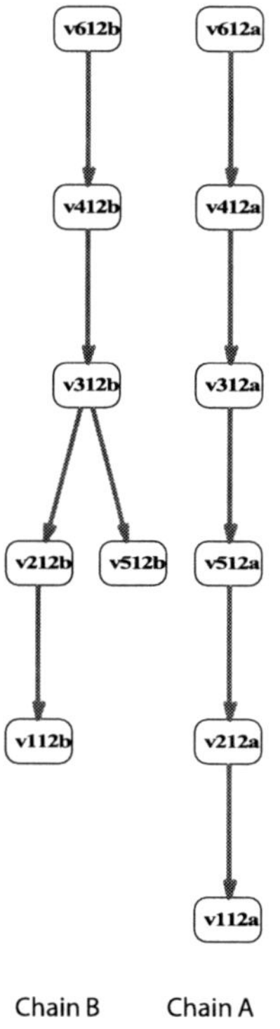

Fig. 7. The implicative diagram among the responses of students of grade 11 to the conversion tasks of Test A and Test B (C.H.I.C. 3.5)

However, in this diagram, the structure of each implicative chain has a more linear form relatively to the diagram of ninth graders, illustrating more explicitly the hierarchical ordering of the conversion tasks with respect to the type of the algebraic relation involved. In particular, the most difficult tasks which implied success to all the other tasks in each test involved the conversion of the function $y = x - 2$ (v612a or v612b) (see Appendix). Less complex conversion tasks in the two tests involved the function $y = -x$ (v412a or v412b) and students who tackled them succeeded at the easier tasks incorporating the algebraic relation $y > x$ (v312a or v312b). Consecutively, students who carried out the latter tasks were successful at the conversion tasks of the constant function $y = 3/2$ (v512a or v512b) in both tests. In Chain B (Test B), success at the conversion task of the verbal representation of $y > x$ implied directly success also at the conversions of another verbally given inequality, that is, $xy > 0$ (v212b), which in turn entailed success at the easiest conversion task of the test, involving the inequality $y < 0$ (v112b). In Chain A (Test A), carrying out the conversion task of the graphic representation of $y = 3/2$ (v512a) implied success at the conversions of the relation $xy > 0$ (v212a) corresponding to a region of points, which in turn entailed successful performance in the conversions of the graphic form of the relation $y < 0$ (v112a) representing a region of points as well.

6 Discussion

This study investigated students' abilities in the conversions of functions and other algebraic relations among representations, as well as, their interrelations. The data, which were collected using two tests involving conversion tasks of the same mathematical content with a different source representation, were analyzed from different perspectives using three distinct statistical methods, each of which is based on a different rationale. CFA is used to test statistically whether a hypothesized connection pattern between the observed variables and their underlying factors exists. The a priori hypothesis in this study stemmed from knowledge of past empirical work and theory which suggested that students encounter difficulties in transferring information gained in one representational context to another related to the concept of functions [1, 3, 9, 10]. The hierarchical clustering of variables aims at bringing to light the consistency among students' responses to the various tasks in a hierarchical manner. The implicative method gives information about whether success at one task implies success at another task and about the relative difficulty of the tasks based on students' performance (Table 2, item 1). A major concern of this study is to compare in detail the findings of these statistical analyses so as to gain insight about the advantages of each method, as well as, about whether their outcomes on the same sample data are congruent and can complement each other.

Tables 2 and 3 summarize and compare the key outcomes of the three statistical methods on the data of the study that concur or have a complementary role amongst them. Whereas the implicative technique and the hierarchical clustering of variables incorporated only observed measurements, CFA allowed the development and validation of a model that involves not only observed but also latent (unobserved) variables, which can not be observed or measured directly (Table 2, item 2). These constructs, which lied behind the corresponding observed measures, were the ability to perform conversions of algebraic relations from the graphic representation to symbolic and to verbal form, and the ability to carry out conversions of algebraic relations from the verbal representation to symbolic and to graphic form. The fact that two factors are needed to account for the effects of the two types of the initial represantation exemined here, i.e. graphic form and verbal form provided a strong case for the role of the source represantation in the conversions of algebraic relations ammong different representations. Another abstract construct of a higher-order level was assumed to underlie these abilities, indicating that despite the discrepancy in students' performance among the different types of conversion, both abilities are still basic components of a common construct, i.e. general ability in transferring an algebraic relation from one representation to others.

The difference in the strength of the relations of the two first-order factors to the second-order factor in the model revealed that the graphic and the verbal representations have a different and an almost autonomous function in the conversions of algebraic relations. The outcomes of the other two methods of analysis were the ones that revealed in a more explicit way students' compartmentalized responses to the conversion tasks with respect to the source representation (Table 2, item 3). The separate grouping of the responses to conversions having as the initial representation the verbal form or the graphic form in the similarity diagrams showed students' inconsistency when dealing with conversions of different initial representations. The implicative diagrams included distinct chains of variables with respect to the initial representation of the conversion indicating that success in one type of conversion of an algebraic relation did not necessarily imply success in another mode of conversion of the same relation. Weak connections or lack of implications among conversions of the same mathematical content with a different starting representation are the main features of the phenomenon of compartmentalization and indicate that students did not construct the whole meaning of the concept of function and did not grasp the whole range of its applications [10]. As Even supports [16], the ability to identify and represent the same concept in different representations, and flexibility in moving from one representation to another allow students to see rich relationships, and develop a deep understanding of the concept.

In comparing the three methods of analysis, the CFA validated the same grouping of tasks as the implicative method and the hierarchical classification method, since the measurement indicators of each factor in the model formed a separate group in the other two analyses. Nevertheless, the implicative and the hierarchical classification methods provided further insight and a more analytic view about the construction and hierarchical structure of these groups and the implicative relations among students' responses to the tasks (Table 4).

For example, the hierarchical clustering of variables revealed some discrepancies in the ways students tackled particular tasks of the two tests, that were not evident in the CFA model (Table 4, item 1). Whereas the conversions of

CFA	Hierarchical clustering	Implicative method
1. Factorial structure of students' abilities in the conversion of algebraic relations from one representation to another	1. Hierarchical classification and consistency of students' responses to the conversions	1. Implicative relations between students' responses to the conversions, relative difficulty of the tasks
2. Development of a model involving two latent (unobserved) factors for the effects of two types of initial representation and a second-order factor standing for the general ability to convert functions from one representation to others	2. Similarity groupings among observed measurements standing for students' responses to the conversions of algebraic relations having the verbal or the graphic mode as the source representation	2. Implications among observed variables standing for students' responses to the conversions of algebraic relations having the verbal or the graphic mode as the source representation
3. Difference in the strength of the relations of the two first-order factors to the second-order factor: the graphic and the verbal form of the same content operate rather autonomously in the conversions	3. Compartmentalization in students' responses to the conversions with respect to their source representation	3. Two implicative chains involving students' responses to the conversions of a verbal representation and the conversions of a graphic one respectively, lack of implications between the variables of the two chains
4. Lower factor loadings of the conversions of inequalities relatively to the conversions of functions: considerable autonomy among the conversion processes of functions and the conversions of algebraic relations not representing functions	4. Separate grouping of the variables of conversions of inequalities and the variables of conversions of functions: inconsistency between them, significant role of the conceptual properties of the algebraic relations on students' consistency	4. The conversions involving functions were more complex than the tasks involving regions of points. Success at carrying out a conversion of a function implied success at a conversion of the same type involving an inequality

Table 2. The congruent and complementary outcomes of the CFA, the hierarchical clustering of variables and the implicative method on the data of the study (Part 1)

5. Lower factor loading of the conversion of the graphic form of a constant function relatively to the corresponding conversions of the other functions	5. Relatively weak similarity of students' conversions of the graphic form of the constant function and of the other functions: distinct ways of approaching the conversion of a graphic representation of this kind of function	5. The conversion of the constant function was the easiest one among the conversion tasks of the other functions.
6. The factorial structure remains invariant across grades.	6. The abilities to carry out conversions of graphic or verbal representations of algebraic relations remain compartmentalized in the two grades	6. Two distinct chains with respect to the type of the conversions in both grades, success at a conversion of a graph does not imply success at a conversion of a verbal representation in neither grade.

Table 3. The congruent and complementary outcomes of the CFA, the hierarchical clustering of variables and the implicative method on the data of the study (Part 2)

the verbal representations of the algebraic relations $y < 0$ and $xy > 0$ were approached similarly to each other, this was not the case in the conversions of the graphic representations of the corresponding relations. This outcome provides evidence to the students' disconnected and distinct ways of using the verbal form and the graphic form as source representations in conversions, giving further support to the phenomenon of compartmentalization. The implicative diagram indicated that the conversion of the graphic representation of the function $y = x - 2$ (v612a) was the most complex task of both tests and students who provided a correct solution at it, succeeded at all of the other conversion tasks of both tests. Moreover, the implicative relations linking responses to tasks of the two tests showed that success at some tasks of Test A entailed success at the corresponding tasks of Test B. The above outcomes of the implicative method provide evidence to the distinct and more difficult character of the conversions starting with a graph relatively to conversions starting with a verbal description. This differentiation and increased difficulty may be due to the fact that the perceptual analysis and synthesis of mathematical information presented implicitly in a diagram often make greater demands on a student that any other aspect of a problem [22]. The graphic register functions effectively only under the conventions of a different mathematical culture. Due to students' poor knowledge of this new culture, graphic modes of representation are difficult to decode and work out. It can be asserted that the support offered by mathematical meta-language is more fundamental than the aid given by the graphic register for carrying out a translation from one mode of representation to another [10].

Furthermore, the outcomes of the three statistical processes uncovered how students dealt with different types of algebraic relations in conversions of the same source representation (Tables 2 and 3, items 4 and 5). The conversions of the algebraic relations that corresponded to inequalities and thus regions of points were found to have considerable autonomy from the conversions of relations involving functions. The separate grouping of the former variables from the latter ones in the similarity diagrams revealed that students tackled the conversions of inequalities differently from the conversions of functions. The lower factor loadings of the variables referring to the conversions of inequalities relatively to the corresponding loadings of the variables standing for the conversions of functions in the CFA model were in line with this outcome. The implicative diagrams revealed additional information to the above findings, suggesting that the tasks involving functions were more complex than the tasks involving regions of points in any type of conversion. Moreover, the implications among the variables showed that students who carried out conversions of function from one representation to another were able to succeed at conversions of the same type involving inequalities. Considering the relations within the responses to the function tasks, a relatively weak similarity was observed between students' response to the conversion of the constant function and their responses to the other functions, revealing their distinct ways of approaching the conversion of a graphic representation of this kind of function. Given that this distinction did not apply in the conversions of the verbal representation (Test B), it is indicated that the interpretation of the graphic form of the particular type of function was the main factor differentiating students' performance. The relatively lower factor loading of the variable referring to the conversion of a constant function from a graphic representation to another representation in comparison to the factor loadings of the corresponding variables of the other functions gave further support to students' different conversion processes when dealing with this special kind of function. The implicative relations revealed that the task involving the particular function was the easiest one among the conversion tasks of the other functions. The above findings highlight the effect of the inherent mathematical foundation of the algebraic relations, i.e. whether they are functions or not and what kind of functions they are, on students' processes, consistency and success in conversions of the same type.

Addressing the effect of age on the abilities to transfer algebraic relations from one representation to another, the outcomes of all of the analyses (Table 3, item 6) indicated that students' compartmentalized ways of using representations and thinking of algebraic relations in general and functions in particular occurs in both grades. Specifically, the factorial structure described above remains invariant, while the abilities to carry out conversions of graphic or verbal representations of algebraic relations remain compartmentalized in the similarity and the implicative diagrams for the two grades involved in the study. These findings, which provide support to the results of the studies by Elia et al [10] and Gagatsis and Christou [11] indicate that relations among translations from one mode of representation to another do not vary as a function of grade levels. Thus the relative inherent nature of difficulties of each type of conversion has age endurance, suggesting that development or regular instruction does not change students' processes while dealing with conversions of functions from one representation to another.

Despite this invariance, some discrepancies in the performance of students of the two grades occurred as regards the inherent mathematical properties of the algebraic relations (Table 4, item 2). Based on the similarity diagrams, eleventh graders were found to respond to the conversion of the graph of the constant function more coherently with the conversions of the other functions relatively to the ninth graders. This suggests that the older students were more competent in recognizing the conceptual components of some algebraic relations, i.e. whether they represented a function or not, and dealt with their conversions more consistently. This competence was also indicated by the outcomes of the implicative analysis on the data of the two age groups respectively. The hierarchical ordering of the tasks with respect to their level of difficulty was clearer in the implicative diagram of eleventh graders compared to the diagram of the ninth graders. Given that the different levels of difficulty of the tasks stemmed from the type of the algebraic relations involved, it can be asserted that the older students identified more efficiently the distinct conceptual properties of each algebraic relation. Thus, the conceptual features of the algebraic relations seemed to have a stronger impact on their processes and success levels relatively to the younger students' responses.

In general, the application of all of the analyses yielded congruent results. However, at the same time given that these statistical processes approached the data from different perspectives, they emphasized different aspects of students' outcomes. This differentiation allowed for the accumulation of a number of new distinctive elements by each analysis that contributed to the unravelling and making sense of students' performance, structure of abilities, difficulties and inconsistencies on the particular subject. The findings of the study suggest that the three statistical methods are open to complementary use and each one does not operate at the expense of the other. CFA provided a means for making sense of the structure of students' abilities in the conversion of functions among different representations. The hierarchical clustering of variables provided a means for classifying students' responses, for identifying students' consistencies and inconsistencies among different conversions and for investigating the factors influencing this behaviour. The implicative method provided a means for examining the implicative relations among the responses to the tasks and the relative difficulty of the different conversions on the basis of students' performance. Provided that applying these methods of analysis is consistent with the objectives of a study, their combination on the same sample data could contribute to overcome some significant limitations of

Hierarchical clustering	Implicative method
1. The conversions of the verbal representations of the algebraic relations $y < 0$ and $xy > 0$ were approached similarly to each other unlike the conversions of the graphic representations of the corresponding relations: disconnected ways of using the verbal form and the graphic form as source representations	1a) The conversion of the graphic representation of the function $y = x - 2$ was the most complex task of both tests; b) Success at the tasks of Test A entailed success at the corresponding tasks of Test B: greater difficulty of the conversion starting with a graph relatively to a conversion starting with a verbal description
2. The conversions of the constant function and of the other functions were tackled more coherently by the 11th graders than by the 9th graders: the older students were more competent in recognizing the conceptual components of some algebraic relations	2. The hierarchical ordering of the tasks with respect to their level of difficulty was clearer in the 11th graders' diagram compared to the 9th graders' diagram: the older students were more able to identify the distinct conceptual properties of each algebraic relation

Table 4. The new additional outcomes of the hierarchical clustering of variables and the implicative method to the outcomes of the CFA

each analysis employed separately, and consequently could enrich and deepen the outcomes of the investigation.

[1]Glossary of initials used in the text:
CFA: Confirmatory Factor Analysis
EQS: Bentler's Structural Equation Modeling program
CFI: Comparative Fit Index
RMSEA: Root Mean Square Error of Approximation
SEM: Structural Equation Modeling
EFA: Exploratory Factor Analysis
CHIC: Classification Hierarchique, Implicative et Cohesitive

References

1. R. Duval. The cognitive analysis of problems of comprehension in the learning of mathematics. *Mediterranean Journal for Research in Mathematics Education*, 1(2):1–16, 2002.
2. J. Kaput. Representation systems and mathematics. In C. Janvier (ed.): *Problems of representation in the teaching and learning of Mathematics*, pages 19–26, Lawrence Erlbaum Associates Publishers, Hillsdale NJ, 1987.
3. A. Gagatsis, M. Shiakalli. Ability to translate from one representation of the concept of function to another and mathematical problem solving. *Educational Psychology*, 24(5):645–657, 2004.

4. A. Sierpinska. On understanding the notion of function. In E. Dubinsky, G. Harel (eds.): *The concept of function: Aspects of epistemology and pedagogy*, pages 25–28, The Mathematical Association of America, United States, 1992.
5. R. Lesh, T. Post, M. Behr. Representations and translations among representations in mathematics learning and problem solving. In C. Janvier (Ed.): *Problems of representation in the teaching and learning of Mathematics*, pages 33–40, Lawrence Erlbaum Associates Publishers, Hillsdale N.J, 1987.
6. A. Gagatsis, I. Elia, A. Mougi. The nature of multiple representations in developing mathematical relations. *Scientia Paedagogica Experimentalis*, 39(1):9–24, 2002.
7. A. Evangelidou, P. Spyrou, I. Elia, A. Gagatsis. University students' conceptions of function. In M.Jonsen Hoines, A. Berit Fuglestad (eds.): *Proceedings of the 28th Conference of the International Group for the Psychology of Mathematics Education*, pages 351–358, Bergen University College, Bergen, Norway, 2004.
8. I. Elia, A. Panaoura, A. Eracleous, A. Gagatsis. Relations between secondary pupils' conceptions about functions and problem solving in different representations. *International Journal of Science and Mathematics Education*, 5:533–556, 2007.
9. I. Elia, P. Spyrou. How students conceive function: A triarchic conceptual-semiotic model of the understanding of a complex construct. *The Montana Mathematics Enthousiast* 3(2):256–272, 2006.
10. I. Elia, A. Gagatsis, R. Gras. Can we "trace" the phenomenon of compartmentalization by using the I.S.A.? An application for the concept of function. In R. Gras, F. Spagnolo, J. David (eds.): *Proceedings of the Third International Conference I.S.A. Implicative Statistic Analysis*, pages 175–185, Universita degli Studi di Palermo, Palermo, Italy, 2005.
11. A. Gagatsis, C. Christou. The structure of translations among representations in functions. *Scientia Paedagogica Experimentalis*, 39(1):39–58, 2002.
12. B. Dufour-Janvier, N. Bednarz, M. Belanger. Pedagogical considerations concerning the problem of representation. In C. Janvier (ed.): *Problems of representation in the teaching and learning of mathematics*, pages 109–122, Lawrence Erlbaum Associates Publishers, Hillsdale NJ, 1987.
13. J. G. Greeno, R.P. Hall. Practicing representation: Learning with and about representational forms. *Phi Delta Kappan*, 78:361–367, 1997.
14. S. Ainsworth, P. Bibby, D. Wood. Evaluating principles for multi-representational learning environments. *Paper presented at the 7th European Conference for Research on Learning and Instruction*, Athens, Greece, 1997.
15. J. Kaput. Technology and mathematics education. In D. A. Grouws (ed.): *Handbook of research on mathematics teaching and learning*, pages 515–556, Macmillan, New York, 1992.
16. R. Even. Factors involved in linking representations of functions. *The Journal of Mathematical Behavior*, 17(1): 105–121, 1998.
17. R. Duval. A cognitive analysis of problems of comprehension in a learning of mathematics. *Educational Studies in Mathematics*, 61(1–2):103–131, 2006.
18. J.P. Smith, A.A. diSessa, J. Roschelle. Misconceptions reconceived: A constructivist analysis of knowledge in transition. *Journal of the Learning Sciences*, 3:115–163, 1993.
19. T. Eisenberg, T. Dreyfus. On the reluctance to visualize in mathematics. In W. Zimmermann, S. Cunningham (eds.): *Visualization in Teaching and Learning*

Mathematics, pages 9–24, Mathematical Association of America, United States, 1991.

20. A. Sfard. Operational origins of mathematical objects and the quandary of reification — The case of function. In E. Dubinsky, G. Harel, (eds.): *The concept of function: Aspects of epistemology and pedagogy*, pages 59–84, The Mathematical Association of America, United States, 1992.
21. Z. Markovits, B. Eylon, M. Bruckheimer. Functions today and yesterday. *For the Learning of Mathematics*, 6(2):18–28, 1986.
22. L. Aspinwall, K. L. Shaw, N. C. Presmeg. Uncontrollable mental imagery: Graphical connections between a function and its derivative. *Educational Studies in Mathematics*, 33:301–317, 1997.
23. R. Duval. Registres de Représentation Sémiotique et Fonctionnement Cognitif de la Pensée. *Annales de Didactique et de Sciences Cognitives*, 5:37–65, 1993.
24. B. M. Byrne. *Structural Equation Modeling with EQS and EQS/Windows: Basic concepts, applications and programming*, SAGE Publications Inc., Thousand Oaks CA, 1994.
25. P. M. Bentler. Causal modeling via structural equation systems. In J. R Nesselroade, R. B. Cattell (eds.): *Handbook of multivatriate experimental psychology*, (2nd ed.), pages 317–335, Plenum, New York, 1988.
26. R. B. Kline. *Principles and practice of structural equation modeling*, Guilford Press, New York, 1998.
27. P. M. Bentler. *EQS structural equations program manual*, Multivariate Software Inc., Encino CA, 1995.
28. P. M. Bentler. Comparative fit indexes in structural models. *Psychological Bulletin*, 107:301–345, 1990.
29. A. Bodin, R. Coutourier, R. Gras. *CHIC: Classification Hiérarchique Implicative et Cohésitive-Version sous Windows, CHIC 1.2.* Association pour la Recherche en Didactique des Mathématiques, Rennes, 2000.
30. R. Gras. Data analysis: a method for the processing of didactic questions. *Selected papers for ICME 7*, La Pensée Sauvage, Grenoble, 1992.
31. R. Gras, et al. *L'implication statistique*, Collection associée à Recherches en Didactique des Mathématiques, La Pensée Sauvage, Grenoble, 1996.
32. I.C. Lerman. *Classification et analyse ordinale des données*, Dunod, Paris, 1981.
33. R. Gras, P. Peter, H. Briand, J. Philippe. Implicative Statistical Analysis. In C. Hayashi, N. Ohsumi, N. Yajima, Y. Tanaka, H. Bock, Y. Baba (eds.): *Proceedings of the 5th Conference of the International Federation of Classification Societies*, pages 412–419, Springer-Verlag, Tokyo, Berlin, Heidelberg, New York, 1997.

Appendix

	Test A			
N=183	Graphic→Verbal		Graphic→Symbolic	
	Occurrence	Mean	Occurrence	Mean
V1: $y < 0$	139	.76	102	.56
V2: $xy > 0$	93	.51	72	.39
V3: $y > x$	67	.37	46	.25
V4: $y = -x$	75	.41	36	.20
V5: $y = 3/2$	65	.36	70	.38
V6: $y = x - 2$	28	.15	13	.07

	Test B			
N=183	Verbal→Graphic		Verbal→Symbolic	
	Occurrence	Mean	Occurrence	Mean
V1: $y < 0$	132	.72	109	.60
V2: $xy > 0$	116	.63	72	.39
V3: $y > x$	83	.45	91	.50
V4: $y = -x$	58	.32	47	.26
V5: $y = 3/2$	68	.37	80	.44
V6: $y = x - 2$	53	.29	44	.24

Table 5. Frequencies of occurrence and means of Grade 9 students to the tasks of Test A and Test B

	Test A			
N=404	Graphic→Verbal		Graphic→Symbolic	
	Occurrence	Mean	Occurrence	Mean
V1: $y < 0$	322	.80	288	.71
V2: $xy > 0$	244	.60	193	.48
V3: $y > x$	171	.42	145	.36
V4: $y = -x$	166	.41	137	.34
V5: $y = 3/2$	205	.51	177	.44
V6: $y = x - 2$	156	.39	107	.26

	Test B			
N=404	Verbal→Graphic		Verbal→Symbolic	
	Occurrence	Mean	Occurrence	Mean
V1: $y < 0$	329	.81	300	.74
V2: $xy > 0$	339	.84	218	.54
V3: $y > x$	250	.62	269	.67
V4: $y = -x$	195	.48	224	.55
V5: $y = 3/2$	257	.64	286	.71
V6: $y = x - 2$	204	.50	190	.47

Table 6. Frequencies of occurrence and means of Grade 11 students to the tasks of Test A and Test B

Implications between learning outcomes in elementary bayesian inference

Carmen Díaz[1], Inmaculada de la Fuente[2] and Carmen Batanero[3]

[1] Facultad de Psicología, Campus El Carmen, Universidad de Huelva
21071 Huelva, Spain
carmen.diaz@dpsi.uhu.es

[2] Facultad de Psicología, Campus de Cartuja, Universidad de Granada
18071 Granada, Spain
edfuente@ugr.es

[3] Facultad de Educación, Campus de Cartuja, Universidad de Granada
18071 Granada, Spain
batanero@ugr.es

Summary. In this research implicative analysis served to study some previous hypotheses about the interrelationships in students' understanding of different concepts and procedures after 12 hours of teaching elementary Bayesian inference. A questionnaire made up of 20 multiple choice items was used to assess learning of 78 psychology students. Results suggest four groups of interrelated concepts: conditional probability, logic of statistical inference, probability models and random variables.

Key words: Bayesian inference, teaching, conditional probability, undergraduate students, assessment

1 Introduction

There is a tendency nowadays to recommend that teaching of Bayesian inference should be included in undergraduate statistics courses as an adequate and desirable complement to classical inference [22, 25, 26]. Situations where prior information can help to make an accurate decision and software that facilitates the application of these methods are becoming increasingly available. Moreover, top core statistical journals now include an important proportion of Bayesian papers but this does not yet translate into comparable changes in the teaching of statistical inference to undergraduates [6].

Some excellent textbooks whose understanding does not involve advanced mathematical knowledge and where basic elements of Bayesian inference are contextualized in interesting examples (e.g. [7] or [2]) can help follow these recommendations. There are also a wide number of Internet didactic resources

C. Díaz et al.: *Implications between learning outcomes in elementary bayesian inference*, Studies in Computational Intelligence (SCI) **127**, 163–184 (2008)
www.springerlink.com

that might facilitate the teaching of these concepts (e.g. those available from Jim Albert's web page at http://bayes.bgsu.edu/). These and other authors (e.g. [8]) have incorporated Bayesian methods to their teaching and are suggesting that Bayesian inference is easier to understand than classical inference.

This is however a controversial point. On one hand, it is argued [29] that Bayesian inference relies too strongly on conditional probability, a topic hard for undergraduate students in non-mathematical majors to learn. On the other hand, in the past 50 years errors and difficulties in understanding and applying frequentist inference have widely been described (e.g. in [3, 21]). These criticisms suggest researchers do not fully understand the logic of frequentist inference and give a (incorrect) Bayesian interpretation to p-values, statistical significance and confidence intervals. It is then possible that learning Bayesian inference is not as intuitive as assumed or at least that not all the concepts involved are equally easy for students. Moreover, empirical research that analyze the learning of students in natural teaching contexts is almost non-existent.

Consequently, the first aim of this research was to explore the extent to which different concepts involved in basic Bayesian inference are accessible to undergraduate psychology students. A second goal was to compare learning outcomes with our previous hypotheses that there are different groups of related concepts (and not just conditional probability) that are potentially difficult for these students. We finally wanted to explore the implications between concepts included in each of these groups with the aim of providing some recommendations about how to best organize the teaching of the topics.

In this sense, implicative analysis was an essential tool. As suggested by [17], researchers in human sciences are interested in discovering inductive nonsymmetrical rules of the type "if a, then almost surely b". The method provides an implication index for different types of variables; and moreover serves to represent these implications in a graph or an implicative hierarchy as a complex non linear system. This specially suits our theoretical framework [16], where knowledge is seen as a complex system, more than as a linear object, and for this reason, in our research we were interested in finding the implications between understanding the different mathematical objects involved in basic Bayesian inference, that is, what the knowledge contents A that facilitate learning of other different contents B are.

2 Teaching Experiment

The sample taking part in this research included 78 students (18–20 years old) in the first year of the Psychology Major at the University of Granada, Spain. These students were taking part in the introductory statistics course and volunteered to take part in the experiment. The sample was composed of 17.9% boys and 82.1% girls, which is the normal proportion of boys and girls in the Faculty. These students scored an average of 4.83 (in a scale 0–10) in the statistics course final examination with standard deviation of 2.07.

The students were organized into four groups of about 15–20 students each and attended a short 12 hour long course given by the same lecturer with the same material. The 12 hours were organized into 4 days. Each day there were two teaching sessions with a half hour break in between. The first session (2 hours) was dedicated to presenting the materials and examples, followed by a short series of multiple choice items that each student should complete, in order to reinforce their understanding of the theoretical content of the lesson. In the second session (one hour), students in pairs worked in the computer lab with the following Excel programs that were provided by the lecturer to solve a set of inference problems:

1. Program Bayes: This program computes posterior probabilities from prior probabilities and likelihood (that should be identified by the students from the problem statement).
2. The program Prodist transforms a prior distribution $P(p = p_0)$ for a population proportion p in the posterior distribution $P(p = p_0 \mid \text{data})$, once the number of successes and failures in the sample are given. Prior and posterior distributions are drawn in a graph.
3. The program Beta computes probabilities and critical values for the Beta distribution $B(s, f)$, where s and f are the numbers of successes and failures in the sample.
4. The program Mean computes the mean and standard deviation in the posterior distribution for the mean of a normal population, when the mean and standard deviation in the sample and prior population are known.

In Table 1 we present a summary of the teaching content. Students were given a printed version of the didactic material that covered this content. Each lesson was organized in the following sections: a) *Introduction*, describing the lesson goals and introducing a real life situation; b) *Progressive development of the theoretical content*, in a constructive way and using the situation previously presented; c) *Additional examples* of other applications of the same procedures and concepts in other real situations, d) *Some solved exercises*, with description of main steps in the solving procedure; e) *New problems* that students should solve in the computer lab; and f) *Self assessment* items. All this material together with the Excel programs described above was also made available to the students on the Internet (http://www.ugr.es/ mcdiaz/bayes). We added a forum, so that students could consult the teacher or discuss themselves their difficulties, when needed.

3 A-Priori Analysis of the Questionnaire

Two weeks after the end of the teaching, the students were given a questionnaire to assess their understanding of the topic. Students prepared in advance

Lesson	Content	Session 1: classroom	Session 2: computer lab.
1	Bayes theorem in the context of clinical diagnosis	Prior and posterior probabilities; likelihood; Bayes theorem; comparing subjective and frequentist probability; revision of beliefs; sequential application of Bayes theorem	Solving Bayes problems (Program Bayes)
2	Inference for proportion. Discrete case in the context of voting	Parameters as random variables; prior and posterior distribution; informative and non informative prior distribution; credible intervals; comparing Bayesian and frequentists approaches to inference	Computing credible intervals for proportion; assigning non informative and informative prior distributions (Program Prodist)
3	Inference for proportion. Continuous case in the context of production	Generalizing to continuous case; Beta distribution; its parameters and shape; credible intervals; Bayesian tests	Assigning non informative and informative prior distributions; computing credible intervals for proportion; testing simple hypotheses (Program Beta)
4	Inference for the mean of a normal population in the context of psychological assessment	Normal distribution and its parameters; credible intervals and tests for the mean of a normal distribution with known variance; non informative and informative prior distributions	Assigning non informative and informative prior distributions; computing credible intervals for means; testing simple hypotheses (Program Mean)

Table 1. Teaching content and its organization

for the assessment that was part of the analysis course they were following. The BIL (Bayesian Inference Learning) questionnaire (which is included in Appendix) was prepared for this research and is composed of both multiple choice and some open ended items that were developed by the authors with the specific aim to cover the most important contents in the teaching. The aim was to assess learning in the following groups of concepts, which in our a-priori analysis were assumed to be the core content of basic Bayesian inference and might cause different types of difficulties to students. These concepts, as well as the philosophical principles of Bayesian inference had been introduced in the teaching at an elementary level, adequate to the type of students. We also assumed learning of one of these groups of concepts would not automatically assure the learning of the other groups, so in the implicative analysis the three groups could be unrelated.

Conditional probability and the Bayes' theorem

As was argued before, different authors pointed to students difficulties in understanding conditional probability. For example, the students' confusion between the two probabilities $P(A \mid B)$ and $P(B \mid A)$ was termed the *fallacy of the transposed conditional* [13]. [20,34] described the identification of causality and conditioning (causal conception of conditional probability) and the belief that an event could not condition another event that occurs before it (chronological conception); confusion between simple, joint and conditional probability was described by [31]. All these errors might cause difficulties in computing different types of probabilities (item 2), understanding of the differences between prior and posterior probability and likelihood (items 1 and 18), and using the Bayes' theorem as a tool to transform prior into posterior probabilities (item 7 and 18). In addition, students' difficulties with the Bayes' theorem were also described by the afore mentioned and other authors (see [5] for a survey).

Parameters as random variables, their distribution, distinction between prior and posterior distribution

In Bayesian inference, parameters are considered to be random variables with a prior distribution, while in frequentist inference they are assumed to be unknown constants (items 3, 5), a distinction which is not too clear for some students [23]. Moreover, the aim of Bayesian inference is to transform the prior into a posterior distribution via the Bayes' theorem (item 18). A prior distribution provides all the information for the parameter before collecting the data (item 4), non informative priors are given by uniform distributions and are used when no previous information is available for the parameter (item 6).

There are different models to represent prior distributions. The Beta distribution was introduced in the teaching, and students had to learn the meaning of its parameters (item 8, 20) and how to select a specific Beta distribution in a particular inference problem (item 9). Students knew the normal distribution from previous lessons. However, they had to learn the rule to compute the posterior distribution for a mean when the prior distribution is normal (item 13, 14, 15, 16). In managing all these distributions, Bayesian statistics uses the rules of probability to make inferences, and that requires dealing with formulae, but actual calculus used is minimal as students only have to understand that probability is given by different types of areas under a density function [8]. However, the extent to which all of this is grasped by psychology students has still to be assessed.

Logic of Bayesian inference

The aim of Bayesian inference is updating the prior distribution via the likelihood to get the posterior distribution, which provides all the information for the parameter, once the data have been collected [9]. However, it is also possible to carry out procedures similar to those used in frequentist statistics, although the interpretation and logic is a little different [7, 16].

Credible intervals provide the epistemic probability that the parameter is included in a specific interval of values, for the particular sample, while confidence intervals provide the frequentist probability that in a percentage of samples from the same population the parameter will be included in intervals of values computed in those samples. Credible intervals are computed from the posterior distribution (item 17) and students should be able to compute them by using the tables of different distributions (items 10, 16); they should understand that the interval width increases with the credibility coefficient and decreases with the sample size (item 12).

In Bayesian inference we can compare at the same time different hypotheses; in this case we compute the probabilities for those hypotheses given the data by using the posterior distribution and select the hypothesis with higher probability (item 11). In testing only one hypothesis we either compute the probability for the hypothesis or for the contrary event (item 14); acceptance or rejection will depend on the value of that probability.

So, there are some conceptual and interpretative differences between classical and frequentist approaches, but, since both approaches often lead to approximately the same numerical results, students might not understand these differences and confuse both approaches [23].

4 Implications in Learning Outcomes

In order to assess understanding of the three groups of concepts above and see if our previous hypothesis that learning of the three groups of concepts might be unrelated, we gave the BIL questionnaire to the students who participated in the teaching experiment and analyzed their responses[4]. We also were interested in the extent to which each task was easy for the students. In table 2 we present the number and percentage of correct responses to items and sub items. In item 18 we considered three different scores: correct identification of prior probabilities (18.1), correct identification of likelihood (18.2) and correct computation of posterior probabilities (18.3). In item 2 each response was scored independently.

The average number of correct responses per student was $\overline{x} = 16.6$. Given that the maximum possible score was 26; these results show a reasonable result

[4] This assessment was complemented with analysis of self- assessment tests and open tasks solved by the students along the teaching sessions. We are not including here the analysis of this complementary data, due to space limitation.

		Credible interval (95%)		
Item	Percent correct	Lim sup	Lim inf	Content assessed
1	88.7	78.4	94.3	Likelihood, conditional probability
2.1	79.0	67.3	87.2	Simple probability
2.1	38.7	27.6	51.1	Conditional probability
2.2	29.0	19.2	41.2	Conditional probability of contrary event
2.4	51.6	39.4	63.5	Joint probability
3	66.1	53.7	76.6	Comparing parameters in frequentists and Bayesian inference
4	58.1	45.6	69.5	Prior distribution
5	61.3	48.8	72.3	Parameter as random variable; statistics
6	50.0	36.6	60.4	Correct assignment of a non informative prior distribution for proportion
7	93.5	84.5	97.3	Bayes' theorem as a tool to transform prior into posterior probabilities
8	53.2	40.9	65.0	Parameters in Beta distribution, defining prior informative distribution for proportion
9	85.5	74.6	92.1	Parameters in Beta distribution
10	64.5	52.0	75.2	Computing credible intervals for proportion; from Beta tables
11	58.1	45.6	69.5	Testing simple hypotheses for proportion; from Beta tables
12	53.2	40.9	65.0	Properties of credible intervals
13	69.4	57.0	79.3	Posterior distribution of mean; non informative prior; known variance
14	30.6	20.6	42.9	Testing simple hypotheses for means
15	40.3	29.0	52.7	Posterior distribution of mean; non informative prior; unknown variance
16	69.4	57.0	79.3	Credible intervals for means
17	69.4	57.0	79.3	Posterior distribution for mean, informative prior
18.1	85.8	75.4	91.3	Identifying prior probabilities from a problem statement
18.2	74.4	63.9	83.0	Identifying likelihood from a problem statement
18.3	79.0	67.3	87.2	Bayes' theorem as a tool to transform prior into posterior probabilities
19	58.1	45.6	69.5	Meaning of likelihood
20.1	82.3	70.9	89.7	Parameters in Beta curve. Spread
20.2	72.6	58.2	80.0	Parameters in Beta curve. Centre

Table 2. Percent of correct responses and contents assessed in the BLI items

of the teaching experience. A reliability analysis of responses gave a value for the internal consistency coefficient $\alpha = 0.68$, which is reasonable, given the variety of contents (the test is multidimensional). The Pearson correlation coefficients between BIL and final score in the statistics course $\rho = 0.41$: was significant at .01 level, although moderate.

The easiest tasks were those related to distinguishing prior probabilities, posterior probabilities and likelihood and identifying them from a problem statement (items 7 and 1). Correct assignment of an informative distribution for proportions (item 9), interpreting parameters in the Beta curve (items 9, 20.1 and 20.2), computing credible intervals for the mean of a normal distribution with known variance (item 16), distinguishing statistics and parameters in a problem statement (item 17), getting a posterior distribution for the mean in a normal population from uniform prior distribution (item 13), understanding parameters as random variables (item 3 and 5), computing credible intervals for proportions (item 10) were all relatively easy tasks with over 60% correct responses on average.

There were only 4 difficult tasks (mean percentage of correct responses under 50%). These tasks were item 14 (testing hypotheses about the mean), where students either made a mistake in the reasoning by contradiction (choosing distractor c) or did not understand the standardization operation and choose distractor *a*. Of course this is a highly complex item, where the logic of testing hypothesis is mixed with knowledge of probability calculus and standard Normal distribution. Moreover understanding proof the logic of statistical tests and reasoning by contradiction was also difficult in other research related to frequentist statistics (e.g. [35]).

Students also found items 2*b* and 2*c* difficult, where they confused a conditional probability and its inverse, a problem that has been repeatedly reported (e.g. in [13,32]). In comparing these results with those in item 1 and 18 where we found a high percentage of correct responses, we remark that distractors in item 2 are given only by formulas (instead of using a verbal expression).

We conclude that the expressions prior and posterior probabilities and likelihood helped students to better distinguish a conditional probability and its inverse in these items and students possibly did not remember the symbolic expression for a conditional probability. Finding a posterior distribution for the mean (item 15) was difficult because students forgot to divide by the square root of the sample size to find the standard deviation in the posterior distribution. All the other tasks had a medium difficulty (between 50–60% correct responses).

To study the interrelations and implications between learning objectives we carried out several multivariate analyses, using the CHIC software, Classification Hierarchical, Implicative and Cohesive [11]. The implication index between two dichotomous variables *a* and *b* in a population is defined by (1).

$$q(a,\bar{b}) = \left[\frac{\text{card}\left(A \cap \overline{B}\right) - \frac{\text{card}(A)\cdot\text{card}(\overline{B})}{n}}{\sqrt{\frac{\text{card}(A)\cdot\text{card}(\overline{B})}{n}}} \right] \tag{1}$$

Here A and B are the population subgroups where a and b take the value 1 [19,33]. This index follows the normal distribution N(0,1), and from there an intensity for the implication $a \Rightarrow b$ is defined by (2).

$$\varphi(a,\bar{b}) = \Pr\left[\text{card}\left(X \cap \overline{Y}\right) \le \text{card}\left(A \cap \overline{B}\right)\right] \tag{2}$$

In (2) X and $\overline{Y}$ are dichotomous independent random variables having the same cardinal as A and $\overline{B}$ respectively [27,28]. In our study we have a total of $C_{26.2}$ implication indexes among the 26 sub items in the BIL questionnaire. The software CHIC computes these indexes and provides a graph with all the implications which are significant to a given significance level.

The implication $a \Rightarrow b$ in our study is interpreted in the sense that when a student correctly solves item a there is higher probability for him /her to solve item b. In this sense the implicative graph provides a possible order to introduce the different concepts and procedures whose understanding is assessed in the items in the teaching of the topic. Before carrying out the implicative analysis we checked the assumptions of the method; experimental units of variables and independence of responses by different students. We assumed a binomial model for the responses; that is, we assumed each student to have the same likelihood to correctly solve the items [27], as in fact these are the hypotheses assumed in classical theory of tests [30] that was the model used in building the questionnaire.

In Figure 1 we present the implicative graph with all the relationships that were significant at 99% (dashed line) or 95% level (continuous line). We observe that the implication relationship is asymmetrical and the direction of implication is showed by the arrows in the graph.

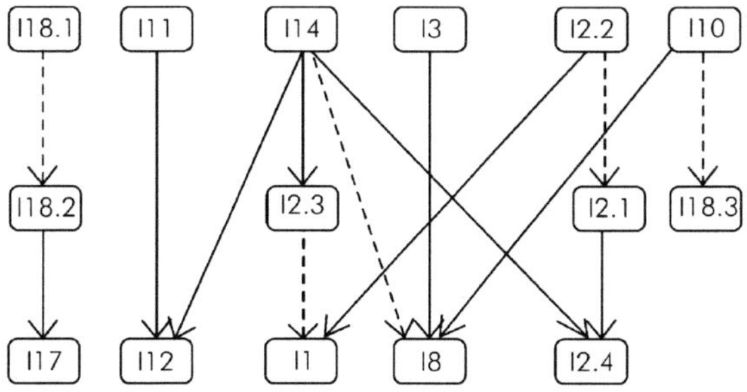

Fig. 1. Implicative graph with significant implications at 99% and 95%

If we study the relationships higher than 99% in the graph (discontinuous line in the diagram), we observe that students who correctly answer item 18.2

(correct identification of likelihood, which is given by a conditional probability) have better likelihood to answer item 18.1 (correct identification of prior probabilities, which are given by simple probabilities). Correct performance in item 10 (computing credible intervals for the proportion, identifying probabilities and critical values from the Beta distribution table and computing credible intervals for a proportion) increases the possibility of a correct computation of posterior probabilities with Bayes theorem (item 18.3). Both tasks involve understanding probability axioms and computing probability, although the first one is more complex. Finally correct computation of conditional probabilities implies correct computation of joint and single probabilities (items 2.1, 2.2, and 2.4).

As regards implications higher than 95% (continuous line in the diagram) we observe that students who correctly perform a Bayesian hypothesis test (items 14 or 11) increase their likelihood to correctly interpret credible intervals (item 12). In fact all the ideas and computations involved in solving the second task are involved in the first one, which adds the need to understand the logic of comparing probabilities for different hypotheses and in the case of item 14 proof by contradiction. Item 14 implies item 2.3, the computation of conditional probability for a contrary event, but, again mastering the idea of proof by contradiction involves correct reasoning on both conditional reasoning and complementation. Students who visualize parameters as random variables (item 3), or compute probabilities for Beta function and credible intervals for proportions (item 10), perform better in correctly assigning a Beta informative prior distribution (item 8), a task that is also facilitated by item 14.

Item 2.3 (computing the conditional probability for the contrary event) or item 2.2 (computing conditional probability) facilitate item 1, distinguishing prior and posterior probabilities and likelihood (all these ideas are supported on correct conditional reasoning). Item 2.2 facilitates computing simple probability (item 1) and both of them together facilitate the computation of joint probabilities (item 2.4), another task which is easier for those who succeeded in Item 14 (testing hypotheses), possibly because the idea of conditional probability is required to test a hypothesis.

5 Implicative hierarchy of learning outcomes

Once the isolate implications between items were studied we carried out an implicative classification analysis to clarify the structure of implication analysed that point to the three groups of concepts in our a-priori analysis, but that in some points is somewhat mixed. This is an algorithm, which uses the implicative indexes in a set of variables to study the internal cohesion of some variables subsets [12, 24]. The cohesion between two variables a y b is defined by $c(a,b) = \sqrt{1 + H^2}$ where H is the entropy for the two variables, and varies between 0 and 1. The cohesion for a class of variables [18] is defined by in (3).

$$C(A) = \left[\prod_{j \in \{2,\ldots,r\}; j>i}^{i \in \{1,\ldots,r-1\}} c(a_i, a_j) \right]^{\frac{2}{r(r-1)}} \tag{3}$$

Finally, given two sets of variables $\underline{A}$ and $\underline{B}$ the strength of implication from $\underline{A}$ to $\underline{B}$ is defined [10] by (4).

$$\Psi(\underline{A}, \underline{B}) = \left[\sup_{i \in \{1,\ldots,r\}; j \in \{1,\ldots,s\}} \varphi\left(a_i, \overline{b_j}\right) \right]^{rs} \cdot \left[C\left(\underline{A}\right) \cdot C\left(\underline{B}\right)\right]^{1/2} \tag{4}$$

The software CHIC builds an implicative hierarchy in the set of variables, taking into account the maximal cohesion into each class and the higher implication from one class to another. In Figure 2 we present the hierarchy produced.

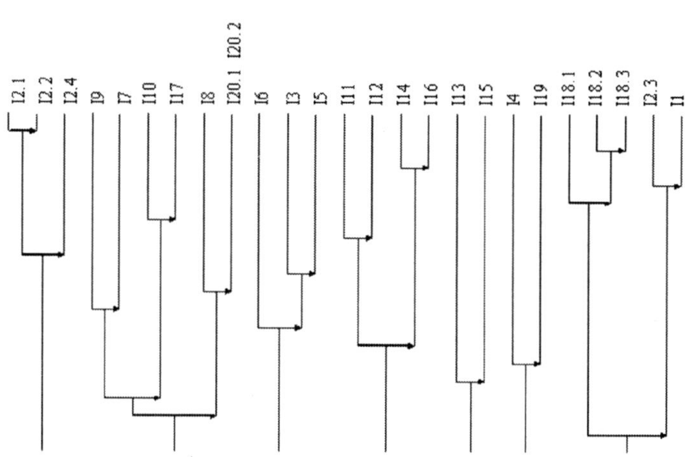

Fig. 2. Implicative hierarchy with 95% node

There are four significant clusters:

1. *Group 1: Conditional probability.* Items 2.2 and 2.1 which are linked to item 2.4, all of them related to probability. The student who correctly computes conditional probabilities (item 2.2), correctly performs simple (item 2.1) and compound probability (item 2.4). The higher difficulty of conditional probability as regards simple and compound is then confirmed as well as our previous hypothesis that conditional probability is one core concept in the learning of Bayesian statistics.
2. *Group 2: Prior and posterior distributions and Beta curves.* Items 9, 7, 10, 17, 8 and the two parts of item 20. Students who are able to interpret

the parameters in the Beta curve (item 9) and understand how posterior distributions are achieved from prior distributions and likelihood through Bayes theorem (item 7) succeeded better in getting a credible interval for proportions in the continuous case, a task that requires interpreting probabilities of Beta curves, and understanding the concept of posterior probability, as well as the concept of credible interval. They also performed better in discriminating prior and posterior distribution of the mean (item 17). All of this leads to better choosing a non informative prior distribution for proportion in the continuous case through the Beta Curve (item 8), and graphically interpreting the parameters in Beta curves (item 20); both tasks are related to understanding the meaning of these parameters. These are a subgroup of the tasks we included in the second group of concepts (parameters, their distribution, prior and posterior distribution) in the a-priori analysis; specifically most of these tasks are related to Beta curves that was a concept new to the students.

3. *Group 3* (items 11, 12, 14 and 16) is a set of the concepts we included in the third group in the a-priori analysis: *Logic of Bayesian inference.* Being able to correctly test a hypothesis for proportions (item 11) increases the likelihood of correctly interpreting credible intervals (item 12); and these two tasks are linked to another group: correctly testing a hypothesis about the mean (item 14), which, in turn increases the likelihood of correctly computing a credible interval for the mean (item 16). All this knowledge is specifically related to the logic of Bayesian methods; understanding the test of hypotheses facilitates that of credible intervals; inference for proportion was easier than inference for mean, possibly because students have to distinguish in the last task the formulas for known or unknown variance.
4. Group 4: Finally there is a second group of tasks related to *conditional probability* (the different parts of Item 18, 2.3 and 1). Correct identification of prior probability (item 18.1) facilitates the correct identification of likelihood from a problem statement (item 18.2) and this leads to correct computation of posterior probabilities (item 18.3). These three abilities lead to better identification of conditional probabilities for the contrary event (item 2.3) and discrimination between prior probability, likelihood and posterior probabilities in the context of a problem (item1). The separation between groups 1 and 4 is explained by the different difficulty of the tasks in the two groups. Tasks in group 4 were easier than those in group 1 where probabilities are only given by formulas.

Other groupings of items that were non significant at the 95% level were as follows:

1. *Group 5*: Items 6 (assigning adequate prior distribution for the non informative case to proportions in the discrete case), 3 (understanding parameters as random variables) and 5 (discrimination between parameters

and statistics); all these tasks are related to understanding parameters from a Bayesian point of view, that is difficult for some students [23].
2. *Group 6*: Items 13 (Posterior distribution of mean when variance is known) and 15 (posterior distribution of mean when variance is unknown) related to specific factual knowledge; so possibly there was no difference between students with good or poor understanding of other concepts.
3. *Group 7*: Item 4 (concept of prior distribution) and item 9 (concept of likelihood); both were easy items, and therefore were unrelated to knowledge of other concepts.

In summary these implications support our a-priori analysis and point to three groups of concepts relevant for students' introduction to the elementary ideas of Bayesian inference and that should be taken into account in planning the teaching, although some of the groups still split in some subgroups:

1. *Conditional probabilistic reasoning* (as shown in groups 1 and 4), a theme where many biases have been described in the literature, but which is basic in defining posterior probabilities and likelihood, as well as in understanding the logic of credible intervals and hypothesis testing. Results also suggested that formulas for different types of probability were harder than verbal expressions for students to understand. Perhaps we should take into account Feller's suggestion ([15] p. 114) that "*conditional probability is a basic tool of probability theory, and it is unfortunate that its great simplicity is somewhat obscured by a singularly clumsy terminology*".
2. *Probability distributions, its parameters* (visualized as random variables), the distinction of prior and posterior distribution of parameters and assignment of prior distributions for informative and non informative cases (Groups 2, 5, 6 and 7). In our teaching we limited to Beta and Normal distributions, since the time available for teaching was restricted, but still so, the understanding of Beta curves appeared as a separated subgroup, as well as remembering the rules for known and unknown variance in inference about normal distributions. The difficulties to understand the different conception of parameters in Bayesian and frequentists statistics [23] also appeared as a separated subgroup.
3. *Logic of Bayesian inference* (Group 3), that is, understanding the logic for computing and interpreting credible intervals and testing simple hypothesis. Performance in these tasks is in fact supported in understanding the previous two groups of concepts, most of which are not specific to Bayesian reasoning. However, limitation of teaching time leads some lecturers to reduce the teaching of the same and to try to pass directly from data analysis to inference. Teaching of Bayesian inference, therefore should only be started when previous groups of concepts are well understood by students.

6 Discussion

In this paper we reported results from assessing students' understanding of elementary Bayesian ideas after a short teaching experiment. The high percentage of correct responses in the questionnaire (even in highly complex tasks, such as computing credible intervals and carrying out hypothesis tests) supports the claims for complementing the teaching of frequentists statistics with some ideas of Bayesian statistics in undergraduate statistics courses (e.g. [1, 25]). Both approaches to inference should be, however based in the teaching of core ideas of probability and conditional probability.

A comparative analysis of the undergraduate teaching of statistics shows a clear imbalance between what it is taught and what it is later needed; in particular, most statistics introductory courses are exclusively frequentist and many students never get a chance to learn some Bayesian concepts which would improve their professional skills ([6]). Our research shows that undergraduate students are able to acquire an intuitive understanding for a number of concepts related in elementary Bayesian inference in a short period of teaching.

The implicative and cohesive classification analyses supported our a-priori analysis of the concepts related to understanding basic Bayesian inference and suggested that possible difficulties are not just related to the understanding of conditional probability. Even when the difficulties in distinguishing a conditional probability and its inverse that have been repeatedly pointed out in the literature [4,13,14] also arose in our students, its influence in their general performance was not so high and moreover the difficulty decreased when tasks included verbal expressions of these probabilities instead of formulas.

However, the study also provides arguments to reinforce the study of conditional probability in the teaching of data analysis to psychologists, not only because of the usefulness of this topic in clinical diagnosis, but as a base for future study of Bayesian inference. This and other concepts that students should have previously mastered (difference between statistics and parameters, use of distribution tables, or operating with standard scores and inequalities) also affected success in some of the tasks and more attention should be paid in introductory statistics courses. At the same time, the classes obtained in the implicative hierarchy provide us with information about the concepts whose understanding is related and their relative difficulty. This is a potential help to prepare didactic materials and to organize the teaching of the topic.

We are conscious this research should continue with new samples of students. However we think we have provided arguments to introduce basic Bayesian statistics in undergraduate courses, whenever we emphasize the elements of statistical thinking; incorporate more data and concepts, and fewer recipes and derivations in the classroom, provide students with automate computations and graphics and foster active learning [8].

Acknowledgement: This research was supported by the project SEJ2004–00789 and grant AP2003–5130, MEC, Madrid, and FQM–126, Junta de Andalucía, Spain.

References

1. J. Albert. Teaching introductory statistics from a bayesian perspective. In B. Philips, editor, *Proceedings of the Sixth International Conference on Teaching Statistics*, CD-ROM, 2002.
2. J.H. Albert and A. Rossman. *Workshop Statistics. Discovery with Data. A Bayesian Approach.* Key College Publishing, 2001.
3. Lecoutre B., Lecoutre M.P., and Poitevineau J. Uses, abuses and misuses of significance tests in the scientific community: Won't the bayesian choice be unavoidable? *ISR*, pages 399–418, 2001.
4. M. Bar-Hillel. *Decision Making Under Uncertainty*, chapter The Base Rate Fallacy Controversy, pages 39–61. North Holland, Amsterdam, 1987.
5. C. Batanero and E. Sánchez. *Exploring Probability in School: Challenges for Teaching and Learning*, chapter What is the Nature of High School Student's Conceptions and Misconceptions about Probability?, pages 260–289. Springer, New York, 2005.
6. J.M. Bernardo. A bayesian mathematical statistics primer. In A. Rossman and B. Chance, editors, *Proceedings of the Seventh International Conference on Teaching Statistics. International Association for Statistical Education*, CD-ROM, 2006.
7. D.A. Berry. *Basic Statistics: A Bayesian Perspective.* Belmont, 1995.
8. W.M. Boldstad. Teaching bayesian statistics to undergraduates: Who, what, where, when, why, and how. In B. Phillips, editor, *Proceedings of the Sixth International Conference on Teaching of Statistics*, CD-ROM, 2002.
9. W. Bolstad. *Introduction fo Bayesian Statistics.* Wiley, 2004.
10. R. Couturier. Subjects categories contribution in the implicative and the similarity analysis. *LMSET*, pages 369–376, 2001.
11. R. Couturier and R. Gras. Chic: Traitement de données avec l'analyse implicative. In S. Pinson and N. Vincent, editors, *Journées Extraction et Gestion des Connaissances (EGC'2005)*, pages 679–684 (Vol. 2), 2005.
12. R. Couturier, R. Gras, and F. Guillet. *Classification, Clustering, and Data Mining Applications*, chapter Reducing the Number of Variables Using Implicative Analysis, pages 277–285. Springer-Verlag, Berlin, 2004.
13. R. Falk. Conditional probabilities: insights and difficulties. In R. Davidson and Swift J., editors, *Proceedings of the Second International Conference on Teaching Statistics*, pages 292–297, 1986.
14. R. Falk. *Studies in mathematics education*, chapter Inference Under Uncertainty via Conditional Probability, pages 175–184 (Vol. 7). UNESCO, Paris, 1989.
15. W. Feller. *An Introduction to Probability Theory and its Applications, Vol. 1.* Wiley, 1968.
16. J. D. Godino. Un enfoque ontológico y semiótico de la cognición matemática. *RDM*, pages 237–284, 2002.

17. R. Gras. Panorama du développement de l'a.s.i. à travers des situations fondatrices. In R. Gras, F. Spagnolo, and J. David, editors, *Troisième Rencontre Internationale A.S.I. Analyse Statistique Implicative. Quaderni di Ricerca In Didattica of G.R.I.M.*, pages 6–24 (Supplemento 2(15)), 2005.
18. R. Gras, Kuntz P., and H. Briand. Les fondements de l'analyse statistique implicative et quelques prolongements pour la fouille de données. *MSH*, pages 9–29, 2001.
19. R. Gras and H. Ratsima-Rajohn. L'implication statistique, une nouvelle méthode d'analyse de données. *RO*, pages 217–232, 1996.
20. R. Gras and A. Totohasina. Chronologie et causalité, conceptions sources d'obstacles epistémologiques à la notion de probabilité conditionnelle. *RDM*, pages 49–95, 1995.
21. L.L. Harlow, S.A. Mulaik, and J.H. Steiger. *What if there were no significance tests?* Erlbaum, 1997.
22. P. Iglesias, J. Leiter, M. Mendoza, V. Salinas, and H. Varela. Mesa redonda sobre enseñanza de la estadística bayesiana. *RSCE*, pages 105–120, 2000.
23. G.R. Iversen. Student perceptions of bayesian statistics. In J. Pereira-Mendoza, editor, *Proceedings of the Fifth International Conference on Teaching Statistics*, pages 234–240, 1998.
24. D. Lahanier-Reuter. Un algorithme de regroupements de modalités de variables en analyse implicative des données. *MSH*, pages 5–8, 2001.
25. B. Lecoutre. Beyond the significance test controversy: Prime time for bayes? In A. Rossman and B. Chance, editors, *Bulletin of the International Statistical Institute: Proceedings of the Fifty-second Session of the International Statistical Institute*, pages 205–208 (Tome 58, Book 2), 1999.
26. B. Lecoutre. Training students and researchers in bayesian methods for experimental data analysis. *JDS*, pages 217–232, 2006.
27. I.C. Lerman. *Classification et analyse ordinale des données.* Dunod, 1981.
28. I.C. Lerman, R. Gras, and H. Rostam. Elaboration d'un indice d'implication pour données binaires i. *MSH*, pages 5–35, 1981.
29. D.S. Moore. *Advances in Statistical Decision Theory*, chapter Bayes for Beginners? Some Pedagogical Questions, pages 3–17. Birkhäuser, Stuttgart, 1997.
30. J. Muñiz. *Teoría Clásica de los Tests.* Pirámide, 1994.
31. A.M. Ojeda. Dificultades del alumnado respecto a la probabilidad condicional. *UNO*, pages 37–55, 1995.
32. A. Pollatsek, A.D. Well, C. Konold, and P. Hardiman. Understanding conditional probabilities. *OBHDP*, pages 255–269, 1987.
33. Gras R. *L'Implication Statistique Nouvelle Methode Exploratoire de Donneés.* La Penseé Sauvage, Grenoble, 1996.
34. A. Totohasina. *Methode Implicative en Analyse de Données et Application à l'Analyse de Conceptions d'Étudiants sur la Notion de Probabilité Conditionnelle. Ph.D. Thesis.* Universidad de Rennes I, 1992.
35. A. Vallecillos. Some empirical evidence on learning difficulties about testing hypotheses. In A. Rossman and B. Chance, editors, *Bulletin of the International Statistical Institute: Proceedings of the Fifty-Second Session of the International Statistical Institute*, pages 201–204 (Tome 58, Book 2), 1999.

Appendix: Questionnaire[5]

Item 1. 10 out of every 100 students in a Faculty study mathematics; 30 out of every 100 students doing mathematics share an apartment with other students. Let S be the event "sharing the apartment" and M the event the student is doing mathematics course. If we pick a student at random and the student is doing mathematics, the probability that he shares the apartment is:

1. A prior probability $P(S)$
2. **A posterior probability $P(S|M)$**
3. A likelihood $P(M \mid S)$
4. A joint probability $P(M \cap S)$

Item 2. Imagine you pick 1000 people at random. You know that 10 out of every 1000 people get depression. A depression test is positive for 99 out of every 100 depressed people as well as for 2 out of every 100 non depressed people. Given that D means depression and + means a positive test, compute the following probabilities:

1. $P(D) =$
2. $P(+ \mid D) =$
3. $P(- \mid D) =$
4. $P(D \cap +) =$

Item 3. The mean value for a variable (for example height) in a population:

1. Is a constant in Bayesian inference
2. Is a random variable in classical inference
3. **Is a random variable in Bayesian inference**
4. Could be constant or variable, depending on the population

Item 4. The prior probability distribution for a parameter:

1. **Provides all the information about the population before collecting the data**
2. Is computed from the posterior distribution by using the Bayes theorem.
3. It can be used to compute the credible interval for the parameter
4. Is an uniform distribution

Item 5. 1000 young Spanish people were interviewed in a survey. On average they spent 3 hours a week in practicing some sports. In Bayesian inference:

1. 3 hours is a parameter in the population of young Spanish people
2. **The average in this population is a random variable; the most likely value is about 3 hours**
3. The average in this population is an unknown constant
4. Each young Spanish person spends 3 hours a week in doing some sport

[5] Correct responses are emphasized in bold.

Item 6. In a factory lamps are sold in boxes of four lamps. We have no information about the proportion of defective lamps. Which of the distributions A, B. C or D better describes the prior distribution for the proportion of defective lamps in a box?

(A)		**(B)**	
Values of Proportion	Probability	**Values of Proportion**	**Probability**
0.00	0.1	**0.00**	**0.2**
0.25	0.1	**0.25**	**0.2**
0.50	0.1	**0.50**	**0.2**
0.75	0.1	**0.75**	**0.2**
1	0.1	**1**	**0.2**

(C)		(D)	
Values of Proportion	Probability	Values of Proportion	Probability
0.00	0.00	0.00	1/4
0.01	0.25	0.25	1/4
0.02	0.50	0.50	1/4
0.03	0.75	0.75	1/4
0.04	1	1	1/4

Item 7. In trying to estimate a proportion a student filled three columns in the Bayes table. He got these data:

Values of proportion	Prior Probability	Likelihood	——	——
0.0000	0.0000	0.0000		
0.1000	0.1000	0.0000		
0.2000	0.1000	0.0233		
0.3000	0.1000	0.1239		
0.4000	0.1000	0.0682		
0.5000	0.1000	0.0065		
0.6000	0.1000	0.0001		
0.7000	0.1000	0.0000		
0.8000	0.1000	0.0000		
0.9000	0.1000	0.0000		
1.0000	0.1000	0.0000		
Sum			0.0222	

The posterior probability that the true value of proportion in the population is 0.4 would be:

1. 0.00682
2. 0.1000
3. **0.3072**
4. 0.00015

Item 8. A clinical survey showed a 15

1. $B(15, 100)$
2. $\boldsymbol{B(15, 85)}$
3. $B(85, 15)$
4. $B(100, 15)$

Item 9. The mean for a *Beta* $B(a, b)$ distribution is:

1. a/b
2. $(a+1)/(a+b)$
3. $(a+1)/(b+1)$
4. $\boldsymbol{a/(a+b)}$

Item 10. In the following table probabilities and critical values for the $B(30, 40)$ distribution are given

Probabilities			Critical values	
p_0	$P(0 < p < p_0)$	$P(p_0 < p < 1)$	$P(0 < p < p_0)$	p_0
0	0.000	1.000	0.000	0.000
0.05	0.000	1.000	0.005	0.296
0.1	0.000	1.000	0.010	0.304
0.15	0.000	1.000	0.015	0.311
0.2	0.000	1.000	0.020	0.316
0.25	0.001	0.999	0.025	0.320
0.3	0.012	0.988	0.030	0.324
0.35	0.090	0.910	0.035	0.327
0.4	0.318	0.682	0.040	0.330
0.45	0.645	0.355	0.045	0.330
0.5	0.886	0.114	0.050	0.333
0.55	0.979	0.021	0.950	0.526
0.6	0.998	0.002	0.955	0.529
0.65	1.000	0.000	0.960	0.533
0.7	1.000	0.000	0.965	0.536
0.75	1.000	0.000	0.970	0.541
0.8	1.000	0.000	0.975	0.545
0.85	1.000	0.000	0.980	0.551
0.9	1.000	0.000	0.985	0.558
0.95	1.000	0.000	0.990	0.567
1	1.000	0.000	1.000	1.000

The 98% credible interval for the proportion in a population described by a posterior distribution $B(30, 40)$ is about:

1. $(0.316 < p < 0.551)$
2. $\boldsymbol{(0.304 < p < 0.567)}$
3. $(0.3 < p < 0.6)$
4. $(0.1 < p < 0.9)$

Item 11. The posterior distribution for the proportion of voters favorable to a political party is given by the $B(30,40)$ distribution. From the above data table, the most reasonable decision is accepting the following hypothesis for the population proportion

1. $H : p < 0.25$
2. $H : p > 0.55$
3. $\boldsymbol{H : p > 0.25}$
4. $H : p > 0.45$

Item 12. For the same posterior distribution of the parameter in a population the $r\%$ credible interval for the parameter is:

1. **Wider if r increases**
2. Wider if the sample size increases
3. Narrower if r increases
4. It depends on the prior distribution

Item 13. In a normal population with standard deviation $\sigma = 5$ and with no prior information about the population mean, we pick a random sample of 25 elements and get a sample mean $\bar{x} = 100$. The posterior distribution of the population mean is:

1. A normal distribution $N(100, 0.5)$
2. A normal distribution $N(0, 1)$
3. A normal distribution $N(100, 5)$
4. **A normal distribution $N(100, 1)$**

Item 14. To test the hypothesis that the mean μ in a normal population with standard deviation $\sigma = 1$ is larger than 5, we take a random sample of 100 elements. To follow the Bayesian method:

1. We compute the sample mean $\bar{x}$ and then compute $P\left(\frac{\bar{x}-5}{0.1} < 5\right)$; when this probability is very small, we accept the hypothesis.
2. **We compute $\bar{x}$ and then compute $P\left(\frac{\bar{x}-5}{0.1} < Z\right)$; when Z is the normal distribution $N(0,1)$; when this probability is very small, we accept the hypothesis.**
3. We compute the sample mean $\bar{x}$ and then compute $P\left(\frac{\bar{x}-5}{0.1} > Z\right)$ when Z is the normal distribution $N(0,1)$; when this probability is very small, we accept the hypothesis.
4. We compute the sample mean $\bar{x}$ and then compute $P\left(\frac{\bar{x}-5}{0.1} > 5\right)$ when this probability is very small, we accept the hypothesis.

Item 15. In a sample of 100 elements from a normal population we got a mean equal to 50. If we assume a prior uniform distribution for the population mean, the posterior distribution for the population mean is:

1. About $N(50, s)$, where s is the sample standard estimation.
2. **About $N(50, s/10)$, where s is the sample standard estimation.**
3. We do not know, since we do not know the standard deviation in the population

4. About $N(0, 1)$

Item 16. The posterior distribution for a population mean is $N(100, 15)$. We also know that $P(-1.96 < Z < 1.96) = 0.95$, where Z is the normal distribution $N(0, 1)$. The 95% credible interval for the population mean is:

1. $(100 - 1.96 \cdot 1.5, 100 + 1.96 \cdot 1.5)$
2. $(100 - 1.96, 100 + 1.96)$
3. $(100 \cdot 1.5 - 1.96, 100 \cdot 1.5 + 1.96)$
4. $\mathbf{(100 - 1.96 \cdot 15, 100 + 1.96 \cdot 15)}$

Item 17. In a survey to 100 Spanish girls the following data were obtained:

	Mean	Standard dev.
Sample	160	10
Prior distribution	156	13
Posterior distribution	158.5	7.9

To get the credible interval for the population mean we use:

1. The normal distribution $N(160, 10)$
2. The normal distribution $N(156, 13)$
3. **The normal distribution $\mathbf{N(158.5, 7.9)}$**
4. The normal distribution $N(160, 0.5)$

Item 18. 20% of boys and 10% of girls in a kindergarten are immigrant. There are about 60% boys and 40% girls in the center. Use the following table to compute the probability that an immigrant child taken at random is a boy.

Events	Prior probabilities	Likelihoods	Product	Posterior probabilities
Sum	1			1

Item 19. In a geriatric center we want to estimate the proportion of residents with cognitive impairment. 2 out of 10 residents taken at random in the residence showed cognitive impairment. The likelihood for the parameter $p = 0.1$ is 0.1937. What is the meaning of this value?

1. $P(\text{data})$, that is, probability of getting this sample.
2. $P(\text{data} \cap p = 0.1)$, that is, probability of getting the sample and that, in addition, the population proportion is 0, 1.
3. $P(p = 0.1 \mid \text{data})$, that is, probability of a population proportion is 0.1 given the sample
4. **$\mathbf{P(data \mid p = 0.1)}$, that is, given than $\mathbf{p = 0.1}$, probability of getting this sample**

Item 20. Observe the following Beta curves:

1. Which of them has a greater spread?

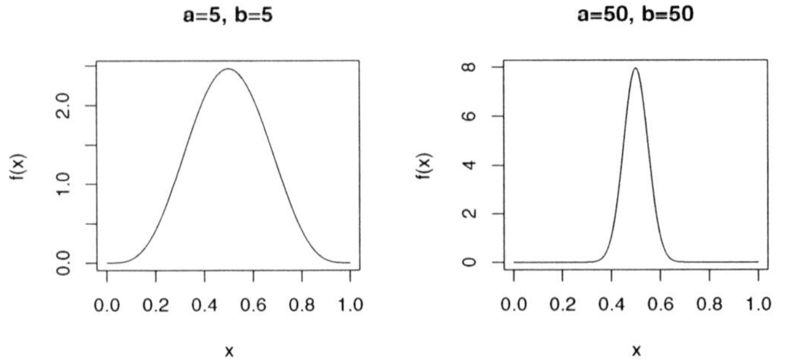

2. Which of them predict a greater value of proportion in the population?

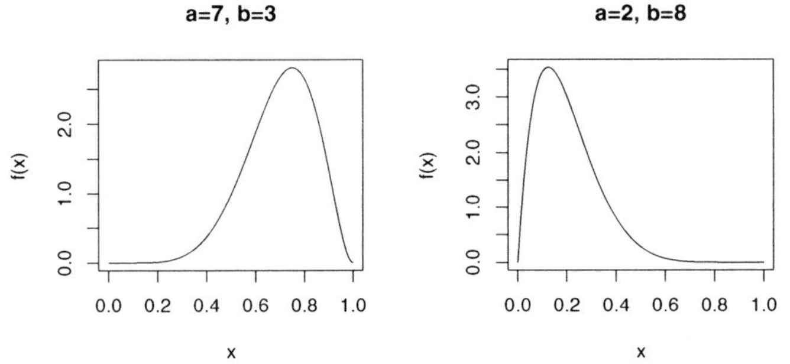

Personal Geometrical Working Space: a Didactic and Statistical Approach

Alain Kuzniak

Equipe Didirem Université Paris 7 France
alain.kuzniak@orleans-tours.iufm.fr

Summary. In this paper, we study answers that pre-service teachers gave in an exercise of Geometry. Our purpose is to gain a better understanding of what we call the geometrical working space (espace de travail géométrique). We first conduct a didactical study based on the notion of geometrical paradigms that leads to a classification of student's answers. Then, we use statistical tools to precise the previous analysis and explain students' evolution during their training.

Key words: Geometry, Didactic, Paradigm, Geometrical Working Space, Teachers Training.

1 Presentation of the study

Various theoretical tools have been developed to study the teaching of geometry and, in the case of teacher training, two of them have been here preferred: geometrical paradigms and geometrical working spaces (GWS; in French: *Espace du Travail Géométrique*). Using these tools, our research focused on the following hypothesis, which our work abundantly supports:

> In education, the sole term *geometry* evokes several distinct paradigms. By and large, these paradigms reflect the breaks observed between the various academic cycles in the teaching and learning of geometry.

In our view, the field of geometry can be mapped out according to three paradigms, only two of which — Geometry I and II — play a part in today's secondary education. Each paradigm is global and coherent enough to define and structure geometry as a discipline and to set up respective working spaces suitable to solve a wide class of problems. Based on these premises, we built a training device designed to make future teachers aware of these paradigms and of their role as cause of certain misunderstandings in a classroom setting. The construction and evaluation of the device requires a precise analysis of

A. Kuzniak: *Personal Geometrical Working Space: a Didactic and Statistical Approach*, Studies in Computational Intelligence (SCI) **127**, 185–202 (2008)
www.springerlink.com

students' spontaneous use of paradigms as they solve geometrical problems. This analysis is meant to understand better the geometrical working space of each student. Existing research provided the elements that lead to distinguishing among four groups of students, each corresponding to a specific approach to the study of geometry.

In this paper, we wish to examine what specific contribution statistical methods can bring to that research. More precisely, we focus on the two following sets of questions:

- The first set bears on the classification resulting from our didactical analysis. Does statistical analysis produce the same outcomes as the initial analysis? Which new elements, if any, emerging from implicative analysis, help better understand the various classes of students and thus predict some of the changes observed during training sessions?
- The second series of questions is concerned with automating the process of sorting students according the classes defined above. Indeed, in addition to being demanding, the didactic analysis calls for advanced knowledge of its theoretical framework, which limits its use by other researchers. To mitigate this problem, we are currently working with Chilean colleagues on developing tools that will enable them to analyze large quantities of data on student performance.

In this paper, we first expand on our theoretical framework in some detail and present the training device. Then we examine the device's key exercise. We will present data on student performance that highlight the role and contribution of the methods we used: a didactic analysis and then a statistical study (factorial and implicative analysis).

2 Object of the study

2.1 Theoretical premises

Work initiated by Bachelard [1] and Koyré [2] and pursued in mathematics by Lakatos [3] showed that the idea of a peaceful scientific evolution of mathematical concepts was an illusion. Kuhn [4] brought the conflicting logic of scientific ideas to a culmination: he sees the transition from one paradigm to another as a revolution whereby a new paradigm replaces the old one.

Our view of the study of geometry is based on an approach asserting that geometry has undergone significant changes of perspectives equivalent to paradigmatic shifts. Following Gonseth [5] who places geometry in relation to the problem of space and applying Kuhn's notion of a paradigm, we consider three geometrical paradigms [6, 7] that organize the interplay between intuition, deduction, and reasoning in relation to space:

- Natural Geometry (Geometry I), which finds it validation in reality and the sensible world. In this Geometry, an assertion is accepted as valid using arguments based upon experiment and deduction. The confusion between the model and reality is great and any argument is allowed to justify an assertion and convince;
- Natural Axiomatic Geometry, whose archetype is classic Euclidean Geometry. This Geometry (Geometry II) is built on a model that approaches reality. Once the axioms are set up, proofs have to be developed within the system of axioms to be valid;
- Formalist Axiomatic Geometry (Geometry III), in which the system of axioms itself, disconnected from reality, is central. The system of axioms is complete and unconcerned with any possible applications in the world.

These various paradigms — and this is originality of our approach — are not organized into a hierarchy, one is not better than the other: their use is different depending on the aim of the problem.

Our theoretical framework is also based on the notion of Geometrical Working Space (GWS) (see 4.4 for some detail) which enables us to analyze how students or experts work when they are involved in a geometrical task.

2.2 The question of the teachers' training

We have examined students' application of geometrical paradigms and use of personal working space in various ways. The approach presented here relies on a relatively complex training device applied in two different phases [8]. Students are primary school teachers in training.

The first phase is based on a written individual questionnaire. Specifically, students are asked to solve geometrical exercises and list the doubts and difficulties they experienced during the résolution. During the second phase, students are asked in particular to participate to a work entitled: "Geometry: Charlotte and Marie, who is right and why? The students do not agree". For the purpose of this work, they look at a selection of solutions and comments written by their peers during phase 1. The solutions and comments were grouped in four categories that reflected the different approaches encountered in students' responses. Then students review their own initial answers.

In the next sections, we present the problem submitted to the students, then we expose different methods used to analyze in depth the solutions given by the students.

2.3 The key problem "Charlotte and Marie"

The following problem (Hachette Cinq sur Cinq 4^e 1998, page 164) exemplifies the kind of geometrical exercises for which the existence of a working space suitable to solve the problem is not obvious.

1.Why can we assert that the quadrilateral OELM is a rhombus? 2.Marie maintains that OELM is a square. Charlotte is sure that it is not true. Who is right?	**1°** Pourquoi peut-on affirmer que le quadrilatère *OELM* est un losange ? **2°** Marie soutient que *OELM* est un carré. Charlotte est sûre que ce n'est pas vrai. Qui a raison ?	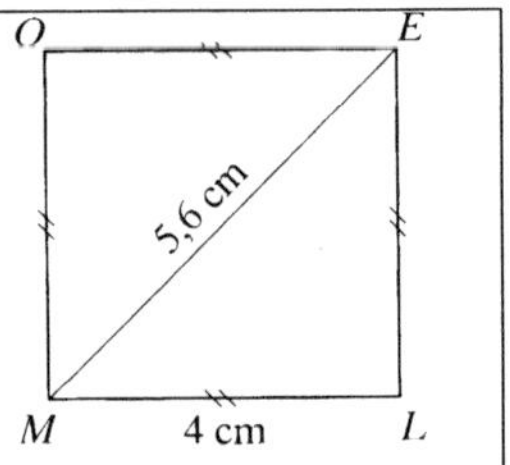

The drawing looks like a square but its status in the problem is not clear. Is the drawing a *real object* the problem suggests to study or does it result from a construction described in a text? And in that case, is the practical achievement essential or does it only serve as a support for reasoning?

The function of the represented object is usually given by the text of the problem: this in turn orients towards a precise geometrical paradigm. Here, the wording gives no such indications and as a student points out: *There are no texts for the wording, only a drawing that can mislead.*

Finally, who is right? Charlotte or Marie? Pythagoras' theorem, which doesn't require the real measurement of the angle, gives a typical way of handling this kind of exercise. But even there, the ambiguity of the choice of the working space reappears. For our purpose, we shall introduce two forms of Pythagoras' theorem, the usual one, an abstracted form, with real numbers and equalities:

If the triangle ABC is right in B then $AB^2 + BC^2 = AC^2$

and the other one, a practical form, using approximate numbers and, in a less common way, approximate figures

If the triangle ABC is "almost" right in B then $AB^2 + BC^2 \simeq AC^2$

The first form leads to work in Geometry which deviates from data of experiment by arguing in the numeric setting. The second formulation appears rather as an advanced form of Geometry I.

If we work in Geometry II by using the abstracted form of Pythagoras' theorem, then we can argue, as one student suggests, giving reason to Charlotte:

We know that if OEM is right in O then we have $OE^2+OM^2 = ME^2$
We verify $4^2 + 4^2 = 5.6^2$ and $32 \neq 31.26$. Thus, OEM is not a right triangle.

If we use the practical Pythagoras' theorem in the measured setting then we shall rather follow the reasoning proposed by another student who concludes:

Marie is right OELM is a square since $\sqrt{32} \simeq 5.6$.

In fact, it would be necessary to conclude that OELM is "almost" a square. But, for lack of a suitable language, students cannot play on these various distinctions. They are faced at the same time with an epistemological and didactical misunderstanding. It seems to us that the interplay between Geometry I and Geometry II explain and work on this problem.

The problem comes from a textbook designed for 14 years old students and, as noticed above, it is especially ambiguous. Its use in a class should be questioned within the framework of high-school Geometry teaching in France [9]. Why give it to seek with the pupils? And with which teaching intentions? In our specific study, we gave it to pre-service teachers with the two main goals of bringing out their knowledge in geometry and making explicit some misunderstandings existing in the teaching of geometry with help of geometrical paradigms.

3 Didactic analysis of students' works

3.1 Towards a classification of the students' answers

From the answers given by the students, we can sketch a classification that takes into account the geometrical paradigm applied in the resolution. It must be clear that only answers and not the students are classified here. But, by doing this, a general understanding of students' behavior is intended.

We have also identified four kinds of answers to the "Charlotte and Marie" problem. This allows us to bring out four main approaches.

We labeled these four groups PII, PIprop, PIperc, and PIexp, for reasons that will be clarify below. In each case, we will give a typical answer (Appendix 1) from the sub-population under study.

First, answers using theorem are common among two groups of students, PII and PIprop.

PII In this case, [St A], the standard Pythagoras' theorem is applied inside the world of abstract figures and numbers without considering the real appearance of the object. Only information which is given by words and signals (code of segments, indications on the dimension of the lengths), is used, and Pythagoras' theorem is applied in its entire formal rigor. To prove that the quadrangle is a rhombus (four sides of the same length) and to show that it is not a square (contrapositive of Pythagoras' theorem), students use minimal and sufficient properties. We shall consider this population as being inside Geometry II.

PIprop. This population groups together students who apply the practical Pythagoras' Theorem, in fact, to be rigorous, the converse. They generally conclude that Marie is right [St B]. In that case, the students recognize the importance of the drawing and of the measurements' approximation. The practical Pythagoras' theorem appears as a tool of Geometry I. We have designated this population as PIprop to insist on the fact that individuals of this group use properties to argue. The question whether these students can play with the differences between Geometry I and Geometry II or if their horizon remains only technological.

An addition to these answers, here are those of students who did not use Pythagoras' theorem.

PIexp. We group together students who use their measuring and drawing's tools to arrive at an answer. They are situated in the experimental world of Geometry I. Generally, this type of students concludes that Marie is right [St C]. But, it is not always the case: a student, using his/her compass, verifies that the vertices of the quadrangle are not cocyclics and s/he can assert that $OELM$ is not a square.

PIper. In this last category [St D], we group together students whose answers are based on perception: Their interpretation of the drawing is the basis for their answer, and they do not give use any information about their tools of investigation. It is not easy to know if this lack of deductive proof is due to a lack of geometrical knowledge or to a real confidence in the appearance of the figure. To answer to this question, we must have a look at their reasoning problems.

3.2 A look at reasoning problems

The typical outcomes presented above are logically quite coherent and do not contain too many reasoning errors and formulation problems. That is not true for all cases and we proceeded by performing an analysis of proofs and the reasoning structure based on the levels of argumentation inspired by Van Hiele.

We classify in level 1 works, which enumerate a non-minimal list of quadrangle properties to justify assertions. In level 2, we place productions, which evoke a correct relation of inclusion between square set and rhombus set. In level 3, we set productions that use minimal and sufficient information to justify assertions.

This analysis allows us to separate two categories of students. In the first one, widely illustrated by our previous examples, students have solid knowledge concerning the figures' properties and use level 3 reasoning. The students of the second category argue with an accumulation of properties and show not very sound knowledge of the geometrical properties. Here are two examples illustrating this second group. [St E]

> 1) The quadrangle $OELM$ is a rhombus. It follows the characteristics of such a figure: four sides are equal; diagonals cut themselves in their middle and form a right angle.
> 2) Both girls are right; $OELM$ is a square, for it has four equal sides and four right angles. It is also a rhombus, even if this figure that is rhombus is not necessarily composed of right angles.

This student justifies his/her first answer by enumerating a list of properties of rhombuses. Thus, we classify his/her production at level 1. The properties employed are partially justified through visual or instrumented indications. This student considers the figure in its material reality and her approach of the problem comes within Geometry I.

The answer to the second question *both girls are right* occurs frequently enough. Its justification shows that the statement *Charlotte is sure that it is not true* is wrongly interpreted as *Charlotte asserts that it is a rhombus* concealing the assertion *It is not a square.* The student focuses on the question of the link between squares and rhombuses. It is a classic question (but not asked here) and the student knows how to answer. That shows that she has level 2 knowledge corresponding to the classification of figures.

With this student, we meet a rather frequent profile. [St F]

> 1) Four sides of the quadrangle are parallel between them and of the same length OE = ML and OM = EL.
> Definition of the rhombus: we can say that diagonals have the same middle and are perpendicular between them.
> 2) Marie is right, $OELM$ is also a square because sides are all of the same length: $OE = ML = EL = OM = 4cm$.
> Let us remember that the square is also a rhombus but which has the peculiarity of having all sides with the same length (thus forming right angles) and having diagonals of the same length.

The employed syntax could refer to level 3: some partially correct implications are evoked. But the body of knowledge is not very reliable. In particular, we find here a rather frequent *pupil's theorem*: Any quadrangle having four equal sides is a square. We are clearly within Geometry I where visual indications are used to support reasoning.

4 The statistical analysis

4.1 Limit of the didactic study

The didactical study leaves some of the questions raised in introduction pending. The classification we obtained is a straight product of our theoretical framework. It is therefore compelling to use statistical tools to test the model and, at the same time, measure the distance between the classes. In phase 2 of the exercise, students had to choose the best explanation among the variety of responses. It turns out that the choices they made depended on the class they belonged to, plus other unknown reasons. To further the analysis, we need to understand what determines students' class membership and define the relevant sub-classes so that we can explain why members of the same class could evolve in different ways during the teaching process.

Moreover, we need to increase our capacity to analyze large amounts of data through automation to learn more about students' personal GWS and to make this kind of research more accessible to researchers with different theoretical backgrounds. Let us note finally that if statistical techniques can put our didactic approach to the test, the latter can in turn help verify whether statistical tools are adequately discerning and explaining the phenomena under study.

We chose two statistical approaches to handle the data; the first is based on factorial analysis, the second uses implicative analysis. To make these different analyses we retain only the productions of two groups of French students, that are 57 subjects.

4.2 The factorial approach

Even if our population is small, we first used principal component analysis to get a global view of data. We encoded and analyzed the answers given by students to three questions set within the framework of the problem "Charlotte and Marie" (Appendix 2).

This encoding, which was performed in association with J.C Rauscher and Chilean colleagues of the University of Valparaiso, allows for the eight components dimension (named here aspect) that underlie the universe of answers, yielding 13 binary variables.

We introduced two components to describe the first question on the rhombus. Aspect 1 represents the sources of information used by the student: does s/he take information directly from the drawing or not? The various justifications the student uses to prove that the figure is a rhombus are represented in Aspect 2: accumulation of arguments, characteristic property.

The answer given by the student to the question 2 — who is right? Charlotte or Marie — is kept in aspect 3. Let us note that 20 students answered Charlotte, 28 Marie, 7 of them both and two students asserted *we could not know*. In order to deal with this aspect, we need two 0–1 variables CHA and MAR. The first one CHA takes value 1 if the student answered Charlotte; the second MAR is 1 when the answer is Marie. With this coding, the answer both gives so values 1 to CHA and MAR at the same time. The arguments given by the student are listed in aspect 4: reference to a theorem, use and type of calculations, remarks on angles or sides, correction and coherence of the reasoning.

The possible use of the triangle congruence cases is taken into account by aspect 5 (suggested by the Chilean colleagues, this use never appeared in France). Aspect 6 was introduced in order to check if students consider a relation between rhombuses and squares to support their argumentation.

Some students add marks on the drawing submitted in the problem (aspect 7). The variable FIG takes the value 1 as soon as a mark or a construction is made on the drawing.

The last question about the doubts of the students is taken into account by aspect 8. This aspect is divided in three variables depending on the content the students mention: properties, drawing and estimation.

These eight aspects are then shaped into disjunctive (Yes/No) variables to allow the statistical analysis: that gives 14 characters. We keep neither aspect 5 nor aspect 4c in this analysis.

The study was made with the program Statistica, we give here (fig. 1) the representation of variables in the prime factorial plane.

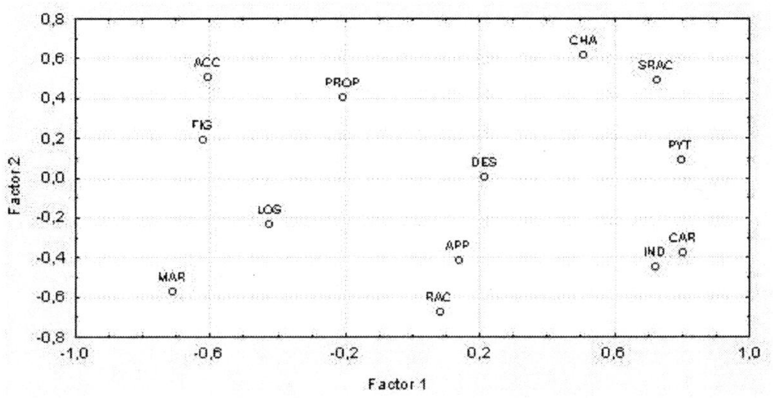

Fig. 1. Factorial analysis

Expressing the answer to the problem, Charlotte (CHA) and Marie (MAR) are obviously the most determining variables and it could have been interesting to consider them as supplementary variables [10, 11]. With the statistical study, we can correlate these two variables with the others which seem in our view to be important, as the use of square roots (RAC) or of drawings on the figure (FIG). Both variables CAR and ACC are pointing in the characteristic properties use and are determining for a better understanding of students' reasoning.

The graph shows the three variables connected to the doubts described by the students: DES for doubts on drawing, APP for the estimation and finally PROP for the expression of problems linked to the properties. Showing the relation between rhombus and square, the place of variable LOS is also going to be interesting to evaluate student's reasoning level.

We should bear in mind that in the didactic approach the subgroup PIperc includes the students who answered Marie but without our knowing the exact nature of their reasoning: is the conclusion they give based on the sole perception or given by default due to the lack of geometrical knowledge and the neglect of certain properties? We introduced the analysis with levels argumentation to better understand the method used by these students, due to its complexity, we were able to make only a qualitative analysis, the statistical analysis complete the first study in a more systematic way.

4.3 The implicative approach

Thanks to factorial analysis, we can first sketch a map that positions students in the prime factorial plane. To ascertain a better grouping of the determining variables, we used the program CHIC and created similarity trees as well as hierarchic trees that reveal one-way relationships among variables [12, 13]. This approach is essential to studying how students' geometrical thinking works and describe their GWSs in a more dynamic way [14, 15].

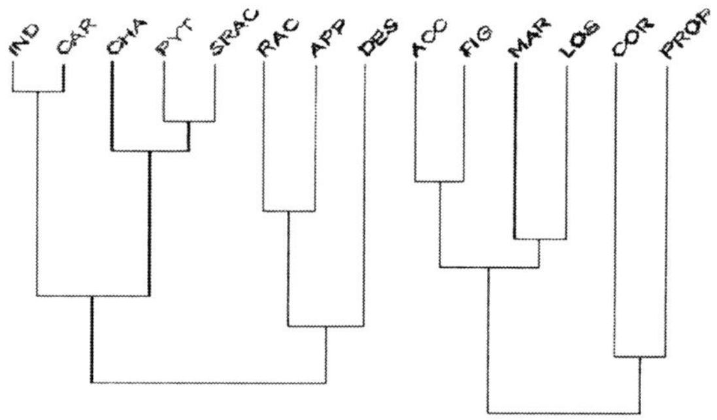

Fig. 2. Similarity tree

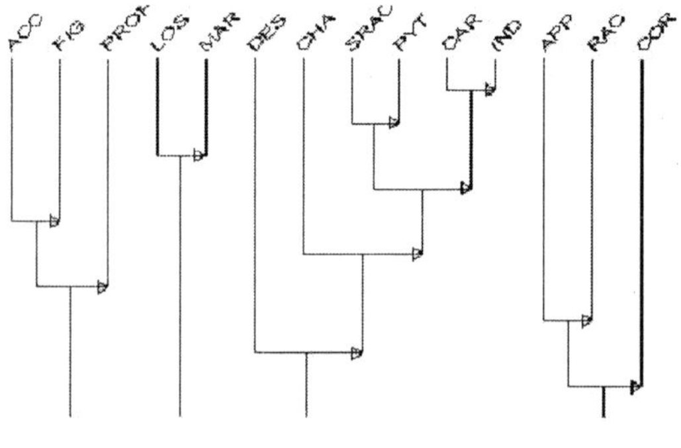

Fig. 3. Implicative tree

The implicative study produces four clearly defined, potentially interesting groupings that we combine with the results of our didactic analysis. Variables

[CHA, SRAC, PYT, CAR, IND] create a first set which shows us how students giving the answer Charlotte (CHA) are reasoning. They use Pythagoras' theorem (PYT) with a calculation without square roots (SRAC), they master the notion of characteristic property (CAR) and use only information given by the problem wording (IND). This group is very close to the one that we identified under the name of PII.

Another group is organized around variables [LOS, MAR]. The implicative analysis confirms that students having told from relations between rhombus and square answered Marie (MAR). This group is close to the one described by variables [ACC, PROP, FIG], but if students of this group lean their reasoning on the figure and give a series of properties, at the same time they indicate their doubts and their difficulties. The students from these two classes argue differently from those belonging to the first group (PII): visual or experimental use of the support of the figure, accumulation of arguments or reasoning based on the global perception of the figure shape.

In a specific way, the statistical analysis shows the coherence of a last group around [APP, RAC, COR]. These answers use the "approximate" form of Pythagoras' theorem. It seems that students from this group are sensitive to the importance of the estimation and to the question of the drawing which looks like a square. The way they argue is close to group PII but their regard to the reality is different.

The combination of both analyses allows the organization of the variables as shown in the graph of the binary variables in the prime factorial plane (fig 4).

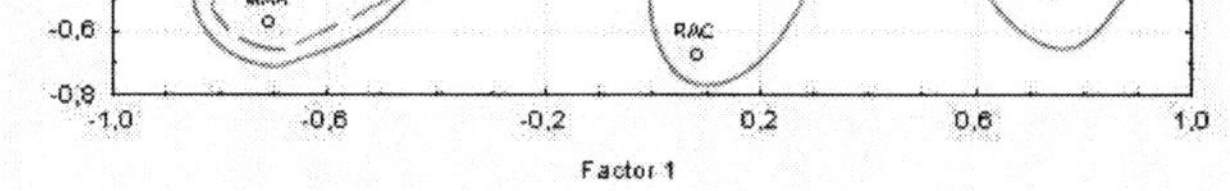

Fig. 4. Factorial analysis

4.4 Precision on the components of the geometrical working space

We must further interpret the results presented above in terms of geometrical working space and for this purpose we first provide some additional elements to clarify a notion so essential to our approach to geometry teaching.

More precisely [16], the Geometrical Working Space (GWS) is the place organized to ensure the geometrical work. It makes networking the three following components:

- the real and local space as material support,
- the artifacts as drawings tools and computers put in the service of the geometrician,
- a theoretical system of reference possibly organized in a theoretical model depending on the geometrical paradigm.

The geometrical working space becomes manageable only when its user can link and master the three components above.

To solve a problem of geometry, the expert has to work with a suitable GWS. This GWS must meet two conditions: its components are sufficiently powerful to handle the problem into the right geometrical paradigm and, depending on the user, its various components are mastered and used in a valid way. In other words, when the expert has recognized the geometrical paradigm involved in the problem, she/he can solve it thanks to the GWS suited to this paradigm. When the problem is set a person (the pupil, the student or the professor), either an ideal expert, this person handles the problem with its personal GWS. This last one will have neither the wealth nor the performance of the GWS of an expert.

This focus on the personal GWS, led us to introduce a cognitive dimension into our GWS approach. For that purpose, we follow Duval [17, p 38] who points out three kinds of cognitive processes:

- visualization process with regard to space representation and the material support,
- construction process depending on the used tools (rules, compass) and on the configuration,
- reasoning in relation to a discursive process.

These three processes are linked in a diagram (fig. 5) we juxtapose with GWS components.

4.5 An interpretation in terms of personal Geometrical Working Space

Now we can interpret the results of the statistical analysis of the students' answers in connection with the notion of personal GWS. In terms of GWS, clearly the study points out two systems of reference: one associated to Geometry I and the other to Geometry II. The new dimension brought by the statistical data is that technical mastery of reasoning and properties knowledge introduces differences between personal students' GWSs. The influence of visualization or artifacts changes according to the student

A first students' group works inside the GWS/GII based on Geometry II system of reference. We can divide this set in two subgroups. Students of the

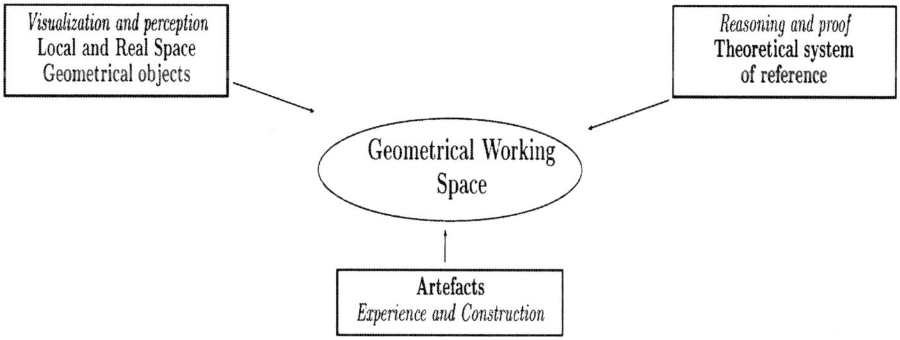

Fig. 5. Global GWS Structure

first groups master enough -at least in the exercise Charlotte and Marie- the theoretical system of reference. This group matches exactly to PII represented above by the production of student A. Within the limits of this exercise, this group masters rules of the geometrical argumentation. When members of this group evoke doubts on the drawing, they underline the misleading aspect of the drawing following the traditional view about figure in French geometry education as soon as Geometry II is set up in the curriculum (Grade 7 or 8).

The second subgroup refers always to Geometry II but with an insufficient mastery due either to the neglect of certain properties, or to the superficial understanding of reasoning rules in Geometry II, encountered during their studies. This subgroup is strictly included in PIperc, a group we noticed the heterogeneity. We meet students here whose answers is similar to student D but who express their doubts in a particularly subtle way as this one:

> Could we say that diagonals are really perpendicular? Could we say that the quadrangle has 4 right angles? By using a set square, yes. By calculating with Pythagora, it is not exact, but approximate.
>
> $$5.6^2 = 31.36 \simeq 32$$

The second large population that the analysis points out groups together students who have in mind Geometry I paradigm and work into the working space GWS/GI To them, the figure given into the problem is a real object they have to study. The analysis reveals two subgroups, members of the first use mainly arguments based on visualization and construction to solve the problem, members of the second use the connection between construction and proof. In this population, we meet vague answers close to student D (PIperc) but also to student C who used drawing instruments to verify properties (PIexp). Part of the students who used the "approximate" Pythagoras' theorem (PIprop) belongs to this large group.

Finally, it is necessary to point out a group which seems to play the game (GI/GII). These students look to balance between GI and GII but the usual rules of the didactic contract leave them few possibilities of expressing clearly their opinion. Members of this group give answers close to student B and use like him/her the "approached" form of Pythagoras' theorem but some can also have used the classic Pythagoras' theorem by writing their doubts on the status of the figure drawn in the problem.

Chart summarizing the results:

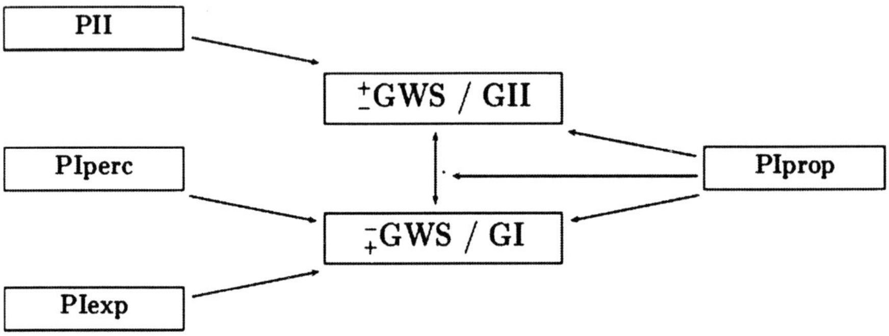

Fig. 6. Relations between GWS and Students types

5 Conclusion

We set out to create a typology of teachers in training (pre-service) who were given a series of geometry problems. The typology is based on students' knowledge of geometry which, in this study, was considered within the context of the paradigms that give its various meanings to geometrical work and thinking. We arrived at the notion that the Geometric Working Space depends on the geometric perspective adopted — in our terminology Geometry I or II — and on the user who adopts it. This observation led us to focus on the personal GWS for each student. Using statistical techniques — in particular implicative analysis — we were able to characterize students' output by exposing how students organize their reasoning within a finite number of categories. We also gained a better understanding of why students evolve, or not, during training.

The study suggests the existence of different categories of students. But it also shows that we should consider these results with caution. We should not assign students to rigid categories but rather be mindful of these classes as useful benchmarks in the training of teachers.

Moreover, personal GWS could depend on national learning curricula as we observed it by comparing Geometry teaching in Chile and France. Explanations and writing of results are not provided on the same way and new arguments based on nature's numbers (irrationality) or on congruent triangles have been used by Chilean students. As the same paradigm could be tackled on different way in various countries and institutions, we need studies of geometrical working spaces in various contexts.

Acknowledgment

This study would not have been possible without the active participation of J.C. Rauscher IUFM d'Alsace.

On the other hand, the project enters a research with Chile supported by ECOS/CONICYT.

References

1. G. Bachelard. La formation de l'esprit scientifique. Vrin Paris (Translation *Formation of the Scientific Spirit (Philosophy of Science)*, Clinamen Press, 1983.
2. A. Koyré. *From the Closed World to the Infinite Universe.* (Hideyo Noguchi Lecture), Johns Hopkins University Press; New Ed edition, 1969.
3. I. Lakatos. *Proofs and Refutations: The Logic of Mathematical Discovery.* Cambridge University Press, 1976.
4. T. S. Kuhn. *The Structure of Scientific Revolutions (Foundations of Unity of Science).* University of Chicago Press, 2Rev Ed edition, 1966.
5. F. Gonseth. *La géométrie et le problème de l'espace.* Griffon Ed, Lausanne, 1945–1952.
6. C. Houdement, A. Kuzniak. Sur un cadre conceptuel inspiré de Gonseth et destiné à étudier l'enseignement de la géométrie en formation des maîtres. *Educational Studies in Mathematics*, volume 40/3, pages 283–312, 1999.
7. C. Houdement, A. Kuzniak. Elementary geometry split into different geometrical paradigms. *Proceedings of CERME 3*, `http://www.dm.unipi.it/~didattica/CERME3/proceedings/Groups/TG7/`, 2003
8. A. Kuzniak, J. C. Rauscher. On Geometrical Thinking of Pre-Service School Teachers. *Cerme IV Sant Feliu de Guíxols Espagne*, `http://cerme4.crm.es/Papers\%20definitius/7/kuzrau.pdf`, 2005.
9. R. Berthelot, M. H. Salin. L'enseignement de la géométrie au début du collège. *petit x*, volume 56, pages 5–34, 2001.
10. P. Orus, P. Gregori. Des variables supplémentaires et des élèves fictifs dans la fouille des données avec CHIC. *Actes des troisièmes rencontres ASI Palerme*, pages 279–292, 2005.
11. A. Scimone, F. Spagnolo. The importance of supplementary variables in a case of an educational research. *Actes des troisièmes rencontres ASI Palerme*, pages 317–326, 2005.
12. R. Gras. *L'implication statistique: nouvelle méthode exploratoire de données.* La pensée sauvage, 1996.
13. P. Kuntz. Classification hiérarchique orientée en ASI. *Actes des troisièmes rencontres ASI Palerme*, pages 53–62, 2005.

14. S. Ag Almouloud. Une étude diagnostique en vue de la formation des enseignants. *Annales de Didactique et de Sciences Cognitives.*, Volume 9, pages 223–246, 2004.
15. M. Bailleul, Ratsimba-Rajohn. Analyse de la gestion des phénomènes d'ostension et de contradiction par l'analyse implicative. In *Colloque Méthodes d'analyses statistiques multidimensionnelles en didactique des mathématiques*, Caen, pages 199–216, 1995.
16. A. Kuzniak. Paradigmes et espaces de travail géométriques. *Canadian Journal of Science and Mathematics*. volume 6.2, pages 167–187, 2006.
17. R. Duval. Geometry from a cognitive point of view in Mammana *Perspectives on the Teaching of Geometry for the 21st Century: An ICMI Study*, Kluwer, pages 37–51, 1998.

Appendix 1 Student's answers

PII [Student A]

1) OELM is a rhombus because its successive sides are equal.
2) If OELM is a square, then MEL is a right-angled triangle L. According to the Pythagoras' theorem we would have then, $ME^2 = ML^2 + LE^2$ as $ML^2 + LE^2 = 16 + 16 = 32$ and $ME^2 = 5,6^2 = 31.36$
Thus, angle ELM is not a right angle.
Consequently, OELM is not a square and it is Charlotte who is right.

PIprop [Student B]

1) OELM is a rhombus, for $OE = OM = ML = LE$ and a rhombus has its four sides of the same length.
2) Marie is right because all the sides of the quadrangle have the same length and there is at least a right angle. We can verify it by Pythagoras' theorem. $ME^2 = ML^2 + LE^2$
$4^2 + 4^2 = 16 + 16 = 32$
$ME = \sqrt{32} = 42 \simeq 5.6$ thus $MLE = 90$

PIexp [Student C]

1) OELM is a rhombus, for its diagonals cut themselves in their middle (measuring) by forming right angles (using a set square).
Remark: the student built the second diagonal on the figure.
2) Marie is right. It is a square, for besides being a rhombus, OELM has its angles right (set square).

PIperc [Student D]

1) Four sides of the quadrangle are parallel between them and of the same length $OE = ML$ and $OM = EL$. According to the definition of a rhombus, we can say that diagonals have the same middle point and are perpendicular between them.
2) Marie is right; $OELM$ is also square because its sides form a right angle.

Appendix 2 Codification of the problem Charlotte and Marie

Question 1 Why is it a rhombus?
Aspect 1: source of the information IND
Code 1 Exclusive use of the information given in the wording.
Code 2 Use of information not given explicitly by the wording.
Aspect 2: Why is it a rhombus? CAR and ACC
Code 1 correct Justification based on a necessary and sufficient condition using the sides.
Code 2 correct Justification based on a necessary and sufficient condition but without using sides.
Code 3 Justification using an accumulation of arguments.
Code 4 Wrong justification
Question 2 Who is right? Why?
Aspect 3: Who is right? CHA and MAR
Code 1: Charlotte
Code 2: Marie
Code 3: we cannot know
Code 4: both
Aspect 4a) Justification of the question 2 Why? PYT
Code 1: Refer to the Pythagoras' theorem
Code 2: does not make reference to the Pythagoras' theorem
Aspect 4b) (Calculations) RAC and SRAC
Code 1: calculations without square roots
Code 2: calculations with square roots
Code 3: without calculations
Aspect 4c) (Arguments)
Code 1: only with angles
Code 2: only with sides
Code 3: with angles and sides
Code 4: with the diagonal
Aspect 4d) COR
Code 1: relevant Arguments
Code 2: wrong Arguments
Other remarks
Aspect 5: Use of the congruence of triangles
Code 1: yes
Code 2: no
Aspect 6: relation between squares and rhombuses LOS
Code 1: reference to the relation between squares and rhombuses
Code 2: no reference to this relation
Aspect 7: Presence of marks or lines on the drawing FIG
Code 1: visible tracks or reference to use of instruments
Code 2: no visible Tracks

Question on the doubts and difficulties

Aspect 8a) (Knowledge) PROP

Code 1: doubts on the knowledge of the definitions and of the properties

Code 2: no doubts expressed on this point

Aspect 8b) (Drawing)

Code 1: doubts about the status or the information given by the drawing

Code 2: no doubts on this point

Aspect 8c) (Estimate) APP

Code 1: doubts on the estimate

Code 2: no doubts on this point

Part III

A methodological answer in various application frameworks

Statistical Implicative Analysis of DNA microarrays

Gerard Ramstein

LINA, Polytech'Nantes
Rue Christian Pauc BP 50609 44306 Nantes cedex 3, France
gerard.ramstein@univ-nantes.fr

Summary. This chapter presents an application of the Statistical Implicative Analysis to microarray gene expression data. The specificity of these data requires an adaptation of the concept of intensity of implication. More specifically, we propose to study the rankings of observations instead of the measurements themselves. This method makes our analysis more robust and insensitive to any monotonic transformation of gene expression. We introduce the concept of rank interval and show that the integration of the implicative method in this framework is more efficient than correlation techniques. Our method is applied to the most challenging problems encountered in gene expression analysis, namely the discovery of gene coregulation, gene selection and tumour classification. We compare our method with performing algorithms that are dedicated to gene expression data or that are well-suited to high-dimensional variable space.

Key words: ranking analysis, feature selection, classification rules, gene coregulation, microarray data analysis

1 Introduction

The microarray (DNA chip) technology [23] monitors a consequent subset of the whole genome on a single chip, so that biologists can observe the interactions among thousands of genes simultaneously. This powerful technique revolutionizes the traditional methods in molecular biology generally consisting in experiments focused on one well-defined gene. Gene expression analysis tends to be a major issue in data mining [21], involving new prospects in drug discovery and disease diagnosis. From a theoretical point of view, microarray analysis provides a deeper insight into the processes involved in the mystery of life. This paper introduces the application of Statistical Implicative Analysis [15] to these challenging data. Our objective is to extract hidden structures from noised data, in the context of unsupervised and supervised analyses. For unsupervised analysis, the discovery process makes no assumption on the observations. For supervised analysis, it uses the class information relative

G. Ramstein: *Statistical Implicative Analysis of DNA microarrays*, Studies in Computational Intelligence (SCI) **127**, 205–225 (2008)
www.springerlink.com

to experimental conditions. The latter corresponds to microarray studies in which the phenotypes (i.e. diseases) are known. Linking clinical phenotypes to genotypes is a major issue for tumour classification and drug discovery. DNA microarrays may be used to discriminate different tumour subtypes by monitoring gene expression profile on a genomic scale.
Owing to the imprecision of the measurements, we consider the relative ranks of the observations instead of their values. In the analysis of microarrays, the study of the ranks presents several advantages. As it is scale-independent, it is not sensible to the natural variation of gene expression. On the other hand, ranking is a useful indicator for the biologist. High ranking (resp. low ranking) corresponds to observations presenting an over-expresssion (resp. under-expression). These expression levels play a crucial role for the analysis of microarray data. The definition of these levels is however very difficult: it is not possible to define absolute thresholds of expression representing a typical expression level. Our rank analysis overcomes this difficulty as it only considers the order of the data without any partitioning. The paper is organized as follows. Section 2 surveys related work in the area. Section 3 introduces some concepts and defines our implicative measure. We notably show that it is more efficient than correlation measures for the analysis of gene expression profiles. In section 4 we address the problem of tumour classification. We apply our method to extract the most informative genes that discriminate different tumour subtypes. We use the implicative technique for the discovery of classification rules and compare our results with different techniques of tumour classification. Finally, section 5 presents an analysis of gene association network, based on the concept of intensity of implication.

2 Related work

Dealing with imprecise and noisy data is an important issue that has already been addressed by the researchers in the area of implicative analysis. Their works put emphasis on the determination of intervals. In [14], an optimal partition on numeric variables is defined and the quality of implication is determined by the union of elements of the partition. Another interesting work [13] introduces fuzzy partitions. We do not use either of these approaches because we prefer avoiding a partioning procedure. We indeed observed that microarray datasets often follow a monomodal distribution and the definition of a partition, fuzzy or not, tends to be arbitrary.

To our knowledge, the implicative analysis has not yet been applied to microarray data. However, the discovery of association rules has been recently proposed in this particular application field.

In [8], association rules are extracted from gene expression databases relative to the yeast genome. A preprocessing retains the genes that are under-expressed or over-expressed according to their expression values. This work is based on the *A priori* algorithm [1] and the usual rule parameters, support and confidence. In [24], the authors present a set of operators for the exploration of comprehensive rule sets. The expression values are discretized according to predetermined thresholds. The rules are filtered with classical support and confidence parameters. One drawback of these two methods is the dependency of the obtained rules to arbitrary thresholds. A similar study [5] incorporates annotation information, combined with over- or under- expression. [20] presents HAMB, a machine learning tool that induces classification rules from gene expression data. FARMER [7] also discovers association rules from microarray datasets. Instead of finding individual association rules, FARMER finds interesting rule groups, i.e. a set of rules that are generated from the same set of individuals. FARMER uses a supervised discretization procedure, based on entropy minimization. A case study on human SAGE data [3] explores large-scale gene expression data using the Min-Ex algorithm, which efficiently provides a condensed representation of the frequent itemsets. The data have been transformed into a boolean matrix by a discretization phase, the logical true value corresponding to gene over-expression. The authors analysed the effect of three different discretization procedures. Our work is closer to the one proposed in [16]. The authors define the concept of emerging patterns, where itemsets are boolean comparison operators over gene expressions. They use an entropy minimization criterion that strongly differs from our approach, since it takes into account all the samples, while we prefer to extract higher quality rules, even if they concern only a subset of observations.

3 Implication over rank intervals

3.1 Definitions

For the sake of generality, we consider a set of m individuals for which n measurements have been performed. In our study, individuals are genes (actually gene products to be more precise) and the measurements correspond to a set $\mathcal{O}$ of n different experimental conditions. A set of experiments generally refers to a biological study involving different tissue samples. We will call *observation* an experiment relative to a particular biological condition and implying the whole set of individuals (genes). Let $M(k,l)$ be the measurement value associated to an indidividual k and an observation l. Note that this value may refer to any ordinal data type. Our analysis relies on this matrix, although the same study could be performed on the transposed matrix. We call *profile* of the individual k the vector $p(k) = (M(k,l), l \in [1,n])$. The profile in our application is usually called the expression profile and concerns the whole set of measurements relative to gene k. We define the operator *rank* that takes a profile $p(k)$ and returns its observation indexes, ranked in increasing order. For example, let us consider the profile $p(k) = (4.1, 12.3, 1.2, 3.7)$. We have $rank(p(k)) = (3, 4, 1, 2)$, which means that the lowest value (i.e. 1.2) has been measured under condition 3, the value 3.7 under 4, and so on. The study of the observation ranks may reveal hidden relationships. In market basket analysis, it can provide interesting relations between amounts of transactions. We could for example discover association rules such as the fact that if a customer buys a lot of pizza, its shopping cart will contain few fresh foods. In microarray data, similar studies will concern associations between expression levels. A rule $A \rightarrow B$ will for instance denote an association between an over-expresssion observed on a gene A and an over-expression observed on a gene B. Note that these types of interactions concern only a subset of observations (e.g. a customer type in market basket data, experimental conditions in microarray data). We then need to precise a rank interval that limits the range of the associations. Let I be the set of all subintervals of [1, n]:

$$I = \{[p,q],(p,q) \in [1,n]^2, p \leq q\} \tag{1}$$

By misuse of language we will simply call *interval rank* the observations belonging to a given interval rank. Thus, the interval rank $r_k(i)$ will refer to the set of observations relative to the interval i of I. This set is defined as follows: $r_k(i) = \{o_j \in rank(k), j \in i\}$. In the previous numerical example, the interval rank $r_k([1,2])$ contains the observations 3 and 4, and corresponds to the two lowest values of the profile. Table 1 gives an example of two profiles. The visual inspection of the values does not permit to detect any relations between the two individuals A and B. Table 2 gives a better insight of a possible hidden association. After ranking, the observations 1, 4 and 9 are close together in both observation sets. The previous definition helps us to precise this similarity as follows: the rank intervals $r_A([4,6]) = \{9,1,4\}$ and $r_B([6,9]) = \{1,3,9,4\}$ are joint sets. This similarity can be verified by using the statistical implicative analysis. This approach gives a quantitative answer to the following question: is this set conjunction really surprising or could comparable results be achieved by chance? The latter hypothesis corresponds to the case where no hidden structure exists between the two profiles. Consequently, the rank operator will not give any further information, which means that the sets $r_A([4,6])$ and $r_B([6,9])$ could be considered as the result of a random selection process. Implicative analysis offers a convenient framework to address this problem. Following [15], we suppose two sets α and β having the same respective cardinals than $r_A(i)$ and $r_B(j)$. The intensity of implication $\varphi(\alpha, \beta)$ represents the quality of the association of $r_A(i) \rightarrow r_B(j)$. In the previous example, the intensity of implication equals to 0.86, which represents an association of little significance.

profile	1	2	3	4	5	6	7	8	9	10
p(A)	6	4	10	7	8	13	3	12	5	2
p(B)	15	12	16	19	10	14	8	7	17	21

Table 1. Measurement values issued from individuals A and B.

rank	o_1	o_2	o_3	o_4	o_5	o_6	o_7	o_8	o_9	o_{10}
r(A)	10	7	2	**9**	**1**	**4**	5	3	8	6
r(B)	8	7	5	2	6	**1**	3	**9**	**4**	10

Table 2. Reordering of the observations using the rank operator. The values represent the observations of the previous table, namely the column indexes.

It is possible that other intervals of I exist for which a better conjunction would be observed. It is therefore preferable to define the quality of a rule $A \rightarrow B$ as the best association that can be found among all the possible intervals. We then adapt the concept of intensity of implication as follows:

$$\varphi_I(A,B) = \max(\varphi(r_A(i), r_B(j)), (i,j) \in I^2) \tag{2}$$

Note that this measure is very robust with respect to the data: it is insensitive to monotonic transformations, an interesting property for microarray data, often prone to various preprocessings. The value of $\varphi_I(A,B)$ indicates the quality of the association. Besides, the relative intervals $imax$ and $jmax$ for which the maximum defined in eq. 2 has been found provides useful information. The rule $A \rightarrow B$ can thus be expressed in a more precise and operational form.

Let us define $t_A = \min(M[A,o], o \in r_A(imax))$, $T_A = \max(M[A,o], o \in r_A(imax))$, $t_B = \min(M[B,o], o \in r_B(jmax))$ and $T_B = \max(M[B,o], o \in r_B(jmax))$. Let o be an observation. The association rule can be expressed as follows:

$$\text{if } t_A \leq M[A,o] \leq T_A, \text{ then } t_B \leq M[B,o] \leq T_B \tag{3}$$

Suppose for example that table 1 concerns two genes A and B issued from the expression matrix M defined at the beginning of this section. The rule $A \rightarrow B$ can then be written as follows: if for an observation o we have $5 \leq M[A,o] \leq 7$, then $15 \leq M[B,o] \leq 19$.

3.2 Numerical and computational considerations

As definition 2 concerns an optimal value, the intensity of implication of coregulated genes often lies in $[0.9, 1]$. For practical reasons, it is easier to express this quality measure as follows:

$$\lambda_I(A,B) = -\log_{10}(1 - \varphi_I(A,B)) \tag{4}$$

This measure presents the advantage to possess an infinite positive range and to also increase with the quality of the rule. The parameter $\lambda_I(A,B)$ is easily interpretable: instead of having for instance $\varphi_I(A,B) = 0.9999$, we will consider $\lambda_I(A,B) = 4$, which means that the risk of observing a comparable situation by chance is equal to 10^{-4}. In this paper we will equally use expression 2 or 4 in our numerical examples.

Definition 2 requires to explore all the subintervals of I. For one profile k, we need to translate a window of size $w_k = q - p + 1$ over the interval $[1, n - w_k + 1]$. For a simple rule $A \rightarrow B$, the time complexity is $O(n^4)$ (without considering the rank operator, in $O(nlog(n))$). This search can be limited by considering the following property: high quality rules necessarily concern joint sets having similar sizes. Indeed, w_A cannot be much larger than w_B, since if $w_A \geq w_B$ the number of counterexamples is always greater than or equal to $w_A - w_B$. Reciprocally, if w_B is much larger than w_A, the intensity of implication of the rule is very low. Therefore, it is more judicious to consider a common window size w and to set $w_A = w$ and $w_B = w + \epsilon$, where $w \in [w_{\min}, w_{\max}]$ and $\epsilon \in [\epsilon_{\min}, \epsilon_{\max}]$. Then, $w_{\min}, w_{\max}, \epsilon_{\min}$ and $\epsilon_{\max}$ are input parameters of the extraction algorithm. A more important algorithmic restriction concerns the interval set I, that will be replaced by the following set:

$$I' = \{[p, q], (p, q) \in [1, n]^2, p \leq q, p = 1 \vee q = n\} \qquad (5)$$

The interval set I' has a particular importance for microarray data, as biologists generally search for expression levels corresponding to either under-expression or over-expression.

This is due to the fact that experimentation is mostly based on differential situations. This means that the measurement defines the gene expression in a tissue sample relatively to a reference sample. For example the study may concern an ill tissue versus a sound one, or experiments on patients having absorbed drug or not. The two extremities of the ranking are then the most interesting observations related to "abnormal" gene activity. For this reason, we propose to replace the computation of $\lambda_I(A, B)$ by $\lambda_{I'}(A, B)$. The rule discovery algorithm using this new interval set is less time consuming ($O(n^2)$ complexity instead of $O(n^4)$) and provides a simpler interpretation of rules. In the case of over-expressions for instance, a rule may be transposed using eq. 3 in the following condensed form: if $M[A, o] \geq t_A$ then $M[B, o] \geq t_B$.

3.3 Comparison of association rules and correlation techniques

The most widely used microarray analysis concerns the study of gene coregulation. A gene A is said to be coregulated with a gene B if the expression profile of both genes is similar. This similarity is generally measured by different metrics, such as the Euclidean distance or the Pearson's correlation. Note that the intensity of implication is oriented and can favour an association from A to B rather than from B to A. The usual metrics do not precise any orientation and has another drawback: the similarity measure is issued from a global estimation that takes into account all the observations of $\mathcal{O}$ while our implicative analysis searches for partial similarity between interval ranks that are subsets of $\mathcal{O}$. A largely used technique in microarray analysis consists in hierarchical clustering based on absolute Pearson correlation. As a clustering method, this measure may not be the most pertinent one, as demonstrated by the example shown in fig. 1. This figure represents two profiles issued from two genes belonging to the *cerevisiae specie*, commonly known as bakers'yeast. The yeast genome is essential for research in molecular biology since it corresponds to a relatively simple organism permitting numerous experiments. As a considerable body of knowledge has been collected regarding the yeast genes, it is a particularly interesting case study. Our example is issued from the microarray dataset selected in [11]. It comprises a set of genes whose functions are known and that do not contain missing values. The gene expression has been measured under 89 different experimental conditions such as heat shock or nitrogen starvation.

Figure 1 represents the implication of gene CHA1 over gene SAM1. The filled triangles indicate that the former is under-expressed in response to a signal of amino acid starvation, and, to a smaller degree, of nitrogen depletion. CHA1 is involved in the catabolism of threonine, an essential amino acid found in peptide linkage in proteins. Filled circles show that SAM1, a gene interfering in the metabolism of the methionine, is over-expressed for the same experiment set.

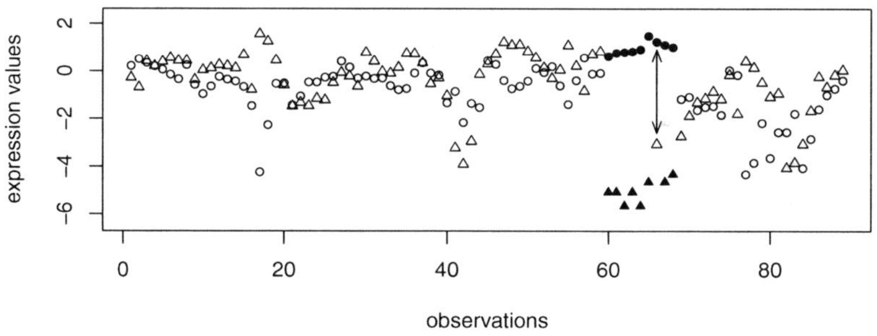

Fig. 1. Implication of CHA1 over SAM1. The absciss axis represents the 89 experimental conditions. The ordinate axis represents the expression measures. Triangles belong to the profile of gene CHA1 (YCL064C) and circles belong to the profile of SAM1(YLR180W). The filled points denote the observations belonging to the rank intervals that maximize the intensity of implication. The two rank intervals are identical and the corresponding rule accepts one exception, indicated by a double arrow. This arrow points out to the fact that there exists an observation (shown by an unfilled triangle) that is less under-expressed and that does not belong to the rank interval of SAM1 while it is included in that of CHA1.

This particular subset of conditions covers only 9% of the set $\mathcal{O}$. Correlation measures are incapable to detect such partial associations. Table 3 summarises the values of different statistic measures, including the intensity of implication for this pair of genes. These results show that the implicative measure is finer than correlation techniques. Indeed the low values obtained from the latter will not permit the gene association to emerge, notably with respect to the large amount of data. In contrast with these results, the implicative value clearly reveals the quality of the rule, stating that the risk to encounter such an association by chance is less than one over a thousand.

method	value
Intensity of implication	0.9992
Pearson correlation	-0.16
Kendall correlation	-0.0089

Table 3. Comparison of the implication and correlation measures.

4 Application to tumour discrimination

The microarray technology being complex and expensive, the vast majority of the expression data correspond to supervised analysis for which the observations are relative to precise phenotypes. We will consider a very important application, namely the classification of tumour samples. This challenging problem refers to the assignment of particular tumour samples to already-defined classes, based on gene expression monitoring by DNA microarrays. These classes define different tumour subtypes. Recognizing these subtypes is crucial to determine the malignity of the cancer. Depending on the subtype, the clinical course can indeed vary from indolence over decades to explosive growth and patient's death.
We suppose the a priori knowledge of a set $\mathcal{C}$ of classes, each observation belonging to only one class of $\mathcal{C}$. In our application study, the observation is relative to a patient for whom a precise diagnosis has been made. Let $L(o_k) = c_j \in \mathcal{C}$ be a class labelling function that associates to an observation o_k its corresponding class c_j. We will address two related problems. The first one concerns the prediction of the class corresponding to an unkown observation from a learning dataset. The second one is the selection of the most discriminating genes, also called informative genes or marker genes, i.e. genes whose expression profile is different from one class to the other. As it has been explained in section 3.2, these differences generally correspond to either under-expression or over-expression. The selection of informative genes is a challenging problem, because of the great number of genes that are analysed (from several thousands to the whole genome on a single chip). One observes that most of the genes on a microarray do not participate to the discrimination of tumour types and that the majority of them present a low amplitude of

variation. The definition of marker genes is crucial for clinical investigations, since it permits to predict the outcome for a patient with a relatively low-cost procedure. We first describe a gene selection method based on implicative analysis and then its application to tumour classification.

4.1 Gene selection

To drastically reduce the number of genes, biologists generally use a filtering technique which is mostly based on statistic tests such as the t-test, ANOVA-F, Cochran, Kruskall-Wallis, Brown-Forsythe and Welsh [6]. The implicative method can be applied to extract informative genes, by the discovery of classification rules of the form:

$$r_g(i) \to c \tag{6}$$

This condensed notation expresses the fact that the observations o of $r_g(i)$ verify $L(o) = c$. In this rule, conclusion c gives the class label associated to gene g. The premise corresponds to the rank interval relative to $i \in I'$. We consider the interval set I' defined in eq.5 to focus our analysis on either under-expression or over-expression. In other words, the classification rule states that the observed expression level on gene g and defined by the interval i concerns observations belonging to class c.

As in eq. 2, the quality of a classification rule is given by the optimal intensity of implication over interval ranks:

$$\varphi_{I'}(g \to c) = \max(\varphi(r_A(i), \mathcal{O}_c), i \in I', \mathcal{O}_c = \{o \in \mathcal{O} \mid L(o) = c\}) \tag{7}$$

The only difference with eq. 2 is that $r_B(j)$ is now replaced by a unique observation set, the set $\mathcal{O}_c$ of observations of class c. As we search for classification rules, we only consider predefined classes. Note however that a more complex analysis could be performed by accepting any subset of $\mathcal{O}$, this issue being similar to the selection of genes in an unsupervised study. When the class partition is unequal, the use of the intensity of implication presents an important advantage. Contrary to the confidence, our measure takes into account the fact that a class c is over-represented. Note that the intensity of implication is always null in the extreme case where $\mathcal{O}_c = \mathcal{O}$.

The selection of informative genes among a gene set $\mathcal{G}$ is performed by the following algorithm:

Selection algorithm
inputs:
- M, $\mathcal{G}$: expression matrix and its gene set
- $\mathcal{C}$, L : the classes and the class labelling function
- K : the number of informative genes per class

outputs:
- igs : the informative gene set

begin

$igs \leftarrow \emptyset$
for each class $c \in \mathcal{C}$ do
    $genelist \leftarrow \emptyset$
    for each gene $g \in \mathcal{G}$ do
        – compute $\varphi = \varphi_{I'}(g \to c)$
        – $genelist \leftarrow genelist \cup \{(g, \varphi)\}$
    end;
    – sort all pairs $(g, \varphi) \in genelist$ in decreasing order of φ
    – add into igs the K first genes of the sorted gene list
end;
end.

Note that the selection process permits to discover informative genes capable of discriminating more than one class among the others. A typical situation will concern genes presenting an over-expression for a given class and an under-expression for another one. A special case is relative to datasets that are partitioned into two classes. One expects that a gene that discriminates one class will automatically be informative for the other one. Actually, this assertion will not necessarily be true, according to the position of counterexamples at each extremity of the ranking. Two different triplets (g, c_1, φ_1) and (g, c_2, φ_2) will indeed be associated to a same gene g. In the case where $\varphi_1 \neq \varphi_2$, the same gene may be retained in one class and rejected in the other.

Application to leukemia subtype discrimination

The study [12] presents a well-known experiment of cancer classification based solely on gene expression monitoring. This paper has demonstrated that microarrays can provide a tool for cancer classification. This work explores the capacity of gene expression analysis to dicriminate between two subtypes of tumour, acute myeloid leukemia (AML) and acute lymphoblastic leukemia (ALL). Differentiating ALL from AML is critical for successful treatment: it has be proven that distincting therapies improves cure rates and diminishes toxicities. We used the initial dataset of the authors, which contains measurements corresponding to ALL and AML samples from bone marrow. These 38 samples concern 27 ALL and 11 AML. The dataset initially comprises 6,817 human genes. We proceeded to a filtering phase for the normalization of the data and we eliminated genes whose official symbol name could not be recovered. After this preprocessing, we obtained a set of 3571 genes. Figure 2 shows the distribution of the intensity of implication $\lambda_{I'}$ relative to the whole set of classification rules that can be extracted. The histogram reveals a non-negligible proportion of genes that are potentially discriminative. Approximatively 10% of the gene set present an intensity of implication φ greater than 0.9997 ($\lambda \geq 3.52$).This means that around 300 genes have differential expression with respect to ALL and AML subtypes.

This property is also implicitly mentioned in [12]. However, the authors select 50 genes to represent the most informative genes for the discrimination

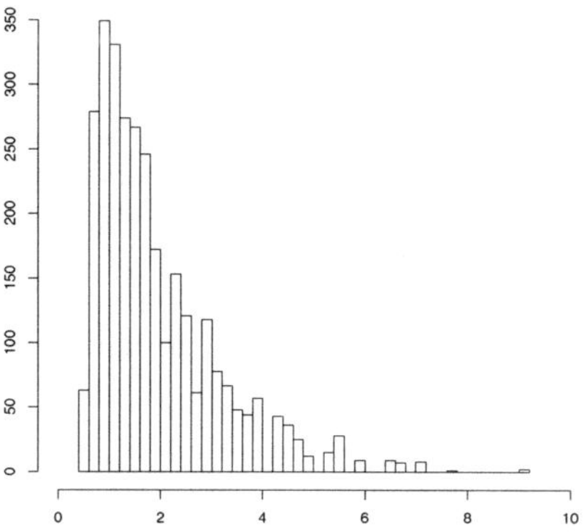

Fig. 2. Histogram of the intensity of implication. The absciss axis represents $\lambda_{I'}$ and the ordinate axis corresponds to the number of genes.

between ALL and AML. As we have two classes, we did the same by setting the input parameter K of our selection algorithm to 25. We obtained a set of classification rules that comprises 14 genes described by the authors as discriminative and included in their list of 50 genes. Table 4 shows that the genes selected by Golub *et al.* are comparable whith our set in terms of intensity of implication. The same minimum has been found in both sets and the mean is almost identical. However, one observes for our selected set that the dispersion is lower, which seems to indicate that the quality of our classification rules is higher. To verify this assumption and to analyse the discriminative power of

gene set	min	median	max	mean	variance
Golub & al.	$8.3\ 10^{-10}$	$3.6\ 10^{-6}$	$2.8\ 10^{-5}$	$7.6\ 10^{-6}$	$6.9\ 10^{-11}$
Our gene set	$8.3\ 10^{-10}$	$3.4\ 10^{-7}$	$3.2\ 10^{-6}$	$1.2\ 10^{-6}$	$1.6\ 10^{-12}$

Table 4. Comparison of gene sets. The statistics indicated in this table refer to the intensity of implication.

these two gene sets, we proceeded to a comparison of their prediction capacity. We applied the K-Nearest-Neighbours algorithm for holdout validation tests ($K = 3$). For each gene set $\mathcal{G}$ that we tested, we used the expression matrix M reduced to the genes belonging to $\mathcal{G}$. We thus obtained two distinct matrices, one corresponding to the gene set of the authors and the other to our own gene set. Table 5 shows that our selection provides better holdout validation results.

% test	Golub & al.	our method	p-value
50%	3.11	0.79	1.2e-7
25%	1.67	0.11	2.7e-4
10%	1.00	0.00	4.5e-02
2.6%	0.00	0.00	1.0

Table 5. Holdout validation in the Golub learning set. The first column gives the percentage of the dataset that has been used as test set. The second and third columns represent the mean error rate expressed in percentage over 100 random sets. The p-value is obtained from a Student's t test. The results indicate that the error rate differences are statistically significant for small learning sets.

4.2 Tumour classification

Cancer remissions highly depend on specific therapies that distinguish the treatments according to distinct tumour types. Cancer classification has historically relied on specific biological insights. DNA microarray technology permits to discriminate between tumour subtypes that present the same morphological appearance. Tumour classification by gene expression monitoring is then a crucial and challenging task.

The previous section has shown that the intensity of implication can be used to determine the most informative genes. We now propose to examine the quality of the classification rules for the prediction of tumour subtypes. Our learning dataset comprises a set $\mathcal{G}$ of genes and a set $\mathcal{O}$ of observations, providing an expression matrix M. All observations of $\mathcal{O}$ are classified by a class labelling function L.

Let $\Gamma(M)$ be the set of classification rules that are extracted from this learning dataset, using the selection algorithm described in the previous section. The cardinal of $\Gamma(M)$ is $K. \mid \mathcal{C} \mid$, where K is the input parameter of the selection algorithm and $\mid \mathcal{C} \mid$ the number of classes. Let s be an unknown tissue sample (i.e. a new observation) for which an expression profile $p(s)$ has been measured. This vector contains the expression values corresponding to the gene set $\mathcal{G}$ of the learning set. We call $p_g(s)$ the expression value associated to gene $g \in \mathcal{G}$. The classification procedure is defined by the following algorithm:

Classification algorithm
inputs:
    M : expression matrix
    $\Gamma(M)$: set of discriminative rules
    $p(s)$: the expression profile of an unknown sample s
outputs:
    c_s : the predicted class of s
begin
parameter:

count : vector of size $\mid \mathcal{C} \mid$, initially set to 0
for each class $c \in \mathcal{C}$ do
for each rule $(r_g(i) \rightarrow c) \in \Gamma(M)$ do
– Let t_g and T_g be resp. the minimal and
the maximal expression value relative to the gene g
and to the observation set $r_g(i)$
if $t_g \leq p_g(s) \leq T_g$ then
$count[c] \leftarrow count[c] + 1$
end;
end;
$c_s < -argmax_{c \in \mathcal{C}}(count)$
end.

Different microarray datasets are publicly available for tumour classification studies. We illustrate our method with the following data:

Brain tumour dataset This dataset contains gene expression profiles from 5 different tumours of the central nervous system. 42 tumour tissue samples are partitioned according to their tumour subtypes: 10 medulloblastomas, 10 malignant gliomas, 10 atypical teratoid/rhabdoid tumours (AT/RTs), 8 primitive neuro-ectodermal tumours (PNETs) and 4 human cerebella. The raw data are available at the web site of the Whitehead Institute Center for Genomic Research (`http://www-genome.wi.mit.edu/cancer`). After prepocessing, 5,597 genes remained.

Colon cancer dataset This dataset contains expression levels of colon tissues. The study concerns 40 tumoural and 22 normal tissues. The expression of 6,500 genes has been measured using the Affymetrix technology. The data are available at the web site of the Colorectal Cancer Microarray Research (`http://microarray.princeton.edu/oncology`).

Leukemia dataset This dataset comprises the expression of 72 tumours relative to acute lymphoblastic leukemia (ALL, 47 cases) or acute myeloid leukemia (AML, 25 cases). The gene expressions were obtained from Affymetrix oligonucleotide microarrays. The data are available at the web site of the Whitehead Institute Center for Genomic Research (http://www-genome.wi.mit.edu/cancer).

Table 6 summarizes the main properties of our test set.

Dataset	Publication	# samples	# classes	# genes	Response
Brain	Pomeroy [22]	42	5	5, 597	tumour subtypes
Colon	Alon [2]	62	2	2, 000	tumoural/normal tissues
Leukemiea	Golub [12]	72	2	3, 571	tumour subtypes

Table 6. Publicly available datasets.

We compare our method with two major contributions in tumor classifications that emphasize the critical importance of feature selection. [9] is based on a supervised clustering of genes and a plurality voting with classification trees. [26] uses Self-Organizing Maps and fuzzy c-means clustering. We also tested the following general-purpose machine learning algorithms:

k-Nearest-Neighbours. This classification algorithm extracts the k nearest neighbours of an unknown sample s, according to a distance function $d(x, y)$. We have used the absolute Pearson coefficient, its associated distance being expressed as follows:

$$d(x,y) = 1 - \mid \frac{\sum_{i=1}^{n}(x_i - \mu(x))((y_i - \mu(y))}{(n-1)\sigma(x)\sigma(y)} \mid \quad (8)$$

where μ and σ are respectively the mean and the standard deviation of the expression profile. The classification algorithm assigns s to the most numerous class within the neighbour set. When many features are bound to be little relevant, feature-weighted distances are preferable to the Pearson distance. However, the standard k-NN method is easy to implement and, compared to more sophisticated techniques, it provides relatively good classification results for microarray data [27].

Random Forest. This classifier [4] consists of many decision trees that deal with a random choice of samples (with replacement). At each selection node, only a random choice of conditions is used. The forest selects the classification having the most votes over all the trees in the forest. Random forest is especially well-suited for microarray data, since it achieves good predictive performance even when the number of variables is much larger than the number of samples, as it has been demonstrated in [10].

Support Vector Machines. A support vector machine [25] is a machine learning algorithm that finds an optimal separating hyperplane between members and non-members of a given class in an abstract space. As random forests, this classifier shows excellent performance in high-dimensional variable space and then is well-adapted to classification of microarray samples [19].

Table 7 presents the results we obtained on a leave-one-out validation test with these classifiers and our three datasets. Our method, although it is based on a simple counting of the rules that s verifies, achieves comparable performances with the most sophisticated techniques.

5 Analysis of gene association networks

In the previous section, we presented the use of implicative analysis for tumour classification and the selection of informative genes. We will now illustrate the application of association rules to the study of gene associations. The selection

method	Brain	Colon	Leukemia
our method	14.3	12.9	2.8
Gene clustering	11.9	16.1	2.8
Fuzzy c-means	14.3	11.4	4.1
random forest	19.0	14.5	2.8
support vector machine	11.9	12.9	2.8
K Nearest Neighbours	23.8	22.6	1.4

Table 7. Comparison of different classifiers. We give the percentage of error rates relative to a leave-one-out validation procedure. The first line concerns the results obtained from our classification algorithm (the classification rules are of course extracted from learning sets that do not contain the unknown sample s).

algorithm will enable us to focus our analysis on the most discriminative genes. We will first propose an original visual representation of the gene space, based on implicative analysis. We will then present a partial view of the gene association network, corresponding to the most informative genes.

5.1 Gene representation based on implicative analysis

We apply our study on tumour samples relative to brain cancer. The dataset is the same as the one presented in the previous section (42 samples, 5 classes). It is the most complex dataset, because of its number of tumour subtypes and its classification error rates. As in [4] the data are preprocessed. This preprocessing comprises thresholding, filtering, a logarithmic transformation and standardisation of each experiment to zero mean and unit variance. Filtering includes a selection of the first thousand of genes by decreasing order of variance.

The gene selection process extracts the $K = 10$ most discriminative genes for each of the five tumour types. Let $\Gamma(M)$ be our new set containing these 50 genes. We compute for each pair (g_i, g_j) of genes in $\Gamma(M)$ the intensity of implication associated to $g_i \rightarrow g_j$. We obtain a gene association network that can be vizualised using any arbitrary layout algorithm. We prefer to position the genes with respect to the quality of their associations with their neighbours. Therefore we define a similarity function $sim(g_i, g_j)$ as follows: $sim(g_i, g_j) = \max(\varphi(g_i, g_j), \varphi(g_j, g_i))$. To visualise our genes, we express the distance between two genes g_i and g_j as follows:

$$distance(g_i, g_j) = ms - sim(g_i, g_j)$$

where ms is the maximal value of the elements of the matrix sim. Multidimensional scaling (MDS [18]) provides a visual representation of the proximities among a set of objects. This method permits to associate to a gene g a point in a plane.

Figure 3a shows the mapping of the genes. It reveals a good clustering of the genes according to their class label. The clusters are well separated and correspond to the predefined classes, without any counterexamples. To compare this result with standard techniques, we replaced our distance function by the absolute Pearson distance defined in eq. 8. We applied it to the same set of genes and performed the MDS algorithm on the new distance matrix. Figure 3b indicates that the Pearson correlation is not capable of separating the genes into distinct clusters. This difference demonstrates that the implicative method can better discriminate the gene similarity observed on a particular subset of observations, contrary to correlation methods that consider the observation set as a whole.
The application of the implicative method has proven to be efficient in the particular context of the unsupervised analysis. This problem refers to class discovery and concerns experiments for which the cancer subtypes are unknown. In this case, the distance function that we propose can be directly used to identify clusters. Therefore, it suffices to apply an efficient clustering algorithm such as PAM (Partionning Around Medioids [17]) that extracts clusters from a dissimilarity matrix.

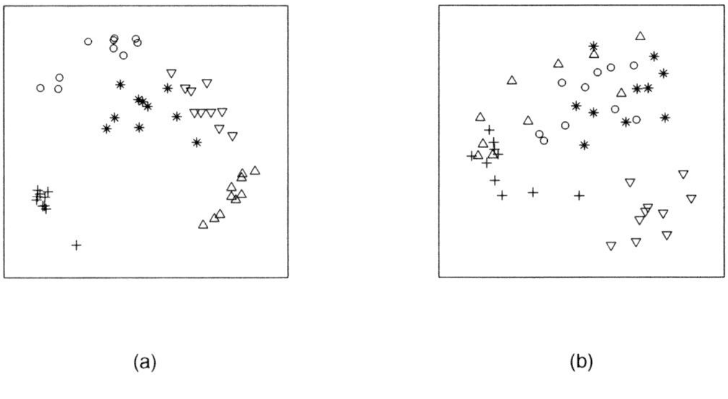

Fig. 3. Representation of the gene similarity. (a) represents the MDS based on implication analysis: the different tumour types associated to our gene selection are well separated, contrary to (b), in which the dissimilarity measure is the absolute value of the Pearson correlation. We obtained a similar mapping by using the Euclidian distance instead of the Pearson coefficient (figure not shown).

5.2 Discovery of gene association using the intensity of implication

The previous model gives an original insight of the similarities between genes. An interesting feature of implicative analysis has been neglected so far: its capability to define the orientation of these similarities. The implication is indeed an oriented relation and reveals that an association $A \rightarrow B$ is more or less pertinent than its counterpart $B \rightarrow A$. It is necessary to apprehend the meaning of implications over genes. An asymmetric relationship between two genes traduces the fact that the knowledge of the expression value relative to one gene better determines the order of magnitude (i.e the interval range) of the expression value on the other than the contrary. This property can somehow be related to gene regulation: the activity of one gene causes the change of expression rate of another. One must however be careful not to interpret implication rules as the direct evidence of a gene regulation. The microarray experiments cannot encompass all the biological mechanisms that take place in the cell. Nevertheless, gene association networks may help biologists to appreciate the polarity of gene coexpressions. Figure 4 presents the 20 genes belonging to the set $\Gamma(M)$ (K being set to 10) issued from the Leukemia dataset presented in section 4.2.

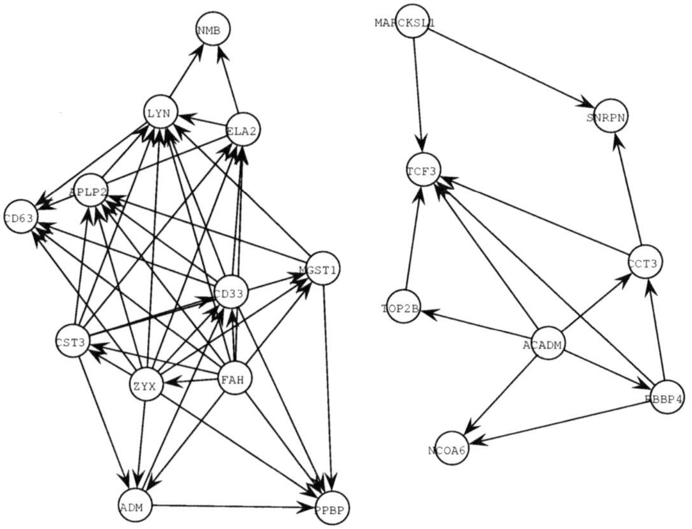

Fig. 4. Gene association network

In a dataset comprising two classes, there exist only two types of gene profiles that discriminate these classes. Following the analysis done by the authors [12], we then consider the two following patterns:

π_1: Genes that present an under-expression for ALL samples and an over-expression for AML samples.

π_2: Genes that present an over-expression for ALL samples and an under-expression for AML samples.

It seems interesting to discriminate these two kinds of situations. Unfortunately, definition 2 authorizes associations over gene profiles that are anticorrelated, as in the example shown in figure 1. This means that we can extract strong associations having either an under-expression in premise and an over-expression in conclusion, or the opposite. In both cases, these associations concern genes belonging to distinct types of profiles. In order to simplify the interpretation of the association network, we then introduce the following restriction: we only retain associations $r_A([p_i, q_i]) \rightarrow r_B([p_j, q_j])$ that respect the condition: $(p_i = p_j = 1) \vee (q_i = q_j = n)$. Note that this simple constraint does not assume the existence of classes.

One can observe in fig. 4 that two clusters emerge from the association network. The left cluster corresponds to the differential expression π_1 previously described, the right cluster to π_2.
The nodes of the network are the informative genes labelled by their official symbol name. The edges of the network represent associations whose intensity of implication is greater than 0.999 ($\lambda(A, B) \geq 3$). One remarks that the right cluster is less connected than the left one. This feature indicates that the expression profile π_1 is more observed than π_2. Although it is only a conjecture, this phenomenon could explain why, in section 4.1, we achieve better performances in holdout validation with our selected genes, compared to [12]: contrary to the authors, we did not impose the predefined expression patterns π_1 and π_2.

The association network presents several genes that are mentioned in biomedical literature. For example, CD33 has been proven to be useful in distinguishing lymphoid from myeloid lineage cells (see [12] for more details). Genes that are the sources of many arrows (e.g. FAH and ZYX) have a profile that is very close to their respective expression pattern. This means that the measurements are well separated into two expression levels, each level approximatively corresponding to one class of observation.

Table 8 gives some examples of rules that have been extracted and represented in fig. 4. The level column indicates whether the rule is relative to an under-expression (interval $([1, q], 1 \leq q \leq n$, noted $\vdash$) or to an over-expression $([p, n], 1 \leq p \leq n$, noted $\dashv$). The subtype defines the class c of observations that is mostly concerned by the rule. The pattern column indicates the type of profile deduced from the two previous pieces of information. The support is the percentage of observations of $\mathcal{O}$ that respect the rule. The class column gives the percentage of observations of class c that are concerned by the rule. The homogeneity h is the percentage of observations belonging to the majority class c that respect the rule. This measure is the percentage expression of the sensitivity (also called the true positive rate). Finally, the rule quality λ is the intensity of implication expressed in its logarithmic form (eq. 4).

The rule $FAH \rightarrow ZYX$ in line 5 is the most pertinent rule, since it concerns all the individuals of class ALL (100 % of class support) and only them (100% of class homogeneity): it is a perfect indicator of the leukemia subtype. Rule $ACADM \rightarrow RBBP4$ (line 1) presents a high intensity of implication, although it does not concern all the observations of class ALL. Indeed, 7.4% of observations of class ALL do not respect the rule (note however that this rule conversely presents a perfect class homogeneity, i.e. all the observations belong to ALL class). The high value of λ is due to its support that is greater than that of rule 5 (i.e. rule 1 concerns more individuals than rule 5). Rule 2 has the same premise than the previous rule. Its class homogeneity is reduced. Almost one third of the individuals do not belong to the majority class of the rule. This rule reflects the fact that coregulation is not always associated to the leukemia subtype. It is indeed possible to encounter a remarkable statistical association between genes that is not necessarily linked to the studied phenotype. That is why one must consider all the parameters expressed in table 8 before interpreting a rule.

line	rule	level	subtype	pattern	support	class	h	λ
1	$ACADM \rightarrow RBBP4$	$\vdash$	ALL	π_1	65.8	92.6	100	4.2
2	$ACADM \rightarrow TCF3$	$\dashv$	AML	π_1	42.1	100	68.7	3.7
3	$CD33 \rightarrow CD63$	$\vdash$	AML	π_2	34.2	100	84.6	4.0
4	$FAH \rightarrow PPBP$	$\dashv$	ALL	π_2	68.4	96.3	100	3.6
5	$FAH \rightarrow ZYX$	$\vdash$	AML	π_2	28.9	100	100	3.8
6	$MARCKSL1 \rightarrow SNRPN$	$\dashv$	AML	π_1	28.9	90.9	90.9	3.2
7	$MARCKSL1 \rightarrow TCF3$	$\dashv$	AML	π_1	23.7	81.8	100	3.0
8	$TOP2B \rightarrow TCF3$	$\vdash$	ALL	π_1	28.9	40.7	100	3.3
9	$ZYX \rightarrow LYN$	$\dashv$	ALL	π_2	57.9	81.5	100	3.3
10	$ZYX \rightarrow PPBP$	$\vdash$	AML	π_2	26.3	90.9	100	3.4

Table 8. Association rules between genes.

6 Conclusion

Microarray data analysis is generally based on the measure of the similarity between gene expression profiles, such as the absolute Pearson correlation. The drawback of this measure is that it assumes a global relationship between genes, while an implication may only concern a particular group of conditions. The intensity of rank implication is more appropriate for the discovery of partial dependencies. Association rule may therefore help to infer gene regulatory pathways. Our method is very robust to noise and, unlike correlation techniques, it provides the direction of the relationship.

We also propose a gene selection method based on implicative analysis. Our model identifies informative genes that have proven very predictive, compared to selection sets existing in the literature. Such a model is potentially useful for medical diagnoses. Indeed, a reliable classification of tumours is essential for successful treatment of cancer. Although simple, our rule-based classifier is an efficient algorithm, which achieves performances comparable to high performing machine learning techniques.
Due to their simplicity and ease of interpretation, association rules are very promising for the analysis of gene expression data. They determine a network of genes that are differentially or coordinately expressed under specific conditions. The implicative analysis can be used to discover such association networks.

References

1. R. Agrawal and R. Srikant. Fast algorithms for mining association rules. In *Proceedings of the 20th Very Large Data Bases Conference*, pages 487–499. Morgan Kanfmann, 1994.
2. U Alon, N Barkai, D A Notterman, K Gish, S Ybarra, D Mack, and A J Levine. Broad patterns of gene expression revealed by clustering analysis of tumor and normal colon tissues probed by oligonucleotide arrays. *Proc Natl Acad Sci U S A*, 96(12):6745–6750, Jun 1999.
3. C. Becquet, S. Blachon, B. Jeudy, J. F. Boulicaut, and O. Gandrillon. Strong-association-rule mining for large-scale gene-expression data analysis: a case study on human sage data. *Genome Biol*, 3(12), 2002.
4. L. Breiman. Random forests. *Machine Learning*, 45:5–32, 2001.
5. Pedro Carmona-Saez, Monica Chagoyen, Andrés Rodríguez, Oswaldo Trelles, José María Carazo, and Alberto D. Pascual-Montano. Integrated analysis of gene expression by association rules discovery. *BMC Bioinformatics*, 7:54, 2006.
6. D. Chen, Z. Liu, X. Ma, and D. Hua. Selecting genes by test statistics. *Journal of Biomedicine and Biotechnology*, 2:132–138, 2005.
7. G. Cong, A. Tung, X. Xu, F. Pan, and J. Yang. Farmer: Finding interesting rule groups in microarray datasets, 2004.
8. C. Creighton and S. Hanash. Mining gene expression databases for association rules. *Bioinformatics*, 19(1):79–86, January 2003.
9. M. Dettling and P. Buhlmann. Supervised clustering of genes. *Genome. Biol. Res.*, 3(12):research0069.1–0069.15, 2002.
10. R. Diaz-Uriarte and S. Alvarez de Andres. Gene selection and classification of microarray data using random forest. *BMC Bioinformatics*, 7, 2006.
11. A. Gasch and M. Eisen. Exploring the conditional coregulation of yeast gene expression through fuzzy k-means clustering, 2002.
12. T. R. Golub, D. K. Slonim, P. Tamayo, C. Huard, M. Gaasenbeek, J. P. Mesirov, H. Coller, M. L. Loh, J. R. Downing, M. A. Caligiuri, C. D. Bloomfield, and E. S. Lander. Molecular classification of cancer: class discovery and class prediction by gene expression monitoring. *Science*, 286:531–537, 1999.
13. R. Gras, R. Couturier, F. Guillet, and F. Spagnolo. Extraction de règles en incertain par la méthode statistique implicative. In *12èmes Rencontres de la Société Francophone de Classification*, pages 148–151, Montreal, 2005.

14. R. Gras, E. Diday, P. Kuntz, and R. Couturier. Variables sur intervalles et variables-intervalles en analyse statistique implicative. In *Société Francophone de Classification (SFC'01)*, pages 166–173, Univ. Antilles-Guyane, Pointe-à-Pître, 2001.
15. R. Gras. *L'implication Statistique.* La Pensée Sauvage, Grenoble, 1996.
16. Xiaoming Jin, Xinqiang Zuo, Kwok-Yan Lam, Jianmin Wang, and Jia-Guang Sun. Efficient discovery of emerging frequent patterns in arbitrary windows on data streams. In *ICDE*, page 113, 2006.
17. L. Kaufman and P.J. Rousseeuw. *Finding Groups in Data: an Introduction to Cluster Analysis.* Wiley-Interscience, New York, 1990.
18. J. B. Kruskal and M. Wish. *Multidimensional Scaling.* Sage Piblications, Beverly Hills, CA, 1978.
19. Yoonkyung Lee and Cheol-Koo Lee. Classification of multiple cancer types by multicategory support vector machines using gene expression data. *Bioinformatics*, 19(9):1132–1139, 2003.
20. Gary Livingston, Xiao Li, Guangyi Li, Liwu Hao, and Jiangping Zhou. Analyzing gene expression data using classification rules. In *CSB'2003: Proceedings of the 2003 Computational Systems Bioinformatics Conference*, pages 1–8. ACM Press, August 2003.
21. Gregory Piatetsky-Shapiro and Pablo Tamayo. Microarray data mining: facing the challenges. *SIGKDD Explorations*, 5(2):1–5, 2003.
22. Scott L Pomeroy, Pablo Tamayo, Michelle Gaasenbeek, Lisa M Sturla, Michael Angelo, Margaret E McLaughlin, John Y H Kim, Liliana C Goumnerova, Peter M Black, Ching Lau, Jeffrey C Allen, David Zagzag, James M Olson, Tom Curran, Cynthia Wetmore, Jaclyn A Biegel, Tomaso Poggio, Shayan Mukherjee, Ryan Rifkin, Andrea Califano, Gustavo Stolovitzky, David N Louis, Jill P Mesirov, Eric S Lander, and Todd R Golub. Prediction of central nervous system embryonal tumour outcome based on gene expression. *Nature*, 415(6870):436–442, Jan 2002.
23. A Schulze and J Downward. Navigating gene expression using microarrays–a technology review. *Nat Cell Biol*, 3(8):190–195, Aug 2001.
24. Alexander Tuzhilin and Gediminas Adomavicius. Handling very large numbers of association rules in the analysis of microarray data. In *KDD*, pages 396–404, 2002.
25. Vladimir N. Vapnik. *The nature of statistical learning theory.* Springer-Verlag New York, Inc., New York, NY, USA, 1995.
26. Junbai Wang, Trond Hellem Bø, Inge Jonassen, Ola Myklebost, and Eivind Hovig. Tumor classification and marker gene prediction by feature selection and fuzzy c-means clustering using microarray data. *BMC Bioinformatics*, 4:60, 2003.
27. C H Yeang, S Ramaswamy, P Tamayo, S Mukherjee, R M Rifkin, M Angelo, M Reich, E Lander, J Mesirov, and T Golub. Molecular classification of multiple tumor types. *Bioinformatics*, 17 Suppl 1:316–322, 2001.

On the use of Implication Intensity for matching ontologies and textual taxonomies

Jérôme David, Fabrice Guillet, Henri Briand, and Régis Gras

Laboratoire d'Informatique de Nantes Atlantique
Equipe COnnaissances & Décision
Site Ecole Polytechnique de l'Université de Nantes
La Chantrerie — BP 50609 — 44306 Nantes cedex 3
{jerome.david,fabrice.guillet,henri.briand}@univ-nantes.fr,
regisgra@club-internet.fr

Summary. At the intersection of data mining and knowledge management, we shall hereafter present an extensional and asymmetric matching approach designed to find semantic relations (equivalence and subsumption) between two textual taxonomies or ontologies. This approach relies on the idea that an entity A will be more specific than or equivalent to an entity B if the vocabulary (i.e. terms and data) used to describe A and its instances tends to be included in that of B and its instances. In order to evaluate such implicative tendencies, this approach makes use of association rule model and Interestingness Measures (IMs) developed in this context. More precisely, we focus on experimental evaluations of IMs for matching ontologies. A set of IMs has been selected according to criteria related to measure properties and semantics. We have performed two experiments on a benchmark composed of two textual taxonomies and a set of reference matching relations between the concepts of the two structures. The first test concerns a comparison of matching accuracy with each of the selected measures. In the second experiment, we compare how each IM evaluates reference relations by studying their value distributions. Results show that the implication intensity delivers the best results.

Key words: Ontology alignment, ontology matching, association rule, interestingness measure

1 Introduction

In our information society, large databases and data warehouses have become widespread. This huge amount of information has led to the increasing demand for mining techniques for discovering knowledge nuggets. To meet this demand, the Knowledge Discovery in Databases (KDD) [16] community proposed the association rule model [1].

Initially motivated by the analysis of market basket data, the task of association rule mining aims at finding relations between items in datasets [9].

J. David et al.: *On the use of Implication Intensity for matching ontologies and textual taxonomies*, Studies in Computational Intelligence (SCI) **127**, 227–245 (2008)
www.springerlink.com

Association rules are propositions of the form "If antecedent then consequent", noted *antecedent* → *consequent*, representing implicative tendencies between conjunctions of valued attributes or items. Association rules have the advantage of being an easy and meaningful model for representing explicit knowledge. Furthermore, this unsupervised learning technique does not need particular information about knowledge to be discovered contrary to classical supervised techniques (such as decision trees). These advantages have motivated a great deal of research and the publication of association rule extraction algorithms such as Apriori [1, 2]. Nevertheless, if only minimal support and confidence values are used, these algorithms typically produce many rules and it is hard to only select those which may interest the user. One way to face this problem is to use Interestingness Measures (IMs). IMs aim at assessing the implicative quality of association rules but also some useful characteristics such as novelty, significance, unexpectedness, nontriviality, and actionability [9,19]. IMs allow to rank and reduce the amount of rules, and consequently to help the user to choose the best ones according to his/her preferences.

The information society has also led to the development of the Web and then a great increase in the available data and information. In this vast Web, resources, often in textual form, tend to be organised into hierarchical structures. This hierarchical structuring of web contents ranges from large web directories (e.g. Yahoo.com, OpenDirectory) to online shop catalogs (e.g. Amazon.com, Alapage.com). Furthermore, with the arrival of the Semantic Web, such a hierarchical organisation is also used through OWL ontologies which aim at providing formal semantics of the web contents. Even if the use of hierarchies helps to structure web information and knowledge, the Web remains heterogeneous. Data exchanges and communications between software programs or software agents using hierarchical-organised data is consequently difficult. In order to address such interoperability problems, one must be able to compare such data structures and find matches between them. Thus, many matching methods have been proposed in the literature [27, 34, 35]. These methods aim at finding semantic relations (i.e. equivalence, subsumption, etc) between entities (i.e. directories, categories, concepts, properties) defined in different hierarchical structures (filesystems, schemas, ontologies). Even if the proposed approaches are issued from different communities, they mostly use similarity measures and as a consequence, a majority of them are restricted to finding equivalence relations only.

At the intersection of these two research fields, we proposed to use the association rule paradigm for matching ontological structures [11]. Our original approach, named AROMA (Association Rule Ontology Matching Approach), heavily relies on the asymmetric nature of association rules, which allows it to match not only equivalence relations but also subsumption relations between entities. The consideration of subsumption between entities helps to characterise more precisely the matching relations between hierarchical structures regarding only similarity based approaches. Also, it allows enhancements the output matches. Furthermore, unlike most approaches designed for matching schema or ontologies, AROMA relies heavily on extensional data provided with structures. This type of matcher, named an instance-level matcher [34], is especially designed to work on structures with limited schema information (i.e. only concept or element names and a partial order relation between them). For example, AROMA can deal with textual hierarchies, such as Web directories or semi-structured data.

The main objective of this chapter is to show and explain the relevance of the Implication Intensity measure relative to other IMs in the context of the problem of matching text hierarchies.

This chapter is organised as follows: in a first section, we present related work concerning ontology/schema matching and propose a classification. Then, we focus on IMs proposed in the literature. First, we classify measures according to three criteria. Then, we use this classification to justify the selection of the best IMs according to our matching context. In the second section, we detail the two stages of AROMA methodology and, we describe a criterion for reducing rule redundancy and enhancing the accuracy of matching results. The last section reports the results of two experiments made on a well-known benchmark provided with two catalogs and a set of reference matching relations. The first experiment evaluates the selected IMs according to a classical information retrieval accuracy measure: the F-measure. In the second experiment, we compare the value distributions of IMs obtained on relevant and non-relevant relation sets.

2 Related work

2.1 Textual taxonomy matching

For the last six years, ontology and schema matching have been widely studied and many approaches have been proposed in the literature. These methods come from different communities such as artificial intelligence [7,18,20], databases [12,29,32], graph matching [23,30], information retrieval [28], machine learning [13,31,36], natural language processing and statistics [24]. Although they are heterogeneous and consequently difficult to compare, preliminary afforts have been result in surveys of matching techniques [27,34,35].

One survey [34] focuses on database schema matching techniques and proposes a classification which discriminates the *extensional or element-based approaches* from the *intensional or only-schema-based approaches.* While many efforts are concentrated around the intensional matchers, few extensional approaches have been proposed in the literature.

A survey of intensional matchers can be found in [35]. The authors propose two classifications of this type of matchers. The first one permits us to distinguish methods according to their granularity and their interpretation of the input information (it distinguishes element-level and structure-level techniques and then the syntactic, external and semantic methods). The second classification distinguishes three classes:

1. the *terminological approaches* (T) based on string similarity measures (TS) or based on measures which use terminological resources (TL) like WordNet;
2. the *structural approaches* (S) which compare two concepts from their internal structure (SI) (shared attributes or properties) or from their external structure (SE), that is to say their respective position within their taxonomy;
3. the *semantic approaches* (SEM), which use formal semantic model.

To sum up, most approaches use terminological matchers with string-similarities (Anchor-PROMPT [32], Coma [12], CMS [26], Cupid [29], S-MATCH [20]) or/and external oracles such as Wordnet ([20], H-match [8]). They can also use structural matchers (Similarity Flooding [30], [8], GMO [23], Omap [12,18,29,32,36]). Some methods such as OLA [15], combine the various terminological and structural criteria. Only a few approaches use formal semantic models [18, 20].

Extensional matchers

In this section, we focus on extensional matchers since the methods designed for matching textual taxonomies (i.e. web directories, catalogs, etc.) heavily rely on textual content. Beacause, such data structures typically have poor schema information, extensional matchers are relevant. We briefly present four approaches which are explicitly designed to work on such data structures. Then, we propose a synthetic classification of these approaches based on their differences.

GLUE [13]. This conceptual hierarchy matching tool uses machine learning techniques. It combines two strategies: the estimation of joint probabilities between concepts and the relation labeller. In the first strategy, for the joint probability estimation, the two hierarchies must share the same set of textual documents. As it is often not the case, the authors propose to classify the documents associated to the concepts of the first hierarchy to the concepts of the second one. This classification process uses several machine learning classifiers: a naive Bayesian classifier on the textual content of the documents, and a naive Bayesian classifier on the names of concepts concatenated with the names of ascendant concepts (concept path). These two classifiers are firstly trained on the documents associated with each hierarchy. The different classifier predictions are then combined by using the meta-learner. The joint probability is evaluated using Jaccard similarity.

oPLMap [31]. This tool is based on a logical and probabilistic model: DATALOG. It aims at finding the matching set which maximises the alignment probability. This method considers rules of the form $S_i \rightarrow T_j$ and evaluates the confidence of the rules by combining several classifiers. It relies on terminological classifiers for concept names (concept name identity, Jaccard measure on words composing concept names, Jaccard measure by considering the concept path name) and machine learning classifiers for textual documents (kNN classifier and Naive Bayes text classifier on the documents). The oPLMap approach also uses some constraints for taking into account the structure of the taxonomies. In the output, this method provides a set of n-to-n mapping elements valued by a probability measure.

Hical or SBI [25]. This method uses a statistical test, named κ-statistic [10], on shared documents for determining matching. κ-statistic tests the null hypothesis: "$\kappa = 0$". A relation between concepts holds if the null hypothesis can be dissmissed with a significance level of 5%. Hical proposes a top-down approach in order to reduce the computing time.

CAIMAN matching service [28]. This matching method is enclosed in the CAIMAN system for facilitating the exchange of relevant documents between geographically dispersed people within their communities of interest. In the CAIMAN matching service, each document is represented by a document vector composed of words and word frequencies weighted by the TF/IDF measure. Then, the characteristic vectors of concepts are computed from the document vectors by using the Rocchio classifier. Finally, similarities between concepts are deduced by evaluating the cosine value between characteristic vectors of concepts. The output matching is a one-to-one matching: for each source concept, the method retains only the target concept for which the measure is maximised.

Classification of textual taxonomy matchers

	Kind of relations	**Measure**	**Comparison level**	**Pre-processing step**
GLUE	equivalence	Jaccard	document level	classification of documents
oPLMap	equivalence	confidence (conditional probability)	document level	classification of documents
Hichal	equivalence	k-statistic	document level	none
CAIMAN	equivalence	cosines	term level	characteristic vectors computation
AROMA	subsumption, equivalence	Implication Intensity	term level	selection of sets of relevant terms

Table 1. Comparison of hierarchy matchers

Table 1 compares of four matchers and AROMA method discussed in Section 3. We analyse them according to four criteria:

- The *kind of relation* that the method considers. It can be equivalence or subsumption relations.
- The *measure* used by the method for evaluating a potential matching between concepts. The measures used in the five approaches are Jaccard similarity, the conditional probability (called confidence in the association rule community), κ-statistic, cosine and the Implication Intensity.
- The *comparison level* shows how concepts are represented and then explains on which basis the similarity or measure is calculated. We denote two modalities: (1) the document level for methods comparing shared documents between entities; and (2) the term level for methods which represent entities by a set of terms.
- The *kind of pre-processing step* refers to the kind of preprocessing applied to the extension in order to be able to calculate similarities. This characteristic is partially determinated by the type of comparison level. In the case of document level approaches, documents can be classified from each structure into the other one so that they share the same set of documents. Term level approaches use linguistic processing in order to extract terms and then select and/or weight these terms according to their relevance to the studied entities.

We can see that only AROMA considers the subsumption relation. The others are restricted to the equivalence. Concerning the measures used, three methods use similarities (which are symmetric) while only oPLMap and AROMA use asymmetric measures. Nevertheless, oPLMap does not seem to use its asymmetric measure to find subsumption relations between concepts. We can also note that Hichal and AROMA use measures based on a statistical model. For the comparison level, three methods use the documents for representing the extension of concepts from which they calculate the measure values. Only CAIMAN and AROMA work at the terminological level by representing concepts by sets of terms (characteristic vectors or relevant term sets). Finally, in order to work on a common document base, the GLUE and oPLMap methods use a combination of classifiers for the documents.

2.2 Interestingness Measures

In the framework of association rule discovery, and in order to select the most interesting rules, many Interestingness Measures (IMs) have been proposed and studied (see [3, 19, 22, 37] for a survey). In this context, some researchers are interested in principles and properties defining a good IM [3,17,22,33,37], while others work on the comparison of IMs from a data-analysis point of view.

According to our objective of hierarchy matching, we selected some IMs that may be relevant for our work. In the context of AROMA, unlike association rule discovery, an IM is not used for ranking rules in a post-processing step, but during the rule extraction process. In order to be able to choose a threshold value more easily, we retained only IMs respecting the principle of minimal and maximal value [22]. IMs respecting this behaviour are more intelligible to a user.

In this section, we firstly introduce notations used for representing association rules and their characteristics. Then, we classify selected IMs according to three main criteria proposed in [4,6]. The resulting taxonomy of IMs shows their main properties and permits understanding of their behaviours and semantics.

Definition of association rule and notations

In this section we use the following notation: A finite set T of n individuals is described by a set I of p items. Each transaction t can be considered as an itemset, so that $t \subseteq I$. $A = \{t \in T; a \subseteq t\}$ is the extension of itemset a and $\overline{B} = T - \{t' \in T; b \subseteq t'\}$ is the extension of $\bar{b}$. An association rule [1] is an implication of the form $a \rightarrow b$, where a and b are disjoint itemsets. In practice, it is quite common to observe some transactions which contain a and not b without in spite of a general trend to have b when a is present. Then, we introduce the quantities $n_a = card(A)$, $n_{\bar{b}} = card(\overline{B})$ and $n_{a \wedge \bar{b}} = card(A \cap \overline{B})$.

Taxonomy of IMs

The taxonomy classifies IMs according to three criteria. The first one concerns *subject* of IMs (deviation from independence or equilibrium) and the second one, the *nature* of IMs (descriptive or statistical). The last one, the *scope* of the IMs (quasi-implication, quasi-conjunction, quasi-equivalence), explains the semantics of the measure according to logical operators.

- *Subject.* A family of IMs evaluate the *deviation from the independence* situation where the number of counter-examples is equal to those expected in a random case ($n_{a\bar{b}} = n_a . n_{\bar{b}}/n$). These measures have a fixed value at the independence. The other family of IMs evaluate the *deviation from the equilibrium.* The equilibrium situation is reached when the number of counter-examples and examples are equal in number ($n_{a\bar{b}} = n_{ab}$).
- *Nature.* The nature of an IM can be *descriptive* or *statistical.* The descriptive measures are not influenced by a proportional expansion of the cardinalities taken into account. A descriptive IM m satisfies $m(n_a, n_b, n_{a\bar{b}}, n) = m(\alpha.n_a, \alpha.n_b, \alpha.n_{a\bar{b}}, \alpha.n)$ with $\alpha. > 0$. Conversely, the statistical measures vary with the expansion of cardinalities. According to the authors, this type of IM allows the validity of rules to be statistically assessed. Some of these measures are also particularly effective at detecting rules that have novel consequents, i.e. consequents that have not been seen in previous rules. For example, Implication Intensity decreases with an increase

of n_b, and thus gives preference to statistically valid rules having novel consequents.

- *Scope*. Finally, this last way of distinguishing IMs relies on the idea that IMs may evaluate a proximity between the rule and a logical configuration such as an implication, a conjunction, or an equivalence [4]. To qualify the scope of an IM, we will use the terms *quasi-implication*, *quasi-conjunction* and *quasi-equivalence* indexes because rules are not strict logical propositions since they may have counter-examples. Furthermore, some IMs only evaluate the tendency to verify the consequent when the antecedent is true (i.e. they only consider the examples of a rule). Such measures (e.g. the Confidence or IPEE [5] measures) are not classified in the Table 2.

For each modality of the scope characteristic of an IM, Table 2 shows their symbol, counter-examples, equivalent linkage and property that the IM must respect. By comparing this table and the semantic relations defined by [20], we can see that based on their scope, measures could be more or less adapted for mining certain types of semantic relations. Subsumption relations must be evaluated by quasi-implication measures, overlapping relations by quasi-conjunction measures and equivalence relations by quasi-equivalence measures. In schema or ontology matching, methods often rely on quasi-conjunction measures for evaluating equivalence relations of the approaches considered in Section 2.1, only Hichal uses an index (κ or kappa) that is a quasi-equivalence measure. GLUE uses a quasi-conjunction index (Jaccard) and oPLMap uses a rule index (confidence or conditional probability).

	quasi-implication	quasi-conjunction	quasi-equivalence
Symbol	$\Rightarrow$	$\leftrightarrow$	$\Leftrightarrow$
Counter-examples	$a \wedge \overline{b}$	$a \wedge \overline{b}$ $\overline{a} \wedge b$ $\overline{a} \wedge \overline{b}$	$a \wedge \overline{b}$ $\overline{a} \wedge b$
equivalence	$a \supset b \equiv \overline{b} \supset \overline{a}$	$a \leftrightarrow b \equiv b \leftrightarrow a$	$a \leftrightarrow b \equiv \overline{a} \leftrightarrow \overline{b}$
IM's property	$I(a \rightarrow b) = I(\overline{b} \rightarrow \overline{a})$	$I(a \rightarrow b) = I(b \rightarrow a)$	$I(a \rightarrow b) = I(\overline{a} \rightarrow \overline{b})$

Table 2. Scope of IMs

Table 3 shows, for each selected IM, its scope (rule ($\rightarrow$), quasi-implication ($\Rightarrow$), quasi-conjunction ($\leftrightarrow$), quasi-equivalence ($\Leftrightarrow$)), its nature (statistical (S) or descriptive (D)), its subject (deviation from independence (I) or equilibrium (E)), its fixed value in the independence or equilibrium situation (depending on its subject) and its formula.

Measure	Scope	Nature	Subject	Fixed value	Formula
II	$\Rightarrow$	S	I	0.5	$P(n_{a\bar{b}} < Poisson(\frac{n_a.n_{\bar{b}}}{n}))$
Loevinger	$\Rightarrow$	D	I	0	$1 - \frac{n.n_{a\bar{b}}}{n_a.n_{\bar{b}}}$
IPEE	$\rightarrow$	S	E	0.5	$P(n_{a\bar{b}} < Binomial(n_a, 1/2))$
Confidence	$\rightarrow$	D	E	0.5	n_{ab}/n_b
Likelihood Linkage Analysis	$\leftrightarrow$	S	I	0.5	$P(n_{ab} \geq Poisson(\frac{n_a.n_b}{n}))$

Table 3. Selected IMs and their properties

3 AROMA methodology

AROMA (Figure 1) was designed to find matching between conceptual hierarchies populated from textual documents. This method permits the discovery of a set of significant association rules holding between concepts obtained from two hierarchical structures and evaluated by the Implication Intensity measure. AROMA takes, as input, two conceptual hierarchies $\mathcal{H}_1$ and $\mathcal{H}_2$, each defined as a tuple $\mathcal{H} = (C, \leq, D, \sigma)$, where C is the set of concepts, $\leq$ is the partial order organising concepts into a taxonomy, D is the set of textual documents, and σ is the relation associating a set of documents to each concept (i.e. for a concept $c \in C$, $\sigma(c)$ represents the documents associated to c). Thanks to the first part of the method concerning the acquisition and the selection of relevant terms for each concept, we are able to redefine each hierarchy as a tuple $\mathcal{H}' = (C, \leq, T, \gamma)$ where T is the set of relevant terms selected. In order to consider the partial order, we assume that a term associated with a concept is also associated with its parent concepts, and thus, we extend γ to the relation γ' as follows: $\gamma'(c) = \bigcup_{c' \leq c} \gamma(c')$.

3.1 Association rules discovery between hierarchies

The second stage of AROMA consists of the discovery of implicative matching relations between concepts by evaluating association rule between their respective sets of relevant terms. The algorithm takes in two pre-processed hierarchies $\mathcal{H}'_1$ and $\mathcal{H}'_2$ and considers only the terms shared by the two structures. The set of common terms for the two hierarchies $\mathcal{H}'_1$ and $\mathcal{H}'_2$ is noted $T_{1\cap 2} = T_1 \cap T_2$. The relation $\gamma'_{1\cap 2}$ associates a subset of $T_{1\cap 2}$ for each concept $c \in C_1 \cup C_2$:

$$\gamma'_{1\cap 2}(c) = \begin{cases} \gamma'_1(c) \cap T_2 \; if \; c \in C_1 \\ \gamma'_2(c) \cap T_1 \; if \; c \in C_2 \end{cases} \quad (1)$$

The extracted rules are of the form $a \rightarrow b$ and are valued by a $\varphi(a \rightarrow b)$ value. A valid rule $a \rightarrow b$ (i.e. rule having a $\varphi(a \rightarrow b)$ value greater than or equals to a chosen threshold) represents a quasi-implication (i.e. an implication allowing some counter-examples) from the set of relevant terms of the concept

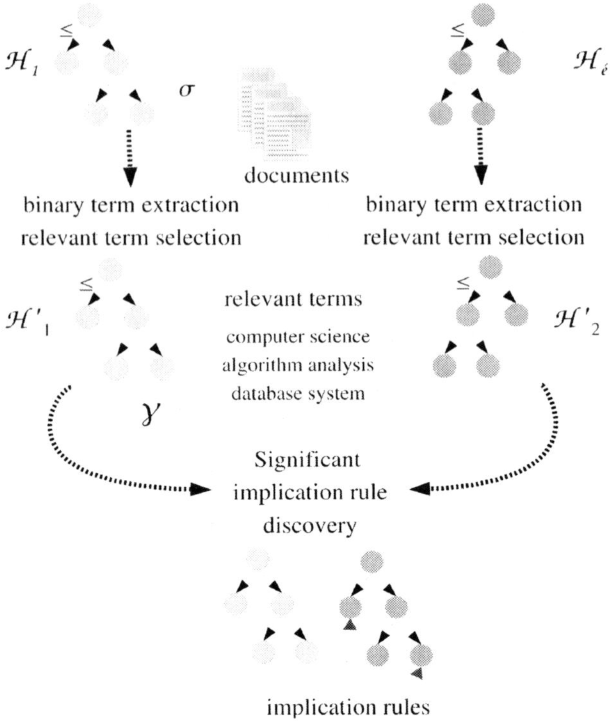

Fig. 1. The AROMA approach

a into the set of relevant terms of the concept b. The existence of such a valid rule means that the concept a (issued from $\mathcal{H}_1$) is probably more specific than or equivalent to the concept b (issued from $\mathcal{H}_2$).

3.2 Selection of significant rules

The algorithm provides a top-down search of association rules. It uses two criteria for selecting significant rules and reduce redundancy. Then, a rule $a \rightarrow b$ (between the concepts $a \in C_1$ and $b \in C_2$) will be significant if it respects the two following criteria:

$$\varphi(a \rightarrow b) \geq \varphi_r \tag{2}$$

$$\forall x \geq a,\ \forall y \leq b,\ \varphi(x \rightarrow y) \leq \varphi(a \rightarrow b) \tag{3}$$

The first criterion (Equation 2) guarantees the **quality** of the implication tendency between the two concepts for a given threshold φ_r. The Implication Intensity of the rule $a \rightarrow b$ is explained in Figure 2 and defined as follows:

$$\varphi(a \rightarrow b) = 1 - \Pr\left(N_{a \wedge \bar{b}} \leq n_{a \wedge \bar{b}}\right) \tag{4}$$

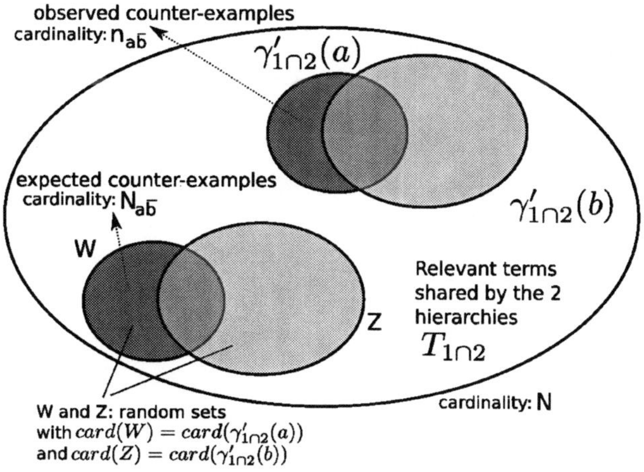

Fig. 2. Implication Intensity of a rule $a \rightarrow b$

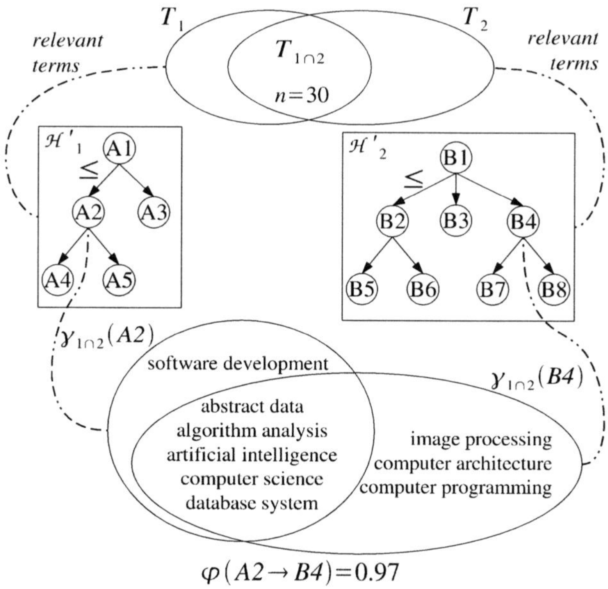

Fig. 3. Evaluation of rules

where $n_{a \wedge \bar{b}} = card(\gamma_{1 \cap 2}(a) - \gamma_{1 \cap 2}(b))$ is the number of relevant terms for concept a that are not relevant for concept b. $N_{a \wedge \bar{b}}$ is the random number of relevant terms for concept a that are not relevant for concept b.

For example (Figure 3), the rule $A2 \rightarrow B4$ has $n_{A2 \wedge \overline{B4}} = 1$ counter-examples. Its Implication Intensity value is calculated using a Poisson law (which is a possible model for the Implication Intensity [21]):

$$\varphi(A2 \rightarrow B4) = \sum_{k=0}^{n_{A2 \wedge \overline{B4}}} e^{-\lambda} \times \frac{\lambda^k}{k!} = 0,97$$

where $\lambda = n_{A2}.n_{\overline{B4}}/n = 6.(30 - 8)/30$.

The second criterion (Equation 3) verifies the **generativity** of the rule and thus allows the redundancy to be reduced in the extracted rules set. Indeeds, a valid rule (i.e. a rule satisfying the first criterion) is significant if there does not exist a more generative rule having an Implication Intensity value greater than or equals to it. A rule $x \rightarrow y$ is more generative than a rule $u \rightarrow v$ if $u \leq x$ and $y \leq v$ (with $x \rightarrow y \neq u \rightarrow v$). For example (Figure 3), the rules $A2 \rightarrow B7$, $A2 \rightarrow B8$, $A1 \rightarrow B4$, $A1 \rightarrow B7$, and $A1 \rightarrow B8$ are more generative than the studied rule $A2 \rightarrow B4$. The rule $A2 \rightarrow B4$ will be significant and thus selected if none of its generative rules have a φ value greater than or equals to its φ value.

4 Experimental results

The experiments presented in this section concern only the second part of the AROMA (i.e. the rule selection phase). After describing the data used for the experiments, we first compare the performance of the measures in terms of the F-Measure, which aggregates precision and recall. Next, we describe an analysis of the distribution of the measure values on two sets of matching relations: a set of hand-made reference matching relations, noted $R+$ and a set of irrelevant relations $R-$.

4.1 Analysed data

The experiments used the "Course catalog" benchmark [14]. This benchmark is composed of two catalogs of courses descriptions which are offered at the Cornell and Washington universities. The courses descriptions are hierarchically organised into schools and colleges and then into departments and centers within each college. These two hierarchies contain respectively 166 and 176 concepts to which are associated 4360 and 6957 textual course descriptions. The benchmark data also include a set of 54 manually matched relations from concepts of the Cornell catalog to the Washington catalog. Only equivalence relations are included in the manually matched set.

4.2 Evaluation of IMs

Here, we describe the evaluation performed by the AROMA algorithm with each selected IM. For each measure, we varied the rule selection threshold φ_r from 0 to 1 with a step of 1 percent. For each threshold value, the rule selection algorithms was executed twice, once to select implications from the Cornell concepts to the Washington concepts and the second time to select implications from the Washington concepts to the Cornell concepts. From the two implicative matching sets, we retained only equivalence relations by following this rule: if $A \rightarrow B$ and $B \rightarrow A$, then $A \leftrightarrow B$.

In order to evaluate the relevance of results according to the reference set $R+$, we use two standard metrics from information retrieval: the precision and the recall. These metrics are defined as follows: let F be the set of matching pairs found using AROMA and $R+$ be the set of "reference" matching pairs. The **precision** ($precision = card(F \cap R+)/card(F)$) measures the ratio of the number of good matching pairs (i.e. matching pairs that are both in our result set and in the reference matching set) over the number of matching pairs found by AROMA. The **recall** ($recall = card(F \cap R+)/card(R+)$) measures the ratio of good matching pairs over the number of reference matching pairs. Finally, these two measures are aggregated into the **F-measure** (Dice's similarity between the sets F and R): $F - measure = 2.precision.recall/precision + recall$.

Figure 4 shows a high correlation between Confidence and Loevinger in terms of efficiency. Their best F-measure scores (around 0.45) are obtained around a threshold of 0.3. This threshold value is greater than those of independence situation for Loevinger but it is less than those of the equilibrium situation for Confidence. The IPEE index has tendency to have the same trend as Confidence and Loevinger. It has lower maximum F-measure values but it is more robust to the increase of the selection threshold value. Likelihood Linkage Analysis (LLA) don't have good F-measure scores. It obtains nearly constant recall values just above 0.5 but it has very bad precision values. Possibly the symmetric nature of this index is not well adapted to our algorithm. Finally, the best F-measure scores are obtained by the Implication Intensity index. Its best value is a little greater than 0.5 for a rule selection threshold fixed at 0.9. We can also notice a stagnation of F-measure before the independence situation (rule selection threshold of 0.5 for II).

These results show that Implication Intensity is the most relevant measure in this context. The two descriptive indexes, Confidence and Loevinger, tend to have similar trends. The selected quasi-conjunction index, LLA, is not relevant for AROMA.

4.3 Distributions of IMs

In this experiment, we studied and compared how the IM's evaluated matching relations independently of the AROMA rule selection algorithms and their

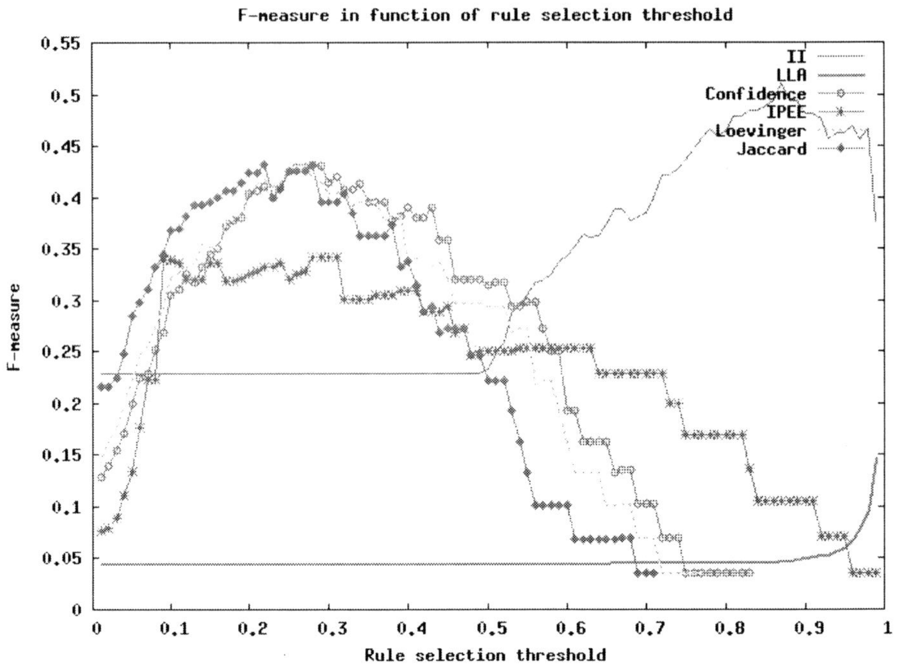

Fig. 4. Evolution of F-measure

selection criteria. Then, we propose to draw the values distributions of IM in two cases. The first case consisted of evaluating the relations from the manually matched set $R+$ and in the second case, we have tried the measures on the set of irrelevant relations $R-$. The set $R-$ was built manually and contains the same number of relations as the reference one $R+$. For these two evaluations, we performed the selection of relevant terms with the Implication Intensity measure for a threshold of 0.9.

In the first test, for each manually matched relation represented by a pair (A, B), we evaluated the rules $A \rightarrow B$ and $B \rightarrow A$. For each pair, we kept only the best value, in the cases where the studied measure was not symmetric.

Regarding Figure 5, the Confidence measure yields a value under 0.5 for the majority of the relations. According to the equilibrium situation, these relations are not relevant because their antecedents is more concomitant with the negation of their consequents than with their consequents. The results obtained with IPEE confirm this situation since many rules have a quality value near 0. Nevertheless, the results found with Loevinger show that matching relations are good regarding the independence case. The majority of rules have a Loevinger value greater than 0 (only 4 relations have smaller values).

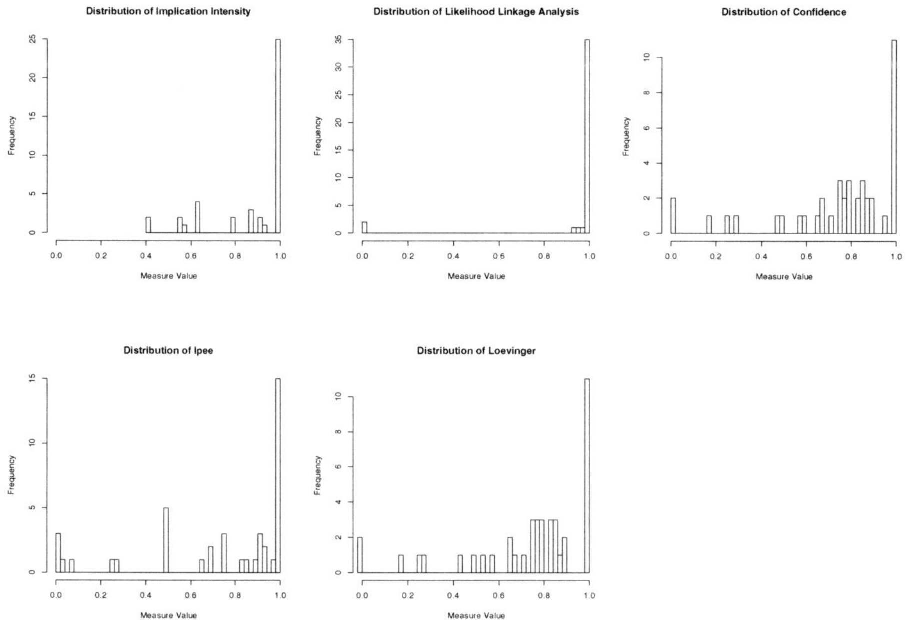

Fig. 5. Measures distributions on manual matching relations $R+$

Thus, we can say that their number of observed counter-examples is negatively correlated to the expected one under the independence hypothesis.

The two probabilistic measures of deviation from independence, the Implication Intensity (II) and the Likelihood Linkage Analysis measures, evaluate the majority of matching relations with good values. The second measure seems to work a little better on this benchmark. This result is not surprising because this second measure is symmetric and designed for evaluating quasi-conjunction relations. But, unlike Implication Intensity, such a measure is not adequate in the case of mining quasi-implication relations between concepts.

Then, after the study of distribution of measures values on good matching relations, we performed a second test on a set of non-relevant relations. In this case, we consider the minimal value obtained for each pair of concepts. In Figure 6, all Confidence and IPEE values are near 0. Loevinger yields less than 0 for 30 rules, that is to say under the independence situation. II confirms this tendency because only 2 rules have a value greater than or equal to 0.5. Nevertheless, LLA evaluates a majority of rules with good values (i.e. greater than 0.5 obtained at the independence situation). Regarding Figures 5 and 6, only Loevinger and II clearly distinguish the two sets of rules. These two sets of distributions show that IMs of deviation from independence work better

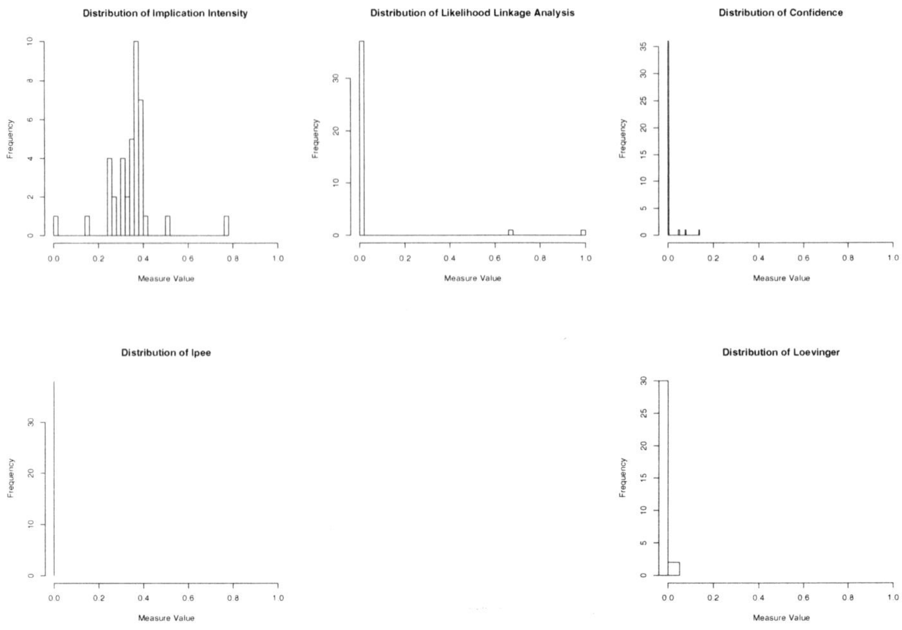

Fig. 6. Measures distributions on irrelevant matching relations $R-$

on this type of rules. We can also notice that the quasi-conjunction measure, LLA, does not distinguish good rules from bad ones.

From these experiments, we conclude that matching relations are better evaluated by IMs of deviation from independence. In such cases, the number of counter-examples needed to reach the equilibrium situation is less than the number needed to reach the independence situation. We also found that the statistic measure of quasi-implication, II, is well suited for distinguish good rules from bad ones.

5 Conclusion

In this paper, we proposed an original use of the association rule model and interestingness measures in the context of schema/ontology matching. More precisely, we described the AROMA approach, which is an extensional matcher for hierarchies indexing text documents. A novel feature of AROMA is that it uses of the asymmetrical aspect of association rules in order to discover subsumption matches between hierarchies or ontologies. Based on studies of IMs, we selected several IMs according to three criteria (subject, nature and scope) and we evaluated them on a matching benchmark. The two experiments show that deviation from independence measures are the best adapted IM family for such an application since the evaluated rules are good regarding the independence situation, but bad in terms of equilibrium deviation. From these results, we can also argue that the two descriptive indexes used, Confidence and Loevinger, tend to have the same behaviour. Due to its deviation from independence subject, its statistical nature and its quasi-implication scope, the Implication Intensity obtains the best scores on this benchmark. In this paper, we analysed measure behaviour on the process of rule extraction between concepts. We did not study the terminological step of AROMA, which consists of extracting and selecting concept relevant terms. Such an evaluation would be interesting since this terminological extraction process significantly influences the accuracy of the results.

References

1. R. Agrawal, T. Imielinski, and A. Swami. Mining association rules between sets of items in large databases. In *Proceedings of the 1993 ACM SIGMOD International Conference on Management of Data*, pages 207–216. ACM Press, 1993.
2. R. Agrawal and R. Srikant. Fast algorithms for mining association rules. In J.B. Bocca, M. Jarke, and C. Zaniolo, editors, *Proceedings of the 20th International Conference Very Large Data Bases (VLDB'94)*, pages 487–499. Morgan Kaufmann, 1994.
3. Jr. Bayardo, J. Roberto, and R. Agrawal. Mining the most interestingness rules. In *Proceedings of the 5th ACM SIGKDD International Conference On Knowledge Discovery and Data Mining (KDD'99)*, pages 145–154, 1999.
4. J. Blanchard. *A visualization system for interactive mining, assessment, and exploration of association rules.* PhD thesis, University of Nantes, 2005.
5. J. Blanchard, F. Guillet, H. Briand, and R. Gras. Assessing rule interestingness with a probabilistic measure of deviation from equilibrium. In *Proceedings of the 11th international symposium on Applied Stochastic Models and Data Analysis (ASMDA-2005)*, pages 191–200. ENST, 2005.
6. J. Blanchard, F. Guillet, R. Gras, and H. Briand. Using information-theoretic measures to assess association rule interestingness. In *Proceedings of the fifth IEEE international conference on data mining ICDM'05*, pages 66–73. IEEE Computer Society, 2005.
7. S. Castano, V. De Antonellis, and S. De Capitani Di Vimercati. Global viewing of heterogeneous data sources. *IEEE Transactions on Knowledge and Data Engineering*, 13(2):277–297, 2001.
8. S. Castano, A. Ferrara, and S. Montanelli. Matching ontologies in open networked systems: Techniques and applications. *Journal on Data Semantics*, 3870(V):25–63, 2006.
9. A. Ceglar and J. F. Roddick. Association mining. *ACM Computing Surveys*, 38(2):5, 2006.
10. J. Cohen. A coefficient of agreement for nominal scales. *Educational and Psychological Measurement*, 20(1):37–46, 1960.

11. J. David, F. Guillet, R. Gras, and H. Briand. Conceptual hierarchies matching: an approach based on discovery of implication rules between concepts. In *Proceedings of the 17th European Conference on Artificial Intelligence (ECAI-2006)*, pages pages 357–361, 2006.
12. H.H. Do and E. Rahm. Coma - a system for flexible combination of schema matching approaches. In *Proceedings of the 28th International Conference on Very Large Data Bases (VLDB '02)*, pages 610–621, 2002.
13. A. Doan, J. Madhavan, P. Domingos, and A. Halevy. Learning to map between ontologies on the semantic web. In *Proceedings of the 11th International WWW Conference (WWW'02)*, pages 662–673. ACM Press, 2002.
14. A. Doan, J. Madhavan, P. Domingos, and A. Halevy. Ontology matching: a machine learning approach. In S. Staab and R. Studer, editors, *Handbook on Ontologies in Information Systems*, pages 397–416. Springer-Velag, 2004.
15. J. Euzenat and P. Valtchev. An integrative proximity measure for ontology alignment. In *Proceedings of the Semantic Integration Workshop, 2nd International Semantic Web Conference (ISWC-03)*, 2003.
16. U. M. Fayyad, G. Piatetsky-Shapiro, P. Smyth, and R. Uthurusamy, editors. *Advances in Knowledge Discovery and Data Mining.* AAAI/MIT Press, 1996.
17. Alex Alves Freitas. On rule interestingness measures. *Knowledge-Based Systems*, 12(5-6):309–315, 1999.
18. F. Fürst and F. Trichet. Axiom-based ontology matching. In *Proceedings of the 3rd international conference on Knowledge capture (K-CAP '05)*, pages 195–196. ACM Press, 2005.
19. Liqiang Geng and Howard J. Hamilton. Interestingness measures for data mining: A survey. *ACM Comput. Surv.*, 38(3):9, 2006.
20. F. Giunchiglia, P. Shvaiko, and M. Yatskevich. S-match: an algorithm and an implementation of semantic matching. In *European Semantic Web Symposium*, LNCS 3053, pages 61–75, 2004.
21. R. Gras et al. *L'implication statistique, une nouvelle méthode exploratoire de données.* La pensée sauvage, 1996.
22. Robert J Hilderman and Howard J Hamilton. *Knowledge Discovery and Measures of Interestingness.* Kluwer Academic Publishers, 2001.
23. W. Hu, N. Jian, Y. Qu, and Y. Wang. Gmo: A graph matching for ontologies. In *Proceedings of the K-CAP 2005Workshop on Integrating Ontologies*, pages 41–48, 2005.
24. R. Ichise, M. Hamasaki, and H. Takeda. Discovering relationships among catalogs. In E. Suzuki and S. Arikawa, editors, *Proceedings of the 7th International Conference on Discovery Science (DS'04)*, volume 3245 of *LNCS*, pages 371–379. Springer, 2004.
25. R. Ichise, H. Takeda, and S. Honiden. Integrating multiple internet directories by instance-based learning. In G. Gottlob and T. Walsh, editors, *Proceedings of the eighteenth International Joint Conference on Artificial Intelligence (IJCAI-03)*, pages 22–30. Morgan Kaufmann, 2003.
26. Y. Kalfoglou and B. Hu. Cms: Crosi mapping system - results of the 2005 ontology alignment contest. In *Proceedings of the K-CAP 2005 Workshop on Integrating Ontologies*, pages 77–85, 2005.
27. Y. Kalfoglou and M. Schorlemmer. Ontology mapping: the state of the art. *Knowledge Engineering Review*, 18(1):1–31, 2003.

28. M. S. Lacher and G. Groh. Facilitating the exchange of explicit knowledge through ontology mappings. In *Proceedings of the 14th International Florida Artificial Intelligence Research Society Conference (FLAIRS'01)*, pages 305–309. AAAI Press, 2001.
29. J. Madhavan, P. A. Bernstein, and E. Rahm. Generic schema matching with cupid. In *Proceedings of the 27th International Conference on Very Large Data Bases (VLDB'01)*, pages 49–58, 2001.
30. S. Melnik, H. Garcia-Molina, and E. Rahm. Similarity flooding: A versatile graph matching algorithm and its application to schema matching. In *Proceedings of the 18th International Conference on Data Engineering(ICDE'02)*, pages 117–128. IEEE Computer Society, 2002.
31. H. Nottelmann and U. Straccia. A probabilistic, logic-based framework for automated web directory alignment. In Zongmin Ma, editor, *Soft Computing in Ontologies and the Semantic Web*, Studies in Fuzziness and Soft Computing, pages 47–77. Springer Verlag, 2006.
32. N. Noy and M. Musen. Anchor-prompt: Using non-local context for semantic matching. In *Proceedings of the Workshop on Ontologies and Information Sharing at the 17th International Joint Conference on Artificial Intelligence (IJCAI'01)*, pages 63–70, 2001.
33. G. Piatetsky-Shapiro. Discovery, analysis and presentation of strong rules. In G. Piatetsky-Shapiro and W. Frawley, editors, *Knowledge Discovery in Databases*, pages 229–248. AAAI Press/MIT Press, 1991.
34. E. Rahm and P. A. Bernstein. A survey of approaches to automatic schema matching. *The VLDB Journal*, 10(4):334–350, 2001.
35. P. Shvaiko and J. Euzenat. A survey of schema-based matching approaches. *Journal on Data Semantics*, 4(LNCS 3730):146–171, 2005.
36. U. Straccia and R. Troncy. omap: Combining classifiers for aligning automatically owl ontologies. In *Proceedings of the 6th International Conference on Web Information Systems Engineering (WISE-05)*, number 3806 in LNCS, pages 133–147. Springer Verlag, 2005.
37. Pang-Ning Tan, Vipin Kumar, and Jaideep Srivastava. Selecting the right objective measure for association analysis. *Information Systems*, 29(4):293–313, 2004.

Modelling by Statistic in Research of Mathematics Education

Elsa Malisani and Aldo Scimone and Filippo Spagnolo

G.R.I.M. (Gruppo di Ricerca sull'Insegnamento delle Matematiche), Department of Mathematics, University of Palermo, via Archirafi 34, 90123 Palermo (Italy)
http://www.math.unipa.it/~grim
e.malisani@alice.it, aldo.scimone@libero.it, spagnolo@math.unipa.it

Summary. The aim of this paper is to study the quantitative tools of the research in didactics. We want to investigate the theoretical-experimental relationships between factorial and implicative analysis. This chapter consists of three parts. The first one deals with the didactic research and some fundamental tools: the *a priori* analysis of a didactic situation, the collection of experimental data and the statistic analysis of data. The purpose of the second and the third section is to introduce the experimental comparison between the factorial and the implicative analysis in two researches in mathematics education.

Key words: research in didactics, theory of didactic situation, statistics, implicative analysis, factorial analysis

Introduction

Modelling, by means of a statistical argumentation, supplies research in the didactics of mathematics with a greater possibility of transferability of the made experience.

It is evident, as it has been widely debated by La Casta-Brousseau [7], Gras [14] and Spagnolo [11,32,34,36], that the statistical argumentation would have not any valence without a theoretical remark from the view-point of didactics and therefore of the epistemology of the mathematical contents. Only a parallel study of all the possible argumentative paths of a research can bring us to results which are considered reliable.

1 The research in didactics, some tools

The research in Didactics places itself as a goal-paradigm with respect to other research paradigms in education science by using both the paradigm of

E. Malisani et al.: *Modelling by Statistic in Research of Mathematics Education*, Studies in Computational Intelligence (SCI) **127**, 247–276 (2008)
www.springerlink.com

the discipline subject of the analysis- and the paradigm of the experimental sciences. The research in Didactics can be considered a sort of "Experimental Epistemology".

The fundamental tool for the research in didactics is the *a priori* analysis of a didactic situation.

What does "a-priori analysis" of a didactic situation mean?

It means the analysis of the "Epistemological Representations", "Historic-epistemological Representations" and "Supposed Behaviours", correct and not, to solve a given didactic situation.

1. *The epistemological representations are the representations of the possible cognitive*[1] *paths regarding a particular concept. Such representations can be prepared by a beginner subject or by a scientific community in a specific historic period.*
2. *Historic-epistemological representations are the representations of the possible cognitive paths regarding the syntactic, semantic and pragmatic*[2] *reconstruction of a specific concept.*
3. *The supposed behaviours of students in facing the situation/problem are all the possible strategies*[3] *of solution of it both correct and not. Among the erroneous strategies, those which can become correct strategies will be taken into consideration.*

The a-priori analysis of the didactic situation allows:

1. the identification of the "*space of the events*"[4] regarding the particular didactic situation with respect to the professional knowledge of the researcher- teacher in a specific historic period;
2. the identification, by means of the possible space of the events, of the "good problem[5]" and therefore of a "fundamental didactic situation" for the set of problems which the didactic situation refers to;

---

[1] The cognitive paths permit the highlighting of the conceptual networks regarding the didactic situation.

[2] The semiotic perspective for the analysis of the disciplinary knowledge allows the management of the contents with reference to the problems of "communication" of the contents. This position is not particularly new with respect to the human sciences, but it represents a real innovation for the technical and scientific disciplines.

[3] In any case, a didactic situation poses a "problem" for the student to solve, either as a traditional problem (i.e. in the scientific or mathematical framework) or as a "strategy" to organise the best knowledge to adapt oneself to a situation.

[4] For "space of the events" we mean the set of the possible strategies of solutions, correct or not, and always supposed within a specific historic period by a specific community of teachers.

[5] The "good problem" is the one which, with respect to a given knowledge, allows the best formulation in ergonomic terms.

3. the identification of the variables of the situation/problem and of the didactic variables[6];
4. the identification of the hypotheses of the research in Didactics of a more general type with respect to those analysable by a first analysis of the situation/problem.

So, the a-priori analysis represents the basic element of an educational research and it takes into account both the epistemology of the discipline and its history. Let us briefly summarise those parts of the research in didactics which can be more significative, by pointing out other possible deeper studies (Table 1) [6, 11].

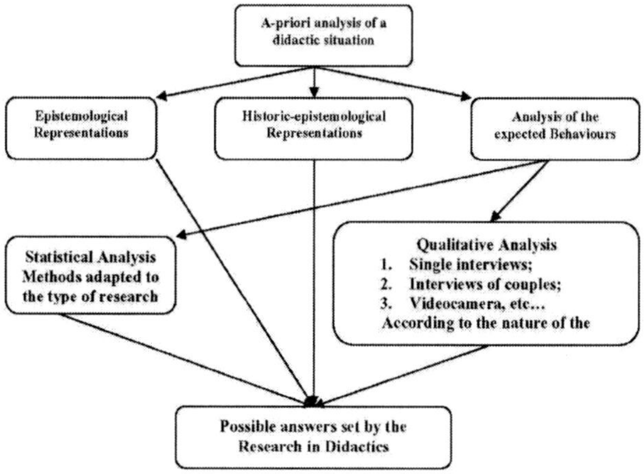

Fig. 1. The diagram summarizes the relationship between research in didactics and collection of experimental data.

1.1 The data

Each didactic research inevitably causes us to collect some data which can be considered formed by a collection of elementary informations. Each piece

6 The "variables of a didactic situation" are all the possible variables which happen into the situation. The "didactic variables" are those which permit a change of the pupils' behaviours. So, the didactic variables are a sub-set of the variables of the didactic situation.

What is the Research in Didactics?	• Research with its Paradigm; • Research with its own language; • Theoretical Research: Epistemological and historic-epistemological analysis of the discipline with respect to a specific Knowledge; • Experimental Research: 1. A-priori analysis of the situation/problem; 2. Identification of the hypotheses of Research; 3. Falsification of the hypotheses[7]; 4. Analysis of the experimental data with respect to small samples by means of appropriate statistical tools. 5. A-posteriori analysis of the experimental data.
Which is the use of the research in Didactics of mathematics?	• Prevision of "didactic phenomena" through "Reliable models" • with respect to Theoretical-Experimental Research. For "Reliable models" we mean those Models which allow the possibility of making forecasts about didactic phenomena; • Communication of the results of the Research to the community of Teachers by means of a strong argumentation such as the a-priori analysis and the statistical tools.
What is Research in Didactics concerned with?	•Problems regarding the "communication of a specific discipline" • through: - Preparation of appropriate a-didactic situations; - Analysis of the errors and obstacles derived from the communicative processes; - Study of didactic and epistemological [8] obstacles such as: • Tools for reflection on the construction of didactic curricula; • Tools for a deeper and better understanding of the communicative processes; • Tools for the preparation of a-didactic situations.

Table 1.

of elementary information reports, in general, a behaviour of a pupil in a given situation. A statistics, therefore, will be a set composed by: a student, a situation and the behaviour of the student.

The student belongs to an observed sample E, assumed to be extracted from a larger population, either by chance or following a system of control situations (for example: scholastic level, gender, previous personal knowledge...).

The situation is chosen in a set S (of questions, exercises...) generated and structured by conditions and parameters of various natures (the knowledge in play, material conditions, didactic conditions...).

The behaviours (typical of knowledge or of aimed knowledge) are taken in a set C of the student's possible answers in the conditions which he is placed into.

A *class* can be defined as a set E of students, a mathematics course as a set S of exercises, the results of the students as a certain application of E onto the set $S \times C$ where C is the set of the behaviours of successes or errors, a note as an application of $S \times C$ onto R.

The knowledge of a certain behaviour can be represented by a certain application of a set of questions onto a set of behaviours.

The use of statistics from teachers and researchers

The teacher must take rapid and many decisions and can correct them very quickly if they prove to be inappropriate. He cannot wait for the result of the statistical treatment of all his questions. The teacher must try to utilise these statistical treatments which allow him to arrive quickly at certain conclusions.

The researcher must follow an opposite process:

1. Which hypotheses correspond to the questions that interest us?
2. What data should be collected?
3. Which treatments should be used?
4. What conclusions?

More than the rapidity and the immediate usefulness, it is the consistency, the stability, the pertinence, and the sureness of the responses which interest a researcher.

The research with appropriate statistical methods will allow:

1. the communication between teachers about the information they need and which they collect on the results of the students; the value of the methods used...;
2. the use, also with discernment, of the results of the research in didactics;
3. the knowledge of the possibilities and the limits of the statistical methods and so the legitimacy of the knowledge used in their profession;
4. the discussion about this legitimacy;
5. the formulation of some open conjectures to be put to the test of the experimental contingency;
6. the imagination of the plausibility of these conjectures;
7. to know how to convert their experience into knowledge;
8. the participation in some research.

The teacher must take rapid and many decisions and can correct them very quickly if they prove to be inappropriate. He cannot wait for the result of the statistical treatment of all his questions. The teacher must try to utilise these statistical treatments which allow him to arrive quickly at certain conclusions.

The researcher must follow an opposite process:

1. Which hypotheses correspond to the questions that interest us?
2. What data should be collected?
3. Which treatments should be used?
4. What conclusions?

5. More than the rapidity and the immediate usefulness, it is the consistency, the stability, the pertinence, and the sureness of the responses which interest a researcher.
6. The research with appropriate statistical methods will allow:
7. the communication between teachers about the information they need and which they collect on the results of the students; the value of the methods used. . . ;
8. the use, also with discernment, of the results of the research in didactics;
9. the knowledge of the possibilities and the limits of the statistical methods and so the legitimacy of the knowledge used in their profession;
10. the discussion about this legitimacy;
11. the formulation of some open conjectures to be put to the test of the experimental contingency;
12. the imagination of the plausibility of these conjectures;
13. to know how to convert their experience into knowledge;
14. the participation in some research.

Observations

An observation consists in attributing a value to a variable with reference to an individual: the observed subject.

The statistics principally permits to treat the case when:

1. many observations are collected.
2. and when these observations totally:
   a) regard different individuals with the same property;
   b) refer to different properties for the same individual;
   c) regard both cases i) and ii).

If "24" (value observed) is attributed to student X (the subject under observation), as "result of the mathematical exam" (variable observed), the set of the values or of the possible cases, in our case, is a whole number between 0 and 30.

The variables can be numeric, interval, ordinal, nominal.

1. *Numeric Variable:* when the values are expressed by numbers (belonging to the sets **N**, **Z**, **Q**, **R**) and the operations are significative for the variable.
2. *Interval Variable:* when only the interval among the values is significative while the sum is not. For example, the points obtained in a sport can constitute an interval variable.
3. *Ordinal Variable:* when the values express only an order among the observations. In an ordinal variable the sum of the two values is not a value.
4. *Nominal Variable:* When the values are characters (letters) or attributes. This variable can have two values: its attribute and its negation. Even if it is expressed by numbers, such as 0 and 1, a nominal variable is not a numeric one: the sum between the two characters is not defined and

in general their order too. The only operations are the logical ones (set theory).

It is always possible to transform a numeric variable into an interval, ordinal or nominal variable (losing some information); an interval variable can be transformed into an ordinal or nominal variable; an ordinal variable can be transformed into a nominal variable. The reverse is not true.

1.2 The correspondence factor analysis and the implicative analysis among variables in research in didactics of mathematics: an experimental comparison

The research in didactics uses quantitative and qualitative tools. In this paper we are trying to deal with the quantitative tools, spending more time, overall, on theoretical-experimental relationships between factorial and implicative analysis. Some significant experimental situations will be analysed in the parts 2 and 3.

Implicative analysis

The problem faced by R. Gras [12] arose from the attempt to answer the following question: "Given some binary variables $\boldsymbol{a}$ and $\boldsymbol{b}$, how can I be sure that into a population, from each observation of $\boldsymbol{a}$, there necessarily follows the observation of $\boldsymbol{b}$? " Or in an even more succinct manner: "Is it true that if $\boldsymbol{a}$ then $\boldsymbol{b}$? "

The answer is not generally possible and the researcher must be satisfied with an "almost" true implication. By the implicative analysis by R. Gras, one tries to measure the degree of validity of an implicative proposition between binary and not binary variables. This statistical tool is used in the Didactics of Mathematics[9].

Some observations on Factorial Analysis

The approaches to factorial analysis are of two types; the first one through the study of the self-values of equations and the second one through a geometric interpretation (vectors) and some contents of rational mechanics.

The approach presented here is the second one.

Let us consider the Cartesian product $\mathbf{E} \times \mathbf{V}$ (E constituted in general by $\boldsymbol{n}$ students, $\boldsymbol{n}\in\mathbf{N}$; and V by $\boldsymbol{m}$ variables, $\boldsymbol{m}\in\mathbf{N}$). This is a typical situation to find data in didactics. The problem is to represent geometrically the distribution of the two sets in a space of $\boldsymbol{n} \times \boldsymbol{m}$ dimensions. Factorial analysis interprets the geometrical representations. This fact, in the sphere of Human Sciences, has had many applications in the field of Psychology, but allowing an

[9] All the information related to the mathematical theory are found in this volume.

analysis of small samples in the field of nonparametric Statistics significantly contributes to the interpretation of didactic phenomena. See Bastin et al. [3] for a geometrical approach to factorial analysis and Escofier and Pages [10] for the interpretation of the graphic representation of the data.

1.3 Experimental comparison between Statistic Implicative Analysis (SIA) and the Correspondence Factor Analysis (CFA)

The comparison between the two statistical tools has a prejudicial epistemology:

1. Factorial analysis is in the field of descriptive statistics: averages, distances by geometric methods. The measure used among variables is symmetrical. This distance accentuates the role of rare observations. The geometrical representations are given on different planes.
2. Implicative analysis is in the field of inferential statistics. The measure used among variables is asymmetrical and highlights the not-commonplace observations. The geometrical representations are carried out on a single plane.

The types of variables treated are more varied in the ASI than in the CFA, and one thinks they may be also fuzzy variables [37].

As regards the control of the data, the information:

1. in the ASI is complete and it is controlled by the indices of implication and by the level of significance of the hierarchical levels:
2. in the CFA is controlled by the explicit inertia. There is a series of indications to keep in mind to accept the level of the explicit inertia: for example, regarding the number of observed variables or the number of individuals.

Notwithstanding these two great epistemological differences, the two statistical tools are used for the research in the didactics of Mathematics as appropriate tools for the multi-varied analysis of small samples.

It is clear that the Correspondence Factor Analysis (CFA) can lead to the study of variables (factors) which can be used to analyze experimentally the conceptions, and the relationships among the conceptions referred to the identification of the factors. One has another thing, instead, when supplementary variables are introduced. In the case of research in didactics the supplementary variables are individuals made up in function of the hypotheses and of a-priori analysis. On the transposed matrix the variables become the students with the addition of the supplementary variables (theoretical students who correspond to well defined characteristics by the nature of the problem of the research). The consequent CFA, by the information with regards to the possible identified factors, gives information on how many individuals group themselves onto one supplementary variable and this leads us to infer a reasoning of the implicative type: *If n individuals group themselves onto a supplementary variable, then the supplementary variable identifies a significant conception.*

The Statistical Implicative Analysis is concerned exactly with implications and this activity is supplied without the introduction of supplementary variables. However, when the supplementary variables are introduced, implicative analysis turns out to be much clearer and evident.

The comparison between the two methods can be significative only in the case of the introduction of supplementary variables for what has been said up to now.

Two experimental researches, on the same sample and with the same supplementary variables, separately analysed, are compared.

Supplementary variables and Goldbach's conjecture

In the doctoral thesis by Aldo Scimone, through the introduction of supplementary variables regarding some reasoning schemes on the solution of Goldbach's conjecture, the same data were compared by CHIC and by SPSS for the factorial analysis[10].

The case of the passage from arithmetic to algebraic language

Elsa Malisani's doctoral thesis compared the aspects of variable as unknown and the functional relation in problem-solving, by considering the semiotic contexts of algebra and analytical geometry. The goal was to investigate if the notion of unknown interferes with the interpretation of the functional aspect, and if the procedures in natural and/or arithmetic language prevail as solving strategies for want of an adequate knowledge of algebraic language. Also for this work we are trying to analyse the differences between factorial and implicative analysis[11].

1.4 Conclusions

Both Correspondence Factor Analysis and Statistic Implicative Analysis do analyses on small samples with the difference that CFA is a descriptive statistics while ASI is of inferential type. This first difference places ASI in a different situation to infer from the sample of the population. However, as already said in paragraph 2, the introduction of supplementary variables can allow us an experimental comparison between the two statistical tools.

The measure used by CFA is symmetrical while the measure introduced by ASI is asymmetrical. The asymmetry is due to the introduction of the relationship of inclusion and so for its inferential nature (from the cause to the effect).

[10] See section 2.

[11] See section 3.

The numerous data collected in experimental work regarding degree theses and doctoral theses bring us to affirm that implicative analysis, when supplementary variables are introduced, proves to be more incisive in the literature of the contingencies. Instead, factorial analysis can give a useful argumentative contribution as a support to implicative analysis when it must analyse only the observed variables. This fact perfectly agrees the epistemological analysis of the two statistical tools. So, ASI allows us a better approach to the statistical analysis of the evolution of the conceptions in the dynamics of the classes.

2 The importance of supplementary variables in a case of an educational research

2.1 The framework of the research

The theoretical framework of this educational research is the theory of didactical situations in mathematics by Guy Brousseau [6].

It is known that this theory is based on the conception of the didactic situations, and in particular, this paper concerns an a-didactic situation, namely that part of a didactic situation which teacher's intention respect to pupils is not clear into. An a-didactic situation is really the moment of the didactic situation in which the teacher does not declare the task to be reached but he gets the pupil to think about the proposed task which is chosen in order to allow him to acquire a new knowledge which is to be looked for within the same logic of the problem.

So, an a-didactic situation allows a pupil to appropriate and to manage the staking dynamics, to get him to be a protagonist of the process, to get him to perceive the responsibility of it as a knowledge and not as a guilt of the sought result.

2.2 The historical context of the research

The research concerns some conceptions of pupils facing a conjecture, and in particular the famous historical Goldbach's conjecture. Goldbach's conjecture was chosen because it has a long historical background allowing an efficient a-priori analysis, which is an important phase for the experimentation in order to foresee the possible pupils' answers and behaviours in front of the conjecture. Moreover, it has a fascinating formulation allowing pupils to mix many numerical examples, and to discuss fruitfully about its validity and some possible attempts for a demonstration of it. So, the historical context is important because it suggests an interplay between the history of mathematics and the mathematics education.

2.3 Using implicative analysis

This research was carried out by a quantitative analysis along with a qualitative analysis. The statistical survey for the quantitative analysis was made by two phases: in the first experiment, which was realized by a sample of pupils attending the third and fourth year of study (16–17 years) of secondary school, the method of individual and matched activity was used; the second experiment was carried out in three levels: pupils from the first school (6–10 years), pupils from primary school (11–15 years) and pupils from secondary school.

The quantitative analysis of the data drawn from pupils' protocols was made by the software of inferential statistics [14] CHIC 2000 (*Classification Hiérarchique Implicative et Cohésitive*) and the factorial statistical survey S.P.S.S. (*Statistical Package for Social Sciences).*

The research pointed out some important misconceptions by pupils and some knots in the passage from an argumentative phase to a demonstrative one of their activity which need to be deepened.

2.4 The experimentation

The research was realized on different levels by two experiment. The first experiment, was realized with pupils attending the third and fourth year of study (16–17 years) of secondary school, the method of individual and matched activity was used. Pupils working individually were expected, within two hours, to answer the following question:

a) Using the enclosed table of primes, the following even numbers can be written as a sum of two primes (in an alone or in a manner more)? 248; 356; 1278; 3896.

b) If you have answered the previous question, are you able to prove that it occurs for every even number?

The pupils working in couples were expected, within an hour, to answer this question (in a written form and only if they agreed):

Is it always true that every even natural number greater than 2 is a sum of two prime numbers? Let argue about the demonstrative processes motivating them.

In both cases the procedure was acoustically recorded and the transcript of those records with comments was made.

The second experiment was carried out in three levels: pupils from the primary school (6–10 years), pupils from middle school (11–15 years) and pupils from higher secondary school. The experiment was carried out on the lowest level in two phases: In the first phase the pupils could answer this question:

How can you obtain the first 30 even numbers by putting together prime numbers of the table you have just made?

In the second phase, the pupils created small groups and tried to answer the following question:

Can you derive the even numbers obtained by summing always and only two primes? If it is so, can you state this is always the case for an even number?

The pupils from lower secondary school solved the following problem within 100 minutes:

Is the following statement always true? "Can an even number be resolved into a sum of prime numbers?" Argue your claims.

The procedure had four phases:

a) a discussion about the task in couples (10 min.)

b) an individual written description of a chosen solving strategy (30 min.)

c) the division of the class into two groups discussing the task (30 min.)

d) the proof of a strategic processing given by the competitive groups (30 min.)

Pupils from higher secondary school solved the same problem like the pupils from middle school in the same way and within the same time limit.

Individual works were analyzed (a-priori analysis), the identification of parameters was carried out and those were subsequently used as a basis for the characteristics of pupils' answers. It enabled to do a quantitative analysis of the answers, to establish an implicative graph (graph of functionality), a hierarchical diagram, a diagram of similarities and also an analysis of data. The analyses, graphs and diagrams (or trees) were part of the evaluation of each experiment together with conclusions.

2.5 The first experiment and its analysis by CHIC

The first statistical survey was made by using a sample of 88 pupils attending the third and fourth year of study of secondary school in Palermo (Sicily). The students worked in pairs for the part relating to interviews and individually for the production of solution protocols related to the proposed conjecture. The variables used for the a-priori analysis were 15 and they were settled in the following manner:

1. He/she verifies the conjecture by natural numbers taken at random. **(N-random)**
2. He/she sums two prime numbers at random and checks if the result is an even number. **(Pr-random)**
3. He/she factorizes an even number and sums its factors, trying to obtain two primes. **(Factor)**
4. Golbach's method 1. He/she considers odd prime numbers lesser than an even number, summing each of them with successive primes. **(Gold1)**
5. Golbach?s method 2 (letter to Euler). He/she writes an even number as a sum of more units, combining these in order to get two primes. **(Gold2)**
6. Cantor?s method. Given an even number 2n, by subtracting from it the prime numbers $x \leq 2n$ one by one, by a table of primes one tempts if the obtained difference $2n - x$ is a prime. If it is, then $2n$ is a sum of two primes.**(Cant)**

7. The strategy for Cantor's method. He/she considers the primes lower then the given even number and calculates the difference between the given number and each of primes. **(S-Cant)**
8. Euler. He/she is uneasy to prove the conjecture because one has to consider the additive properties of numbers. **(Euler)**
9. Chen Jing-run?s method (1966). He/she expresses an even number as a sum of a prime and of a number which is the product of two primes.**(Chen)**
10. He/she subtracts a prime number from an any even number (lower then the given even number) and he/she ascertains if he/she obtains a prime, so the condition is verified. **(Spa-pr)**
11. He/she looks for a counter-example which invalidates the statement of the conjecture. **(C-exam)**
12. He/she considers the final digits of a prime to ascertain the truth of the statement.**(Cifre)**
13. He/she thinks that a verification of the statement by some numerical examples needs to prove the statement. **(V-prova)**
14. He/she does not argue anything for the second question. **(Nulla)**
15. He/she thinks the conjecture is a postulate. **(Post)**

2.6 The implicative graph

The analysis of the implicative graph shows, with percentages of 90%, 95% and 99%, that pupils' choice to follow some of the strategies is strictly linked to a relevant strategy, namely Gold 1, or the one according which the pupil considers odd prime numbers summing each of them with successive primes. Hence the basis of pupils' behaviours is the sequential thinking.

2.7 The factorial analysis by S.P.S.S.

The graph shows that a part of pupils is inclined either to proceed by a sequential manner or by preferring a method based on a random choice. On the other hand, the second component shows that the real strong characterization of most pupils is Gold1-Chen which is nearer to the intersection of the two components. So, this is the winning strategy among pupils to pass from an argumentation to a possible demonstration of the conjecture. This is a kind of a photo of the more frequent approaches to the conjecture by students.

2.8 Supplementary variables and pupils' profiles

A further step was made by introducing three supplementary variables to get other informations about the obtained data. They were the following ones:

a) **Abdut**: this is the pupil proceeding by *abduction*, which indicates (Peirce) the first moment of an inductive process, when a pupil chooses a hypothesis by which he can explain determined empirical facts. On the base of

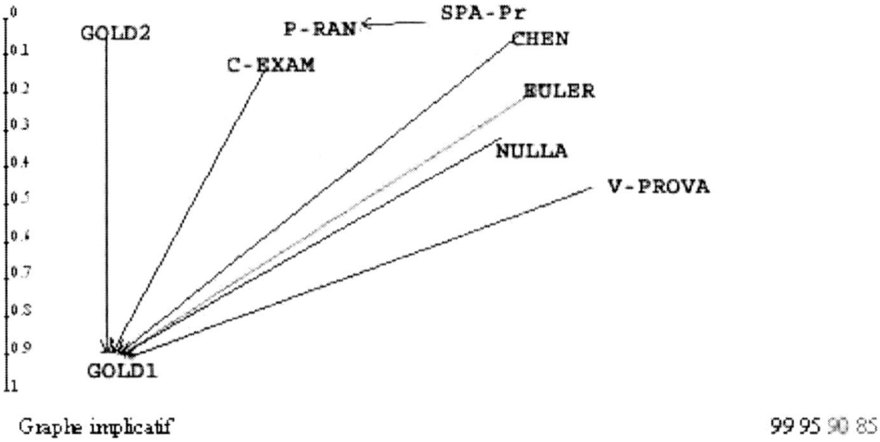

Fig. 2. The implicative graph 90%.

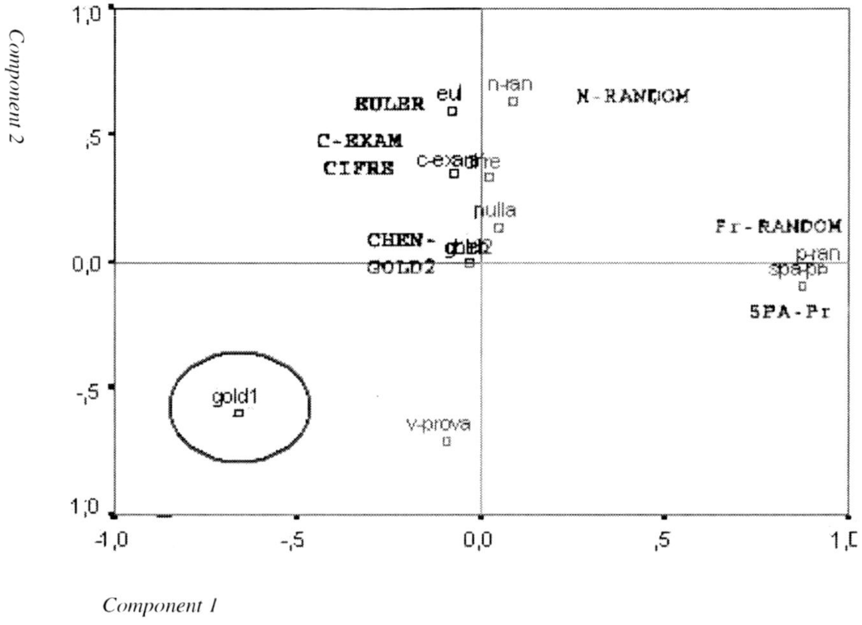

Fig. 3. Factorial analysis variables

such a definition, the pupil named Abdut is who observes how Goldbach's conjecture to be verified in a large number of cases, so he supposes it is also valid for any very large even number, and this fact leads him to the final thesis, namely the conjecture to be valid for every even natural number.

b) **Intuitionist**: this is the pupil using the *N-random* and *Euler* strategies in common with *Abdut*, but thinking that the demonstration of the conjecture can be deduced by a simple numerical evidence, because he is convinced that what happens for the elements of a small finite set of values can be generalized to the infinite set which the small set belongs to. So, he uses the *V-prova* strategy. In short, in an inductive argumentation used by the intuitionist pupil the statement is deduced as a generic case after specific cases.

c) **Ipoded**: this is just the pupil using a deductive argumentation which can be directly transposed into a deductive demonstration.

With these new additional variables a transposed matrix (changing rows by columns) was made by Excel and interpreted by CHIC. The more interesting results are displayed in figure 4.

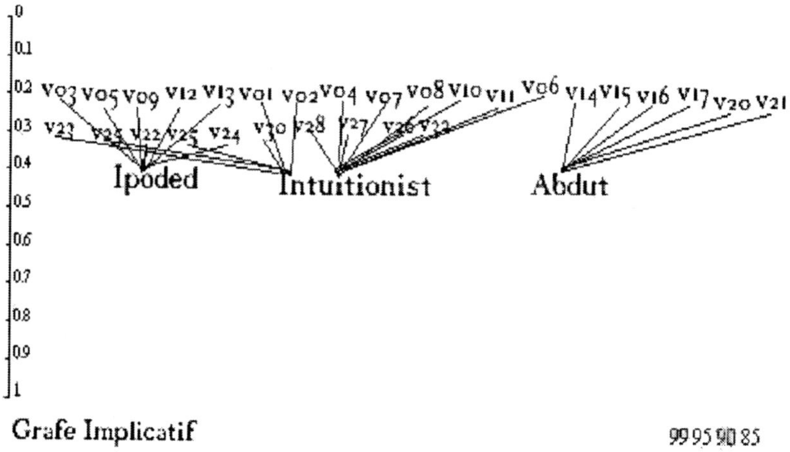

Fig. 4. The implicative graph with supplementary variables

2.9 The implicative graph

It is evident that the three profiles corresponding to the additional variables are significative as much as they *catalyze* the outlines of reasoning of the pupils. The supplementary variables play in this case the same role played by an equivalence relation when we put it into a set. In fact, what the relation does in this case? It orders the elements of the set, so the supplementary variables, in this case, give much more order to the data. They get the interpretation of data more effective. Really, they begin attractors for pupils' behaviours.

2.10 Factorial Analysis

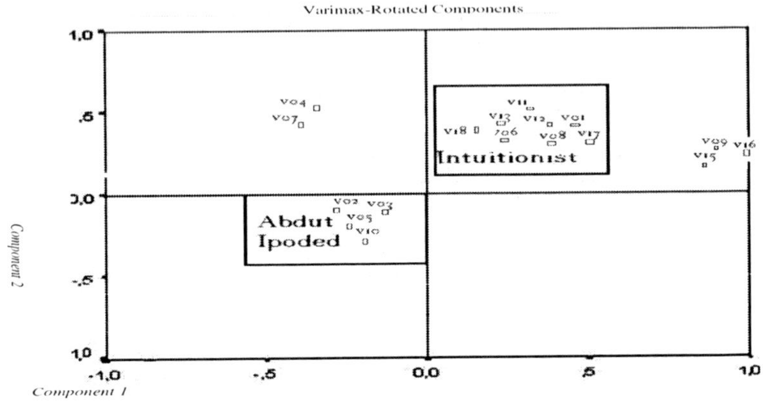

Fig. 5. Factorial analysis with supplementary variables. *Component 1* is refered to abscissae, *Component 2* is refered to ordinata

From the viewpoint of the horizontal component the variable *Intuitionist* characterizes it weakly, while the variables *Abdut* and *Ipoded* with a lot of other variables characterize it much more. On the other hand, this is a paradigmatic situation which has its historical counterpart in the attempts made along centuries by different mathematicians facing the conjecture. So, *Abdut* and *Ipoded* profiles are winners, while the intuitive method of approach is less productive. This characteristic situation is stationary also when one observes the graph from the viewpoint of the second component. This means that in any way Abdut and Ipoded methods are more interesting for pupils.

2.11 Some final observations

The experimentation about Goldbach's conjecture has pointed up that in general most pupils, while facing an unsolved historical conjecture (without knowing it is yet unsolved), start at once with an empirical verification of it which can support their intuition, but after they distinguish themselves along three different solving typologies:

1. a part of pupils bites off more than one can chew with the following conclusion: since the conjecture is true for all of these particular cases, then it has to be true anyway. These are pupils who have a strong faith in their convictions, but who do not know clearly enough how to pass from an argumentation to a demonstration, by using the achieved data.
2. a part of pupils proceeds at the same time by an empirical verification and by an attempt of argumentation and demonstration ending to a mental statement. They try to clear a following hurdle: how can I deduce a general statement from the empirical evidence? These are pupils who before making any generalization want to be sure of the made steps, therefore they tread carefully.
3. few pupils, after a short empirical verification, look at once for a formalization of their argumentations, but if they are not able to do that, they are not diffident about claiming they are in front of something which is undemonstrable. These pupils have a high consideration for their mental processes therefore they think that if they are not able to demonstrate anything, then it has to be undemonstrable anyway.

By this experimentation we argue that the argumentation favoured by pupils facing a historical conjecture like Goldbach's is the abductive one. Some questions arise from the results which would be advanced by other experimentations:

1. Is this result generalizable?
2. To what extent is it generalizable?

But the fundamental kernel of this experimentation about the interplay between history of mathematics and mathematics education is that such results could not be pointed out if the a-priori analysis had not been made by the historical-epistemological remarks which have inspired it.

3 The Statistic Implicative analysis and the correspondence factor analysis in a research in Mathematics Education: unknown or "thing which is varying"

3.1 Introduction

Modelling through statistic argumentation gives to the research in mathematics education a greater possibility of transferability of the experience. However, the statistic argumentation would not have any weight without an accurate theoretical reflection from the viewpoint of the didactic and the epistemology of the mathematical contents [11].

In the a-priori analysis of a didactic situation it is necessary to consider the epistemological representations, the historical-epistemological representations and the hypothetical behaviours, correct and not correct. Besides the a-priori analysis allows us to individualize the variables of the situation-problem and the hypotheses of research. These hypotheses can be falsified through the statistic analysis and/or the qualitative analysis of the data.

In the last decade two statistic methods have been very used: the implicative analysis (ASI) of Regis Gras [13, 14] and the correspondence factor analysis (CFA). The implicative analysis is a powerful tool. It allows a clear visualization of the relations of similarity and implication among variables or classes of variables of the situation-problem, through the graphs elaborated by the software CHIC. The correspondence factor analysis represents geometrically, in a multi-dimensional space, a distribution of two set: the individuals and the variables of the situation [13]. Since it allows an analysis on small samples in the field of the not parametric Statistics, it contributes to interpret meaningfully the didactic phenomena. In the last decade the aim of some studies was to improve the tool in the field of the didactic research and, chiefly, to create some models *ad hoc* [35].

This paper is a contribution to the studies on the application of the Statistic Implicative Analysis (SIA) and the Correspondence Factor Analysis (CFA) in different fields, particularly, in Mathematics Education. This research puts in evidence the relations between the Implicative Analysis and the Factorial Analysis to falsify hypotheses of research in mathematics education. We want to analyze too the type of information obtained by the application of the two statistic methods[12].

3.2 A condensed theoretical framework

There are a lot of studies on the obstacles which pupils meet during the passage from the arithmetic to the algebraic thought. Some of them reveal that the introduction of the concept of variable represents the critical point of transition [25, 39, 40].

This is a complex concept because it is applied with different meanings in different situations. Its management depends precisely on the particular way of its use into the activity of problem-solving.

The notion of variable could take on a plurality of conceptions: *generalized number* (it appears in the generalizations and in the general methods); *unknown* (its value could be calculated by considering the restrictions about the field of existence of the solutions of a problem); *"in functional relation"* (relation of variation with other variables); *sign totally arbitrary* (it appears in the study of the structures); *register of memory* (in informatics) [38].

In Malisani and Marino [23] and Malisani and Spagnolo [24] we observed that the pupils spontaneously evoke the different conceptions of variable as: numerical value, unknown, "*thing which is varying*", also in absence of an adequate mastery of the algebraic language.

[12] This part of the paper is based on [22]

It is possible that many difficulties in the study of algebra derive from the inadequate construction of the concept of variable [8]. An opportune approach to this concept should consider its principal conceptions, the existing interrelationships between them and the possibility to pass from one to the other with flexibility, in relation to the exigencies of the problem to be solved.

The historical analysis emphasizes that the notions of unknown and the one of variable as "*thing which is varying*" have a totally different origin and evolution. Even if both the concepts deal with numbers, their processes of conceptualizations seem to be entirely different [26].

In Malisani [20, 21] we studied the relational-functional aspect of the variable in problem-solving, considering the semiotic contexts of algebra and analytical geometry. We showed that there is a certain interference of the conception of unknown on the functional aspect, in the context of a situation-problem and in absence of visual representative registers. We also demonstrated that the students find some difficulties to interpret the concept of variable in the process of translation from the algebraic language into natural one.

This paper belongs of the statistic analysis of that experimentation, by which we wanted to verify if the conception of variable as "*thing which is varying*" is evoked, when the notion of unknown prevails in the context of a situation-problem.

To carry out this research we chose the linear equation in two variables for two reasons: firstly, because it represents a nodal point from which the students derive the conceptions of the letters as unknowns or "*things which are varying*". Secondly, this kind of equation is well known by the pupils, because they studied it from different viewpoints: linear function, equation of a straight line and component of the linear systems.

3.3 Methodology of the research

One hundred eleven students — aged 16–18 — of the Experimental High School of Ribera (AG, Italy) participated to the research.

The questionnaire presentes four questions, but in this paper we introduce only the resolution of the first problem which is the following:

Charles and Lucy win the total sum of a Euro 300 in the lottery. We know that Charles wins the triple of the betted money, while Lucy wins the quadruple of her own.

1. *Determine the sums of money that Charles and Lucy have betted. Comment on the procedure that you have followed.*
2. *How many possible solutions are there? Give reasons for your answer.*

In this problem the variable takes on the relational-functional aspect in the context of a concrete situation-problem. We also asked the pupils to think over the solution set. With this question we wanted to analyze the solving strategies used and if the unknown's notion interferes with the interpretation of the functional viewpoint.

We carried out a-priori analysis of the problem. The aim was to determine all the possible strategies that the pupils could use. Some errors that students could possibly make in the application of these strategies were also identified.

The pupils worked individually, we did not allow them consulting books or notes. The where given time was sixty minutes.

In a table we filled in with a double input "pupils/strategy", and we indicated for every pupil the strategies he used by the value 1 and those he didn't apply by the value 0. The data were analyzed in a quantitative way, by using the statistic implicative analysis of Regis Gras [13, 14], the software CHIC 2000 and the factorial statistical survey S.P.S.S. (Statistical Package for Social Sciences).

3.4 The a-priori analysis

We determined the principal experimental variables from an a-priori analysis. They were the following ones:

AL1: The pupil answers the question.

AL2: He/she shows a procedure in the natural language.

AL3: He/she shows a procedure by trial and errors in natural language and/or in a half-formalized language.

AL4: He/she adds a datum.

AL4.1: He/she adds a datum, but he/she considers that the winnings are divided in half.

AL4.2: He/she adds a datum, but he/she considers that the winnings of the two teenagers are equal to Euro 300.

AL4.3: He/she adds a datum, but he/she considers that the bets are equal.

AL5: He/she translates the problem into a first degree equation of two unknowns.

AL7: He/she translates the problem into a first degree equation with two unknowns and he/she uses the algebraic method of "substitution into the same equation"[13].

AL9: He/she abandons the pseudo-algebraic procedure and he/she tries another method.

AL11: He/she considers, in an explicit or implicit way, that the problem represents a functional relation.

AL13: He/she makes some errors in the resolution of the equation and he/she finds (or he/she tries to find) the only solution.

AL14: He/she considers that a relation of proportionality exists between x and y.

[13] We called "procedure of substitution into the same equation" the incorrect method where he/she writes one variable in function of the other. Then he/she replace this variable in the original equation, and thus he/she obtain an identity. In short, the pupil applies the method of substitution used to solve the systems of equations to a single equation.

ALb1: The pupil calculates the solution set.

ALb2: He/she shows a particular solution verifying the equation.

ALb3: He/she shows several solutions verifying the equation.

ALb4: He/she considers the infinite solutions expressly.

ALb5: He/she explicitly considers the data are insufficient to determine only one solution.

ALb6: He/she considers multiple solutions (it includes ALb4 and ALb5).

The hypothesis

If the conception of variable as unknown prevails in the context of a situation-problem	**then** the relational-functional aspect is not evoke
AL4, ALb2	~AL3, ~AL11, ~AL14, ~ALb3, ~ALb4, ~ALb5, ~ALb6

Table 2. Hypothesis and experimental variables

The conception of variable as an unknown is highlighted by the experimental variable AL4 *"he/she adds a datum"*. Precisely, "adding a datum" is equivalent to introducing a new equation and thus to forming a system of two linear equations with the equation of the problem or part of it. The solution of the system is a *"particular solution verifying the equation of the problem"* (ALb2).

The relational-functional aspect of the variable is evoked when the pupil exhibits a *"procedure by trial and errors"* (AL3), through which he recalls the notion of dependence among the variables. In this way, the pupil *"considers implicitly or expressly that the problem represents a functional relation"* (AL11) or he manifests, incorrectly, that *"a relation of direct proportionality exists among the variables"* (AL14). Therefore the pupil *"shows some solutions veryfing the equation"* (ALb3) or *"he considers that the problem has multiple solutions"* (ALb6)[14] Accordingly *"not to evoke the relational-functional aspect of the variable"* is equivalent to the negation of the experimental variables above described: ~AL3, ~AL11, ~AL14, ~ALb3, ~ALb4, ~ALb5, ~ALb6.

[14] In this study we prefer to use the term "multiple solutions" rather than the one of "infinite solutions", because we have not considered the possible connotations of the word "infinite". However, we defined two experimental variables ALb4 to take into account the cases in which the pupil explicitly considers the existence of infinite solutions.

3.5 Statistic Implicative Analysis (ASI)

Implicative graph

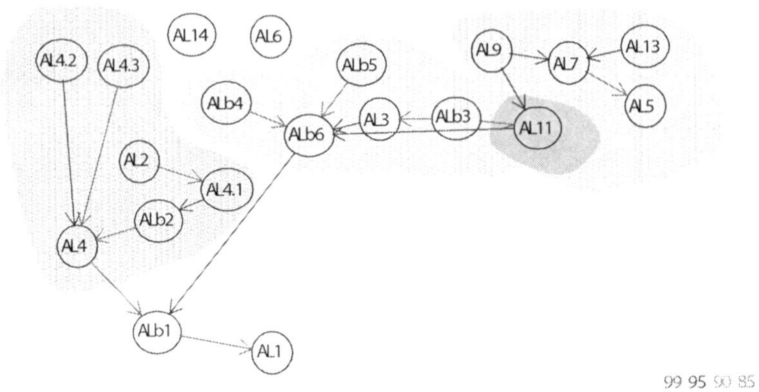

Fig. 6. Implicative graph

The implicative graph on Figure 6 (carried out with the software CHIC 2000) shows three well defined groups of experimental variables with statistic percentages of 95% and 99%. They are pointed out by the cloud on the left (cloud L), the cloud in the center of the figure (cloud C) and the cloud on the right (cloud R) (the grey cloud (cloud I) around AL11 indicates the intersection between cloud C and cloud R).

The three groups are directly or indirectly connected with the variable ALb1 *"the pupil calculates the solution set"* and AL1 *"the pupil answers to the question"*. Every group corresponds to a different kind of strategy used by the students:

- ***Procedure in natural language*** (cloud L): the pupil adds a datum considering that the winnings are equal (generally dividing Euro 300 in half) or that the bets are equal[15]. In this way, the student transforms the question into a typical arithmetic problem and he resolves it finding only a particular solution verifying the equation. This result is confirmed by

[15] The experimental variable **AL4** *"he/she adds a datum"* considers two possibilities: equal winnings or equal bets (AL4.3). The first case takes into account the two other alternatives: the winnings are divided in half (AL4.1) or both the teenagers win Euro 300 (AL4.2).

"To add a datum" is equivalent to introduce a new equation and to forme (with the equation of the problem $3x+4y=300$ or part of it) a system of two equations into two unknowns. Therefore a system corresponds to each case.

"The winning of Euro 300 are divided in half" (AL4.1): it is equivalent to the system $3x+4y=300$, $3x-4y=150$

the implicative links among the experimental variables AL2, ALb2 and AL4 (with its variations AL4.1, AL4.2 and AL4.3). The procedure in the natural language is the most used by the pupils (Cfr. Table of frequencies in the *Appendix*), and it leads to the single solution. So the predominant conception of variable is that of unknown.

- ***Method by trials and errors in natural language or in half-formalized language*** (cloud C): the pupil generally assigns several values to a variable (e.g. Charles' bet) and he finds the corresponding values in the other variable (Lucy's bet). In this way, the student shows some solutions verifying the equations and/or he considers that it has multiple solutions. That is, he generally considers in an implicit way that the problem represents a functional relation. This result result is obtained by the implicative links between the variables AL3, ALb3, AL11 and ALb6. This method leads to many solutions, allows evoking the dependence between the variables, but a strong conception of the relational-functional aspect does not appear yet.

- ***Pseudo-algebraic strategy*** (cloud R): the pupil translates the text of the problem into an equation of first degree with two unknowns and applies the method of "substitution into the same equation", that is the incorrect procedure where he/she writes one variable in function of the other. Then he/she replaces this variable it the original equation and thus he/she obtain an identity. Since the pupil does not succeed in interpreting the identity, either he/she changes his/her resolving procedure abandoning the pseudo-algebraic one or he/she resumes the resolution of the equation and he/she makes some errors trying to find only one solution. This result is deduced by the implicative links among the experimental variables AL9, AL13, AL7 and AL5.

It is interesting to observe that, if the pupil abandons this strategy, then he/she considers, in an implicit or explicit way, that the problem represents a functional relation. This result is confirmed by the implicative link between the experimental variables AL9 and AL11. This link allows the connection between the two procedures: by trials and errors and pseudo-algebraic (cloud I).

However, the pseudo-algebraic strategy is rarely used and it leads to the correct solution of the problem only in some cases.

Falsification of the hypothesis

We are considering:

"The winnings are equal to Euro 300 for both the teenagers" (AL4.2): it corresponds to the system $3x = 300$, $4y = 300$

"The bets are equal" (AL4.3): is equivalent to the system $3x + 4y = 300$, $x = y$

p: in the context of a situation-problem, the conception of variable as unknown (experimental variables AL4 and ALb2) prevails;

q: the relational-functional aspect is evoked (experimental variables AL3, AL11, AL14, ALb3, ALb4, ALb5 and ALb6).

The hypothesis 1 is equivalent to:

$p \rightarrow \sim q$ that, from the logical viewpoint, is equivalent to $\sim (p \wedge \sim (\sim q))$ or $\sim (p \wedge q)$.

Therefore, to falsify this hypothesis it is sufficient to demonstrate the empty intersection between the experimental variables of p and q, in other words:

p corresponds to the *procedure in the natural language* which the conception of variable as unknown prevails into;

q is equivalent to the *method by trials and errors* which the relational-functional aspect of the variable predominates into.

From the implicative graph we deduce that the sets of experimental variables, corresponding to p (cloud L) and to q (cloud C), are disjoined. This result allows us to falsify the formulated hypothesis.

Profile of the pupils

From the previous analysis an important aspect emerges: the existence of a certain correspondence between the solving strategies used by the pupils and the conceptions of variable as unknown and "*thing which is varying*". To examine carefully these results we apply a particularly effective methodology by introducing some supplementary variables in the "students" component. In other words, the researcher defines a profile of student that satisfies some characteristics he considers important. In the specific case of this analysis we define five profiles:

1. **NAT:** this profile corresponds to the pupil performing a procedure in the natural language. Then he/she adds a datum considering that the winnings are equal (generally dividing it in half) or that the bets are equal and he/she resolves the problem finding only a particular solution verifying the equation. This profile is characterized by the presence of the followings experimental variables: AL2, AL4 (AL4.1, AL4.2 or AL4.3) and ALb2.
2. **FUNZ:** it corresponds to the pupil applying a strategy by trials and errors in natural language and/or in half-formalized language. He/she generally assigns several values to a variable and he/she finds the corresponding values in the other variable. So the student shows some solutions verifying the equations and/or he/she considers that it has multiple solutions. The experimental variables describing this profile are: AL3, AL11, ALb3, ALb4, ALb5 and ALb6.

3. **PALG1:** it corresponds to the student who uses the pseudo-algebraic procedure. He/she translates the text of the problem into a linear equation with two unknowns and he/she applies the method of "substitution into the same equation". When the student reaches the identity he/she does not succeed in interpreting it. Then he/she resumes the resolution of the equation, he/she makes some errors of syntactic kind trying to find only one solution. This profile is characterized by the presence of the experimental variables: AL5, AL7, AL13 and ALb2.
4. **PALG2:** it is a variation of the profile PALG1. In this case, when the pupil arrives to the identity he/she changes the solving procedure abandoning the pseudo-algebraic one. The experimental variables which describe this profile are the followings: AL3, AL5, AL7, AL9, AL11 and ALb3.
5. **ALG:** it corresponds to the pupil who applies an algebraic procedure. He/she translates the problem into an equation of first degree with two unknowns, he/she considers, in an implicit or explicit way, that it represents a functional relation and so that it is verified by multiple solutions. The experimental variables of this profile are the followings: AL5, AL11, ALb4 and ALb6.

3.6 The correspondence factor analysis (CFA)

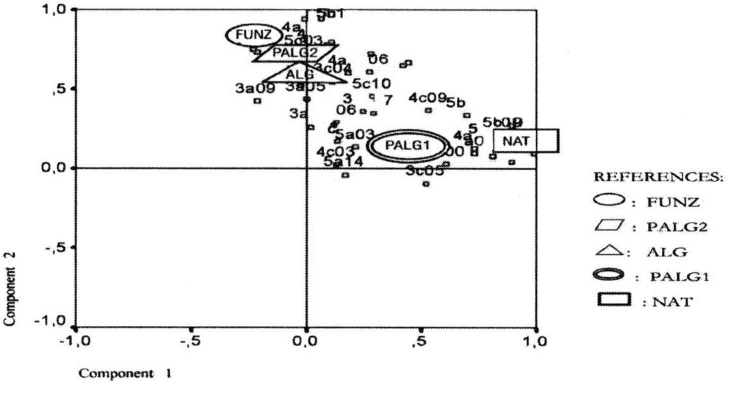

Fig. 7. Factorial analysis with supplementary variables

The graph shows that the first component (horizontal axis) is strongly characterized by the pair of supplementary variables: NAT and PALG1.

The profiles ALG, PALG2 and FUNZ form a cloud that strongly characterizes the vertical component. The supplementary variable PALG2 is very near to FUNZ, because the student who abandons the pseudo-algebraic procedure generally adopts the profile described in FUNZ.

The winning strategies are precisely those described in the profiles ALG, PALG2 and FUNZ which lead to multiple solutions, while NAT and PALG1 lead to the oneness of the solution. This finds a strong correspondence with the different conceptions of "variable". Therefore, the horizontal axis represents the conception of variable as *unknown*, the vertical axis, instead, reproduces its *relational-functional* aspect. These results allow us to falsify the hypothesis again.

3.7 Conclusions

The implicative graph shows the solving strategies applied by the students to solve the problem:

1. *procedure in natural language*: it is the most used by the pupils and it leads to the single solution. The predominant conception of variable is that of unknown.
2. *methods by trials and errors in natural language and/or in half-formalized language* (generally arithmetic): it gets to several solutions. The dependence of the variables is evoked, but a strong conception of the relational-functional aspect does not appear yet.
3. *pseudo-algebraic strategy*: it is little used by the pupils and it leads to the correct solution of the problem only in some cases.

To examine carefully these results we introduced some supplementary variables in the "students" component. These profiles represent the supplementary individuals putting out the fundamental characteristics of the a-priori analysis. They are displayed in Table 3.

SUPPLEMENTARY VARIABLES	**STRATEGIES**
NAT	***in natural language***
FUNZ	***by trials and errors in natural language and/or in half-formalized language***
PALG1	***pseudo-algebraic*** + *resolution of the equation with some errors of syntactic kind*
PALG2	***pseudo-algebraic*** + *other strategy*
ALG	***algebraic***

Table 3. Correspondence between supplementary variables and strategies

The hierarchic tree shows that the profile NAT is the most meaningful because it represents the strategy the pupil used most. We observe a small set of pupils connected to this group. They followed the procedure described in NAT; but, afterwards, they effected the passage from the single solution to multiple solutions [22, pp. 96].

From the factorial analysis we observe that the horizontal component is characterized by the profiles NAT and PALG1, while the vertical component by FUNZ, PALG2 and ALG. Therefore we note a strong correspondence among the principal components and the conceptions of variable: the horizontal component represents the notion of *unknown*, while the vertical one denotes the aspect of *"thing which is varying"*.

The obtained results, by the implicative analysis and the factorial analysis, allow us to falsificate the formulated hypothesis, namely: "If the conception of variable as unknown prevails in the context of a problematic situation, then the relational-functional aspect is not evoked".

It is interesting to observe that, in some cases, we verify the passage from the single solution to multiple solutions in the linear equation, even if the notion of unknown prevails. From here an important question emerges: *"This passage coincides (or not) with the passage from the conception of unknown to that relational-functional one"?* In this experimentation we have not found the answer. To study carefully this matter we carried out a new experimental research of qualitative kind, submitting the same questionnaire to pairs of pupils.

This research puts in evidence the relations between the Statistic Implicative Analysis (ASI) and the Correspondence Factor Analysis (CFA) to falsify hypotheses of didactic research in mathematics. The implicative analysis puts in evidence the strategies of the students, the factorial analysis puts in contrast the correspondence between the principal components and the conceptions on the variable.

References

1. R. Agrawal, T. Imielinski, and A.N. Swami. Mining association rules between sets of items in large databases. In P. Buneman and S. Jajodia, editors, *ACM SIGMOD International Conference on Management of Data*, pages 207–216, 1993.
2. F. Arzarello, L. Bazzini, and G. Chiappini. L'algebra come strumento di pensiero. analisi teorica e considerazioni didattiche. *Progetto Strategico CNR-TID*, (6), 1994.
3. Ch. Bastin, J.P. Benzecri, Ch. Bourgarit, and P. Cazes. *Pratique de l'Analyse des Données*, volume 1–2. Dunod, 1980.
4. A. Bodin. Improving the diagnostic and didactic meaningfulness of mathematics assessment in france. In *Annual Meeting of the American Educational Research.* Association AERA, New York, 1996.
5. G. Brousseau. *Theory of didactical situations in mathematics.* Kluwer Academic Publishers, 1997. Edited and translated by N. Balacheff, M. Cooper, R. Sutherland and V. Warfield.
6. G. Brousseau. *Théorie des situations didactiques. Didactique des mathématiques 1970–1990.* La pensée sauvage, 1998. Textes rassemblés.

7. E. La Casta and G. Brousseau. *Méthodes d'analyses statistiques multidimensionnelles en didactique des mathématiques*, chapter Utilisation de la contingence par l'analyse factorielle. Traitement d'un cas: Le graphique, pages 53–90. ARDM, Rennes, 1995.
8. I. Chiarugi, G. Fracassina, F. Furinghetti, and D. Paola. Parametri, variabili e altro: un ripensamento su come questi concetti sono presentati in classe. *L'insegnamento della Matematica e delle Scienze integrate*, 18B(1):34–50, 1995.
9. R. Couturier. Traitement de l'analyse statistique implicative dans chic. In *Actes des Journées sur la "Fouille dans les données par la méthode d'analyse implicative"*, 2001.
10. B. Escofier and J. Pages. *Analyses factorielles simples et multiples (objectifs, méthodes et interprétation)*. Dunod, Paris, 1990.
11. F.Spagnolo. *Insegnare le matematiche nella scuola secondaria*. La Nuova Italia, Firenze, 1998.
12. R. Gras. *L'implication statistique (Nouvelle méthode exploratoire de données)*. La Pensée Sauvage, Grenoble, 1996.
13. R. Gras. Metodologia di analisi di indagine. *Quaderni di Ricerca Didattica*, 7:99–109, 1997.
14. R. Gras. I fondamenti dell'analisi statistica implicativa. *Quaderni di Ricerca Didattica*, 9:189–209, 2000. Text available at: `http://dipmat.math.unipa.it/grim/quaderno9.htm`.
15. R. Gras, R. Couturier, F. Guillet, and F. Spagnolo. Extraction de règles en incertain par la méthode statistique implicative. In *Comptes rendus des 12e Rencontres de la Société Francophone de Classification*, pages 148–151, 2005.
16. R. Gras, E. Diday, P. Kuntz, and R. Couturier. Variables sur intervalles et variables-intervalles en analyse implicative. In *Actes du 8e Congrès de la Société Francophone de Classification*, pages 166–173, 2001.
17. J.B. Lagrange. Analyse implicative d'un ensemble de variables numériques; application au traitement d'un questionnaire aux réponses modales ordonnées. *Revue de Statistiques Appliquées*, 46(1):71–93, 1998.
18. I.C. Lerman. *Classification et analyse ordinale des données*. Dunod, 1981.
19. I.C. Lerman, R. Gras, and H. Rostam. Elaboration et évaluation d'un indice d'implication pour des données binaires, i et ii. *Mathématiques et Sciences Humaines*, 74 and 75:5–35 and 5–47, 1981.
20. E. Malisani. The notion of variable in semiotic contexts different. In *Proc. of the Int. Conf. "The Humanistic Renaissance in Mathematics Education"*, pages 245–249, University of Palermo-Italy, 2002. Text available at: `http://dipmat.math.unipa.it/~grim/21project.htm`.
21. E. Malisani. The notion of variable: some meaningful aspects of algebraic language. In A. Gagatsis, F. Spagnolo, G. Makrides, and V. Farmaki, editors, *Proc. of the 4th Mediterranean Conf. on Mathematics Education (MEDCONF 2005)*, volume 2, pages 397–406, University of Palermo-Italy, 2005.
22. E. Malisani. *The concept of variable in the passage from the arithmetic language to the algebraic language in different semiotic contexts*. PhD thesis, Palermo, Italy, 2006.
23. E. Malisani and T. Marino. Il quadrato magico: dal linguaggio aritmetico al linguaggio algebrico. *Quaderni di Ricerca in Didattica*, 10, 2002. Text available at: `http://dipmat.math.unipa.it/~grim/quaderno10.htm`.
24. E. Malisani and F. Spagnolo. Difficulty and obstacles with the concept of variable. In *Proc. of CIEAEM 57*, pages 226–231, Piazza Armerina-Italy, 2005.

25. M. Matz. *Intelligent Tutoring Systems*, chapter Towards a Process Model for High School Algebra Errors, pages 25–50. Academic Press, London, 1982.
26. L. Radford. *Approaches to Algebra. Perspectives for Research and Teaching*, chapter The roles of geometry and arithmetic in the development of algebra: historical remarks form a didactic perspective, pages 39–53. Kluwer, 1996.
27. A. Scimone. Following goldbach's tracks. In *Proc. of the Int. Conf. "The Humanistic Renaissance in Mathematics Education"*, University of Palermo-Italy, 2002. Text available at: `http://dipmat.math.unipa.it/~grim/21project.htm`.
28. A. Scimone. La congettura di goldbach tra storia e sperimentazione didattica. *Quaderni di Ricerca in Didattica*, 10:1–37, 2002. Text available at: `http://dipmat.math.unipa.it/~grim/quaderno10.htm`.
29. A. Scimone. *Conceptions of pupils about an open historical question: Goldbach's conjecture. The improvement of Mathematical Education from a historical viewpoint.* PhD thesis, Palermo, Italy, 2003. published on Quaderni di Ricerca in Didattica 12, Text available at: `http://dipmat.math.unipa.it/~grim/tesi/_it.htm`.
30. A. Scimone. An educational experimentation on goldbach's conjecture. In *Proc. CERME 3, Group 4*, pages 1–10, Bellaria-Italy, 2003.
31. A. Scimone. How much can the history of mathematics help mathematics education? an interplay via goldbach's conjecture. In *Zbornik, Bratislavskehoseminara z teorie vyucovania matematiky*, pages 89–101, Bratislava, 2003.
32. F. Spagnolo. *Obstacles Epistémologiques: Le Postulat d'Eudoxe-Archimede.* PhD thesis, Universiy of Bordeaux I, 1995.
33. F. Spagnolo. L'analisi a priori e l'indice di implicazione di regis gras. *Quaderni di Ricerca in Didattica*, 7:110–117, 1997. Text available at: `http://dipmat.math.unipa.it/~grim/quaderno7.htm`.
34. F. Spagnolo. A theoretical-experimental model for research of epistemological obstacles. In *Int. Conf. on Mathematics Education into the* 21st *Century*, 1999. Text available at: `http://dipmat.math.unipa.it/~grim/model.pdf`.
35. F. Spagnolo. L'analisi quantitativa e qualitativa dei dati sperimentali. *Quaderni di Ricerca in Didattica*, 10, Supplemento, 2002. Text available at: `http://dipmat.math.unipa.it/~grim/quaderno10.htm`.
36. F. Spagnolo. La modélisation dans la recherche en didactiques des mathématiques: les obstacles épistémologiques. In *Recherches en Didactiques des Mathématiques*, volume 26. La Pensée Sauvage, Grenoble, 2006.
37. F. Spagnolo and R. Gras. Fuzzy implication through statistic implication: a new approach in zadeh's framework. In S. Dick, L. Kurgan, W. Pedrycz, and M. Reformat, editors, *Proc. of Annual Meeting of the North American Fuzzy Information Processing Society (NAFIPS 2004)*, volume 1, pages 425–429, Banff,Canada, 2004.
38. Z. Usiskin. Conceptions of school algebra and uses of variables. In A.F. Coxford and A.P. Shulte, editors, *The ideas of Algebra*, pages 8–19. NCTM, Reston-Va, 1988.
39. S. Wagner. An analytical framework for mathematical variables. In *Proc. of the Fifth PME Conference*, pages 165–170, Grenoble, France, 1981.
40. S. Wagner. What are these things called variables. *Mathematics Teacher*, 76(7):474–479, 1983.

Appendix: Table of frequencies

Variable	Absolute frequency	Relative frequency	Percentage	Rest
AL1	106.00	0.95	95	0.21
AL2	44.00	0.40	40	0.49
AL3	34.00	0.31	31	0.46
AL4	71.00	0.64	64	0.48
AL4.1	53.00	0.48	48	0.50
AL4.2	11.00	0.10	10	0.30
AL4.3	13.00	0.12	12	0.32
AL5	27.00	0.24	24	0.43
AL6	4.00	0.04	4	0.19
AL7	13.00	0.12	12	0.32
AL9	9.00	0.08	8	0.27
AL11	34.00	0.31	31	0.46
AL13	8.00	0.07	7	0.26
AL14	3.00	0.03	3	0.16
ALb1	99.00	0.89	89	0.31
ALb2	63.00	0.57	57	0.50
ALb3	33.00	0.30	30	0.46
ALb4	25.00	0.23	23	0.42
ALb5	13.00	0.12	12	0.32
ALb6	36.00	0.32	32	0.47

Didactics of Mathematics and Implicative Statistical Analysis

Dominique Lahanier-Reuter

Université Charles-de-Gaulle,
Equipe THEODILE E.A. 1764
59653 Villeneuve d'Ascq, France
dominique.reuter@univ-lille3.fr

Summary. People working in Didactics of Mathematics have constantly regarded statistical implicative analysis as a profitable and heuristic method of data analysis. First we intend to show the reasons for this interest: implicative links that S.I.A. has pointed out may be interpreted as rules and regulations connecting actions, discourses, ..., or as a group's characteristics. We develop some examples showing how S.I.A. can be used and what special research results it can provide. We insist upon some points that may be interesting methodologically to focus on: asymmetric links, nodes and separate implicative ways.

Key words: Mathematical didactics, rules, regulations, school subjetc, geometric task, geometric skill

Theorizing the relations between these two fields of research, didactics of mathematics and Statistical Implicative Analysis, in terms of producers of models and techniques on one hand and in the field of application on the other is without doubt too reductive, from a historical point of view. In fact the still very brief history of the emergence of these two scientific domains shows us some more complex connections: the coincidence of the time of emergence and recognition, that of their geographical and institutional situations- some universities and associations of researchers in France —and finally, and above all, the presence of common actors — Regis Gras particularly [4–6]— imply a dynamic which is specific to these relations. Thus the implicative analysis of data has been able to be identified as a preferred method of analysis in mathematical didactics and reciprocally in some way, some problems of didactics have been able to raise questions in implicative analysis.

First we seek to clarify the reasons why we see it as a fruitful cooperation, in seeking to understand how the clarification of rules (or quasi-rules) that implicative analysis allows proves to be a valuable and pertinent tool for mathematical didacticians. Then we will explain two of the main problems in which these rules have meaning in mathematical didactics: that of the

D. Lahanier-Reuter: *Didactics of Mathematics and Implicative Statistical Analysis*, Studies in Computational Intelligence (SCI) **127**, 277–298 (2008)
www.springerlink.com

regulation of observable behaviours of students in a situation and that of controls, in this instance understood as effects of teaching devices.

1 Rules and Regulations

The implicative analysis of data allows us to show rules (or quasi rules) that structure a set of data from calculations of the co-occurrences of some modalities of variables. These rules can be generically represented by an expression of the type "if A then B". They are consequentially hierarchical, that means that they operate an asymmetry between the regulated modalities. In this, the implicative analysis of data (henceforth S.I.A) is distinguished from other modes of statistical analysis that, if they are equally based on calculations of co-occurrences of modalities of variables, only exhibit symmetrical rules that therefore do not discriminate between the variables studied. Studying the relations between S.I.A. and the didactics of mathematics consequently questions the theoretical status that mathematics didacticians can grant to these models of rules as well as the nature or the status of data from which the modalities of variables subject to the S.I.A. are constructed.

If one can define mathematical didactics as a plan of scientific study of the phenomena tied to the transmission of disciplinary knowledge, this means that the didacticians' preferred field of observation is that of the mathematics class: the mathematics class in the sense of the material space, certainly (as the one can explore the posters, the students' notebooks...) but also in the sense of symbolic space (the mathematics class still exists when the teacher prepares his classes, when the student learns his lessons, at home, at study hall...). This class thus exists once the interaction between subjects places them, one as teacher, and the other as student, in relation to an object of disciplinary knowledge. Very schematically, one of the main objects of study in of didactics is that of the manifestations of this relationship between these three interdependent elements, the teacher, the student and the knowledge of the discipline, of its establishment and its maintenance over time. Two consequences can be drawn from this modelling. Firstly the observables are consequently constructed as interactions between master and student, master and knowledge, student and knowledge. Secondly, this study requires the analysis of the *regulations* that are simultaneously going to be generated by this system of interactions and assure its functioning. Thus, for example, one of the most fruitful problems in mathematical didactics is that of the regulations which affect the interactions between the student and the knowledge at hand if the observables are in this case actions in which the student is engaged (linguistic or not), the regulations that govern these actions (the engagement of procedures, some choices made...) or that are produced by these actions (the abandonment of certain ways of acting, certain controls...)

Some of the rules revealed through SIA are thus interpretable in mathematical didactics in terms of regulation of interactions. Two positions can then be adopted: either these rules have the status of hypotheses for the didacticians, it being his responsibility to invalidate them or to confirm them by other methods of analysis (as for example the undertaking of interviews) or they have the status of facts of experience (thus allowing the rules to contradict or to not be able to invalidate the analysis *a priori*). Some of the rules revealed through SIA are thus interpretable in mathematical didactics in terms of regulation of interactions [8, 9].

The asymmetry that these rules present is also to be taken into account in the interpretation that mathematical didactics can make of them. It poses the problem of an explication of the asymmetries by the modelling in terms of a system of interactions. The didactic system that we have summarily evoked (a triplet of interactions between student, teacher and knowledge) is a system in which the asymmetries of the characteristics tied by a rule can be explained in a number of different ways.

Let us begin with the most classic case, in which the established rules are from data corresponding to observables (the actions or the effective declarations of students or teachers gathered on site). An asymmetry between observables must correspond to the asymmetry between modalities of variables linked by a rule which has resulted from S.I.A. "all the subjects having the characteristic A have the characteristic B". This asymmetry leads to question didactically the fact that very few students have done B without having done A, have succeeded in B without having succeeded in A, have answered B without having answered A, that very few teachers have done B without having done A, etc. These regulations of doing, saying and of their effects can be the effects of temporality, of differences between the tasks proposed, of organization of knowledge...

Another case can be envisioned, that in which the established rules are from data corresponding to observables on site, but also from data perpetuated from these observables. The stability of the latter therefore reveals groups of "fixed" subjects (the students of a same socio-cultural milieu, "novice" vs. "experienced" teachers, etc.). S.I.A. can then either provide rules linking actions, statements, the effects of these actions, of these words and these constituted groups, or, by the study of the contributions of subjects to rules, establish tendencies shared by subjects from the same group, or on the contrary, equally characteristic avoidances.

We will present several cases of studies in mathematical didactics exemplifying these different uses.

2 Regulations of Situated Actions, Rules Established from Observable Modalities.

2.1 Asymmetries of Rules Established and Chronology of Tasks

The example that we develop first is that of the study of responses of students of CM1-CM2 level (9 to 10 years) to an exercise that is composed of two successive tasks. Firstly students are asked to put into order written decimals and fractions 1.2; 5.9; 7.5; 4; 9.5; 12; 5.15; ½; 2.5 secondly to place them on a graduated line. The question is: "range par ordre croissant 1,2 – 5,9 – 7,5 – 4 – 9,5 – 12 – 5,15 – ½ – 2,5". In French the word numbers associated with 5.15 and 5.6 are pronounced 'five, comma fifteen" and "five, comma six". This way of pronouncing the numbers explains a frequent error at this school level, which consists of placing 5.6 before 5.15, by only comparing the decimal parts of these numbers. However the numbers have been chosen so that the reproduction of this classification error in the second part of the task leads to a contradiction that the students —still normal for students at this school level— can comprehend. In fact, to put a point corresponding to 5.6 on the graduated line, then to put one that corresponds to 5.15, in moving back the first one by a space of "9" (the space between 15 and 6) leads the student to place 5.15 erroneously on the point that should correspond to 6.5 $(5.6 + 0.9)$. This placement can seem contradictory with that which corresponds to 6.2 and other points. We will say in this case that the information given to the student by the erroneous placing of 5.15 is an element of the *environment* with which the student interacts.

If consequently we expect certain students to commit errors in the ordering of written numbers, on the other hand, we wonder about the effects of the consequences of these errors during the execution of the second task. Two types of common considerations in mathematical didactics allow us to anticipate them. Firstly, the information that the erroneous placement of the points on the line provides is not "naturally" interpreted in terms of a contradiction. The reading and the comprehension of this information by the student require that he uses certain knowledge: in fact it concerns considering the placement of 5.15 and 6.2 as "strange" and sees to consider them as a consequence of the classification error of 5.6 and 5.15. Previous research done on this error or this problem in a school situation lead us to differentiate a student's recognition of an error from his doing of that error. Or to put it simply, the perception of a contradiction in his results is often insufficient to lead a student in a class to invalidate the latter because he still does not feel invested with the responsibility to resolve the problem raised [1,2,11]. The question of the study of students' behaviour is therefore a legitimate question.

The study of the corpus of written productions of the students makes apparent the diverse strategies used to respond to the two questions of the exercise. To order the numeric writing, some students used a classification strategy by 'types of writing', in classifying first the written fractionals, then the written entire numbers, that have no decimal point, then the written decimals, in separating those that only have one figure after the decimal point from those that have two. The pupils adopting that classification take into account only the length of the numbers as they are written out. Others, as we could have expected, classified the written figures according to their entire part (visible or calculated in the case of ½), then according to their decimal part, which was also considered as an integer number (5.15 is then placed after

5.6). Finally, certain students 'neglected' to use the reference points of the graduated line and instead used the line as a 'writing line' without placing any points. This study also allows us to decide if, in the end, the student produces two different orders which are consequently contradictory, or if on the contrary he produces two coherent orders, even if they are erroneous.

If S.I.A. is applied to data it may result in an association group as in Graph 1 that allows us to see the following rules:

1. "Adopting, definitively, a classification or writing by types of writing" implies, 'accepting a lack of accord between the two orders produced" (99%). (Graph 1, 3 $\Rightarrow$ 13)
2. "Working, definitively, on the graduated line, as a writing line" implies obtaining "two coherent orders, even if they are erroneous" (95%) (Graph 1, 7 $\Rightarrow$ 12).
3. "Producing, finally, an exact classification of written numbers" implies obtaining "two coherent orders" (95%) (Graph 1, 4 $\Rightarrow$ 12).

These three rules are interpreted as regulations of student behaviour when faced with these two tasks. It is possible to read in rule (1) the fact that certain of these students see —or decide to see— the two tasks as distinct. For instance, one of the students answers (at first question): "4; 12; 1,2; 2,5; 7,5; 9,5; 5,15; 1/2". However, he puts marks on the graduated line for "1,2", next for "2,5", next for "4", etc. We can consider that they do not understand (in the situation explored) the articulation between the order that the linear arrangement of the written numbers "shows" and that which the arrangement of points on the graduated line "shows". Secondly, rule (2) can be interpreted, in the context of the situation, as an "avoidance" of the second task. The student copies the preceding list of writing onto the graduated line, and thus avoids taking into account the eventual difficulties that he will face in assuring coherence between the two orders. Finally rule (3), in establishing the asymmetry between the two modalities[1], leads us to surmise that checking the coherence of the two orders allows, for certain students, a rectification of the classification of the written figures.

Thus the didactic organization of these two tasks, and particularly the conception of an environment by the interpretation of its retro-actions, to reveal an incoherence of the results, is insufficient: it is necessary in fact, that the student accepts to link these two tasks in order for him to accept to interpret the results of one according to the other.

2.2 Asymmetry of Rules and Representations of Subjects.

The use of S.I.A. in didactics of mathematics goes beyond the problematic that we have just mentioned. Another field of investigation uses S.I.A. as well.

[1] That could not particularly be assumed by a test of χ^2

We will call usually, the field of reconstruction of observed subjects' representations. Indeed, teaching and learning situations can be considered as social situations defined by the stakes, positions and specific roles of those involved. The reconstructions the authors of these situations make of these stakes, of these positions and of these roles have consequences on effective actions. As a matter of fact, these representations can be considered as knowledge networks. In that case, S.I.A. can contribute to the recognition of such networks. This time, implicative rules of the type "if A then B" can be interpreted as follows: factor A is predominant compared to B in the creation of the representation.

The example that we are presenting here is the study of representations of some school subjects' organization for high school students. The notion of "school subject" is a complex one, even though it can be more or less naturalized within the school system institution.

What is "French", for example, or "Physique-Chimie"
(Physics-Chemistry)? Can we describe a school subject by the organization of knowledge which makes it specific or should we deal with it according to the teaching and learning techniques that constitute it? If these theoretical issues are far from being solved, the few studies done with students confirm the interest researchers have for them.

In fact, it seems that a large number of students have trouble identifying the different subjects: for example, some of them cannot identify the terms of a French exercise from the terms of a History exercise. It also seems that the identification criteria are very often material in primary school pupils (from 7 to 11). Those who best identify the different disciplines say, e.g. they do so by using material clues, such as a notebook's colour. However, the point is that the difficulties have consequences on how well students do in school. Identifying a discipline's boundaries and being able to recognize some of its functioning is a factor of success.

We are presenting here a study about these different issues that deal with high school students' representations of scientific disciplines. A questionnaire was given to four scientific junior classes (Première S in French–), the students are 17 years old), and scientific senior classes (Terminale S in French –the students are 18 years old). The questionnaire's aim was to ask students at the end of their high school years how they perceive the different scientific fields they are being taught.

The different questions they were asked are pertinent to the following intensive subjects: mathematics, analysis and statistics identification. Students of this level recognize analysis and statistics as "parts" of mathematics. However, what "parts" means is not as clear as it seems to be. Indeed, analysis and statistics can be considered as separated mathematical fields from an epistemological point of view: even if some of their objects and methods are undoubtedly common, knowledge projects, special applications fields, and symbolic representations contribute to specify these scientific areas. But we may suppose that this approach is not the one the students have. One of our theoretical hypotheses is that the partition between analysis and statistics, and the relationships between mathematics, analysis and statistics, elaborated by students, are generated by their school practices. Therefore, we intend to focus especially on differences that can be related to experiences of mathematics, analysis, and statistics teaching and learning.

The first questions are about the identification of the school level where the students think the teaching of math (analysis, statistics) started. Students can usually say in a coherent way when mathematics and statistics started to be taught (in kindergarten and elementary school for math, in Junior high school for statistics). Students have much more trouble locating when they were first taught analysis; they usually avoid the question. Then they are asked how useful they think the preceding disciplines are. Here again the answers are pretty clear cut: they see a cognitive use in mathematics, but more rarely any use in real life (future jobs, etc.). As for statistics as a school subject, it is quite the opposite, pupils see it as useful outside school, in the real world. The next questionnaire item asks students to show how mathematics is used in other disciplines. All students answer that maths are used in physics but very few of them point to other school subjects in which analysis could be useful. The last question in this part of the questionnaire, about the usefulness of these three fields of knowledge, is the following: the student is asked if he remembers or not his teacher talking about how useful these disciplines can be. The four last questions concentrate more particularly on school work habits: how they identify a class (math, analysis, statistics), how they organize themselves (do they use separated class folders or not), how they identify exercises of the same subjects, and how much they think they have learnt in these subjects throughout the school year.

A student's answer can be considered as an indicator of the way he puts back together the school subject referred to and the way it is organized for him, through his memories of it, how useful he thinks it is, and what definition he gives to it. We therefore consider the answer as a trace of what Yves Reuter calls "subject awareness" [13].

To understand the graphic (Graph 2) we have only kept the characteristics of related answers.

The organization into a hierarchy of the different items shows us how the students deal with the three different disciplines. They have most trouble identifying analysis whereas they identify mathematics and statistics more clearly. One of the first results of this study is therefore to show that students dealing with closely related fields of knowledge, in the same physical space of the classroom, have some trouble defining the boundaries of the different fields involved. This result is important for us to know, because in the French educational system, even in junior high school, students often face classes like "Histoire-Géographie" (History- Geography), "Physique-Chimie" (Physics-Chemistry), etc.

Looking at the graph (Graph 2) shows that these identifications are interrelated.

(1) "Students recognize the characteristics of an analysis lesson" implies "Students recognize the characteristics of an analysis exercise" (85%) (Graph 2 22 ⇒ 25)

(2) "Students recognize the characteristics to an analysis exercise" implies "Students recognize a list of the knowledge learnt in analysis class" (90%) (Graph 2 25 ⇒ 31)

The two combined rules can be interpreted as the traces of a complex network of knowledge which keeps the representations of a specific subject. They suggest several levels to these representations of "analysis": the highest level would be determined by the ability to identify, in what is taught, the subject's characteristics.

Other implicative ways are also to be considered:

(3) "Students recognize the characteristics to an analysis class" implies "Students recognize the characteristics of a mathematics exercise" (85%) (Graph 2 22 ⇒ 24).

(4) "Students recognize the characteristics to a mathematics exercise" implies "Students recognize the characteristics of a statistics lesson" (85%) (Graph 2 24 ⇒ 23).

The degrees of subject awareness are therefore probably not independent from one another: the graph nodes in particular (here the identification of an Analysis class specificities) show the interrelation of these networks.

Lastly, the *graphs' separation* also suggests that factors of subject' identification are different and not linked. As a matter of fact, the items in the central graph point back to everything that has to do with reconstruction within work tasks.

The items on the right end side refer back mainly to what has to do with reconstruction involving other parties, here the teacher.

(5) "To remember a teacher's presentation of the usefulness of analysis" implies "Remembering the presentation of statistics usefulness by a teacher" (85%) (Graph 2 16 ⇒ 17).

This separation among implicative ways and therefore among networks making up the representations studied here, is interesting.

What seems to be important here, apart from the fact that what the teacher says can help define the subjects taught, is on the contrary to find out that there is no link between subject practices within a classroom and what the teacher says about them, no link between learning and its social usefulness.

Subject awareness as it comes up in what students say is not, therefore, the sum of the different perceived characteristics, but rather the delicate elaboration of links between these different characteristics. One of the interesting aspects of S.I.A. is the way it allows the description of knowledge networks, since this possibility is directly linked to one of the problematics in didactics, the study of students' and teachers' representations and conceptions.

2.3 Rules Interpreted as Traces of Skills.

One of the last S.I.A. fields of applications in mathematical didactics that we will present is that of skills reconstruction, stemming from observations of student behaviours. As we have already seen, rules of the type "if A then B" showed by S.I.A. must be interpreted.

A and B still refer to observable behaviours. The implicative relationship can be interpreted, according to the different cases, such as: B is a consequence of A or A is an explicative factor of B. The reconstruction of students' ability to complete a particular task can be read in the implicit rules which govern students' behaviours.

The example that we have chosen to explore stems from a study carried out in the didactics of mathematics, even though it is only a part of a much wider research which involves different disciplinary didactics [10].

The main problematic in which this study makes sense is in the relationships between teaching and learning, or to be more precise, that of measuring the effects of a particular pedagogical set of devices on an aspect of mathematics learning. Numerous prior studies can be quoted on this theme, amongst which two equally interesting syntheses have been recently and simultaneously published (on the one hand [12] and on the other hand [3]). Their hypothesis is that the particularities of didactic management of teaching and learning situations by the teacher can influence the building of mathematical knowledge and the appropriation of other knowledge of the discipline by the students concerned. We will extend this hypothesis to attitudes and behaviours students present when faced with specific tasks.

The study tries to measure the effects of a pedagogical set of devices, which make it necessary to compare skills developed by students within a particular study set up and the skills of students who are not in the study setup. We will get back to our goal later. However, one of the stages of this undertaking, is first to describe the skills brought forth by the group of students while they were doing the specific subjects tasks.

The example we will bring out here is, as we have mentioned above, in this research perspective[2]. It is supported by the analysis of skills in geometry and the language skills of 9 and 10 years old pupils who had to do a writing geometry task.

The task is the following: "how to draw this figure?" (See Fig. 1).

It was given in seven different classes, by the teachers themselves, without any outside observer. It thus appeared as a more or less ordinary task within the class. 165 texts were collected, of which 163 can be taken into account.

The study is therefore about a writing task taken from "an instruction program to reproduce a complex figure". To be carried out, the writer is required first to identify and to name some of the constructible elements, then to point out the constructible relationships that exist among the different elements.

[2] These results come from an IUFM Research "Effet d'un mode de travail pédagogique Freinet en Z.E.P" R/RIU/04/007

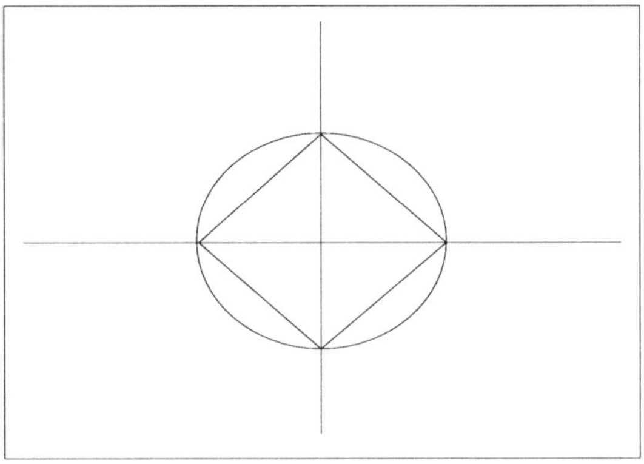

Fig. 1. "how to draw this figure?"

We analyze these texts as pupils' works, that is to say we try to take into account the context in which they were created and the different pupils' status. It is not the same, for instance, for a nine years old or a ten years old pupil to produce, as a construction program, the following: "draw two perpendicular lines of which centre is the meeting point of lines and link the intersection of the two lines and of the circle" or "trace a square, its diagonals and its centre, trace the circle whose centre is the same as the square's and which touches the summits of the square" or "draw a circle and a square inside the circle". As a matter of fact, the first text points to objects and geometrical relationships between the objects that are constructible by pupils at this school level. On the other hand, the second one needs the drawing of a square, which isn't as easy for them. The last one avoids the necessary drawings to be able to fit a square in a circle. Taking into account the pupils' school level and considering the outcomes as productions, we have chosen not to rank them according to how correct the answers are, but according to the choices pupils made.

We have kept as first indicators of the way pupils write, the chosen elements and their designations, the geometrical relationships mentioned and their designations. The corresponding indicators are the numbers of geometrical terms used, the chosen elements -circles, lines, summits, ... -, relationships such as perpendicular, topological positions, etc. Therefore, the point is to determine the figure analysis pupils have chosen to make through what they say. The way of analyzing can be quite different from one pupil to the other. Some of them only dealt with the lines that shape the diagram: the two lines, the circle and the square, O or the four summits of square A, B, C, D. From a theoretical point of view, these two ways of looking at it are linked to different analytical skills. As a matter of fact, looking at a geometrical figure as a punctual structure requires going beyond immediate perception, which only shows lines entangled. The use of S.I.A. makes it possible to test the theoretical hypothesis of the different analytical stages in constructing a geometrical diagram.

2.3.1 Split implicative ways: the example of different geometric skills.

Studying the graph (Graph 3) showing the implicative links between the different items kept makes it possible to dissociate two implicative ways. One of the networks links the items which take into account points A,B, C, D to those that determine the diagram's elements in relation to the others and lastly to a central item, the one that shows that the pupil not only takes into consideration point O but also gives it at least two different status: for example from being the centre of the circle, it becomes the lines' intersection or the square's centre. A second network links items which, on the contrary, indicate that the pupils didn't take into account the different points ("ONM" and "ABCDNM") to those that mark the absence of constraints between the different elements of the diagram.

The first network groups writing productions where the diagram's analysis is rather an analysis in terms of punctual structures. The writing productions communicate certain constraints, which heavily influence the drawing of the different elements. The second network gathers texts which are more like descriptions, in which pupils only need to mention visible lines and sometimes their respective topological positions (inside, on). The first type of output is, for us, an indication that the skills needed to move on to "a construction program task" are met, whereas the second type of output is more the indication of an interpretation of a task in terms of "description of a regular drawing". So, the graph S.I.A. provided, because it can differentiate implicative ways, allows displaying connected geometric skills separately. We provide here two texts that are more or less representative of these positions: "Tracez un carré de 2,9cm. Tracez deux lignes qui se croisent au milieu du carré. Le croisement des deux droites vient former un centre. Celui-ci (centre) permettra de tracer un cercle qui touche tous les sommets du carré " (Valérian, CM2) and " Il faut faire comme une croix puis le cercle de 4,2cm et enfin le losange " (Tiffany, CM2)[3].

2.3.2 The networks' nodes: crucial points.

If the dissociation of implicative nodes in a graph is interesting, the identification of nodes in the different networks is no less interesting. What we call nodes here are the items that take part into several links.

In the first network (Graph 4) we notice how important the node is "O appears and goes through a change of status". As a matter of fact, pupils can mention the centre of the circle or the middle of two lines without having to get into a perspective of linking both lines (or even saying it).

[3] Original spelling has been changed.

Therefore it is indeed the change of status (in turn, centre of the circle and intersection of lines or centre of the circle and intersection of the diagonals) which makes up the decisive criterion for classifying the pupils' production and its grading.

Trusting the reader with this change and finding the discursive ways to do so might be a crucial stage. It involves indeed being able to find the means to bring back, to recall an element already present in the text (therefore a handling or the implementation of anaphora), it also implies being able to get away from visual evidence of the figure, in short being able to go beyond an immediate visual contact to a written account of an invisible change. Point O doesn't move but its function changes.

So S.I.A. allows us proving the deciding role played by some analysis criteria and the necessity of their presence in skills evaluation.

2.3.3 Univocal implicative links: the case of some linguistic characteristics.

The complexity of some of the graphs studied above shouldn't mask the fact that in some cases they are in fact extremely simple. Their "simplicity" is nonetheless a source of information that shouldn't be overlooked. Here we will focus on the study of links between the items corresponding to the linguistic characteristics of the texts produced by pupils (see Graph 3).

These characteristics have to do with the length of text produced, the different modes used (infinitive, imperative, indicative), subjects ("I", "we", "you"). We have also kept the signs of planning: as a matter of fact the task that is proposed can be interpreted as one of the writing of a series of actions aimed at putting the figure back together. Some pupils write in an orderly list of actions. Others note the temporality by using adverbs (now, then, ...). Others conclude their texts by an indication of the type "here it is, the figure is done" or simply by using the word "end". Some disorders can be caused by planning operations. Thus, some pupils refer to elements that have not yet been introduced in their text; others add constraints that "they had forgotten".

Lastly, one of the last characteristics of the produced writings is that of the inadequate use of definite articles — "the..." and indefinite — "a, some..." As a matter of fact, the presented elements can be undetermined by what precedes them or on the contrary entirely determined. For instance, if the pupil has said how to build the four points (and give name to them ABCD) on two perpendicular lines which intersection is 0, the circle he is then going to talk about (0 being its centre and going through the 4 points) is entirely determined. Thus these determinations are not of a linguistic order: it isn't because one element has already been quoted in the text that it is therefore determined, but because the geometrical constraints define it in a unique way. There is no doubt then, that the tension between the two orders of determination explains the numbers of disorders in the use of articles.

Even though there are quite many linguistic characteristics, on the other hand, the graph isn't really complex. Three main rules stand out:

- (1) "Writing "I" implies "using the indicative mode" (99%) (Graph 3 12 ⇒ 11).
- (2) "Using the infinitive mode" implies "Building a generic subject "one" "(99%) (Graph 3 9 ⇒ 13).
- (3) "Using the imperative mode" implies "building a subject "you" " (99%) (Graph 3 10 ⇒ 14).

The links that come up are expected for the most part since the very use, even partial, of the imperative and infinitive modes is indeed linked to the pronouns which signal what the reader puts back together: a "peer" for the imperative mode, signalled by the pronoun "tu" (informal you) and "vous" (formal you), "a generic reader" for the infinitive mode, an "evaluative reader" signalled by "I" and the indicative mode. However, the presence of univocal relationships between the chosen pronouns and the modes used, gives an unexpected rigidity since it is possible, in the indicative mode to use "on" (people)" and "tu" (informal you)". Unlike classical tests suggesting symmetrical links between studied variables, the S.I.A. allows questioning of these strong constraints, stemming from the fact that the writing is produced in a school situation.

This makes it possible to think that the rules students use define them as actualisations of discursive genres. Thus, we suppose that building the reader as a "peer" is characteristic of a school writing genre in math class, and therefore can be perceived as legitimate by pupils for several reasons. It could be that the pedagogical and didactic devices make such positions possible, because help and cooperation are principles put into practice in the language used in the different subjects taught. It could also be that the ways exercises are written in school books define such a reader. Building a "generic reader" is also an identifiable characteristic. However, this characteristic is not as frequent in geometric construction exercises in elementary school schoolbooks. On the other hand, it is frequent in "description of recipes" schoolbooks.

At this level, this genre, like those of "construction programs" or "users' manuals", take up a rather important part in school activities, and can be found in places other than schools. Lastly, the discursive ways of using "I" with which a pupil shows the reader what he can do or manages to do is a characteristic of school evaluation situations.

Therefore, S.I.A. interpretation of results appears to take into account univocal links. Unlike in traditional analysis, links can be interpreted as constraining rules governing the actions observed.

3 Regulations Relative to Groups of Subjects, Rules Established in Observable Modalities and in Contributory Variables Modalities.

An important problematic in mathematics didactics is, as we have shown, that of the interpretation of regulations in pupils' actions seen in a group's characteristics. It is a matter of trying to know if we can give the status of "results of teaching devices used" to certain behaviours and to certain skills. These issues are what trigger experiments as well as comparisons of ordinary practical school experiences. It is about sorting out what is specific to a group of pupils, whichever methodology is used: in the case of experiments, we are trying to compare the performance or competence of pilot groups and those of experimental groups, in the case of observations, the groups constituted are all of the classes.

The study mentioned above addresses this question, since it questions some teachers' demands to set up specific teaching devices in their classes. The study attempts to describe the effects of such a way of functioning, from the point of the pupils' performance. Its goal is to show the results strengthening or invalidating the hypothesis according to which, particular effects, read in pupils' behaviours, can be linked to specific devices used in classes. It is this question that brings us to compare the readings done by the researcher who studied the pupils' group activities, according to whether the students received one kind of teaching or another. Remember that there are seven classes studied. They are elementary classes respectively 9 year olds and 10 year olds, all located in the suburbs of Lille.

Taking up again the different characteristics studied, relative to geometrics and linguistics skills, we are now studying links connecting the different items, showing to what class they belong.

This graph (see Graph5) has been completed keeping only the paths leading or getting to one of the explored "classes": the different CM1 (9 year olds) or CM2 (10 year olds) classes. We are thus trying to bring forth characteristics in groups of pupils.

3.1 "Isolated" Groups

Methodologically, we can first look at the isolated groups. In the case of the example we are developing, one of the CM2 class ("CM2 Wa") is isolated. As it would be the case in "classical" analysis, the absence of implicative links marking this group of pupils is interpreted as a sign of diversity in this group of pupils' written productions (according to the chosen criteria).

3.2 Characteristic Abilities

S.I.A. makes it possible to perceive the cases where characteristics are those of a part of the group of pupils studied, compared to a group where all the pupils of the group were studied. They "almost" all share the same characteristics, contrary to the modes of analysis which produce symmetrical links. We will start by presenting a case where a characteristic comes up as specific to one of the groups.

The example is that of the characteristic "writing a text using "I" which is the characteristic of one of the classes coded CM1 Wb.

- (1) "Writing a text using "I" " implies "belonging to class CM1 Wb" (99%) (Graph 5 12 $\Rightarrow$ 2).

Only the pupils of this class chose a particular writing behaviour, which indicates that pupils read into a suggested situation an evaluation situation: the expected reader is the teacher and pupils show what they can do. But what seems even more important is the fact that the implicative link of maximal intensity means that no pupil "or rather almost none" in other classes have reacted in that way. To explain this specificity, we support the hypothesis that the way the task was done has been singularized in class: maybe only this CM1 teacher has presented the exercise as an evaluation or at least as an exercise that he would check on.

3.3 Groups Characterized by Capacities

On the other hand, other implicative links show that almost all pupils of a same class share some skills and also make the same mistakes, etc. Let's keep in mind some of them:

- (1) "Belonging to CM2 Hb1" implies "using "tu" " (90%) (7 $\Rightarrow$ 14).
- (2) "Belonging to CM2 Wa" implies "Mentioning point 0 and changing its status" (95%) (5 $\Rightarrow$ 29).

This time, these characteristics are met in almost all the pupils of a same class.

We think that they are the results — sometimes indirect — of the didactic or pedagogical approaches used in class. What is left is to interpret these different rules. If almost all the pupils of the first particularized class address a reader who acts like "a peer" in their writings, it is no doubt because help and cooperation are legitimized and encouraged in these classes or because these forms of communication are used. If almost all the pupils of the second class show good geometric skills, it is no doubt thanks to the teaching techniques used.

But this geometry skill cannot be understood without the linguistic skills which make it possible to communicate it. Using "tu" (informal "you") is not always the required school form. Now, these two classes have pupils from different social backgrounds: in the second class the pupils come from more privileged families than the other studied classes. Relationships between social classes and linguistic strategies are certainly complex and not mechanical. However, the results we are getting are coherent with those of other studies on these relations. As a consequence, we cannot neglect the explicative factor of cohesion between observed behaviours. We are bumping here into a recurrent problem: the principles of subjects' categorisation are obviously never unique or uniform. The gatherings of pupils in this particular case cover both institutional groups (school classes) and "social" groups (coming from more or less privileged families). This remark makes it very clear that evaluating the effects of teaching methods cannot take the shape of simple cause to affect relationships.

Another way for bringing up class characteristics is provided by S.I.A. We may have access to contributions of every subject or of additional variables to implicative links [7]. These methods allow us to count how many pupils of each class contribute to these significant links, or to exhibit which group of subjects has more weight in these established links. So doing, we can confirm some of the previous exposed results and, on the other hand, bring to the fore some new ones. An example of confirmation of previous results is given by looking at the subjects that contribute to the link "using "I" " and "using indicative mode". Not surprisingly, all these subjects belong to the CM1 Wb class. Considering the additional variables "belonging to CM1 Wa", "belonging to CM1 Wb", "belonging to CM1 Hb1", ..., S.I.A reveals that "belonging to CM1 Wb" is the most contributive variable to the link "using 'I" and "using indicative mode". Analyzing contributions may also produce some new results. For instance, the implicative link between "square located" and "ABCD not mentioned" (see Graph 4) may be interpreted as the following: some of these pupils regard the given figure as a network of lines and not as a structured set of points. Analyzing contributions of additional variables mentioned above affords no determining difference between them. Nevertheless, it appears that 66% of the optimal group are CM1 pupils, considering the subjects' contributions to this link. So, a more adequate additional variable to this part of the study is: "belonging to CM1 (or not)", which, actually contributes mostly to this link.

This being said, it is in any case obvious that using S.I.A., even if it doesn't help to find cause and effect relationships, makes it possible to accumulate, little by little, shared behaviours, common capacities, specific mistakes, ... linked to groups of subjects. What is left to find, then, is global coherence to the groups' characteristics.

4 Conclusion

Using a central problematic in mathematical didactics, we have been able to show the efficiency of dealing with S.I.A. techniques. The input of this method to analyze data cannot be disregarded for several reasons. The almost-rules established by S.I.A. from observables can easily lend themselves to interpretation in terms of action regulation. The implicative paths can be read in terms of networks. Lastly, the asymmetry of links seems essential in exposing any explicative hypothesis of certain phenomenon pertinent to teaching and learning. We have also been able to sketch, through the detailed relationships of research examples, particular methodological behaviours: the attention to give to the interpretation of the internal cohesion of implicative graphs, but also to the separation of these graphs, as well as to the graphs' nodes and the univocal links. These are paths of thinking to be pursued.

References

1. G. Brousseau. Le contrat didactique : le milieu. *Recherches en didactique des mathématiques.* Volume 3, pages 309–336, La Pensée Sauvage, Grenoble, 1990.
2. G. Brousseau. *Théorie des situations didactiques.* La Pensée Sauvage, Grenoble, 1998.
3. M. Bru, M. Altet, C. Blanchard-Laville A la recherche des processus caractéristiques des pratiques enseignantes dans leurs rapports aux apprentissages. *Revue Française de pédagogie.* Volume 148. INRP, Paris, 2005.
4. R. Gras. L'analyse des données : une méthodologie de traitement de questions de didactique. *Recherches en didactique des mathématiques.* Volume 12:1, pages 59–72, La Pensée Sauvage, Grenoble, 1992.
5. R. Gras, A. Totohasina, S. Almouloud, H. Ratsimba-Rajohn, M. Bailleul. La méthode d'analyse implicative en didactique. Applications. In : M. Artigue, R. Gras, C. Laborde, P. Tavignot (eds.): *Vingt ans de didactique des mathématiques en France.* Pages 349–363, La Pensée Sauvage, Grenoble, 1994.
6. R. Gras. *L'implication statistique, Nouvelle méthode exploratoire de données.* La Pensée Sauvage, Grenoble, 1996.
7. R. Gras, J. David, J.C. Régnier, F. Guillet. Typicalité et contribution des sujets et des variables supplémentaires en Analyse Statistique Implicative. *Extraction et Gestion des Connaissances (EGC'06).* Volume 2, pages 359–370, Cépaduès Editions, 2006.
8. D. Lahanier-Reuter. *Conceptions du hasard et enseignement des probabilités et statistiques.* P.UF., Paris, 1999.
9. D. Lahanier-Reuter. Exemple d'une nouvelle méthode d'analyse de données : l'analyse implicative. *Carrefours de l'éducation.* Volume 9, pages 96–109, CRDP Amiens, 2000.
10. D. Lahanier-Reuter. Enseignement et apprentissage mathématiques dans une école Freinet. *Revue Française de Pédagogie,* Volume 153, pages 55–65, INRP, Paris, 2005.
11. C. Margolinas. *De l'importance du vrai et du faux.* La Pensée Sauvage, Grenoble, 1993.
12. A. Mercier, C. Buty C. Evaluer et comprendre les effets de l'apprentissage de l'enseignement sur les apprentissages des élèves : problématiques et méthodes en didactique des mathématiques et des sciences. *Revue Française de Pédagogie.* Volume 48, pages 47–59, INRP, Paris, 2004.
13. Y. Reuter. Les représentations de la discipline ou la conscience disciplinaire. *La Lettre de la DFLM.* Volume 32, pages 18–22, 2003.

Appendix

a) S.I.A Graph 1.

Chronology of tasks, two types of number classifications.

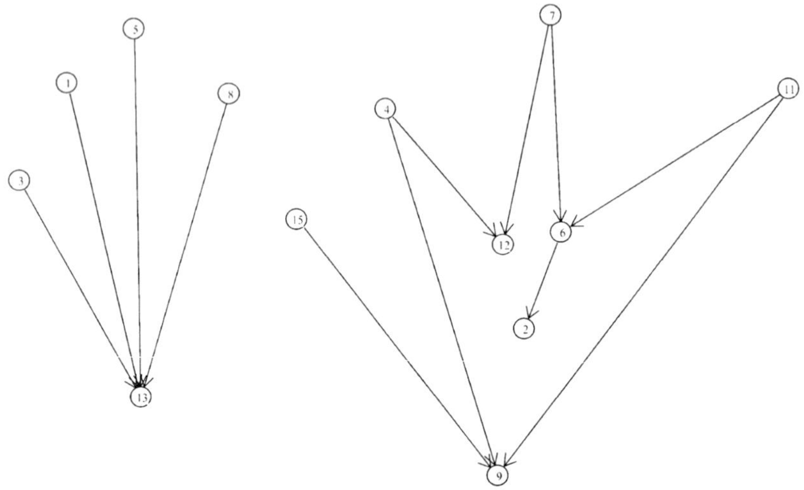

Fig. 2. 1 Writing classification: No answer; 2 Writing classification based on numerical relationship with errors; 3 Writing classification based on length; 4 Writing classification exact; 5 Writing classification incomprehensible; 6 Writing classification with inversion 5,15 and 5,9; 7 Linear classification based on 'line'; 8 Linear classification No answer; 9 Linear classification based on points; 11 Linear classification with inversion 5,15 and 5,9; 12 Adequation between the two classification; 13 Non adequation; 15: Linear classification with 5,15 pointed on 6,5.

b) S.I.A Graph 2.

Questionnaire and students representations of disciplines.

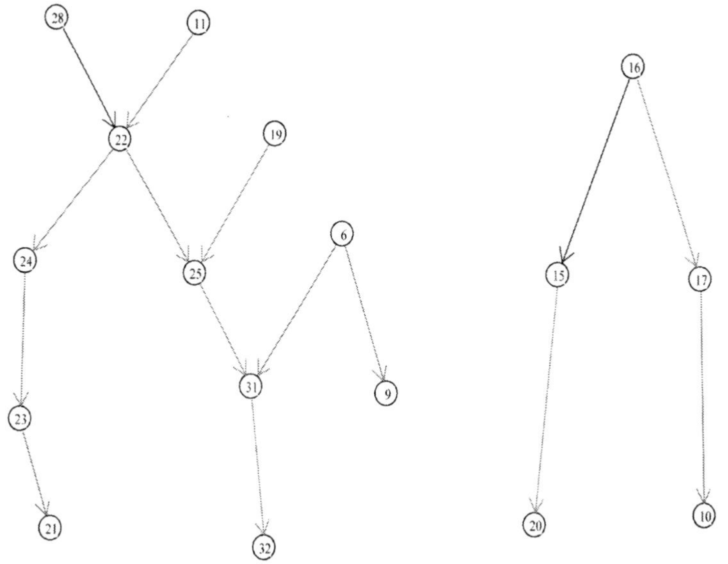

Fig. 3. 6 Analysis first teaching identified; 9 Mathematics cognitive use identified; 10 Mathematics social use identified; 11 Analysis cognitive use identified; 15 Remembering teacher speech about Mathematicsmathematics usefulness; 16 Remembering teacher speech about Analysis usefulness; 17 Remembering teacher speech about Statistics usefulness; 19 Identification of Analysis use in other disciplines; 20 Identification of Statistics use in other disciplines; 21 Identification of Mathematics lessonmathematics class characteristics; 22 Identification of Analysis lesson classcharacteristics; 23 Identification of Statistics lessonclass characteristics; 24 Identification of Mathematicsmathematics exercises characteristics; 25 Identification of Analysis exercises characteristics; 28 Using a special notebook for Analysis; 31 Analysis knowledge identified; 32 Statistics knowledge identified.

c) S.I.A Graph 3.

Writing in geometry (all items included).

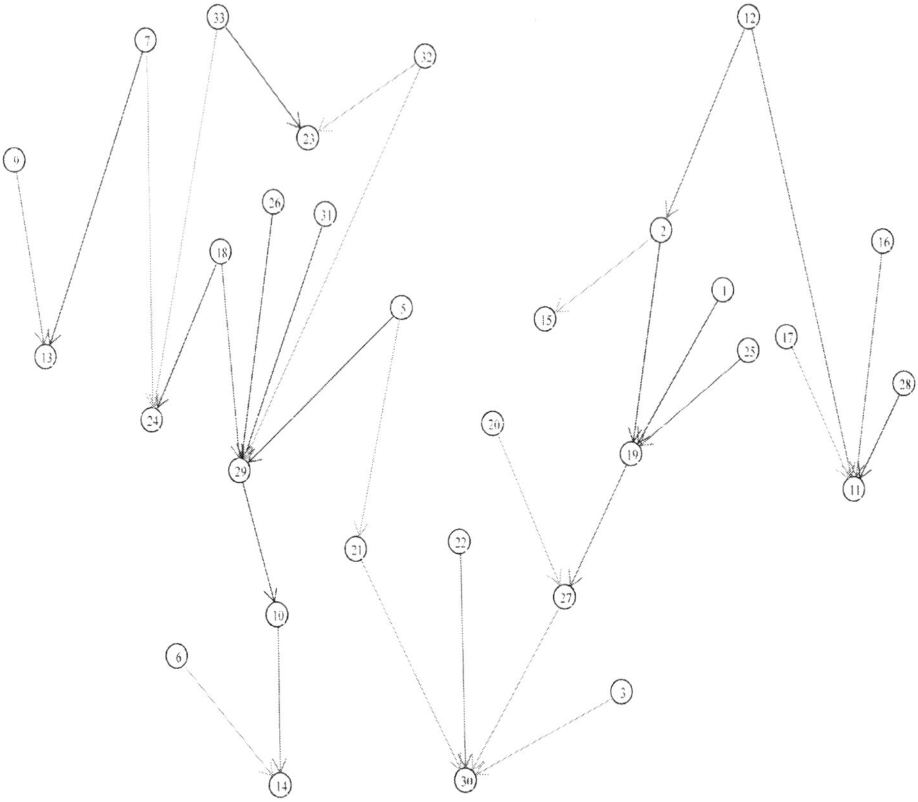

Fig. 4. 1 CM1 Wa; 2 CM1 Wb; 3 CM1 Hb; 5 CM2 Wa; 6 CM2 Hb1; 77 CM2 Hb2; 9 Using infinitive mode; 10 Using imperative mode; 11 Using indicative mode; 12 Using 'Je' (I); 13 Using 'On' (one); 14 Using 'Tu or Vous' (informal you or formal you); 15 Planning marks; 16 End marked; 17 Error on 'le' (the) or on 'un' (a); 18 Circle determined; 19 Circle independent; 20 Circle located; 21 Square independent; 22 Square located; 23 Square determined; 24 Lines independent; 25 Lines located; 26 Lines determined; 27 O No mention; 28 O mentioned; 29 O mentioned, two status; 30 ABCD No mention; 31 ABCD mentioned; 32 ABCD mentioned, two status; 33 ABCD constructed.

d) S.I.A Graph 4.

Writing in geometry, geometrical abilities.

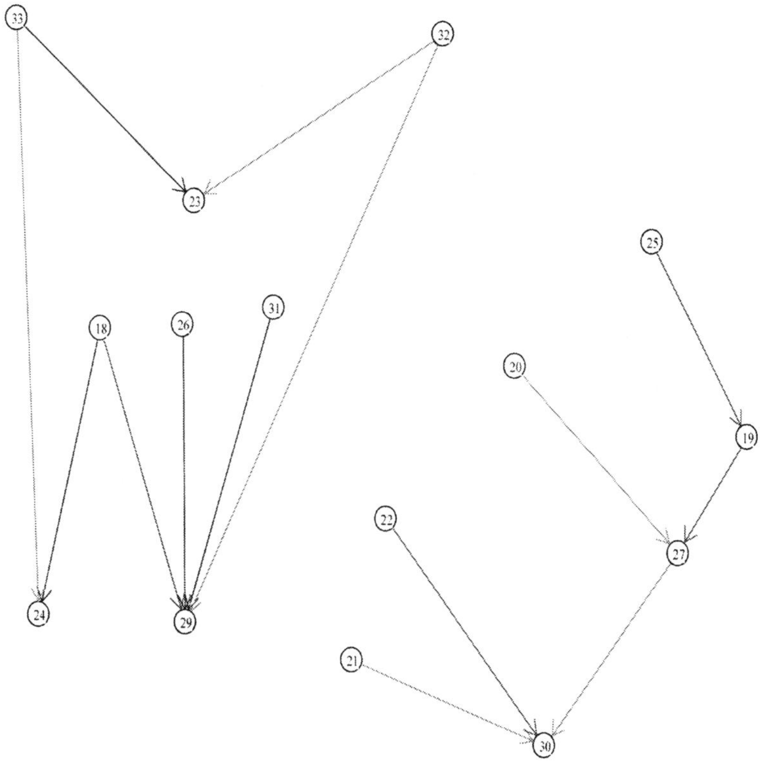

Fig. 5. 18 Circle determined; 19 Circle independent; 20 Circle located; 21 Square independent; 22 Square located; 23 Square determined; 24 Lines independent; 25 Lines located; 26 Lines determined; 27 O No mention; 29 O mentioned, two status; 30 ABCD No mention; 31 ABCD mentioned; 32 ABCD mentioned, two status; 33 ABCD constructed

e) S.I.A Graph 5.

Writing in geometry, groups characteristics.

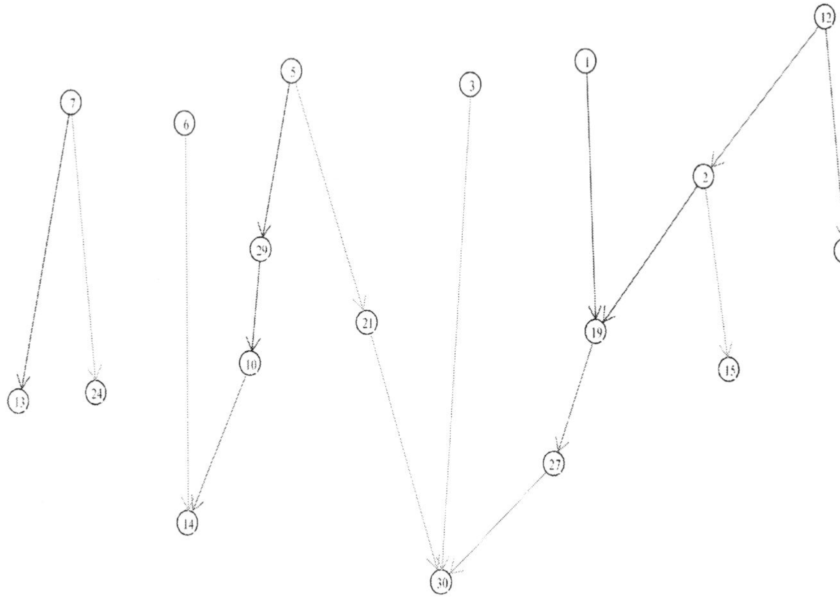

Fig. 6. 1 CM1 Wa; 2 CM1 Wb; 3 CM1 Hb; 5 CM2 Wa; 6 CM2 Hb1; 77 CM2 Hb2; 10 Using imperative mode; 11 Using indicative mode; 12 Using 'Je' (I); 13 Using 'On' (one); 14 Using 'Tu or Vous' (informal you or formal you); 15 Planning marks; 19 Circle independent; 21 Square independent; 24 Lines independent; 27 O No mention; 29 O mentioned, two status; 30 ABCD No mention.

Using the Statistical Implicative Analysis for Elaborating Behavioral Referentials

Stéphane Daviet[1,2], Fabrice Guillet[2], Henri Briand[2], Serge Baquedano[1], Vincent Philippé[1], and Régis Gras[2]

[1] PerformanSe SAS, Atlanpole La Fleuriaye, 44470 Carquefou,
{stephane.daviet, serge.baquedano, vincent.philippe}@performanse.fr
http://www.performanse.com

[2] LINA-École Polytechnique de l'Université de Nantes,
La Chantrerie – BP 50609 – 44306 Nantes CEDEX 3
{stephane.daviet, fabrice.guillet, henri.briand, regis.gras}@univ-nantes.fr

Summary. Various informatic assessment tools have been created to help human resources managers in evaluating the behavioral profile of a person. The psychological basis of those tools have all been validated, but very few of them have follow a deep statistical analysis. The PerformanSe Echo assessment tool is one of them. It gives the behavioral profile of a person along 10 bipolar dimensions. It has been validated on a population of 4538 subjects in 2004. We are now interested in building a set of psychological indicators based on Echo on a population of 613 experienced executives who are 45 years old and more, and seeking a job. Our goal is twofold: first to confirm the previous validation study, then to build a relevant behavioral referential on this population. The final goal is to have relevant indicators helping to understand the link between some behavioral characteristics and current profiles that can be categorized in the population. In the end, it may provide the foundation for a decision support tool intended for consultants specialized in coaching and outplacement.

Key words: Statistical Implicative Analysis, Assessment tool, Behavioral referentials, Decision support system, Validation study

1 Introduction

Human resources managers have been early users of computer tools. The need for evaluating the behavioral profile of a person in human resources has led to the creation of personality assessment tools. Initially, the first tools were paper based ones. Then, the first developed computer ones were expert systems (e.g. Human Edge [12]). Then more complex decision support tools were created: MBTI (Myers-Briggs Type Indicator) [5, 14], PerformanSe Echo. Meanwhile,

S. Daviet et al.: *Using the Statistical Implicative Analysis for Elaborating Behavioral Referentials*, Studies in Computational Intelligence (SCI) **127**, 299–319 (2008)
www.springerlink.com

great strides were accomplished in the field of knowledge discovery in data (KDD) [6], enabling the study of the huge bulk of data collected by those assessment tools.

Today, validation has become a crucial stake for those tools. Till now, they were only validated by the relevance of their results and how they matched the a priori knowledge of psychologist experts on specific assessed stereotyped people. Probing them with less subjective methods is yet crucial for both the assessor firms and the assessed people. Very few studies were driven to confront personality assessment tools with the reality they are meant to model. The results of these studies on the MBTI tool was conflicting: [2, 19] versus [5, 11, 20]. Very few studies have been driven on the Big Five theory, and on the Echo PerformanSe tool (that we are interested in). Yet, the PerformanSe tool is widely used and we have at our disposal of a huge population to manage some relevant statistical studies to validate it. Mining the sample of assessed population for association rule discovery [1] seems to be a right way to achieve this validation.

A previous validation study has been realized over a scope of 4538 evaluations [16,17]. In our case, we are interested in the data collected by the APEC[3]. More specifically, we have carried out a study on a lower sample of population: the executives who are 45 years old and more and seeking a job, extracted from a higher population of 2788 people. We have used the CHIC [4] software and the Statistical Implicative Analysis to conduct our study. We have used classical statistical measures like mean and standard deviation, and the tools of SIA [9]: similarity trees [13], implicative trees and cohesitive graphs [10]. Study targets four objectives. First, combined with the previous validation study of the Echo tool, a statistical survey of this population could confirm the last collected results. Then, we want to draw a deep statistical analysis of this specific population to build a referential on which we can establish a decision support system. The final goal is to have relevant indicators helping to understand the link between some behavioral characteristics and current profiles that can be categorized in the population. In the end, this may provide the foundation for a decision support tool intended for consultants specialized in coaching and outplacement.

First, we describe the data we have studied. Then, we present our methodology to qualify behavioral indicators. We use CHIC to highlight some combinations of characteristic behavioral dimensions. Then, we focus on some of these combinations to bring out the indicators and to associate an appropriate meaning based on expert evaluation to each of them. Second, we present our results in terms of relevant behavioral indicators. Third, we discuss the possibility of completing this approach with a temporal analysis. Finally, we open some possible paths to improve this work.

---

[3] APEC stands for Agence pour l'Emploi des Cadres, in English: Job Center for Executives

2 Applicative Context

Various personality assessment tools are widely used in human resources management for profiling people in job oriented decision support. A personality assessment tool intends to draw up the behavioral profile of a person from the results of a questionnaire. The goals of those types of tools are multiples: support for recruitement, support for vocational guidance, behavioral checkup accompanying a competence checkup. Those tools are not intended to be used in a discriminative way to select among applicants for a job, but more as a basis to help a human ressource manager for instance when receiving people for an interview. We have two types of questionnaires. The first one is composed of open questions where the person is free to enlarge. The answers are examined by a psychological expert to draw up the behavioral profile. It is highly questionnable due to the subjectivity and variability of the interpretation of the expert from one person to another. There are very few questionnaires of this type. To name but one, Phrases [18] has the attendee complete 50 phrases in 30 minutes, under the scrutiny of the examiner. Both the answers and the behavior of the person during the test are evaluated. This type of questionnaire is poorly studied due to the difficulty to build statistical analysis on open questions.

The second type of questionnaires is composed of closed questions and is the most widespread. Those questionnaires could be handwritten or computerized. It generally consists of a set of questions (also named *items*) with 2 or more answers. A set of rules, like those we can encounter in expert systems, have beforehand been established and give a behavioral profile along a predetermined number of personality traits (also named *dimensions*). There is a great number of those types of tools. Here we show a set of the computerized ones:

- Sosie (from ECPA): 20 personality traits evaluated through 98 groups of 4 assertions,
- PAPI (PA Preference Inventory from Cubiks):
  - the classic test: a choice between 90 pairs of sentences,
  - the normative test: 126 assertions with a choice from "totally disagree" to "totally agree",
- MBTI (from Myers and Brigg): 126 questions with a choice between two answers and a profile chosen among 16 predefined ones,
- PerformanSe Echo: 70 questions with two answers and a profile on 10 bipolar dimensions determined through a set of rules,
- Assess First: 90 questions with two choices drawing a profile over 20 behavioral dimensions and 5 families.

Among all these tools, few of them have undergone a real statistical validation study. Indeed, most of those products are based on well-grounded psychological basis: Jungian theory [5, 15], Big Five model [8, 21], or study of motivations [7]. But statistical studies are a necessary counterpart of the psychological validation. Studying the distribution of the population over the behavioral dimensions could, for instance, be an interesting type of validation. Several studies have been driven on the MBTI tool, but the results are conflicting. Some studies [2, 3, 19] have shown that the MBTI is a valid and reliable instrument, others have demonstrated several drawbacks [5, 11, 20].

The PerformanSe Echo tool is the one we have studied. Previous validation studies have already been realized on this tool every 5 or 6 years since 1985. The last one [17] dates from 2004 and it consists in analysing and fine tuning the distribution of 4538 assessments over the 10 behavioral dimensions of the PerformanSe model. It results from the collaboration of the KOD (KnOwledge and Decision) laboratory of Polytech'Nantes, the DPL (Development Psychology Laboratory) laboratory of Rennes 2, the LRI CNRS laboratory from Orsay and the PerformanSe company. The first goal of this study was to get a global overview of the repartition of the population among the 10 dimensions. Indeed, with the time passing and the evolution of certain factors (environment, vocabulary, conceptual references, etc.), it becomes necessary to control that the questionnaire and its results are always up to date and relevant, and to make it evolve if needed. This is the second goal of this study: to recalibrate if needed the tool over this reference sample of 4538 assessments. We will explain in the next chapter concerning the data studied how the tool is calibrated.

In this paper, we have driven a second and more specific study on a subsample of the population. We focus on a cross section of the 4538 evaluations which concerns the executives who are 45 years old and more and seeking a job. The data have been collected by the APEC and anonymized.

3 The PerformanSe Echo Tool

The personality assessment tool, Echo, developed by PerformanSe is a questionnaire with 70 items. Each item is a question with two possible answers, but it is not a Yes/No questionnaire: it is called an ipsative questionnaire. For instance, Fig. 1 shows a question of Echo and its two answers.

Once all the items are answered, the tool draws up the behavioral profile of the person. This profile is described along 10 bipolar dimensions which are detailed in Tab. 3. Each pole of a dimension is called a trait and is valued on a scale from 0 to 35. Each answer ascribes a set of points to one or more traits. From the scores of two opposite traits, one calculates the score of the corresponding bipolar dimension. Each of these dimensions, initially graduated from 0 to 100, has been then discretized in 3 zones: low values under 40 (marked -), medium values between 40 and 60 (marked 0) and high values above 60 (marked +). For instance, for the extroversion, we distinguish EXT-, EXT0 and EXT+. Fig. 2 gives an example of a behavioral bipolar profile.

Each dimension matches a personality trait which has no intrinsic reality. This is the interaction between several traits that trigger an observable

Introversion (INT) *Express:* reserve, modesty, discretion, risk of looking cold, difficulty to communicate, ability to concentrate	**Extroversion (EXT)** *Express:* expansion of self, desire to be noticed, ease of expression, risk of attention scattering, tendency to be invasive
Relaxation (REL) *Express:* maintenance of a state of relaxation, cold-bloodedness	**Anxiety (ANX)** *Express:* pressure, worry, emotive power, maintenance of a waking state, concern, state of tension
Questioning (QUE) *Express:* concern for improving, level-headed opinions	**Assertion (ASN)** *Express:* self-confidence, innermost conviction, firm opinions
Determination (DET) *Express:* distance regarding others, emotions, passive resistance	**Receptiveness (REC)** *Express:* opening towards others, taste for listening to others, understanding (empathy)
Improvisation (IMP) *Express:* taste for the unforeseen and adaptation, spontaneous reaction to events, impulsiveness	**Rigor (RIG)** *Express:* structure of work and environment, sense of method and planning, sense of hierarchy
Intellectual conformism (INC) *Express:* reference to well-tried solutions, analytical approach, sense of precision,difficulty to take a global view of situations, expert knowledge	**Intellectual dynamism (InD)** *Express:* creativeness, social relationships, intellectual curiosity (new ideas), overall understanding of situations, quick-wittedness, risk to neglect details, overall views
Conciliation (CCL) *Express:* patience, search for serene relationships, spirit of consensus, ability to act as an arbiter	**Combativeness (COM)** *Express:* reactive behaviour,search for stakes, sense of competition, offensiveness, impatience
Motivation for faciliation (FAC) *Express:* immediate pleasure *Main fear:* having too much work *Stimulus:* easiness, short missions *Satisfaction:* achieving easy success *Money rewards for:* jumping at the opportunity *Relationship to time:* short term projects *Voc.:* to save time, to use short cuts, to give greater importance to the present...	**Motivation for achievement (ACH)** *Express:* persevering and succeeding *Main fear:* being obliged to give up *Stimulus:* difficult projects *Satisfaction:* making efforts *Money rewards for:* merit *Relationship to time:* long term projects, feels guilty when losing time *Voc.:* to build, to persevere, to deserve, tenacity...
Motivation for independence (IND) *Main fear:* being overwhelmed by the group *Stimulus:* personal freedom *Satisfaction:* having one's own territory *Money rewards for:* individual results *Relationship to time:* protects one's personal time *Voc.:* to take the consequences of one's own choices...	**Motivation for belonging (BEL)** *Express:* influence *Main fear:* being expelled *Stimulus:* the community *Satisfaction:* living in good relationships with people *Money rewards for:* common results *Relationship to time:* dedicate time for the group *Voc.:* consensus, solidarity...
Motivation for protection (PRO) *Main fear:* not having any guarantee *Stimulus:* maintenance of what has been acquired *Satisfaction:* peacefulness of mind *Money rewards for:* an acquired right *Relationship to time:* is provident *Voc.:* to stay in a known environment, to avoid surprises, to perpetuate organisation	**Motivation for power (POW)** *Express:* risk-taking *Main fear:* having no influence *Stimulus:* challenge; deciding and leading *Satisfaction:* initiating events *Money rewards for:* risk and responsibility-taking *Relationship to time:* wants to leave a mark *Voc.:* to be in a dominant position, to be ambitious...

Table 1. The 10 behavioral dimensions

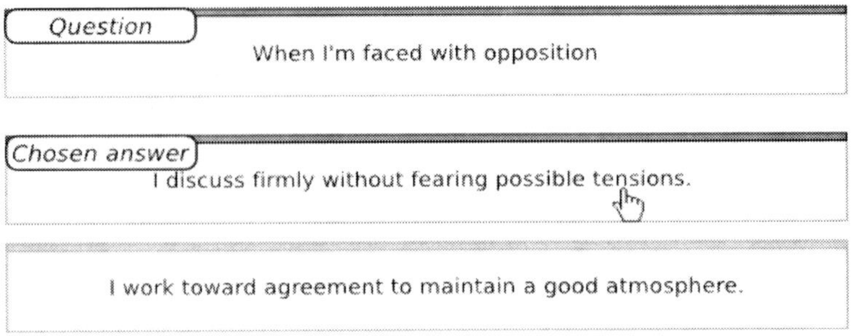

Fig. 1. Screenshot of an ipsative question

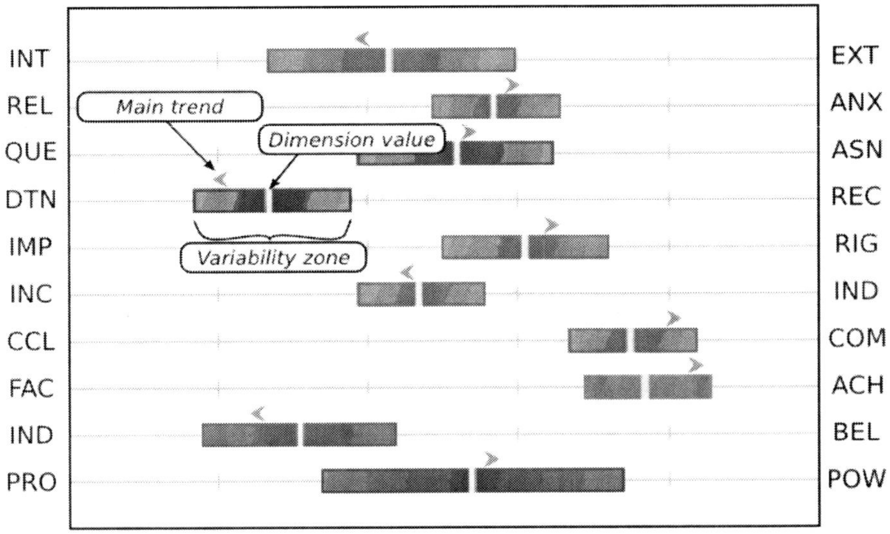

Fig. 2. Screenshot of the behavioral bipolar profile along the 10 dimensions

behavior. Such a set of dimensions is called a factor: a meta-concept linked to a general theory of the personality widely validated and recognized (for instance, Agreeableness in the Big-Five theory). The tool consists of 27000 rules that draw up a human readable text report from a text base containing 2500 pages of text. This text report considers the combination between each dimension to give the real behavioral explanation of the profile. Fig. 3 shows a schema explaining how the system works.

This model is based on the so-called "Big Five" model which describes personality according to five dimensions: Extroversion, Conscientiousness, Agreeableness, Openness and Emotional stability. It is the result of more than forty years of work led by dozens of researchers: Cattell, Fiske, Eysenck, Gulford, Tupes, Christal and Norman, and more recently, Smith, Borgotta, Goldberg, Mc Crae and Costa. This well-tried model has been enhanced by:

- the study of motivations and what leads the individual to act,
- the systemic and behavioral approach that takes an individual and its interactions with the environment as a whole.

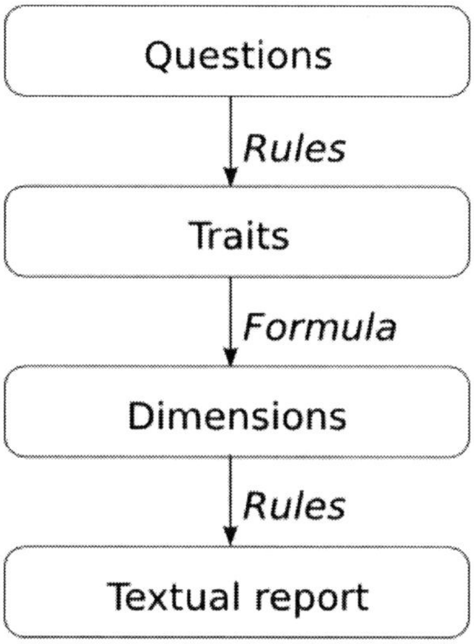

Fig. 3. Explanation of the system of rules

4 Problematic and Goals

This study has been governed by a need of the APEC to get relevant behavioral indicators for supporting their everyday task: to help people finding the right job. The APEC is a national organization that provides job guidance to people. It is somewhat comparable to the ANPE[4], but especially intended for executives. They provide an assistance for all that is related to job orientation, training, reemployment, skill validation and skill assessment. Their major difficulty is to get the good clues to determine at best the function that matches a given person. The PerformanSe tools have been purpose-built for the employment field. The assessment provides a set of recommandations with support/vigilance points. But, those conclusions are quite general and the need of the APEC is more specific to each position. The stake is twofold: first, to determine the main characteristics that promise the better chance of success for reemployment, then the personal profile that better matches a

[4] ANPE stands for Agence Nationale Pour l'Emploi, in English: National Job Center

given job. Finding those characteristics boils down to building a job referential which is our main goal.

To meet those needs, we have organized our study in two steps: the first step consists in bringing out the specificities of this population of executives in respect to the global population. To succeed in this task, both classical statistical tools and more advanced SIA tools are used. In the end, we want to determine both the dimensions and the factors (combination of multiple dimensions) that characterize this population, and thanks to a psychological expert give a meaning to the discovered factors. As previously said, it is not the dimensions themselves but the factors that can be interpreted. The SIA is a good way to obtain those combinations of dimensions, notably via similarity trees and cohesive graphs.

This first study would bring some indicators that may differentiate the studied population from the global population, but also discriminate some subgroups in this studied population. If global indicators may characterize the main part of the sample, there might be some subpopulation that is not fully characterized by those indicators and could be interesting to study. To summarize, the first step may bring the discovery of subgroups that we will analyze in a second more local step. We will use the same tools of usual statistics and SIA to drive this study.

5 Data

5.1 The Reference Population

The reference population is the one that has been used for validating the tool. The data has been collected through a partnership between PerformanSe and a large sample of clients, who have communicated their assessments. It is composed of 4538 people with wide-ranging backgrounds:

- companies, national and international groups and SMEs in all sectors of the economy,
- consultancy firms,
- business schools and engineering schools,
- public organizations for professional mobility and orientation, governed by the Ministry of Employment or the Ministry of Education.

People within this sample could be of any age, employed or not, from miscellaneous socio-cultural origins. Each person of the sample is described with the 20 traits of the Echo questionnaire. For the computation, it is the values from 0 to 35 for each trait that have been used, not those of the bipolar dimensions. On each of the 20 traits of the Echo model, the average score values of this sample teeters from 17.12 to 17.98 on a scale from 0 to 35. The standard deviation values are spread between 5.83 and 7.42. The population follows a normal distribution (centered Gaussian) over the 20 traits. People are divided out: 25% in low values, 50% in medium values and 25% in high values. Tab. 5.1 shows the results of this study over the 20 traits. On every trait, the population follows a centered Gaussian distribution. It is important to specify that this distribution has been obtained directly from the gross results without any curve fitting. That shows the relevance and accuracy of this personality assessment tool. This study has also shown that the tool did not need to be recalibrated.

Traits	Mean	Standard deviation	Minimum	Maximum
EXT	17.3642	7.0355	0.0000	35.0000
INT	17.4916	6.8063	0.0000	35.0000
COM	17.8043	6.8395	0.0000	35.0000
CCL	17.1670	6.0825	0.0000	35.0000
ANX	17.4746	7.2266	0.0000	35.0000
REL	17.2259	5.9921	0.0000	35.0000
ACH	17.9843	7.3375	0.0000	35.0000
FAC	17.8944	6.7562	0.0000	35.0000
InD	17.9006	6.1809	0.0000	35.0000
InC	17.3321	7.0624	0.0000	35.0000
RIG	17.3944	7.2871	0.0000	35.0000
IMP	17.6307	6.2754	0.0000	35.0000
ASN	17.5410	7.4265	0.0000	35.0000
QUE	17.4169	7.1501	0.0000	35.0000
POW	17.6234	6.7130	0.0000	35.0000
PRO	17.1214	6.9842	0.0000	35.0000
BEL	17.2952	6.7044	0.0000	35.0000
IND	17.8821	5.8345	0.0000	35.0000
REC	17.6166	6.7729	0.0000	35.0000
DTN	17.4786	6.6286	0.0000	35.0000

Table 2. Results of the study

5.2 The Studied Population

Thanks to a partnership with the APEC, we get access to the data collected by this national organization. It means a large sample of people having passed the behavioral assessment: 2788 people. In our case, we have restricted our study to a particular cross-section of population: the experienced executives who are 45 years old and more and seeking a job. This restriction stems from a need of the APEC to get a more specific analysis on this particular part of the population. Indeed, the average behavioral profile of this sample may be different from the one of the overall population. In our case, we get a cross-section of the population that contains 613 assessments (one assessment per subject), in other words 20% of the global sample. This may be interesting for characterizing some specificities of this population in respect to the reference population, and then inside this population between some particular profiles typical of some subgroups of the population.

To study the data in CHIC, we have chosen to transform it into binary data. In the previous validation study, the computation was made on the 20 traits valued from 0 to 35. In this study, we have used the 10 bipolar behavioral dimensions discretized in +, 0, − (i.e. for instance: EXT-, EXT0 and EXT+). We have then transform this data into binary data as usually done in this type of case (i.e. 1 if the characteristic is present, 0 if not). A sample illustration is shown in Tab. 5.2. The first reason is that CHIC and the SIA were initially designed to study this type of binary data, and it so also seems to be the simplest way. The second and most important reason is that we want to make some indicators appear, in other terms some factors (or combinations of multiple dimensions). But these indicators may be big trends more than precise values, because it is more likely to give some consistent and meaningful classes than discreet values between 0 and 35. That is why we have driven our study on discretized values.

	Ind1	Ind2	Ind3	Ind4	Ind5	...
EXT-	0	0	1	0	0	
EXT0	1	0	0	0	1	
EXT+	0	1	0	1	0	
ASN-	1	0	1	0	0	
ASN0	0	0	0	1	0	
ASN+	0	1	0	0	1	
...						

Table 3. Transformation to binary data

Finally, the data is anonymized, but we also have further information on those people: their gender, their age, their (previous) activity, etc. It could be interesting in a second step to use this data to refine our study, but now we have not used them in this contribution.

6 Why We Used the Statistical Implicative Analysis and CHIC

For this study, we have used CHIC and the Statistical Implicative Analysis for multiple reasons. The first one is the ability of CHIC to handle the primary functions of classical statistics. This is very useful because the first steps of our study are really simple and classic. Being able to complete this steps and the rest of the study with the same tool is valuable. The second reason is that we cannot just rely on a classical statistical study. CHIC gives advanced tools to study data and can perform hierarchical analysis of data. In our case, it is crucial to dichotomize the population to isolate interesting subgroups. Those groups can then be analyzed by the expert according to their descriptive factors and strongly support the building of a referential. CHIC also provides at once similarities, implicative and cohesitive analyses. Those 3 types of analyses are fully complementary. Finally, the tool is visual and quite simple to use. This is a strong advantage for the expert and the dialog with him. Then, if we want re-use this analysis process in the future, CHIC is easy to use and can be used by the expert himself with a short explanation.

7 Global Study

7.1 Goal

We have firstly dealt with the whole studied population of experienced executives and driven our research along this axis. First of all, we have made a comparative study between the studied population and the reference population thanks to the classical statistical measure: mean. Our goal is to highlight some relevant indicators that differentiate those executives from the common individual. Then, we have more deeply studied the inner characteristics of this population with the tools made available with CHIC. Our goal here is to find out some specific significant subpopulations both in the statistical and semantical meaning, and with the expert support, to isolate the behavioral dimensions implied in this dichotomy and their explanations.

This first step of our study will also help us for the second step. The subpopulation found in this first global study will be more locally analyzed in a second phase. It will be the basis of the second study to reveal indicators. The goal is to complete the characterization realized in the global study and confirm the first draft of the indicators with a set of complementary dimensions.

7.2 Study of Deviations

Occurrence and frequency (also called *mean* in CHIC) give apposite information so that the expert characterizes the studied sample. The standard deviation has not been used in this first step of the study because it is not meaningful with binary variables. With these first measures, we have established the most marked dimensions comparatively to the standard profile (available in the appendix A). In Tab. 7.2, ASN+, ASN0, COM+ and POW+ (highlighted in the table) are significantly more important for this population of executives than for ordinary people. In the light of those results, the significance of ASN+ and POW+ confirms the a priori knowledge of the expert. Indeed, Assertion and Motivation for power are known features of experienced executives. The importance of Communication also matches what was expected by the expert.

This first step is not sufficient because the information brought by this analysis is really poor. The conclusion that an experienced executive is characterized by strong affirmation, communication and motivation for power would

	Occurence	Frequency	Ordinary frequency	Absolute gap
EXT-	138	0.23	0.23	0.00
EXT0	254	0.41	0.54	0.13
EXT+	221	0.36	0.23	0.13
COM-	101	0.16	0.22	0.06
COM0	266	0.43	0.55	0.12
COM+	246	0.40	0.23	0.17
ANX-	206	0.34	0.23	0.11
ANX0	281	0.46	0.54	0.08
ANX+	126	0.21	0.23	0.02
ACH-	132	0.22	0.23	0.01
ACH0	265	0.43	0.54	0.11
ACH+	216	0.35	0.23	0.12
InD-	129	0.21	0.22	0.01
InD0	302	0.49	0.55	0.06
InD+	182	0.30	0.22	0.08
RIG-	177	0.29	0.23	0.06
RIG0	292	0.48	0.55	0.07
RIG+	144	0.23	0.22	0.01
ASN-	107	0.17	0.24	0.07
ASN0	217	0.35	0.53	0.18
ASN+	289	0.47	0.23	0.24
POW-	110	0.18	0.23	0.05
POW0	273	0.45	0.55	0.10
POW+	230	0.38	0.23	0.15
BEL-	218	0.36	0.23	0.13
BEL0	264	0.43	0.54	0.11
BEL+	131	0.21	0.22	0.01
REC-	221	0.36	0.23	0.13
REC0	277	0.45	0.53	0.08
REC+	115	0.19	0.24	0.05

Table 4. Classical statistical measures

be overhasty. Nothing indicates that these characteristics does not split the global population into two subpopulations or more. To delve into this analysis, we have completed this study with a more advanced tool of CHIC: similarity trees.

7.3 Analysis with Similarity Trees

We have used similarity trees with the entropic implication and the Poisson distribution. Indeed, we have a population of more than one hundred people and the classical method is not recommended because less discriminatory. With the same restrictive goal, we have chosen the Poisson distribution.

This second analysis with similarity trees provides a way to determine the essential classes that partition our population of executives. As we can see on

Fig. 4, this analysis confirms the significance level of the ASN+ and POW+ dimensions. Indeed, the pair (ASN+, POW+) forms the first significant node of the tree (marked in bold) with a similarity coefficient of 0.954724. The dimension EXT+, combined with the (ASN+, POW+) pair, appears to be relatively significant and discriminative of the population of senior executives. Those three dimensions underlie most of the significant nodes (similarity of (EXT+ (ASN+, POW+)) = 0.876296). Therefore, this triplet could be considered as a good candidate to partition our population and to be a relevant behavioral indicator.

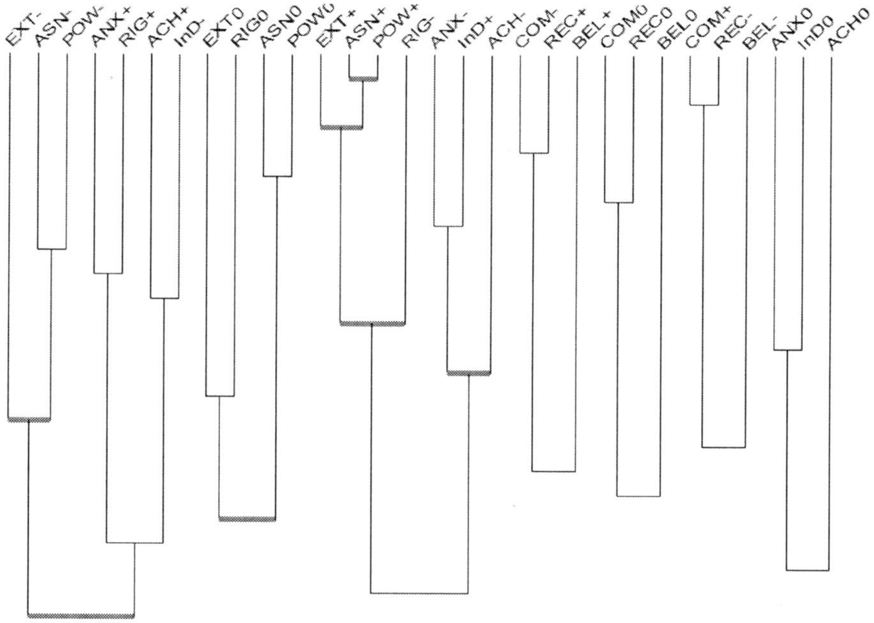

Fig. 4. Similarity tree

The COM+ dimension appears to be also significant because it forms the second level node of the tree (similarity = 0.927691). But this node is not marked as a significant one. Moreover, contrary to the triplet (EXT+ (ASN+, POW+)), COM+ is not part of a strong partition.

This second analysis has confirmed the weight of ASN+ and POW+. Moreover, we can distinguish three classes where ASN and its three modalities (ASN-, ASN0 and ASN+) are discriminative. Then, we have categorized three groups corresponding to each of these modalities. According to the expert, the ASN- seems to be a defeat factor for the re-employment of executives. Therefore, the ASN- class may be an interesting basis for building a relevant indicator showing the ability of an experienced executive to be reemployed.

8 Local Study

8.1 Data Studied and Goal

On the basis of the results discovered through similarity trees, we have studied each of the three discovered subclasses, discriminated with the ASN dimension. Indeed, ASN is not sufficient to build a relevant indicator. We need more clues on what are the main trends of each of these three groups. Thus, we have used implicative and cohesitive trees to detect link between dimensions. Those links could then be used to build our indicators as a combination of multiple dimensions. In the following studies, we have discarded the ASN dimension that is no more discriminative for each of the 3 subpopulations: ASN-, ASN0 and ASN+. We will only present the statistical details of the study concerning the ASN- subpopulation, and our conclusions on the 2 other populations. The conclusions are based both on the comparison between the subpopulations and the global population of executives, and between the subpopulations and the ordinary population.

8.2 Analysis of the Subpopulations

ASN- Subpopulation

This subpopulation contains 107 individuals (18% of the global population of executives). The analysis of the frequencies Tab. 8.2 shows an important offset of this population compared to the original population of experts for the following dimensions: POW-, EXT-, ANX+ and RIG+. For the psychologist, POW- and EXT- are revealing a debasement of self-image and ANX+ and RIG+ indicate an attempt to balance a feeling of insecurity with extra planning and organisation.

Considering the implicative graph Fig. 5, we can see the two pairs (ACH+, InD-) and (InD-, ANX+) may be particulary interesting to contribute to a relevant indicator. But we cannot yet conclude for the (ACH+, InD-, ANX+) triplet. We need to study the cohesitive tree to know if we can really group those three dimensions.

The cohesitive tree Fig. 6 and its values Tab. 8.2 give us valuable information on the set of dimensions that can be eligible to relevant indicators. By combining those results with the previous ones, we see that the dimensions found in the first step (POW-, EXT-, ANX+ and RIG+) are not all eligible for building an indicator. Indeed, POW- appears at the fifteenth level in a group with a cohesion value equal to 0.247. Moreover, it can hardly be combined with EXT- as presumed with a simple statistical analysis: the ((BEL0 EXT-) POW-) has a very low cohesitive level.

We can see here the benefit of using an advanced statistical method like the Statistical Implicative Analysis to prevent eroneous analysis. The POW- dimension may be, according to the expert, interesting on its own because

	Occurence	Frequency	Executives frequency	Ordinary frequency
EXT-	75	0.70	0.23	0.23
EXT0	32	0.30	0.41	0.54
EXT+	0	0	0.36	0.23
COM-	41	0.38	0.16	0.22
COM0	52	0.49	0.43	0.55
COM+	14	0.13	0.40	0.23
ANX-	1	0.01	0.34	0.23
ANX0	27	0.25	0.46	0.54
ANX+	79	0.74	0.21	0.23
ACH-	17	0.16	0.22	0.23
ACH0	54	0.50	0.43	0.54
ACH+	36	0.34	0.35	0.23
InD-	58	0.54	0.21	0.22
InD0	43	0.40	0.49	0.55
InD+	6	0.06	0.30	0.22
RIG-	2	0.02	0.29	0.23
RIG0	30	0.28	0.48	0.55
RIG+	75	0.70	0.23	0.22
POW-	82	0.77	0.18	0.23
POW0	25	0.23	0.45	0.55
POW+	0	0	0.38	0.23
BEL-	39	0.36	0.36	0.23
BEL0	42	0.39	0.43	0.54
BEL+	26	0.24	0.21	0.22
REC-	17	0.16	0.36	0.23
REC0	49	0.46	0.45	0.53
REC+	41	0.38	0.19	0.24

Table 5. Classical statistical measures and gap

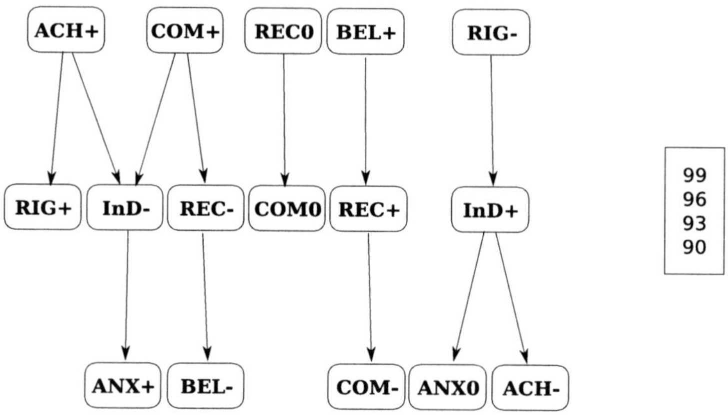

Fig. 5. ASN- implicative graph

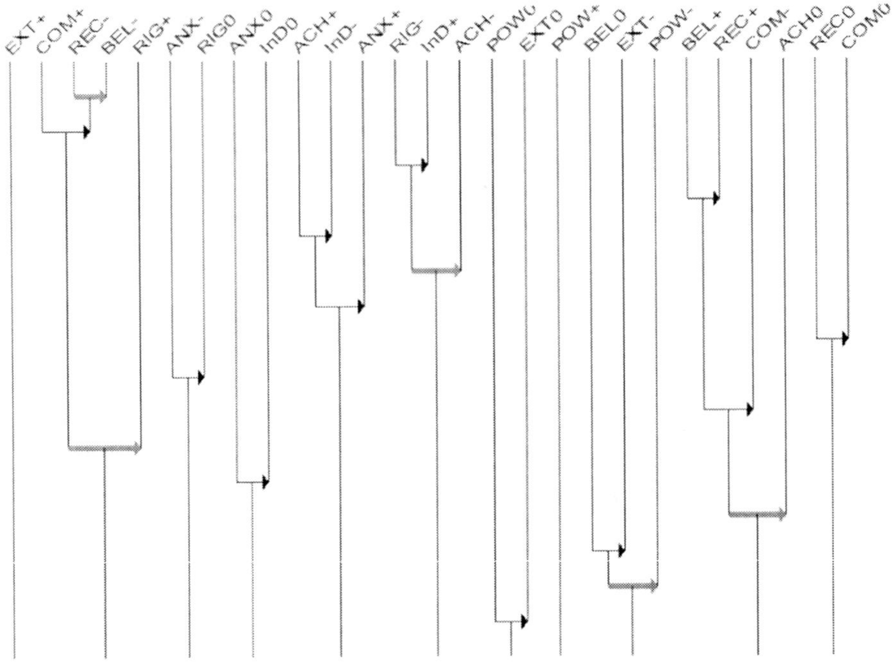

Fig. 6. ASN- cohesitive graph

	Levels	Cohesion
1	(REC- BEL-)	0.993
2	(COM+(REC- BEL-))	0.991
3	(RIG- InD+)	0.966
4	(BEL+ REC+)	0.962
5	(ACH+ InD-)	0.948
6	((RIG- InD+)ACH-)	0.944
7	((ACH+ InD-)ANX+)	0.939
8	(REC0 COM0)	0.893
9	(ANX- RIG0)	0.876
10	((BEL+ REC+)COM-)	0.868
11	((COM+(REC- BEL-))RIG+)	0.856
12	(ANX0 InD0)	0.559
13	(((BEL+ REC+)COM-)ACH0)	0.366
14	(BEL0 EXT-)	0.313
15	((BEL0 EXT-)POW-)	0.247
16	(POW0 EXT0)	0.155

Table 6. Cohesitive values

it reveals the loss of leadership and can explain the difficulty of this subpopulation to reintegrate the working world. However, the expert has found interesting those combinations of dimensions:

- (((BEL+ REC+) COM-) ACH0): relies on others by accepting concessions, so as to lighten his/her work load,
- ((COM+ (REC- BEL-)) RIG+): hides behind an inflexible, aloof and even strongly opposed behavior.

In the light of those results for the ASN- subpopulation, the expert has characterized a set of dimensions that is meaningful according to its psychological knowledge. Hereunder, you can see the indicators built:

- Indicator of adaptation: REC0/COM0 (17% of the sample),
- Indicator of illusion: RIG-/InD+/ACH-,
- Indicator of cry for help: BEL+/REC+/COM-,
- Indicators of autistic withdrawal:
  - passive: EXT-/POW-/BEL0,
  - offensive: COM+/REC-/BEL-,
- Indicator of strictness by:
  - obstinacy: ACH+/RIG+,
  - nervous tensing up: ACH+/InD-/ANX+.

If we consider the implicative graph, we can see that almost all the combinations over the 0.90 threshold have been kept and construed by the expert.

8.3 ASN0 Subpopulation

This subpopulation contains 217 individuals (35% of the global population of executives). The study of frequencies shows that this subpopulation has a highly lower offset with the ordinary population than the previous ASN- subpopulation. The highest offset is on POW0 that is a quite neutral dimension and, according to the expert, not really significant on its own. Once more, the classical statistical tools are not sufficient to build our indicators.

Using the SIA, both similarity and cohesitive trees and implicative graphs show that the (REC-, COM+) pair is a strong indicator of this subpopulation. According to the expert, it can reveal a reject of others. Again, almost all the combinations discovered by the implicative graph have been kept and interpreted by the expert. However, we had to lower the thresholds to 0.80, because this subpopulation is poorly characterized. Hereunder, you can see the basis of indicators built by the psychologist expert:

- Indicator of influence: POW+/ANX0 and POW+/RIG0,
- Indicator of open-mindedness as:
  - curiosity: InD+/ANX-/POW0,
  - intellectual adaptability: InD+/RIG-/POW0,
  - cleverness: InD+/RIG-/ACH-,

- Indicator of interpersonal skill: REC0/COM0 by:
  - sharing: REC0/COM0/POW-,
  - compliance: REC+/COM-/POW0,
  - conviviality: REC+/BEL+/POW0,
  - benevolence: BEL+/REC+/COM-,
- Indicator of autistic withdrawal: EXT-/BEL-,
- Indicator of reject of others: REC-/BEL-/COM+,
- Indicator of strictness and nervous tensing up: ANX+/RIG+/InD-/ACH+.

8.4 ASN+ Subpopulation

This subpopulation contains 289 individuals (47% of the global population of executives). This subpopulation is better characterized than the ASN0 population. Indeed, according to the study of frequencies, the offset is well marked on multiple dimensions: EXT+, COM+, ANX-, RIG- and POW0. Those five dimensions seem to validate the apriori knowledge of the expert that the ASN+ population is the best able to return to the working world. Representing almost fifty percent of the overall executive population, this subpopulation follows pretty much the same trend, and the classical statistical tools are not sufficient to get relevant specific information. As usual, CHIC has been used to go into detail.

The SIA has brought new information on combination of interesting dimensions. This subpopulation is well characterized and many indicators have been found:

- Indicator of enthusiasm: ACH-/RIG-/InD+/ANX-/EXT+/POW+,
- Indicator of interpersonal skill by:
  - involvement: BEL0/REC0/COM0,
  - conviviality: BEL+/REC+/COM-,
- Indicator of vigilance: ANX+/EXT0,
- Indicator of strictness:
  - interpersonal: EXT-/BEL-/REC-/COM+,
  - intellectual: InD-/ACH+,
  - organizational: RIG+/POW0.

9 Results and Outlooks

This study has led to interesting discoveries according to the psychologist expert. The method previously used was designed only over single dimensions with classical statistical tools. As our study has shown, this could give imprecise results, and in some particular case, erroneous ones. Moreover, it is not the results but the expert's interpretation that can be biased by statistical information that do not match those he/she expects to be working with. Thus, this study has proven the interest of using more advanced statistical tools and data analysis methods in the field of psychology, more accustomed to classical statistics.

Thanks to this study, the expert has been able to build a set of relevant indicators on our initial population of experienced executives seeking a job. He has identified three main groups on the assertion (ASN) dimension. On each group, numerous indicators have been designed by the expert. Because it was a prospective study, the results have then been confronted with other data of the APEC on the studied subjects: it appears that the indicators found were quite relevant when looking at the global behavior of each of these three groups. Indeed, each of these groups has a particular behavior considering its working world reintegration. The group in weak assertion is mainly characterized by a longer reintegration period and by a feeling of defeat relative to their unemployment situation, whereas the group in strong assertion has a higher rate of success in reemployment and shows a more positive behavior towards their situation. The group in medium assertion is less clearly defined than the other two and the behavior of its subjects is less uniform. Some of them follow the trend of the ASN- group, others that of the ASN+. The indicators built for each group match those observations. Most of the indicators of the ASN- population (indicators of illusion, cry for help, autistic withdrawal, strictness) can be considered as vigilance points denoting potential difficulties for some people of this population. This does not mean that a person of the ASN- subpopulation is bound to fail, it means that people of this group must be watched more closely. This is exactly the goal of the APEC and it is also a success of our study. Indeed, it shows that CHIC can be used as a decision support tool, combined with the psychological assessment tool Echo.

The results of this study are really hopeful. First, we have been able to show the coherence of the PerformanSe Echo tool over the studied population: the experienced executives, the study of deviations match the knowledge of the expert. Then, we have been able to dichotomize three characteristic subpopulations and highlight the meaningful collusions of several dimensions on which to build our indicators with the expert's support. Third, it seems that the Statistical Implicative Analysis is a means for semi automatically building behavioral referentials. From a study on a large sample of behavioral assessments, we have built a decision support tool that can be used for counseling, training or access to a job. For the moment, the study is still in progress and we trying to take advantage of supplementary variables available in the sample. We are also thinking of studying the historical evolution of these populations thanks to Statistical Implicative Analysis or other multi-dimensional data mining methods.

References

1. R. Agrawal and R. Srikant. Fast algorithms for mining association rules. In J.B. Bocca, M. Jarke, and C. Zaniolo, editors, *20th International Conference on*

Very Large Data Bases, VLDB'94, pages 487–499. Morgan Kaufmann, 1994.
2. J. G. Carlson. Recent assessment of the mbti. *Journal of Personality Assessment*, 49(4), 1985.
3. M. Carlyn. An assessment of the myers-briggs type indicator. *Joumal of Personality Assessment*, 41:461–473, 1977.
4. R. Couturier. Traitement de l'analyse statistique implicative dans chic. In *Journées sur la fouille des données par la méthode d'analyse implicative*, pages 33–55, 2001.
5. D. Cowan. An alternative to the dichotomous interpretation of jung's psychological functions: Developing more sensitive measurement technology. In *Journal of Personality Assessment*, volume 53, pages 459–471, 1989.
6. U. Fayyad, G. Piatetsky-Shapiro, P. Smyth, and R. Uthurusamy, editors. *Advances in Knowledge Discovery and Data Mining*. MIT Press, Cambridge, MA, 1996.
7. J. George and G. Jones. *Organizational Behavior*. Prentic Hall, Upper Saddler River, NJ, 3rd ed. 2004 edition, 2002.
8. L. R. Goldberg. Language and individual differences: The search for universals in personality lexicons. *Review of Personality and Social Psychology*, 2:141–165, 1981.
9. R. Gras. *L'implication statistique : une nouvelle méthode exploratoire de données*. La Pensée sauvage. 1996.
10. R. Gras, H. Briand, P. Peter, and J. Philippe. Implicative statistical analysis. In *Proceedings of International Congress I.F.C.S.*, Kobe, Tokyo, 1997. Springer-Verlag.
11. R. Harvey, W. Murry, and S. Markham. Evaluation of three-short-form versions of the mbti. *Journal of Personality Assessment*, 63(1):181–184, 1994.
12. J.H. Johnson and T.A. Williams. Using a microcomputer for on-line psychological assessement. *Behavior Research Methods & Instrumentation*, 10:576–578, 1978.
13. I.C. Lerman. *Classification et analyse ordinale des données*. Dunod, 1981.
14. I. Myers. The myers-briggs type indicator. *Educational Testing Service*, 1962.
15. P. Myers. *Gifts Differing. Understanding Personality Type*. Davies-Black Publishing, 1995.
16. T. Patel. Comparing the usefulness of conventional and recent personality assessment tools: Playing the right music with the wrong instrument? *Global Business Review*, 7(2):195–218, 2006.
17. V. Philippé, S. Baquedano, R. Gras, P. Peter, J. Juhel, P. Vrignaud, and Y. Forner. étude de validation : Performanse echo, performanse oriente. Technical report, Study realized with the collaboration of PerformanSe, Laboratoire COD de l'École Polytechniquede l'Université de Nantes, Laboratoire de Psychologie Différentielle de l'Université de Rennes 2, 2004.
18. B. S. Stein and J. D. Bransford. Constraints on effective elaboration: effects of precision and subject generation. *Journal of Verbal Learning andVerbal Behavior*, 18:769–777, 1979.
19. O. Tzeng, D. Outcalt, S. Boyer, R. Ware, and D. Landis. Item validity of the mbti. *Journal of Personality Assessment*, 48(3), 1984.
20. T. Vacha-Haase and B. Thompson. Alternative ways of measuring counselees' jungian psychological-type preferences. *Journal of Counselling and Development*, 80, 2002.

21. J. S. Wiggins, editor. *The Five-Factor Model of Personality: Theoretical Perspectives.* Guilford, New York, 1996.

Fictitious Pupils and Implicative Analysis: a Case Study

Pilar Orús and Pablo Gregori

Universitat Jaume I, Castellón E-12071, Spain
{orus, gregori}@mat.uji.es

Summary. We present a case study, in the context of Didactics of Mathematics, in which we adopt the methodology of using fictitious data in the Statistical Implicative Analysis. On the one hand, unlike supplementary variables, the fact of adding fictitious data to the sample does modify analyses results, so caution is needed. On the other hand, *fictitious students* are a tool for better understanding the data structure resulting from the analyses.

Key words: Contribution, entropic implication, fictitious subject, intensity of implication, quasi-implication, statistical implicative analysis, typicality.

1 Introduction

The use of multivariate analysis in the field of Didactics of Mathematics (DM) has already got a long tradition in the frame of fundamental didactics. Important references can be found among the contributions of Journées de Caen (1995 & 2000), such as [3] and [10], in which new statistical tools are provided, motivated in the context of DM as in many other occasions, but fruitful for both the fields of DM and multivariate statistics.

In the usual multivariate methods (generally factor analysis and principal component analysis), Brousseau [3] uses some supplementary individuals (*fictitious individuals*) in his data in order to be able to compare the *a priori* and *a posteriori* analysis of a questionnaire. The *a priori* analysis of the questionnaire leads to certain criteria of characterization of its questions (the variables). In this way, two matrices are obtained: one coming from the pre-experimental analysis (the *a priori* matrix of the questionnaire: criteria × questions) and the empirical matrix, made of the collected data, where questions remain characterised by the present sample (answers × questions).

Fictitious individuals allow for the simultaneous consideration of both the pre-experimental criteria and those provided by the sample in a single matrix. Fictitious individuals, as features of the variables involved in the

P. Orús and P. Gregori: *Fictitious Pupils and Implicative Analysis: a Case Study*, Studies in Computational Intelligence (SCI) **127**, 321–345 (2008)
www.springerlink.com

experiment, contribute both to the improvement of the knowledge on the variables —enhancing the information management furnished by the sample and thoroughly analysed *a priori*—, and to the comparison of the *a priori* and *a posteriori* behavior of the same variables.

The study of dependences between pieces of knowledge, approached in [9], highlighted the existence of non symmetric relations between variables —in a context of didactics— and motivated the search of new tools to analise such relations. Therefore the Statistical Implicative Analysis (SIA) is introduced [10–12] and mainly developed in research in DM [4,13,15,19], although it has also been used in other fields, such as in [7]. This theory is located among other procedures of data mining [1,20,21].

Our contribution intends to be located in the intersection of both contexts. On the one hand, we support the didactical interest of the fictitious individuals in the multivariate analysis of data as Brousseau has done in [3]. On the other hand, we consider SIA as a very appropriate and powerful tool for the processing of data obtained in the DM research. In this frame, the contribution [18] meant an empirical and naive approach to the use of *fictitious individuals* within SIA, following the direction of [22], that explored the contrast between the a priori analysis of a didactical situation and the contingence found in the experimentation.

Contribution [18] was an exercise of experimentation and observation, where the notion of *fictitious individual* was used just as any other resource in the several data analysis provided by the SIA, by means of the statistical software *CHIC* (acronym of *Classification Hiérarchique Implicative et Cohésitive*, see [5]). The role we intended to assign to *fictitious students* was similar to the one played by supplementary variables with respect to ordinary variables. That is, to use their relative placement in the final structure of data, but leaving the computation of ordinary variables untouched [6]. Obviously the nature of SIA —and its implementation in *CHIC*— do not offer this possibility, unlike other techniques such as factor analysis or principal component analysis [3].

Following the classical theory and using the Poisson distribution, several implicative analysis were performed on a binary valued matrix containing the results of a Mathematics test taken by 690 first-year students of University Jaume I (Spain). Five profiles were also used to classify the items *a priori*, and were considered as supplementary students, with their answers to each item forming the a priori matrix of the test.

The presence of these five *fictitious students* in procedures such as the *typicality* and *contribution* of individuals to the formation of similarity, implicative and cohesion classes of items show their potential use in the didactical interpretation of results in [18]. But new questions concerning this methodology arise, such as the ones regarding the conditions under which their use leads to reliable conclusions.

This work shows the instrumental role of *fictitious students* in didactical analysis performed through SIA, which could be transferred to other domains of application, according to the available data, by the consideration of fictitious extra data.

2 Case Study Description

2.1 Aim of the Study

The aim of this chapter is the search of new potential uses of SIA, within the context of DM, in order to promote the development of this theory and its applications. In this way, we present and study a conjecture on the introduction of fictitious individuals in the sample data matrix and in the subsequent processing through *CHIC* [5]. On the one hand, this procedure offers assistance in the didactical interpretation of the results. On the other hand, the final results of the analysis are perturbed by this artificial data, so that a non negligible size of this perturbation would invalidate the advantages achieved with respect to the interpretation. Therefore, this work evolves in two directions: one in DM, and the other one in the methodology of SIA, as a corollary of the first one.

We show, in the several analyses performed, not only results but also the philosophy concerning the use of *fictitious students*: when, where and what they are introduced for. Firstly, we present data. Then, in Sect. 3, we show the results of the different SIA procedures applied to data under the classical version [18], with and without fictitious data, keeping track on the differences, checking that they are reasonably small, and stressing on the gain of information through the procedure. Next, in Sect. 4, we take profit of the conclusions of the previous section to design a new set of *fictitious students* with which we analyse the structure of the same dataset (using the entropic version of the SIA), improving the obtained information, as well as the appreciated differences with respect to the previous results (with classical implication).

Having posed the question on how the structure of SIA results do vary when adding a small number of individuals to an existing sample, we show, in Appendix A, and for the interested reader, a short introduction to quasi-implications and their intensity, and a result on its variation under addition of new individuals to the sample. In the case study shown in this chapter, the number of new individuals are less than 1% of the sample size.

Finally, the questionnaire used to obtain the data of the study is shown in Appendix B.

2.2 Description of the Case

A test on the initial skills in Mathematics has been conducted over the population of first-year students of University Jaume I (UJI) of Castellón (Spain) since 2001 [14, 17, 19]. For the first time, it was part of the development of a DM research project of Bosch and collaborators (see [4]) and a PhD thesis [8].

The items of the test were selected in order to ascertain some given didactics hypotheses on the didactical discontinuities between the mathematics at the pre-university and university levels. Adapted versions of that test have been conducted in the subsequent years, being part of several Educational Improvement Projects promoted by the institution. They have been used to help the UJI Mathematics Department professors to get to know the skills of the students they are going to work with. As a consequence,they would assure what is expected for them to master, and it would allow them to freely modify their didactical strategies of education.

Data.

The present study is based on data obtained through the experimentation of the questionnaire (test) on the initial skills on mathematics of first-year students at University Jaume I, more precisely belonging to the High School of Technology and Experimental Sciences, at the beginning of the academic year 2003–04.

The questionnaire consists of 17 questions corresponding to 21 single items whose answers are coded as 0 (fail or unanswered) or 1 (success). Then we manage 21 binary variables labelled as P1, P2, P3, P4, P5, P6a, P6b, P7, P8, P9, P10, P11, P12, P13a, P13b, P14a, P14b, P15, P16, P17a and P17b. Additional supplementary variables have been taken into account, such as the kind of degree, attendance of mathematics preparation lectures, origin of the student and result in the State University Access Test, adding up to 16 of them. Contribution [18] focused on the role of these variables. In the present work, we manage a Boolean contingence table of 21 variables and 690 students.

The *a priori* matrix (MAP) of the questionnaire: *fictitious students.*

A first *a priori* classification of the items in the questionnaire, according to their type of knowledge and the task involved is shown in Table 1.

Type of knowledge	Type of task	Item
Algebra (A)	*problem* (p)	P1, P6a, P6b, P8, P10, P12, P13a, P13b, P15, P16, P17b
	graphical (g)	P17a
	exercise (e)	P2, P7
Calculus (C)	*problem* (p)	P8, P14b, P17b
	graphical (g)	P3, P17a
	exercise (e)	P3, P4, P5, P9, P11, P14a

Table 1. Type of knowledge and task involved in the items of the questionnaire.

The item features in Table 1 have been used in our first SIA analyses, under the classical theory, in order to define *fictitious students*. Let us note that this classification is uneven in the sense that the cardinals of classes are very different. Fictitious students are *Algebra* (A) and *Calculus* (C), scoring 1 only in items classified under those respective type of knowledge, and *problem* (p), *graphical* (g) and *exercise* (e), scoring 1 only in items classified under those respective types of task.

3 Application of SIA using the Classical Implication

In this section we present our methodology, based in the analysis of the optimal groups of individuals regarding the *contribution* to the formation of classes and their *typicality*, in order to improve the interpretation of the rules and the quantity of information about the sample. We shall present what will be shown in particular cases, without the intention to cover all aspects.

3.1 Classification Analysis

The original data processed through software *CHIC* 3.7 [5] leads to a classification tree, after the similarity index defined by I.C. Lerman [16]. The classification tree for the data containing the supplementary *fictitious students* is so close to the one for the original data, that eye inspection cannot distinguish among them. This fact allows us to go beyond the analysis result, examining the role of new students (item features) in the constitution of classes and rules.

If we relabel items, reflecting the criteria of the *a priori* analysis, and details on the type of task, we display the dependence on the variable class formation in a better way, and we can then compare it to the *fictitious students* appearing in the optimal groups of classes (see Fig. 1). Codes 'm', 't', 'i' do represent *mathematical modelisation* (m), *algorithmic technique* (t), *interpretation or judgment* (i). The relation between former and new labeling is expressed in Table 2.

Fictitious students taking part in optimal groups for the *contribution* to the formation of classes as well as for *typicality* are actually the same (however we display it in Fig. 1 only for significant knots).

Classification tree shows significant knots at levels 1, 6, 9, 12, 15, 17, et 19, being level 12 the most significant.

For instance, items P4, P5 of class 1 are, a priori, calculus exercises, and those features do appear as *fictitious students* in the optimal group of individuals contributing to the formation of that class. On the other side, level 9 items (((P15, P17a), P16), P17b) are algebra problems, but only the fictitious student *Algebra* do appear at the *contribution* optimal group of students.

Item	Relabeled as	Meaning
P1	1Apt	Algepra problem involving an algorithmical technique
P2	2Aet	Algebra exercise involving an algorithmical technique
P3	3Cgt	Calculus graphical representation involving an algorithmical technique
P4	4Cet	Calculus exercise involving an algorithmical technique
P5	5Cet	Calculus exercise involving an algorithmical technique
P6a	a6Apj	Algebra problem focused on the interpretation of results
P6b	b6Apj	Algebra problem focused on the interpretation of results
P7	7Aet	Algebra exercise involving an algorithmical technique
P8	8Xpm	Algebra and Calculus problem requiring modelisation
P9	9Cet	Calculus exercise involving an algorithmical technique
P10	10Apj	Algebra problem focused on the interpretation of a situation
P11	11Cet	Calculus exercise involving an algorithmical technique
P12	12Apj	Algebra problem focused on the interpretation of results
P13a	a13Apm	Algebra problem requiring modelisation
P13b	b13Apm	Algebra problem requiring modelisation
P14a	a14Cet	Calculus exercise involving an algorithmical technique
P14b	b14Cpm	Calculus problem requiring modelisation
P15	15Apm	Algebra problem requiring modelisation
P16	16Apgm	Algebra problem with graphical representation requiring modelisation
P17a	a17Xgt	Algebra and Calculus graphical representation involving an algorithmical technique
P17b	b17Xpm	Algebra and Calculus problem requiring modelisation

Table 2. Relabeling of test items: Each item is described as (1) item number (now with prefix a or b if it had that suffix in the initial labeling) (2) series of letters expressing features of the item, (A: *Algebra*, C: *Calculus*, X: both *Algebra* and *Calculus*, e: *exercise*, p: *problem*, g: *graphic*, m: *mathematical modelisation*, t: *algorithmic technique*, j: *interpretation or judgment*).

This kind of arguments can be drawn at each class formation. To summarise, the analysis of the *contributions* of *fictitious students* in the formation of the classes in the similarity tree shows that, in a first step, the types of knowledge (*Algebra* and *Calculus*) arise as determinants in the class formations whereas the types of task arise associated to type of knowledge (*exercise* attached to *Calculus* and *problem* attached to *Algebra*).

3.2 Implicative Analysis

The results of the implicative analysis driven through *CHIC* on our questionnaire, using the classical implication and the Poisson distribution, depict rules among the questions coming from the answers given by the students sample. Figure 2 (left and center) represents those quasi-implication rules using, respectively, the real sample and the enlarged one (with *fictitious students*).

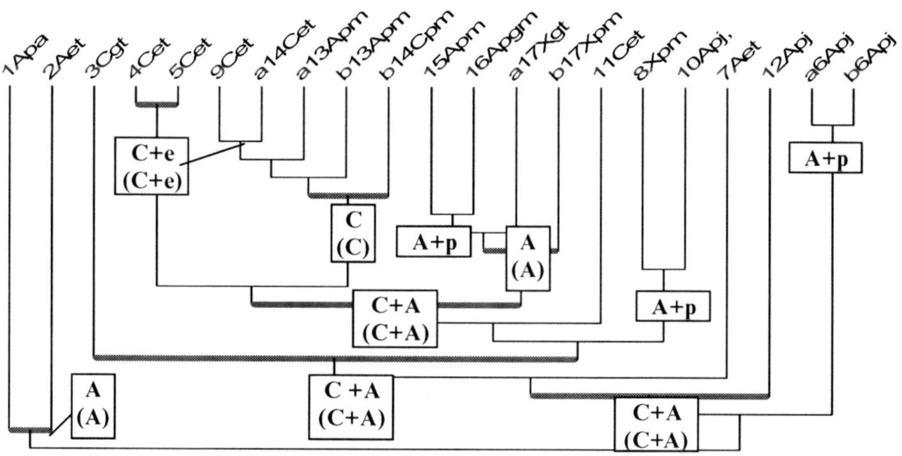

Fig. 1. Similarity tree of the questionnaire using new *fictitious students* (C: *Calculus*, A: *Algebra*, e: *exercise*, p: *problem*, g: *graphic*) and highlighting them when belonging to optimal groups regarding the *contribution* and *typicality* (between parenthesis). Items are labeled as shown in Table 2.

If we focus on the implicative graph regarding items belonging to class 12, we can perceive three subgraphs A, B and C (see Fig. 2, right). They represent rules among the items that constitute themselves complete classes at significant knots 1, 6 and 9, respectively. The implication P4→P5 (between calculus exercises) in subgraph A, show that students who perform correctly the derivative of a rather complex function ($f(x) = 5/(3x-2)^2$), also generally do perform correctly a definite integral of a very simple function ($\int_1^3 2axdx$).

The rules among algebra problems (P17a → P17b → P15 → P16) of subgraph C, specifies a relation between translation activities in two registers: a graphical one and an algebraic one.

The subgraph B gathers the different levels of aggregation conforming class 6. Rule P9 → P14a points out the relation between two calculus exercises concerning limits. The chain (P14b → P14a →P13b → P13a) rapports two function problems, and what we can observe is the following: firstly, the performance of the b-part of a problem implies the performance of the a-part, and secondly, success in problem P14 (involving an exponential function) implies success in problem P13 (involving an affine function). Therefore, B specifies rules between abilities and more complex knowledge, i.e., non algorithmic knowledge.

Analysing the role of *fictitious students* in the *contribution* to the formation of rules, we find that *problem* and *Algebra* play it in (a17Xgt → b17Xpm → 15Apm → 16Apgm) and (b14Cpm → a14Cet → b13Apm → a13Apm → 15Apm → 16Apgm). On the other side *exercise* and *Calculus* play it in (9Cet → a14Cet → 5Cet) and (4Cet → 5Cet). Similarly, only *Calculus* appear in

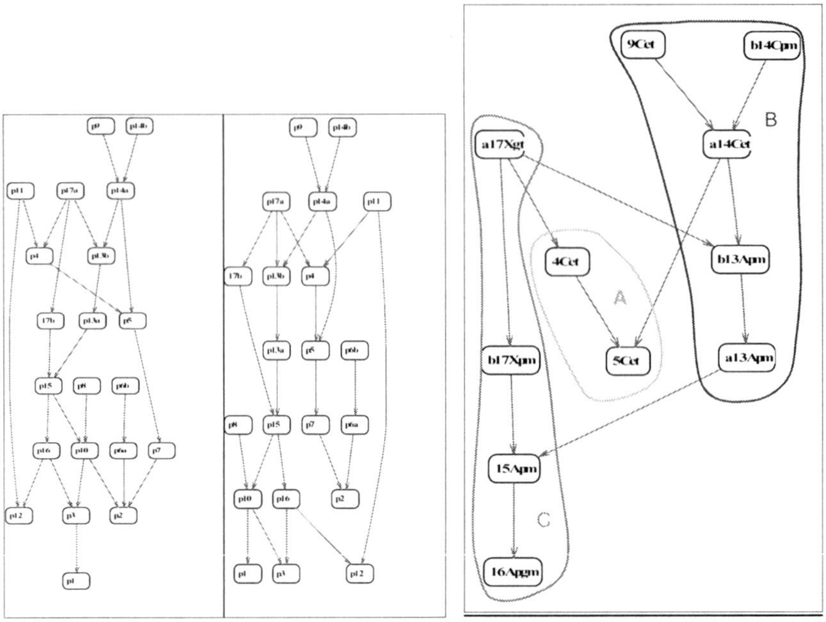

Fig. 2. Implicative graph of data without (left) and with (center) *fictitious students*. Implicative graph of the subset of items forming class 12 in the classification tree (right). Items are labeled as shown in Table 2.

(b14Cpm → a14Cet) and (a17Xgt → 4Cet → 5Cet), and only *Algebra* in (b14Cpm → a14Cet → b13Apm → a13Apm) and (a17Xgt → b17Xpm → 15Apm).

We have experienced that the search of *fictitious students* in *contribution* optimal groups of the implication chains of the implicative graph brings on the features of involved items, completing and tuning the information of the *a priori* categorisation, which can in turn fail to explain the observed data. We confirm again the linked features *exercise-Calculus* and *problem-Algebra*.

Even when observing only slight differences between implicative analysis with and without *fictitious students*, we don't think it is appropriate to proceed with subsequent analysis using *fictitious students* until some stability or robustness of implicative analysis results under addition of new data are proved.

3.3 Cohesion Analysis

The cohesion tree of the original data shows significant knots at levels 1, 3, 8, 14 et 17, being the most significant the one at level 1 (see Fig. 3).

Here we discover again that algebra problems P17a, P17b, P15 and P16 are linked in the same class, but now, the chain of implications given in the implicative analysis shows new nuances in the cohesion tree: [P17b → (P17a

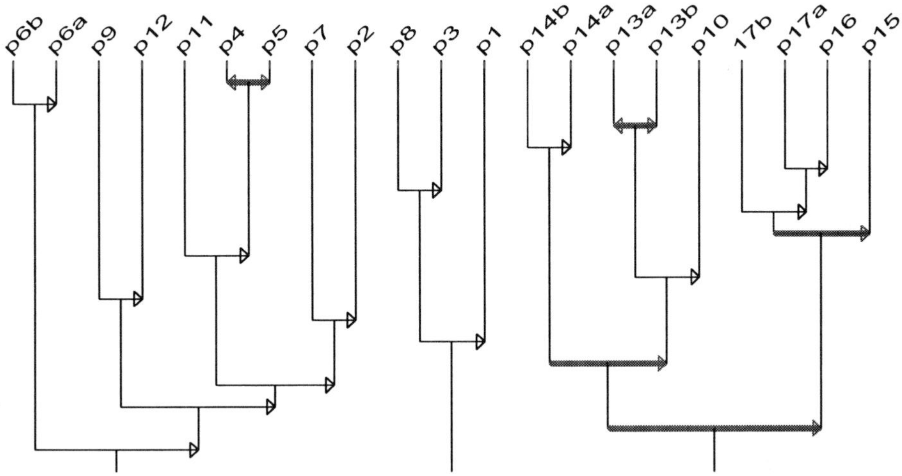

Fig. 3. Cohesion tree of items in the questionnaire.

→ P16) → p15]. This class puts together, not symmetrically, two translation tasks, one in graphical language and another one in formal language.

In level 14 we find items from implicative subgraph B. The cohesion analysis shows a stronger relation between a- and b-parts of each problem P13 and P14, than the implicative analysis did.

Now, the introduction of *fictitious students* lead to a similar cohesion tree, where only a new level of aggregation (level 11) is significant. Cohesion is strong: it equals 1 until level 15 and greater than 0.99 in the lowest significant level. We display in Fig. 4 the cohesion tree with relabeling of items (see Table 2), and *fictitious students* appearing in the *contribution* and *typicality* optimal groups of each significant knot.

If we consider Fig. 4 as the compilation of all the provided information, we can assess the existence of three classes, T, O and M, being M the most significant, not only because of the signification of the meta-rule relating all the items, but also because of the significance of the included rules.

Class M establishes a meta-rule characterised by the need of modelisation. It comprises complex questions, which require the interpretation of the context and the knowledge of mathematical models in several frameworks (graphical, algebraic, functional, . . .). It establishes a dissimetry among two meta-rules, also significant and already analysed [P14b →P14a] → [(P13a → P13b) → P10)], which specifies rules between abilities and high level knowledge (non algorithmic), and [((P17a → (P17b → P16)) → P15], which reveals relations between graphical and algebraic registers. We have found fictitious students *Algebra* and *problem* in the optimal groups for the *contribution* to the formation of classes. Also, the presence of student *Calculus* in [P14b → P14a] allows us to formulate that the ability in mathematical modelisation of

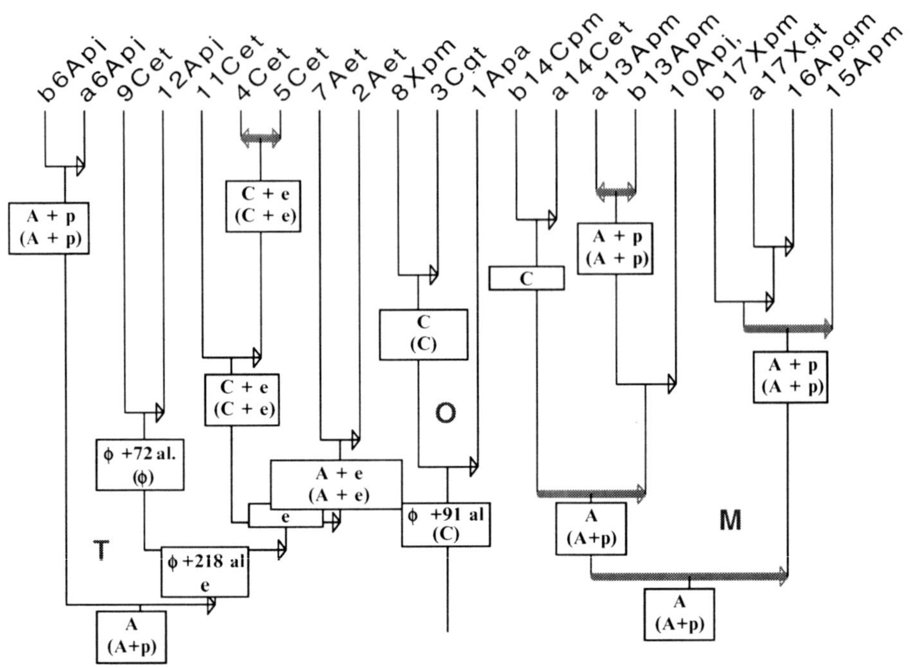

Fig. 4. Cohesion tree of the questionnaire using *fictitious students* (A: *Algebra*, C: *Calculus*, e: *exercise*, p: *problem*, g: *graphic*). We point them out into squares when they belong to optimal groups regarding the *contribution* and *typicality* (the last one between parenthesis). ϕ is used to indicate no *fictitious student* belong to the optimal group. Items are labeled as shown in Table 2.

problems implies the ability in its resolution technique. And it confirms the characterisation that we have done to class M, strengthened by the presence of students *Algebra* and *problem* in the optimal group of *typicality* of the class, and all its significant subclasses.

Class T groups together practically all items which can be solved with algorithmic techniques. It also includes some rules determined by the presence of *Calculus* and *exercise* as students in the optimal groups for the *contribution*. Here the type of knowledge of the item is not relevant but the applied technique used to solve it.

Class O is determined by the contingency: no *fictitious student* appear among the 91 individuals contributing to the formation of this class. *Calculus* appears only in the optimal group for the *contribution* to the rule P8 → P3, together with 93 more students, but it disappears at the following level.

3.4 Conclusions of SIA with Classical Implication

Classification and cohesion analyses based on *fictitious students* bring on complementary information, mainly related to the set of variables (P4, P5, P13a, P13b, P14a, P14b, P17a, P17b, P16, P15) of our questionnaire. They have been used as tools for the mining of our dataset, investigating the nature of the formation of classes to the *a priori* emphasised features of items. However, special attention has to be paid to the size of changes that *fictitious students* produce over global results.

We should highlight the strong dependence existing between the features *exercise* and *Calculus* (in the questionnaire, exercises were mostly related to the systematic application of a technique, generally dealing with functions, and mainly identified with the type of knowledge *Calculus*) and similarly, between *problem* and *Algebra*. This fact explains they usually appear in pairs.

The inspection of the modifications suffered by SIA results after the introduction of *fictitious students* has allowed us to detect weakness in our a priori analysis, motivating a new a priori feature classification of items, then new *fictitious students*, and new possibilities of explanations of the SIA analyses.

4 New Fictitious Students and the Application of SIA with Entropic Implication

We have chosen to use the results of previous hierarchical trees of similarity and cohesion of variables to design new criteria of classification of items (that is, new *fictitious students*, see Table 3), by keeping the cardinal small, 5, and being aware of the eventual changes that this new *a priori* matrix could generate in the structure of original data.

Under the new criteria, item codes are updated as shown in Table 4.

The new 5 selected criteria (F, E, t, i, m) do not represent a disjoint partition of the questionnaire, but allow us a different way of characterising items. That is to say, new *fictitious students* that we shall use in the following analyses. These analyses also need to use a convenient labeling of items that keeps the codes of the former classification as *problem* or *exercise*, allowing us to tune our didactical analysis.

In this section we proceed through a new SIA of our data that has been conducted using *CHIC* 3.7, choosing the entropic version and the Poisson distribution.

4.1 Similarity Tree

Taking into account the new *fictitious students* and the according item labels, we observe that the similarity tree is practically the same one as the one built from original data, therefore only one tree is shown in Fig. 5.

		Item
Type of task	Requires *modelisation* (m)	P8, P10, P12, P13a, P15, P17b
	Requires *interpretation* (i)	P6a, P6b, P8, P10, P12, P13b, P14b, P16
	Application of *technique* (t)	P1, P2, P3, P4, P5, P7, P9, P11, P14a, P17a, P17b
Type of knowledge	*Function* (F)	P2, P6a, P6b, P8, P10, P12, P15, P16, P17b
	Equation (E)	P3, P4, P5, P8, P9, P10, P11, P13a, P13b, P14a, P14b, P16, P17b

Table 3. Redefinition of types of knowledge and task of items in the questionnaire (new *fictitious students*).

Then we use the analysis including the fictitious data in order to get new conclusions. Three classes are identified, respectively dubbed C_1, C_2 and C_3. Class C_1 contains the highest statistical significance levels of aggregation, and shows a class of items characterised by applicability tasks in particular context (what is commonly referred as *problem* and labelled by 'p') even though if the nature of the application varies: use of algorithm (P1, P3, P14a), contextual interpretation (P10, P13b, P14b), mathematical modelisation (P13a) or a combination of them (P8). Class C_2 may be the result of chance, since it does not contain significant levels. Finally, class C_3 gathers items concerning the graphical nature of functions or their operations.

The inspection of optimal groups of students for both the *contribution* and the *typicality* leads to the information shown in Fig. 5.

We can state that, in the classification analysis, *fictitious students* only contribute to the formation of most of subclasses along the first levels of aggregation, because of the heterogeneity of the final classes. However, they belong to *typicality* optimal groups along all subclasses and classes.

4.2 Cohesion Tree

The cohesion tree of the original data resulting under the entropic theory (Fig. 6, left) shows a weaker structure: isolated and binary rules are abundant, and only one meta-rule is present. Nevertheless, it is practically a simple rule, since the implication of the premise involves two items which are parts 'a' and 'b' of a double question of the questionnaire. This cohesion tree does not seem to show relevant information. However, the consideration of *fictitious students* (Fig. 6, right) adds up structure to data: it reduces the number of isolated items whereas the number of meta-rule raises, involving items belonging to class C_1 of similarity tree.

Item	Relabeled as	Meaning
P1	1pt	Problem involving an algorithmical technique
P2	2Eet	Equation-related exercise involving an algorithmical technique
P3	3Fgt	Graphical representation of a function involving an algorithmical technique
P4	4Fet	Function-related exercise involving an algorithmical technique
P5	5Fet	Function-related exercise involving an algorithmical technique
P6a	a6Epi	Equation-related problem focused on the interpretation of results
P6b	b6Epi	Equation-related problem focused on the interpretation of results
P7	7et	Exercise involving an algorithmical technique
P8	8Xpmi	Equation- and function-related problem requiring a modelisation and the interpretation of results
P9	9Fet	Function-related exercise involving an algorithmical technique
P10	10Xpi	Equation- and function-related problem focused on the interpretation of a situation
P11	11Fet	Function-related exercise involving an algorithmical technique
P12	12Epi	Equation-related problem focused on the interpretation of results
P13a	a13Fpm	Function-related problem requiring the modelisation of a situation
P13b	b13Fpi	Function-related problem requiring the modelisation of a result
P14a	a14Fet	Function-related exercise involving an algorithmical technique
P14b	b14Fpi	Function-related problem focused on the interpretation of a result
P15	15Epm	Equation-related problem requiring the modelisation of a situation
P16	16Xpgi	Equation- and function-related problem involving the interpretation of a graphical representation issue
P17a	a17Fgt	Function-related graphical representation solvable by an algorithmical technique
P17b	b17Egt	Equation-related of a graphical representation situation solvable by an algorithmical technique

Table 4. Relabeling of items following the new *fictitious students*: Each item is described as (1) item number (now with prefix a or b if it had that suffix in the initial labeling) (2) series of letters expressing features of the item, (E: *Equation*, F: *Function*, X: both *Equation* and *Function*, e: *exercise*, p: *problem*, g: *graphic*, m: *mathematical modelisation*, t: *algorithmic technique*, i: *interpretation or judgment*).

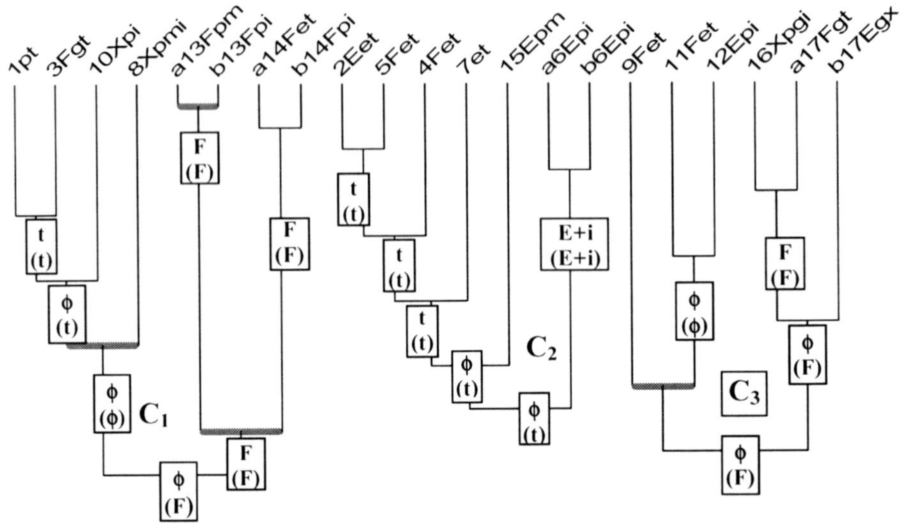

Fig. 5. Similarity tree of the questionnaire using new *fictitious students* (F: *Function*, E: *Equation*, t: *technique*, i: *interpretation*, m: *modelisation*), pointing them out into squares when they belong to optimal groups regarding the *contribution* and *typicality* (the last one between parenthesis). ϕ is used to indicate that no *fictitious student* belong to the optimal group.

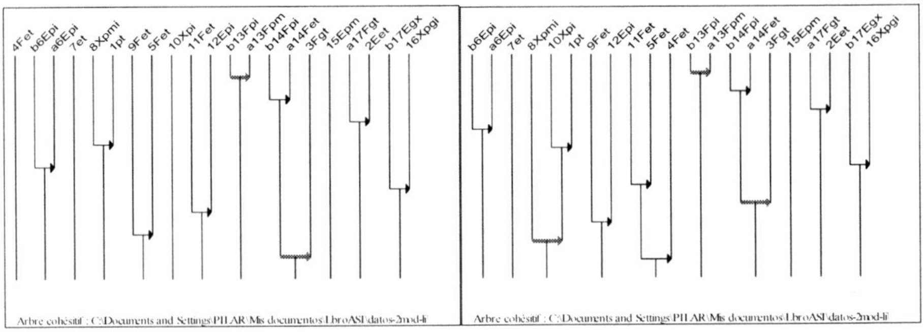

Fig. 6. Cohesion tree under entropic version without (left) and with (right) *fictitious students*.

We display the presence of *fictitious students* in the optimal groups of *contribution* and *typicality* in Fig. 7.

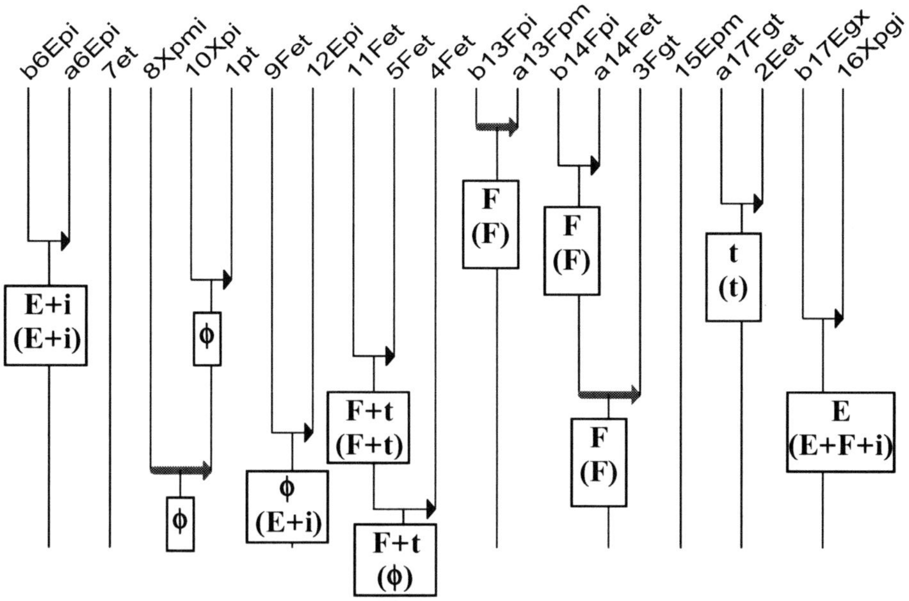

Fig. 7. Cohesion tree of the questionnaire using new *fictitious students* (F: *Function*, E: *equation*, t: *technique*, i: *interpretation*, m: *modelisation*). We point them out into a square when they belong to optimal groups regarding the *contribution* and *typicality* (the last one between parenthesis). ϕ is used to indicate no *fictitious student* belong to the optimal group. Items are labeled as shown in Table 2.

We find again the information previously acquired through the similarity tree and the cohesion tree under the classical theory: relation between parts 'a' and 'b' of double questions but, at least, confirms and strengthen the role of *fictitious students* as elements helping to display the features of the structures underlying the variables according to the sample.

4.3 Implicative Graphs

We show the comparative implicative graphs in order to finish the study. Entropic formulation is more strict than the classical one regarding the formation of rules. We switched the originality threshold in *CHIC* to 0.90 (see Fig. 8).

We observe that both graphs keep a low number of implications at a threshold of 0.99 (the one used under the classical version). The comparison with the classical setting, in which variations in the implication significances and

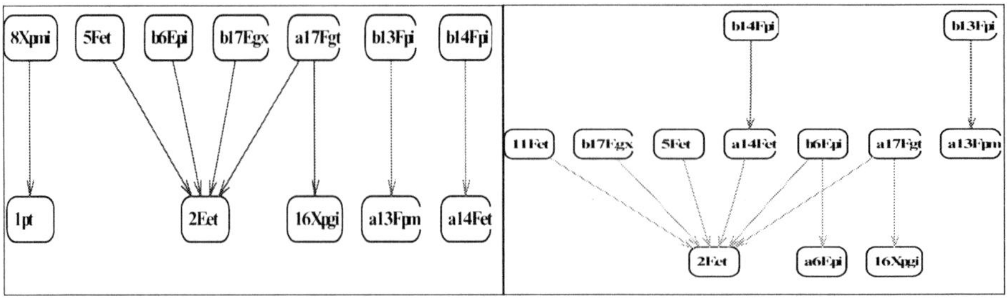

Fig. 8. Implicative graphs under entropic version without (left, threshold=0.95) and with (right, threshold=0.90) *fictitious students*. Arrows are drawn normal (90%), barred (95%) and doubled barred (99%) after the intensity the implications exceed.

an arguable choice of the originality threshold (we need to set it up to 0.90 if we want to keep a similar number of implications) dissuades us from using this analysis.

5 Conclusions

The description of the method involving *fictitious students* in the different analyses performed with software *CHIC*, have permitted us to point out and show pieces of information which are complementary to the ones resulting from the classical descriptive statistics and to SIA, that we summarise as follows:

1. We have ascertained, rather generally, that the introduction of 5 *fictitious students* in the initial data matrix has not basically altered the structures that the different SIA analyses resulted in, just slight modifications showing a logical but low sensibility in front of a low number of added individuals (5 over 690 in our case). Then we warn the potential users to check the size of changes in global analyses before proceeding to the search of conclusions. This stability in SIA results legitimate, up to our knowledge, the use the extended matrix in order to interpret results of the original data through the *fictitious students*. In that sense, the slight variations between the analyses mean a positive feedback to the used methodology.
2. *Fictitious students*, playing the role of students belonging to the optimal group of students regarding either the *contribution* to the formation of classes or the *typicality* within each class, help to explain features of significant classes issued from the classification and cohesion analysis.
   In the first part, using the classical theory of implication, version 3.7 of software *CHIC* and the *a priori* matrix made of fictitious students *Calculus*, *Algebra*, *exercise*, *problem*, *graphic*, we have been able to see that:

- In the similarity tree, *fictitious students* representing the type of knowledge (*Algebra* and *Calculus*) characterise significant classes.
- Features *problem* and *Algebra* do appear together and they characterise several classes of variables involving items P17a, P17b, P16, P15, P13a, P13b, whereas features *Calculus* and *exercise* also get together for characterising items P4, P5, P14b and P14a.
- Fictitious student *Algebra* characterises the implication of P13a, P13b, P14a, P14b (items included in the same significant class) in the classification and cohesion analyses. Note that in both cases the implication involves part 'a' and 'b' of the same question.
- Similarity and cohesion analyses, with the help of *fictitious students*, seem to provide us with right and additional information, mainly about the pack of items P4, P5, P13a, P13b, P14a, P14b, P17a, P17b, P16, P15, in the questionnaire. They have proved to be useful in the mining of our sample.

In the figures displayed throughout the chapter, at the formation of each class, one can see the codes of items forming the class and the fictitious students belonging to the optimal groups. As a matter of fact, the high *correlation* between what the intuition would tell from the codification, and what our proposed methodology actually yields, means a support to the validity of this procedure within SIA, which was motivated by similar procedures conducted in other multivariate analysis techniques.
in which one can see the codification of items under the criteria used for the definition of fictitious students, and the presence of fictitious students in the optimal groups, show a high correlation between what the intuition would tell (codification) and the results of the positioning of fictitious students.

3. The conducted analyses have highlighted a certain insufficiency in the ability of explaining our *a priori* analysis (choice of features of items), and in the same way they have given hints for the choice of new convenient features (*fictitious students*): type of contents *Equation* and *Function*, and types of task *technique*, *interpretation* and *modelisation*.
4. We conducted a second set of analyses of our data, using the entropic implication, establishing comparisons with the same analyses conducted over the extended matrix, remarking that:
   - As in the previous case, no significant variations have been produced in the structure of data, allowing us to use again the interpretation of *fictitious students*.
   - This second set of *fictitious students* are not exactly a separation between type of knowledge and type of task, unlike the previous case.
   - They tend to appear alone in the optimal groups, being *Function* and *technique* more present.
   - Feature *Function* emerges related to classes formed by the two parts of questions 13, 14, 17 (stressing the order b $\rightarrow$ a) and question 16,

always significant. Feature *technique* characterises items P1, P2, P3, P4, P5, P7 and P10, and the different relations involving them.
- However, also features *Equation* and *interpretation* do appear related to both parts of question 6, stressing the order P*b $\rightarrow$ P*a.
- The option *Typicality*, in the cohesion analysis, shows the combination of these features with other classes.

5. The feasibility study and potentiality of the use of our *fictitious students* has been approached by *CHIC* with the SIA through a mainly descriptive work. We have intended to highlight the aspect of research tool of this methodology, and then we have proposed a rather technical presentation of the use of this tool on the available data.
The necessary small size of the *a priori* matrix has been determinant in the final choice of features, although other choices were possible. In spite of this size, we have found that a considerable amount of information has been supplied.
In addition, the consideration of two different a priori matrices has contributed to show the changes in SIA results and hence support the idea of the stability of results of SIA under addition of a small number of individuals in the sample, as well as to appreciate nuances in the interpretation of results.

References

1. R. Agrawal, T. Imielinsky, and A. Swami. Mining association rules between sets of items in large databases. In *Proc. of the 1993 ACM SIGMOD international conference on Management of data*. ACM Press, 1993.
2. A. Bodin. Modèles sous-jacents à l'analyse implicative et outils complémentaires. *Cahiers du séminaire de didactique de l'IRMAR de Rennes*, 1996.
3. G. Brousseau and E. Lacasta. L'analyse statistique des situations didactiques. In *Actes du Colloque Méthodes d'analyses statistiques multidimensionnelles en Didactique des Mathématiques, ARDM*, pages 53–107, 1995.
4. C. Fonseca C, J. Gascón, and P. Orús. Las organizaciones matemáticas en el paso de secundaria a la universidad. análisis de los resultados de una prueba de matemáticas a los alumnos de 1° de la uji. In *Actas Jornadas de la CV*, Universitat Jaume I, 2002. Societat d'Educació Matemàtica de la C.V.
5. R. Couturier. Traitement de l'analyse statistique dans chic. In *Actes des Journées sur la Fouille de Données par la Méthode d'Analyse Statistique Implicative*, pages 33–50, IUFM de Caen, 2000.
6. R. Couturier and R. Gras. Introduction de variables supplémentaires dans une hiérarchie de classes et application à chic. In *Actes des 7èmes Rencontres de la Société Francophone de Classification*, pages 87–92, Nancy, 1999.
7. J. David, F. Guillet, V. Philippé, and R. Gras. Implicative statistical analysis applied to clustering of terms taken from a psychological text corpus. In *Conference International Symposium Applied Stochastic Models and data Analysis, AMSDA*, Brest, 2005.

8. C. Fonseca. *Discontinuidades matemáticas y didácticas entre la secundaria y la universidad.* PhD thesis, Departamento de Matemática Aplicada, Universidad de Vigo, 2004.
9. R. Gras. *Contribution à l'étude expérimentale et à l'analyse de certaines acquisitions cognitives et de certains objectifs en mathématique.* PhD thesis, Université de Rennes, 1979.
10. R. Gras. Méthodes d'analyses statistiques multidimensionnelles en didactique des mathématiques. In *Actes du Colloque Méthodes d'analyses statistiques multidimensionnelles en Didactique des Mathématiques. ARDM*, pages 53–107, 1995.
11. R. Gras and S. Ag Almouloud. A implicaçao estatística usada como ferramenta em um exemplo de análise de dados multidimensionais. In *II Rencontres Internationales A.S.I. Analyse Statistique Implicative*, Sao Paulo, 2003.
12. R. Gras, P. Kuntz, and H. Briand. Les fondements de l'analyse statistique implicative et quelques prolongements pour la fouille de données. *Math. & Sci. Hum.*, 154–155:9–29, 2001.
13. R. Gras and A. Larher. L'analyse implicative, une nouvelle méthode d'analyse des données. *Mathématiques. Informatiques et Sciences Humaines*, 120, 1992.
14. P. Gregori, P. Orús, T. Bort, I. Pitarch, A. Pérez, J. Gual J, J. García, and G. Villarroya. Institucionalització departamental d'una prova inicial de matemàtiques. In *IV Jornada de millora educativa i III Jornada d'harmonització europea de la UJI*, Castelló, 2005. Publicacions UJI.
15. A. Larher. *Implication statistique et application à l'analyse de démarches de preuve mathématique.* PhD thesis, Université de Rennes, 1991.
16. I.C. Lerman. *Classification et analyse ordinale des données.* Dunod, 1981.
17. P. Orús, T. Bort, P. Gregori, I. Pitarch, and G. Villarroya. Evaluación inicial de los conocimientos matemáticos de los alumnos de primero de la uji. In *Actas del I Congreso de la red Estatal de docencia universitaria y III Jornadas de mejora educativa*, pages 646–665, Castelló, 2004. Publicacions UJI.
18. P. Orús and P. Gregori. Des variables supplémentaires et des élèves "fictifs" dans la fouille de données avec chic. In R. Gras, F. Spagnolo, and J. David, editors, *Troisièmes Rencontres Internationales A.S.I. Analyse Statistique Implicative*, pages 279–293, Palermo, 2005.
19. P. Orús, P. Gregori, and A. Roig. Observación y producción de conocimientos en didáctica de las matemáticas mediante la estadística exploratoria. In *Actas VII Simposio de Investigación en Educación Matemática*, pages 100–105, Granada, 2003. Universidad de Granada y Valladolid.
20. J. Pearl. *Probabilistic Reasoning in intelligent systems.* Morgan Kaufmann, San Mateo, CA, 1988.
21. R.M. Goodman RM and P.Smyth. The induction of probabilistic rule set. the itrule algorithm. In *Proc. of the* 6th *int. conf. on machine learning*, pages 129–132, 1989.
22. F. Spagnolo. L'analisi a-priori e l'indice di implicazione statistica di gras. *Quad. Ricerca Didat*, 7:111–117, 1997.

Appendix

A Theoretical Introduction to Implicative Analysis and some Inequalities Regarding the Increment of the Intensity under Addition of New Data

We introduce the necessary notation to present a short theoretical result on the variations of the intensity of quasi-implications when working under the classical theory and using Poisson law and a new individual is aggregated to an existing sample.

Let V be a finite set of binary variables or features (that we shall denote by $\mathrm{a}, \mathrm{b}, \ldots$) and E a set of n individuals. Individual $x \in E$ is said to possess (or to be an example of) feature $\mathrm{a} \in V$, if $a(x) = 1$ or $a(x)$ is true.

The rule or implication "$\mathrm{a} \longrightarrow \mathrm{b}$" is logically valid whenever for each individual $x \in E$ the logical implication "$\mathrm{a}(x) \longrightarrow \mathrm{b}(x)$" is true or, equivalently when the inclusion $\{x \in E : \mathrm{a}(x) = 1\} \subset \{x \in E : \mathrm{b}(x) = 1\}$ holds. An individual $x \in E$ for which the implication $\mathrm{a}(x) \longrightarrow \mathrm{b}(x)$ is false, is said to be a counterexample of the implication $\mathrm{a} \rightarrow \mathrm{b}$.

In the real world, samples rarely offer true logical implications. It is easy to find counterexamples for any imaginable rule. With the purpose of obtaining relevant information on the relation among variables, the concept of the rigid logical implication among variables was extended, weakened, to the concept of quasi-implication or quasi-rule, more present in real situations [10].

In the sequel we shall work with the single pair of variables a and b. Let $A = \{x \in E : \mathrm{a}(x) = 1\}$ and $B = \{x \in E : \mathrm{b}(x) = 1\}$ be subsets of E, and denote $n := \mathrm{card}(E)$, $n_{\mathrm{a}} := \mathrm{card}(A)$, $n_{\mathrm{b}} := \mathrm{card}(B)$ and $n_{\mathrm{a} \wedge \overline{\mathrm{b}}} := \mathrm{card}(A \cap \overline{B})$.

Let us now define a random process: let $\mathcal{A}$ and $\mathcal{B}$ be two random subsets of E of respective sizes n_{a} and n_{b}, whose elements are selected completely at random from E and independently for each subset.

For any given small α such that $0 \leq \alpha \leq 1$, it is said that the (quasi-) implication $\mathrm{a} \rightarrow \mathrm{b}$ is admissible at a confidence level of $1 - \alpha$ when $P(\mathrm{card}(\mathcal{A} \cap \overline{\mathcal{B}}) \leq \mathrm{card}(A \cap \overline{B})) \leq \alpha$.

The implication $\mathrm{a} \rightarrow \mathrm{b}$ will be admissible at a high confidence level $(1 - \alpha)$ whenever the chances of finding as many counterexemples or less, in the random process, as the observed ones in the sample are small (α).

There exists several models describing the random process introduce above, according to different considerations, such as whether the sample and population coincides, or whether the size of the sample is fixed or the result of a random process too (see for instance [2]). We have chosen the one for which the random variable $\mathrm{card}(\mathcal{A} \cap \overline{\mathcal{B}})$ is distributed as a Poisson law of parameter $\lambda = \frac{n_{\mathrm{a}} n_{\overline{\mathrm{b}}}}{n}$.

For large values of λ (for instance $\lambda > 5$) it is commonly admitted the approximation by the Gaussian distribution (of mean λ and standard deviation $\sqrt{\lambda}$). Then, for the implication $\mathrm{a} \rightarrow \mathrm{b}$, the standarised variable

$$Q(\mathrm{a}, \overline{\mathrm{b}}) := \frac{\operatorname{card}(\mathcal{A} \cap \overline{\mathcal{B}}) - \frac{n_{\mathrm{a}} n_{\overline{\mathrm{b}}}}{n}}{\sqrt{\frac{n_{\mathrm{a}} n_{\overline{\mathrm{b}}}}{n}}}$$

follows approximately the standarised normal distribution, and its empirical realisation

$$q(\mathrm{a}, \overline{\mathrm{b}}) := \frac{n_{\mathrm{a} \wedge \overline{\mathrm{b}}} - \frac{n_{\mathrm{a}} n_{\overline{\mathrm{b}}}}{n}}{\sqrt{\frac{n_{\mathrm{a}} n_{\overline{\mathrm{b}}}}{n}}},$$

expresses the gap between the theoretical and observed values assuming independence between $\mathcal{A}$ and $\mathcal{B}$. This value is called implication index in spite of being an indicator of the non implication, since it measures the size of counterexamples.

Now, the intensity of implication $\mathrm{a} \to \mathrm{b}$ is denoted by $\varphi(\mathrm{a}, \mathrm{b})$ and defined as

$$\varphi(\mathrm{a}, \mathrm{b}) = 1 - P(\operatorname{card}(\mathcal{A} \cap \overline{\mathcal{B}}) \leq \operatorname{card}(A \cap \overline{B})) = P(\operatorname{card}(\mathcal{A} \cap \overline{\mathcal{B}}) > \operatorname{card}(A \cap \overline{B}))$$

Whenever the Gaussian approximation fits, and therefore the use of the implication index $q(\mathrm{a}, \overline{\mathrm{b}})$, an approximate value of the implication intensity is

$$\varphi(\mathrm{a}, \mathrm{b}) = 1 - P(Q(\mathrm{a}, \overline{\mathrm{b}}) \leq q(\mathrm{a}, \overline{\mathrm{b}})) = \frac{1}{\sqrt{2\pi}} \int_{q(\mathrm{a}, \overline{\mathrm{b}})}^{\infty} \mathrm{e}^{-t^2/2} \mathrm{d}t$$

Let us show how the intensity $\varphi(\mathrm{a}, \mathrm{b})$ varies when a new individual x' is added to the sample, in the four different cases (see Table 5). When necessary, we use subindex 1 with values and random variables concerning the original sample, and subindex 2 to the respective values regarding extended sample.

x'	a b	Δn_{a}	$\Delta n_{\overline{\mathrm{b}}}$	$\Delta n_{\mathrm{a} \wedge \overline{\mathrm{b}}}$	λ_2	$\Delta\lambda$
(i)	0 0	0	1	0	$\frac{n_{\mathrm{a}}(n_{\overline{\mathrm{b}}}+1)}{n+1}$	> 0
(ii)	0 1	0	0	0	$\frac{n_{\mathrm{a}} n_{\overline{\mathrm{b}}}}{n+1}$	< 0
(iii)	1 0	1	1	1	$\frac{(n_{\mathrm{a}}+1)(n_{\overline{\mathrm{b}}}+1)}{n+1}$	> 0
(iv)	1 1	1	0	0	$\frac{(n_{\mathrm{a}}+1)n_{\overline{\mathrm{b}}}}{n+1}$	> 0

Table 5. Values (or their increments) of the quantities involved in the process of computation of the intensity $\varphi(\mathrm{a}, \mathrm{b})$ when a new individual x' is added to the sample.

Then we want to estimate $\Delta\varphi(\mathrm{a}, \mathrm{b}) := \varphi_2(\mathrm{a}, \mathrm{b}) - \varphi_1(\mathrm{a}, \mathrm{b})$, hence analyse the values

$$P(\operatorname{card}(\mathcal{A}_2 \cap \overline{\mathcal{B}_2}) > n_{\mathrm{a} \wedge \overline{\mathrm{b}}}) - P(\operatorname{card}(\mathcal{A}_1 \cap \overline{\mathcal{B}_1}) > n_{\mathrm{a} \wedge \overline{\mathrm{b}}})$$

in cases (i), (ii) and (iv), and

$$P(\text{card}(\mathcal{A}_2 \cap \overline{\mathcal{B}_2}) > n_{\text{a}\wedge\bar{\text{b}}} + 1) - P(\text{card}(\mathcal{A}_1 \cap \overline{\mathcal{B}_1}) > n_{\text{a}\wedge\bar{\text{b}}})$$

in case (iii).

For cases (i) and (iv), where $\lambda_2 > \lambda_1$, using the Poisson probability function and inequalities based on the derivative of the power function, we deduce

$$P(\text{card}(\mathcal{A}_2 \cap \overline{\mathcal{B}_2}) > n_{\text{a}\wedge\bar{\text{b}}}) - P(\text{card}(\mathcal{A}_1 \cap \overline{\mathcal{B}_1}) > n_{\text{a}\wedge\bar{\text{b}}}) =$$

$$\left| \sum_{i=n_{\text{a}\wedge\bar{\text{b}}}+1}^{\infty} \left(\text{e}^{-\lambda_2} \frac{\lambda_2^i}{i!} - \text{e}^{-\lambda_1} \frac{\lambda_1^i}{i!} \right) \right| = \text{e}^{-\lambda_1} \left| \sum_{i=n_{\text{a}\wedge\bar{\text{b}}}+1}^{\infty} \text{e}^{-(\lambda_2-\lambda_1)} \left(\frac{\lambda_2^i}{i!} - \frac{\lambda_1^i}{i!} \right) \right| \leq$$

$$\text{e}^{-\lambda_1} \left| \sum_{i=n_{\text{a}\wedge\bar{\text{b}}}+1}^{\infty} \frac{\lambda_2^i - \lambda_1^i}{i!} \right| \leq \text{e}^{-\lambda_1} \left| \sum_{i=0}^{\infty} \frac{(\lambda_2 - \lambda_1) i \lambda_2^{i-1}}{i!} \right| =$$

$$\text{e}^{-\lambda_1}(\lambda_2 - \lambda_1) P(\text{card}(\mathcal{A}_2 \cap \overline{\mathcal{B}_2}) \geq n_{\text{a}\wedge\bar{\text{b}}})$$

and

$$P(\text{card}(\mathcal{A}_2 \cap \overline{\mathcal{B}_2}) > n_{\text{a}\wedge\bar{\text{b}}}) - P(\text{card}(\mathcal{A}_1 \cap \overline{\mathcal{B}_1}) > n_{\text{a}\wedge\bar{\text{b}}}) =$$

$$\left| \sum_{i=n_{\text{a}\wedge\bar{\text{b}}}+1}^{\infty} \left(\text{e}^{-\lambda_2} \frac{\lambda_2^i}{i!} - \text{e}^{-\lambda_1} \frac{\lambda_1^i}{i!} \right) \right| = \text{e}^{-\lambda_2} \left| \sum_{i=n_{\text{a}\wedge\bar{\text{b}}}+1}^{\infty} \text{e}^{-(\lambda_1-\lambda_2)} \left(\frac{\lambda_2^i}{i!} - \frac{\lambda_1^i}{i!} \right) \right| \geq$$

$$\text{e}^{-\lambda_2} \left| \sum_{i=n_{\text{a}\wedge\bar{\text{b}}}+1}^{\infty} \frac{\lambda_2^i - \lambda_1^i}{i!} \right| \geq \text{e}^{-\lambda_1} \left| \sum_{i=0}^{\infty} \frac{(\lambda_2 - \lambda_1) i \lambda_1^{i-1}}{i!} \right| =$$

$$\text{e}^{-\lambda_2}(\lambda_2 - \lambda_1) P(\text{card}(\mathcal{A}_1 \cap \overline{\mathcal{B}_1}) \geq n_{\text{a}\wedge\bar{\text{b}}})$$

In summary, the (positive) increment $\Delta\varphi(\text{a}, \text{b})$ belongs to interval

$$\Delta\varphi(\text{a}, \text{b}) \in [\text{e}^{-\lambda_2}(\lambda_2 - \lambda_1)(1 - F_1(n_{\text{a}\wedge\bar{\text{b}}} - 1)), \text{e}^{-\lambda_1}(\lambda_2 - \lambda_1)(1 - F_2(n_{\text{a}\wedge\bar{\text{b}}} - 1))]$$

where F_i is the Poisson cumulative distribution function of parameter λ_i for $i = 1, 2$. In case (ii), where $\lambda_2 < \lambda_1$, inequalities yield

$$\Delta\varphi(\text{a}, \text{b}) \in [-\text{e}^{-\lambda_1}(\lambda_2 - \lambda_1)(1 - F_2(n_{\text{a}\wedge\bar{\text{b}}} - 1)), -\text{e}^{-\lambda_2}(\lambda_2 - \lambda_1)(1 - F_1(n_{\text{a}\wedge\bar{\text{b}}} - 1))]$$

getting a negative increment of the intensity. Although new individual means no counterexample to the rule, cardinal of B rises (but not cardinal of A) and then the "surprise" diminishes.

Finally, previous inequalities lead to the following estimation for case (iii):

$$\Delta\varphi(\text{a}, \text{b}) \in [\text{e}^{-\lambda_2}(\lambda_2 - \lambda_1)(1 - F_1(n_{\text{a}\wedge\bar{\text{b}}} - 1)) - \text{e}^{-\lambda_2} \frac{\lambda_2^{n_{\text{a}\wedge\bar{\text{b}}}+1}}{(n_{\text{a}\wedge\bar{\text{b}}} + 1)!},$$
$$\text{e}^{-\lambda_1}(\lambda_2 - \lambda_1)(1 - F_2(n_{\text{a}\wedge\bar{\text{b}}} - 1)) - \text{e}^{-\lambda_2} \frac{\lambda_2^{n_{\text{a}\wedge\bar{\text{b}}}+1}}{(n_{\text{a}\wedge\bar{\text{b}}} + 1)!}]$$

We encourage researchers to work in this direction in order to find general bounds for the increment of the intensity of implications, and then providing SIA users with practical information on how many *fictitious students* could be added to a sample in order to have intensity results modified by less than a given allowed value.

B Questionnaire Driven in the Case Study

Here we show the questionnaire used in the study, translated into English, and its original version in Spanish.

B.1 Questionnaire (Translation from Spanish into English)

P1 You buy a shirt for PTA 4000 with 15% discount. How much should you pay for the shirt?

P2 Find out the solutions of the system of equations $\begin{cases} 2x + y = 1 \\ 3x + 2y = 3 \end{cases}$

P3 Represent the graph for the function $t(p) = 4p - p^2$

P4 Calculate the derivative of the function $f(x) = \dfrac{5}{(3x-2)^2}$

P5 Calculate the definite integral (where x is the variable of integration and a is a constant): $\int_1^3 2ax\,dx$

P6a When solving an equation you get to the expression $0 \cdot x = 8$, how do you interpret this result?

P6b And how do you interpret it when you get to the expression $0 \cdot x = 0$?

P7 Calculate the least common multiple of 280 and 350.

P8 A firm is getting an income of $I(x) = 50x - x^2$ USD, where x represents produced units, and it has expenses of $C(x) = 38x + 20$ USD. How many units should it produce in order to get benefits?

P9 The functions $f(x) = 3x^4 + x$ and $g(x) = x^3 - 100x^2$ tend to zero as x tends to zero. Calculate the limit of the quotient $\dfrac{f(x)}{g(x)}$ as x tends to zero.

P10 How would you compare the following job offers to distribute electoral brochures? (a) You are paid a fixed amount of PTA 50.000 plus PTA 10 per each delivered brochure. (b) You are paid a fixed amount of PTA 30.000 plus PTA 15 per brochure.

P11 Calculate the derivative of the following function with respect to the variable x: $f(x) = 8s^x$ (where s is a real number)

P12 Can we consider both $x = 4$ and $x = 36$ as the solutions of the equation $\sqrt{3x-8} = 4 - \sqrt{x}$? Provide arguments.

P13a The amount $C(t)$ of water springing from a tap (in litres) is expressed by an affine function with respect to time t (in seconds). If the water gathered in the first second is 3 litres, 5 litres in the second one, and 7 litres in the third one, how much water is gathered in a general instant t?

P13b How much water is gathered in one hour?

P14a The sales of a product after t years from its commercial launch, $V(t)$ (in thousands of units), is expressed by the function $V(t) = 30 \cdot e^{-1.8/t}$. Calculate the limit of $V(t)$ as t tends to infinity.

P14b Interpret the previous result in terms of sales of the above referred product.

P15 Express in an algebraic language the following sentence: "*The product of three consecutive even numbers is* 1287".

P16 The graphic of the function $f(x) = (x-1)(x+1)(x+3)$, in which points does it cross the x-axis?

P17a Draw the curves $x^2 + y^2 = 4$, $y = 2 - x$ along the same coordinate axis.

P17b Find out, algebraically, the points where they both intersect.

B.2 Questionnaire (Spanish Original Version)

P1 Compras una camisa que marca 4000 ptas. y te hacen un descuento del 15%. Calcula lo que tendrás que pagar por la camisa.

P2 Busca soluciones del sistema de ecuaciones $\begin{cases} 2x + y = 1 \\ 3x + 2y = 3 \end{cases}$

P3 Representa gráficamente la función $t(p) = 4p - p^2$

P4 Calcula la derivada de la función $f(x) = \dfrac{5}{(3x-2)^2}$

P5 Calcula la integral definida (donde x es la variable de integración y a es una constante): $\int_1^3 2ax dx$

P6a En la resolución de una ecuación llegas a la expresión $0 \cdot x = 8$, ¿cómo interpretas este resultado?

P6b ¿Y si llegas a la expresión $0 \cdot x = 0$?

P7 Calcula el mínimo común múltiplo de 280 y 350.

P8 Una empresa tiene unos ingresos de $I(x) = 50x - x^2$ dólares, donde x representa las unidades producidas, y unos costes de $C(x) = 38x + 20$ dólares. ¿Cuántas unidades hay que producir para obtener beneficios?

P9 Las funciones $f(x) = 3x^4 + x$ y $g(x) = x^3 - 100x^2$ tienden a cero cuando x tiende a cero. Calcula el límite de la función cociente: $\dfrac{f(x)}{g(x)}$ cuando x tiende a cero.

P10 ¿Cómo compararías las siguientes ofertas de trabajo de repartir propaganda electoral? (a) Te pagan una cantidad fija de 50.000 ptas. más 10 ptas. por cada papeleta depositada en un buzón. (b) Te pagan 30.000 ptas. fijes más 15 ptas. por papeleta.

P11 Calcula la derivada de la siguiente función respecto de la variable x: $f(x) = 8s^x$ (donde s es un número real)

P12 ¿Se pueden considerar como soluciones de la ecuación $\sqrt{3x-8} = 4 - \sqrt{x}$ los siguientes valores, $x = 4$ y $x = 36$?. Razona la respuesta.

P13a El volumen $C(t)$ de agua que mana de un grifo (en litros) viene dado por una función afín respecto del tiempo t (en segundos). Si en el primer segundo el agua recogida es de 3 litros, en el segundo es de 5 litros y en el tercero es de 7 litros, ¿cuál es el volumen de agua recogida en un instante cualquiera t?

P13b ¿Cuál es el volumen de agua recogido en una hora?

P14a La cantidad de miles de unidades vendidas de un producto, $V(t)$, después de transcurridos t años de su lanzamiento comercial, viene dada por la función $V(t) = 30 \cdot e^{-1.8/t}$. Calcula el límite de V(t) cuando t tiende a infinito.

P14b Interpreta el resultado anterior en términos de ventas del producto en cuestión.

P15 Expresa en lenguaje algebraico el enunciado siguiente: "*El producto de tres números impares consecutivos es igual a* 1287"

P16 La gráfica de $f(x) = (x-1)(x+1)(x+3)$, ¿en qué puntos corta al eje de las x?

P17a Dibuja las curvas $x^2 + y^2 = 4$, $y = 2 - x$ sobre los mismos ejes de coordenadas.

P17b Encuentra, de manera algebraica, los puntos donde se cortan.

Identifying didactic and sociocultural obstacles to conceptualization through Statistical Implicative Analysis

Nadja Maria Acioly-Régnier[1] and Jean-Claude Régnier[2]

[1] EA 3729 University of Lyon — France
acioly.regnier@wanadoo.fr
[2] University of Lyon — France
jean-claude.regnier@univ-lyon2.fr

Summary. To understand *culture*'s relationship to *cognition*, this field has studied children or adults with little schooling and often alien to well-educated Western culture. Traditionally centered on extra-curricular knowledge, school-based variables must be considered: *written culture* and teaching/learning strategies can generate obstacles to conceptualization. Subjects are adults who studied at least three years at university: some are professionals. Data was from short clinical-style interviews as well as a questionnaire based survey taken from an observational sample. To find regularities linked to conceptual strength, S.I.A. determined implicative rules between responses and pre-ordered structures. Results suggested representations linked to specific conceptual aspects constitute didactical and/or socio-cultural obstacles.

Key words: culture and cognition, conceptualization, obstacles, scientific concepts, *prototypical* figures.

1 Introduction

In 1744, Tatanga Mani, in his autobiography, "Indian Stoney" commented on his education "Oh yes! I went to the white man's school. I learned how to read his schoolbooks, newspapers and the Bible. But I discovered in time that this was not enough. Civilized people depend far too much on the printed page. I turned to the book of the Great Spirit which is present in the whole of creation. You can read most of this book by studying nature. You know, if you take all your books and spread them out them under the sun and leave them for some time, the rain, snow and the insects will do their work, and not much will remain. But the Great Spirit gave you and I the chance to study at the university of Nature: the forests, the rivers, the mountains, and the animals to whom we belong" [20, Mc Luhan, 1971 p.110 in Dasen 2001].

This piece of research is situated within the field that deals with the relationship between *culture* and *cognition* [1,26]. The subject matter takes its

N.M. Acioly-Régnier and J.-C. Régnier: *Identifying didactic and sociocultural obstacles to conceptualization through Statistical Implicative Analysis*, Studies in Computational Intelligence (SCI) **127**, 347–379 (2008)
www.springerlink.com

source in an academic teaching and learning situation, a course in the didactics of psychology given by Nadja Acioly-Régnier. In other words, the aim is to reach a better understanding of the clash between cultural rules and rationality, in particular to grasp more clearly the psychological status of the procedures and the concepts at work among social actors in various situations of work or study; here among students of psychology. Some ten years ago in Brazil, the point of departure was Nadja Acioly-Régnier's psychology course for Masters level students and Doctorate seminars at the UFPE[3]. The following situation problem was put forward to the students, based on the instruction: *"Draw the moon as you see it when it is not full."* These students, being used to beginning their lectures with situation problems concerning the development of concepts in psychology, did not seem destabilized by this request. The fact is that the spontaneously produced drawings caused sharp disagreements between them without reference to the relevant scientific arguments. From the socio-constructivist point of view, the operational concept of socio-cognitive conflict was at work [9] which in addition Acioly-Régnier sought to introduce into her courses. Thereafter, this situation was reproduced in different contexts both geographically and in terms of the public concerned, to which we will return further in this paper. Strikingly, the answers the students provided evoked characteristics similar to those which various studies had highlighted concerning the performance of illiterate subjects (or of a low level of schooling) marked by cultural characteristics of group membership, to the detriment of scientific conceptualization. Thus Luria [21] observed among illiterate farmers of Uzbekistan and of Kirghizie, a dominant tendency to solve logical tasks with procedures of argumentation and deduction from their immediate practical experience, when he studied the relationship between the intellectual abilities of adults and their cultural context.

Scribner [27] observed a propensity in most non-schooled subjects with little or no education to resort to what she called empirical explanation, as opposed to theoretical explanation. That is, solving problems relating to syllogisms, by resorting to non-relevant or unsuitable reasons taken from outside the field corresponding to the problem they had to solve and that these reasons were subject to strong sub-cultural constraints. The data on which Scribner's study is based, amongst other things, relates to subjects of Central Asia, Western Africa, Mexico and the USA. While seeking to measure the operational power of the mathematical skills developed out of school in the sub-culture within which the studied adult subjects worked, Acioly-Régnier had already made some observations in Brazil. Firstly among sellers of a lottery game called *jogo de bicho* [1], and secondly, among the Pernambuco sugar cane workers [2,3]. Compared with well-educated subjects, illiterate subjects showed a stronger propensity to avoid confrontation with mathematical problems, with which they were unfamiliar.

[3] Universidade Federal do Pernambuco Recife (Brazil), Town situated in the tropical zone of the southerly hemisphere. (8.03S, 34.54W)

Historically the development of the field indicated by *culture* and *cognition* has essentially relied on work carried out on children and teenagers as subjects [7] or adults with little or no education [16], relating to cultural spheres often characterized by their differences with the Western dominant culture of reference, that of the well-read. In this direction, Vergnaud [29–32] noticed that the study of the relationship between the cognitive and the social had only just begun. He observed that a great deal of cognitive knowledge which interests us is cognitive-social (knowledge of children is knowledge with individual and social time) and that the process of construction and appropriation is itself deeply social. Work on social interaction does not constitute a contradiction from the *constructivist* point of view according to which the subject builds or rebuilds their knowledge. However they make it possible to better specify the conditions under which this *construction* is achieved. Vergnaud adds: "I prefer, for my part, to speak of the process of appropriation of knowledge by the subject, because the knowledge of which I study the training is socially marked and independent of the subject. A child does not build a scientific discipline, but it does not learn it either without an effort to "rebuild" it, at least partially".

The originality of the approach adopted here is that we were interested in well-educated adult subjects, most of whom are students for at least three academic years and some of who are already professionals in training or teaching. With regard to the field data, this was collected through short clinical style interviews as well as a questionnaire-based survey based on an observation sample obtained by empirical methods. The data resulting from the questionnaire (appendix 1) are treated in this article by Implicative Statistical Analysis. The aim of data treatment is to start from the concepts of obstacles to training and conceptual development, to describe and analyze the various representations of the moon and their effects, as expressed by Brazilian students in Brazil, by French students of kanaka or caldoche origin in New Caledonia, by year 6 pupils (9–10 year olds) from Noumea as well as students in metropolitan France. The existence of different representations and various capacities to treat a situation is a significant indicator to put forth hypotheses characterizing the nature of the obstacles which underlie the representations. From this information, we seek to enhance our understanding about the nature of the obstacles generated by specific contexts of learning likely to facilitate or block conceptual development.

2 Cognitive development, learning and obstacles

Our concern in this paper is not to understand how competences are acquired, but rather to understand how at a given time they are organized by a level of conceptualization, a notion that underlies the idea of cognitive development. Without calling into question Piaget's model, our approach requires a theoretical framework that integrates the idea of life long adult cognitive development. We thus conserve the reference to Piaget's ontogenetic model whenever we need to consider the interactions between task content and performance. In such cases, Piaget's model of cognitive development remains a valuable theoretical reference. Indeed, it can be applied in the adult situations observed here, if we restrain the sense of this model to one which represents a partial competence acquisition, rather than full acquisition between the phases of cognitive development. This reference to cognitive development underlined also that conceptualization is already an internal construct depending upon processes proper to the individual that are revealed in a social context, inducing restrictions and limits as much as facilities for the process of conceptualization.

2.1 Representations, concepts and conceptual development

Vergnaud's theoretical proposals [32,33] contribute to a better analysis of the logical bases of representations and the concepts involved, within the framework of the theory of conceptual fields. We consider that the relationship between reality, signifiers, signification and the concept itself are well summarized by Vergnaud: "Representation is not restricted to a symbolic system reflects the material world, where signifiers directly represent material objects. In fact, signifiers (symbols or signs), represent signification which is itself of a cognitive and psychological nature". For Vergnaud [30], three levels of entities should be considered where representation is concerned: signifiers, signification and reference. The signifiers level consists of various symbolic systems which are organized differently. The level of signification is central to his theory of representation, in the sense that it is on this level that invariables are recognized, inferences drawn, actions generated and predictions made: this level is essentially cognitive. The reference is the real world, as it appears to the subject's experience. The subject acts in and on this environment according to its conscious or unconscious representations. There are three corresponding problems which must be discussed when speaking about representation: the relationships between signifiers and signification; between signification and reference, and that between the various symbolic systems. We hypothesise that these symbolic systems are arbitrary, that they have no direct relationship with the real world, and already represent a cognitive construct, i.e. that of signification. Consequently, cultural differences may be better understood when one considers the problems of the relationships between signifiers and signification, as well as those relating to the various symbolic systems. Indeed, the importance given to linguistic meaning or to symbolic systems differs from one culture to another. Omitting the distinction between signification and signifiers can lead the researcher to take the symbols and the operations involved as the essential part of knowledge and cognitive activity, whereas this knowledge and activity are mainly on a conceptual level.

Vergnaud [31] affirms that one cannot have a psychology of complex cognitive activities without knowing what is a concept as a notion, integrated into that of representation. He defines the concept in the following threefold way: C= (R, I, S)

R (reference): all situations which give meaning to the concept.

I (signification): all operational invariables on which the effectiveness of signifiers rests (concept-in-act and theorem-in-act).

S (signifiers): the whole set of signifiers — linguistic and non-linguistic forms — which makes it possible to represent symbolically the concept, its properties, the situations and the procedures of treatment. However the symbol is only the "directly visible part of the conceptual iceberg": the symbolic system is only the directly communicable part of the field of knowledge which it represents. Lexicon and syntax would be nothing without semantics and pragmatic activities which produce them, i.e. without the practical and conceptual subjective activity in the real world. From this point of view, Bachelard (1938/1996) draws our attention to the fact that: "at the same time, behind the same word, lie so different concepts! What misleads us is that the same word both designates and explains. Designation is one thing; explanation is another".

Concepts are constructs and conceptualization takes place progressively throughout discovery and the mastery of several kinds of concepts and theorems, most of which remain largely implicit. It is this implicit character which leads Vergnaud to introduce the notions of *concept-in-act* and *theorem-in-act* [34] and to add: "But an implicit concept is not completely a concept. It is thus a fundamental theoretical problem to analyze linguistic and non-linguistic meaning which gives the concept its public character, and which makes it possible to discuss its definition, its properties, and the truth of the proposals into which it fits" [33].

A notion which seems to be pertinent to the research presented here is that of *conceptual field*. This is a whole, greater than the sum of the situations, whose mastery requires a specific system of concepts, procedures and closely connected symbolic notations. It is well known that learning situations do not always emphasize the same aspects of a concept. This can result in various levels and various forms of the conceptualization of reality. Such conceptualization can thus appear to be local, which makes it impossible to establish a relationship between all the elements of a situation, or to recognize these elements or entities in other situations.

In the second component, I, the concept of signifiers intervenes, the invariable construct linked to the behaviour of the subject. From a methodological point of view, Vergnaud specifies that two levels are to be carefully distinguished here:

- the surface level which comprises rules of action and expectations, which are fairly easily put into words by the subjects. Consequently, this level remains more readily accessible to the observer.

- an in-depth level, containing operational invariables and a system of operations based on these invariables, and which makes it possible to generate rules of action and expectations. Here, verbalisation by the subject is much more difficult. Consequently, for the observer, this level remains rather inaccessible.

In conclusion, in a given situation, the possibility of locating the various representations which the subjects use to solve a problem can give us important information to understand their cognitive processing and their level of conceptualization. In particular, they can show us the *obstacles* which prevent certain subjects from passing from one type of representation to another, which may be better adapted and more effective.

2.2 Obstacles to learning and conceptual development

It is now recognised that the production of an erroneous explanation is not solely due to ignorance, uncertainty, chance or tiredness as the empiricist or behaviourist theories of learning seem to recommend. These do not recognise the concept of representation as a relevant and operational concept. Indeed, it is the well-known works of Piaget and Bachelard which first showed how errors should be considered as a step towards the acquisition of knowledge, or as the result of previous knowledge adapted to fit particular circumstances, but which is presented in the form of false information or simply does not fit with other circumstances. Thus, in contrast to the empiricist or behaviourist theories of learning, errors are neither sporadic nor unpredictable, but correspond to representations, and should be thus considered as constructs that constitute true obstacles which must be overcome.

Within the framework of the didactics of mathematics, Brousseau [5] specified precisely the notion of obstacle by observing manifestations through errors bound by a common cause, through ways to conceive a characteristic and coherent design, even if it is incorrect, and finally through previous knowledge which functions within a specific field of action. So overcoming an obstacle requires an effort comparable to that of learning. Knowledge as an obstacle is the fruit of the interaction of the individual with his environment, and, more precisely, with a situation which makes this knowledge relevant. The variability of human beings, knowledge and the context of learning, leads inevitably to the construction of erroneous designs which if locally true, are non generalizable. However, it should be observed that these designs are guided by conditions of interaction (individual, medium, knowledge) which function in a certain way. The identification of these conditions allows the use of didactic objectives. However, these conceptions guided by the parameters of the interaction (i.e. individual, environment, knowledge) can be modified if these parameters are well identified and used with didactic aims. It remains that the obstacles have different origins and content. For our purposes, the main sources of these obstacles to learning and conceptual development can be summarized as follows:

- Obstacles of ontogenetic origin linked to the limitation acquisition potential of the subject during a specific period of cognitive development. Piaget's theory offers many examples of the obstacles related to the period of development of the subject.

- Obstacles of epistemological origin. These are the obstacles identified throughout the history of the conceptual development of a discipline. Some learners' difficulties may be obstacles which arise from this history. These obstacles are met for example in the history of science, elements of which may be observed in the spontaneous models of pupils in learning situations. An epistemological obstacle is then constitutive of incomplete learning. This shows that it is not the single result of a chance error which it is sufficient to correct, or ignorance which can be remedied, or of another unspecified incapacity. It can result from cultural, social and economic conditions, but these causes are brought up to date in designs which resist even when the causes disappear.
- Obstacles which have an educational origin. Bachelard [4] speaks about the teaching obstacle and Brousseau [5] introduced the concept of an obstacle of didactic origin. As far as we are concerned, we consider that the conceptualization of reality as well as the representations are constructed through various types of learning which emphasize certain specific aspects of reality, and which are themselves related to the particular culture in which the learning takes place. These obstacles simultaneously come from different sources: teaching, didactic and socio-cultural and depend upon a model of teaching and learning.

2.3 The moon phases: an object of study for astronomy

The phases of the moon correspond to apparent changes of this satellite of the Earth. These changes depend on the respective positions of the Earth and the Moon in relation to the to the sun. Depending on the time of the month and year, the sun's light will shine on one or other part of the moon. When the moon is between the sun and the Earth, the part lit by the sun is invisible, giving the new moon. When the first zone becomes visible from the Earth, the first crescent is visible. When the moon reaches its first quarter, a half-moon is visible. When the moon is opposite to the sun in relation to the Earth, a full circle, the full moon may be seen. When the moon reaches three quarters of its cycle, it is the last quarter, one sees the other half lit by the sun. This part continues to decrease, the last quarter; then the new moon returns. The cycle of the phases of the moon, called the lunar month, lasts twenty-nine and a half days. During the complete rotation of the Earth, the moon also rotates on its own axis. This brief description (Fig. 1), shows the high degree of conceptual complexity implied in this process of celestial and astronomical mechanics. It requires taking into account the moon, the Earth and the sun, but especially the relationship between three elements, which are in motion. The moon turns on itself and around the Earth, which turns around the sun and on itself. The position of the observer on the Earth as well as the time of day should also be taken into account.

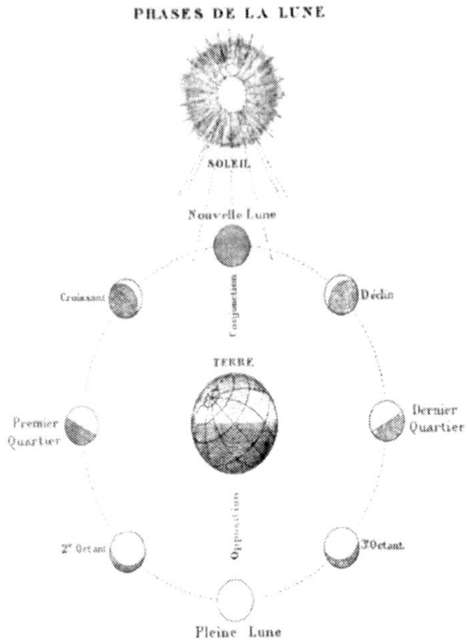

Fig. 1. representation of the moon phases

2.4 The moon and its various socio-cultural representations

Throughout history, human beings have always wondered about the moon [15]. We cannot approach in detail all the questions raised here. A number of works have treated this question, notably, *The Moon, myth and image* by Jules Cashford [8]. One of the dominant representations of the moon appears as the image of a lunar boat crossing the night sky. Figures (2, 3, and 4) show ancient pictorial representations while the following figures (5, 6, 7 and 8) show visual representations of the moon with which subjects may be confronted in their actual daily life. Drawings (6, 7) are taken from comic strips produced in the southern hemisphere respectively by Mauricio de Sousa (Fig. 6), in Brazil and by Bernard Berger in New Caledonia (Fig. 7).

It is also necessary to add all the scientific photographs produced by satellites or even by man's direct visits, on several occasions, to the moon itself. There are also those produced synthetically by computers.

We can find artistic representations of the moon's phases such as figure 8

2.5 Phases of the moon: an object of everyday learning

The moon in all its apparent phases constitutes an object of the perceptive experience of everyday life in early childhood. The observation of this object

Fig. 2. Entry of Venus' Sanctuary, Paphos (Harding 2001 p.92)

Fig. 3. Assyrian winged moon (Harding 2001 p.96)

Fig. 4. Lunar tree surrounded by lattice and torches (Harding 2001 p.93)

Fig. 5. Black/white creation Editions M.D.

Fig. 6. Chico Bento (number 163, 1993) Mauricio de Sousa Editora GLOBO São Paulo Brazil

Fig. 7. Small/large Boat. The Bush in Madness n°10 (1996/2002) Bernard Berger Edition Noumea New Caledonia

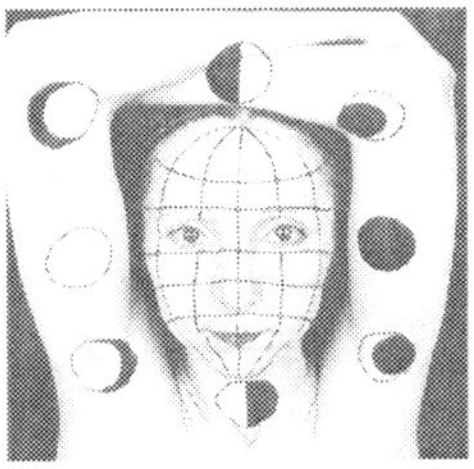

Fig. 8. Lunation 1990 Photographs by Rimma Gerlovina & Valery Gerlovin

constitutes then *a first experience* in the Bachelardian sense. The mental representations that the subjects construct in this way, can constitute obstacles of epistemological origin with which they will be confronted when it is a question of understanding the phenomenon of the phases of the moon. From a Vygotskian understanding, this observation leads to the spontaneous formation of daily concepts relating to the phenomenon of the apparent movement of the moon and of its phases, and its link with the apparent movement of the sun, with a weak use of language. These concepts are isolated from each other and develop apart from any given system. They are temporally or locally relevant, and may also lead to generalizations which can be abusive. Their conceptual weakness is manifested by an incapacity for abstraction or an inaptitude for intentional use. What is characteristic is their incorrect use. Frequently saturated by the rich personal experience of the subject, and as such, socio-cultural obstacles to conceptual development become apparent.

Data resulting from discussions with eight adult illiterate subjects of rural areas of Brazil illustrates our research. These subjects are characterized as all having little contact with a written culture. Questioned on "*how they see the moon when it is not full*", they provide answers giving clues to the nature of the obstacles indicated in this article. Thus Maria, 60 years old, a cleaning lady, from the city of the Sertão, Nordeste of Brazil, draws the moon B (Appendix 1) and explains that "it is like Lampião's hat" (a famous character known in that city). Nen, 40 years old, with the drawing of the moon C (Appendix 1), explains why "it is like a smile", and Neta, 35 years, that "the moon is presented in the form of a hammock". This complementary data consolidate our interpretation of the local aspect of knowledge of the illiterate subjects, but considered further also seem to reveal the effects of what is learned at school, as obstacles to processes of conceptualization.

2.6 Phases of the moon: an object of learning at school

Concerning scientific concepts, Vygotski postulates that they arise from indirect contact with the object and can be acquired only by a continuous process from general experience to the private individual. They are formed under the intentional action of the school, beginning with a teacher's explanation, who exposes a scientific formulation of the concept. Their weakness lies on the one hand in his verbalism, principal source of the shifts which generate obstacles to conceptual development; and on the other hand, insufficient links to concrete experience and knowledge. However, analysis of our research data corroborates with Piaget's idea (1969) where verbalism of the image also exists.

Through a survey carried out over the last five years among primary school teachers still in training and those in secondary schools in initial or professional training at the University Teacher Training Institute (IUFM) of Lyon, we collected data which enabled us to identify three main categories of approaches to teaching the question of the phases of the moon in their courses in France. The first was described as a way of telling a fable or tale to approach the concept of the phases of the moon for young children. The second was based on more technical elements calling upon mnemonic techniques that they themselves had acquired during their own education and more related to learning at secondary school. Finally the third target group was also primary school teachers and was based on a model implying a scientific approach to observation similar to that of astronomers. In an article on Toussaint (1999) we found an extremely relevant and adequate description similar to that which we observed among teachers' replies, related to the first two steps

First approach: "the moon tells lies. . . !"

The concept is introduced through a fable or tale. The following instruction is given to the pupils, and may be thus summarized: "*Don't forget that the moon tells lies: when it looks like the letter C, its not waxing, its really waning; and when it looks like the letter D, its waxing*". (In French, or other Latin languages, C represents the first letter of the verb *croître* — to inCrease. The letter D, *décroître*, means Decrease). This aim of this approach is clearly to help recall and to understand the phases of the moon. The emphasis is on the signifiers the first letters of the verbs, which enables recognition and verbalisation of the different phases. It must be noted that this approach does not in any way reflect the conceptualization of the astronomical movement of the moon. The accent is only on meaning, which makes it possible for the subjects to recognize a position of the moon without taking into account the dynamic relationship of the concept involved. In addition, the concept associated only with only one situation is not restricted here by one meaning to give an account of all the representations which the subjects by will have built elsewhere through early experience by looking at the moon in their childhood. This story-telling approach does not give optimum conditions for learning subjects to go beyond the everyday concepts relating to the phases of the moon, to arrive at scientific concepts with the meaning given by Vygotski [35]. In addition the images given by the forms D and C constitute *prototypical* representations of phases of the moon within Rosch's meaning [25].

Second approach: "the moon is no longer a liar...!"

Here the question of the phases of the moon is introduced with more technical concepts. Toussaint [28] gives an account of it by calling upon its characteristics experienced by a schoolboy.

"*Later, I came across another rule which said that the moon (it was less funny!) was no longer telling lies: by carefully drawing the diameter which goes from one end to the other, one can write in tiny characters a P with the first quarter and a D with the last*". We see that in this case a geometrical concept appears — diameter — which can give the impression of a more learned approach. But the use of diameter does not bring anything more than the mapping of a lunar position with one of the two letters P and D, which are mobilized only as meaning in a way identical to the first step. Surprisingly, the teachers however, seem to regard this approach as requiring a higher level of conceptualization. This is suggested by the fact that this approach was never used with primary school pupils.

Third step: towards a scientific model of the phases of the moon...

This less frequent third approach called upon systematic observations and the use of a model with the pupils. We collected results from a teacher training course on the didactics of physics. In this step, they were asked to identify the representations of the phases of the moon, then to carry out systematic observations of the moon with which the representations are confronted. Finally the trainee teachers are confronted with a model in the shape of a model of the system Earth-moon-sun which reproduces the movements of these bodies. Starting from this mechanical device, they are confronted with problems of the type: in its various phases how does the moon appear from the Earth?

It appears that this model, in spite of its concrete material characteristics, allows only a limited understanding for trainee teachers, insofar as when they approach the phases of the moon they do so infrequently or not at all in their teaching practice. The main argument they use relates to the high cost of such device in a teaching-learning situation. However it is clear that this step, as "costly" as it is, requires an active process of conceptualization.

3 Identifying the conceptualization levels and the associated obstacles

As we said in the introduction to this article, the central question of this study is to identify the conceptual levels and the resulting obstacles that are associated with the conceptualization of moon phases. The starting point of the study was the observation of learning-teaching situations for Masters and Doctorate psychology students in Recife, Brazil. Thereafter, the reference field of these problems was extended to other learning-teaching situations in France and in New Caledonia and involved Bachelor's and Master's students of Educational Sciences, students training to be educational psychologists, social workers in university training in Educational Sciences, and also teacher trainees in initial training and even teachers in professional training. In all the contexts in which Nadja Acioly-Régnier led these teaching sequences, the teaching situation invariable lay in the injunction to draw the moon as each one could see it when it is not full. Interestingly, in all the situations, the main characteristic of the drawings was that they had a closer relationship with the representations of written culture than by direct observations carried out by the subjects. As for the arguments put forward by the subjects, their main characteristic can be summarized as follows: *the drawings are more related to the images we perceive in our socio-cultural environment than what we observe directly by looking at the sky.*

3.1 An initial situation problem focussed on the phases of the Moon

To begin with, we were struck by the fact that the Brazilian students faced with the instruction to draw the moon as they saw it from the subequatorial tropical zone, systematically produced drawings with a typical representation of the moon observed more frequently from mainland French sky. The great majority presented the moon in the shape of a "*crescent*" (Fig. 6) facing the right and a lower proportion in the shape of a "*crescent*" facing the left (Fig. 5). The contradictions of these two categories of response give rise to a genuine situation of socio-cognitive conflict. However by remaining on this level of exchange, the nature of the responses built on the basis of first-hand experience and on the tools provided by the cultural environment and the written culture did not offer the conditions for a significant rise in the level of conceptualization. To modify these conditions, the subjects were then invited to directly observe the moon and to compare their perception with their drawings. The result then seemed to provoke a real cognitive destabilisation and a desire to understand the situation from a conceptual point of view. Within the context of psychology teaching, this didactic situation carried out on the phases of the moon led the students to analyse the school textbooks to pinpoint the role and the place of iconic representations in scientific learning. Piaget himself [23, p.110] considered that: *"image, film and audiovisual processes of which all pedagogy keeps harping on about today and wants to give the illusion of being modern, are invaluable auxiliaries as an addition or as a spiritual crutch, and it is obvious that this is a progress compared to purely verbal teaching. But there is verbalism of the image just as there is verbalism of the word".*

This activity had led the students to verbalise their awareness that the responses initially provided were determined by school learning which, in the name of immediate efficiency, favours excessive simplification and reduces memory based learning of the signifiers without working on the concepts to which they are attached. This perspective is reinforced, in the cultural environment, by the graphic representations of the moon that we find in the media, in comics (Fig. 6 and Fig. 7), in advertisements, etc.

When we transferred the situation in France to the Teacher Training Institute (Institut Universitaire de Formation des Maîtres) in Lyon, as well as at the university (University Lyon2), we modified the procedures. The situation problem given to the students and the trainee teachers was not based on the request of drawing the moon, but on a story told as follows: *"In Brazil, we asked to students to draw the moon as they saw it in their sky. Because they drew the moon as a vertical crescent or a slightly bent one, that faced the right for the most part of them or the left for the others, I asked them to observe it directly by watching in the sky. I told them to increase the stakes by paying for a coffee to anyone that will see it as he drew it".* To end the story we added: "*Nobody came to claim the offer*", and we then addressed the following question to the French students: *"Why was this the case?"*

Using a qualitative approach, the analysis of the responses obtained lead to differentiate five main categories. The subjects were allowed to give several answers in their arguments and also to change the meaning of the answer by exchanging their views with their fellow students.

The first category [CAT1] corresponds to responses overly determined by the socio-cultural dimension. In this category we found the following: "*They did not dare to do so because you are married" or "... you are a professor" or " ... your husband is jealous", and also "Because they watched the moon lying in a hammock" etc.*

The second category [CAT2] corresponds to responses overly determined by the personal characteristics of the subjects questioned. For example: "*They disliked coffee, or pubs" etc.*

The third category [CAT3] corresponds to responses overly determined by local conditions, contexts, or circumstances. We found for example: "*they were living downtown and could not see the sky so well" or "there was fog" etc.*

The fourth category [CAT4] corresponds to answers overly determined by the exact knowledge of the situation-problem. For example: *"I cannot say, I've never been there" or "I went there, and obviously the moon is not like that" etc.*

The fifth category [CAT5] corresponds to responses overly determined by school knowledge. We found for example: *"I believe it has something to do with shadows, and light"* or*"... with the moon, the sun, and the earth and the hemispheres"* etc.

As said in the introduction, Luria [21], studying the illiterate Uzbek farmers without any written culture, observed similar results. The question is then:

Are the facts here collected just anecdotic, representing particular behaviour of subjects confronted with a scholarly situation-problem even though they are at a university level during their initial training?

Alternatively, do these facts reveal socio-cultural obstacles, impeding the rise of the level of conceptualization, with pedagogic and/or didactic origins that could be found in the school institution, or with epistemological origins that are linked to the development of the concept? In order to explore such hypotheses further we designed a questionnaire aimed at investigating mental representations concerning the moon phases and we conducted a survey on subjects from various geographical and socio-cultural contexts.

3.2 Exploration of the mental representations of the phases of the moon

The first questionnaire was drawn up taking into account the observations previously evoked. It comes within the perspective of a structure of data founded on recognition of static shapes proposed *a priori* to the subjects and no longer on the production of a graphic shape supposed to represent the moon for the subjects. The four shapes A, B, C, D (Fig. 3.2) are formulated in relation to question Q1, Q2, Q3, and Q6, reproducing the graphic shapes found in written work in the course of history and which have been able in turn to be prototypical shapes in various eras and contexts. Question Q4 and Q5 repeat the pedagogical approach described by professors in charge of teaching this content. Furthermore, the subject who answers is the subject directly concerned with the instruction in question Q1. In questions Q2 and Q3 the reference point of a particular subject to which the subject who answers must put himself or herself in their place, and aims to introduce the variability of the observer's position on the surface of the Earth. So various forms of the questionnaire were submitted in such a way that in question Q3 it was either the actual geographic location of the subject who answered or it was geographically opposite in question Q2. The aim of question Q6 is to bring out the representations that the subjects have with respect to the perception of the moon from each terrestrial hemisphere and from the equator.

Questionnaire-based survey

This questionnaire-based survey was carried out along with observation obtained through empirical methods located respectively in metropolitan France in Lyon, in Noumea, New Caledonia and in Recife, in north-east Brazil over a long period from 2001–2004. The overall sample is composed of 198 subjects.

Note that the individuals of the Ech_IUFM sample were not submitted that to two questions Q1 and Q2

Representations of the moon

Answers Q1; Q2; Q3; Q6							
A-Yes [1Asim]	A-No [1Anao]	B-Yes [1Bsim]	B-No [1Bnao]	C-Yes [1Csim]	C-No [1Cnao]	C-Yes [1Dsim]	C-No [1Dnao]

Fig. 9. Shapes of the moon

New Caledonia		Metropolitan France	Recife, in North-East Brazil
Adults	Children	Adults	Adults
Students in Master of educational sciences; situation of training	Elementary school students: grade 6 in Noumea	Trainee teachers: 2nd year in IUFM de Lyon	Curso normal superior da UFRPE
Adults professional in the educative system, as teacher or as responsible of teacher's training	Children	Adults in situation of professional training (teachers)	Adults in situation of professional training (future teachers)
EchNC_Ad	EchNC_Enf	Ech_IUFM	Ech_UFRPE
119	22	28	29
Q1; Q2; Q3; Q4; Q5; Q6	Q1; Q2; Q3; Q4; Q5; Q6	Q1; Q2	Q2; Q3; Q4; Q5; Q6

Table 1. Constitution of the overall sample in function of the contexts

Treatment and analysis implicative statistics: modelling and description

The first part of the questionnaire consists of 7 questions which provide information relating to the subjects such as sex, age and the length of professional experience. The variable SEX corresponds to the vector variable (MALE; FEMALE) whose two components are additional binary variables. We also introduced a "place of residence" variable whose detailed form corresponds to the membership of a "usual cultural area: Kanak" in New Caledonia. However we restricted it here to the couple of additional binary variables (HN; HS) indicating respectively the northern and southern hemispheres. The variable "Age" was modelled by the couple of additional binary variables (Child; Adult)

The question Q1 concerning the recognition or not of the shapes and apparent positions of the moon is represented by a variable vector with 12 binary components (1Asim; 1Anao; 1Anr; 1Bsim; 1Bnao; 1Bnr; 1Csim; 1Cnao; 1Cnr; 1Dsim; 1Dnao; 1Dnr). We coded xxnr the absence of answer.

Sample		EchNC_Ad		EchIUFM		EchUFRPE		EchNCEnf	
SEX		Male	Female	Male	Female	Male	Female	Male	Female
Number of responses SEX		50	68	4	24	4	25	15	6
Number of responses SEX and AGE		47	60	4	24	2	25	15	6
AGE in Years	Min.	23	27	24	22	18	17	10	10
	Max.	49	51	33	45	32	40	12	12
	Mean.	38.5	37.6	29	26.4	25	23.0	10.7	10.6
	St.dv	5.2	5.6	3.7	5.9	7	6.2	0.68	0.70
Number of responses SEX and LENGTH		48	66	X	X	2	18	X	X
Length of professional experience in Years	Min.	2	3	X	X	5	0	X	X
	Max.	27	31	X	X	10	20	X	X
	Mean	14.5	14.3	X	X	7.5	2.85	X	X
	St.dv	5.57	6.54	X	X	2.5	4.81	X	X

Table 2. Description of sample (Numbers of responses, measures of location and dispersion)

The questions Q2 and Q3 which urge the subject to be located from another point of view, each one are represented by a variable vector with 15 binary components:

$$\begin{array}{l} Q2 = (2sim; 2nao; 2nr; 2Asim; 2Anao; 2Anr; 2Bsim; \\ 2Bnao; 2Bnr; 2Csim; 2Cnao; 2Cnr; 2Dsim; 2Dnao; 2Dnr) \\ Q3 = (3sim; 3nao; 3nr; 3Asim; 3Anao; 3Anr; 3Bsim; \\ 3Bnao; 3Bnr; 3Csim; 3Cnao; 3Cnr; 3Dsim; 3Dnao; 3Dnr) \end{array} \quad (1)$$

The questions Q4 and Q5 relate to judgements of effectiveness and adaptation of a teaching approach based on the correspondences of the phases of the moon with letters C, D, p or D. They are both modelled by a binary variable vector of dimension 4

$$\begin{array}{l} Q4 = (4ADAP; 4INAD; 4EFFI; 4INEF) \\ Q5 = (5ADAP; 5INAD; 5EFFI; 5INEF) \end{array} \quad (2)$$

Finally the last question Q6 relates to taking into account the three places: southern hemisphere, northern hemisphere and equator. It is modelled by a binary variable vector of dimension 15:

$$\begin{array}{l} Q6 = (6sim; 6nao; 6nr; 6A\_HN; 6A\_E; 6A\_HS; 6B\_HN; 6B\_E; 6B\_HS; \\ 6C\_HN; 6C\_E; 6C\_HS; 6D\_HN; 6D\_E; 6D\_HS) \end{array} \quad (3)$$

The frequency distributions of these variables are provided in appendices (Appendix 2, Appendix 3)

The modeling of the questions by variable vectors with binary components enables us to place ourselves in the context of statistical implicative analysis, SIA developed by Gras Régis and his collaborators [11, 12, 24] starting from prospects released by I. C. Lerman [19] and processed by CHIC software (Couturier). The table of the statistical series is a table of 67 columns made up of values 0 or 1 describing a realization of the 67 binary variables.

Seven open questions formulated from the interrogative one: *Why?* give place to textual answers and are thus modelled by textual variables. At this stage of our work in this article we do not proceed to a refined computer-assisted analysis of contents using software such as SPAD_T[4] [17] which could be also supplemented by the SIA approach.

Questions		Yes	Not	Not-Reference mark	Total	Yes(%)	Not (%)	Not-Reference mark (%)
Q1	1A	171	21	6		86.36	10.61	3.03
	1B	58	116	24		29.29	58.59	12.12
	1C	65	110	23		32.83	55.56	11.62
	1D	160	25	13	198	80.81	12.63	6.57
Q2	2	163	34	1		82.32	17.17	0.51
	2A	40	120	38		20.20	60.61	19.19
	2B	109	48	41		55.05	24.24	20.71
	2C	112	40	46		56.57	20.20	23.23
	2D	54	107	37		27.27	54.04	18.69
Q3	3	114	51	5		62.18	34.45	3.36
	3A	31	102	37		18.49	63.87	17.65
	3B	71	61	38		38.66	42.86	18.49
	3C	65	67	38	170	38.66	42.86	18.49
	3D	30	102	38		18.49	63.03	18.49

Table 3. Frequency Distributions of answers to Q1, Q2, Q3.

Results (Table 3) relating to Q1 (*Did you already see the moon like that in reality?)* confirm the use of prototypical memory (within the meaning of Rosch [25]) in the evocation and the recognition of these lunar forms. According to Eleanor Rosch, among all the levels of abstraction possible, one is psychologically more accessible than the others: called "basic level", level which makes it possible for the individual to obtain the maximum information with the minimum of cognitive effort. Compromised between the most abstract possible level but which offers, in same time, a sufficient number of concrete attributes. Thus, the figures 1A (86.36%) and 1D (80.81%) "*growing directed vertically*", seem figures *prototypic* independently of the group studied (Test of χ^2 Appendix 4) with a prevalence of 1A "*horns directed towards the line*". On the other hand the figure 1B (29.29%) "*horizontal crescent turned*

[4] SPAD_T Système Portable d'Analyse des Données Textuelles CISIA-France

downwards" and the figure 1C (32.83%) "*horizontal crescent turned upwards*" are quoted little by the subjects. These two variables depend on the variable "groups" (Appendix 4) with one *attraction* towards 1B and 1C for EchNC_Ad and one *repulsion* for the three other groups.

It may be noted that no significant dependence (with the test of χ^2 with $\alpha = 0.05$) is detectable between the variables "Sex", "Age" "Hemispheres" and the variables 1A, 1B, 1C and 1D.

	Q1 (A)			Q1 (B)			Total
SEX	1Asim	1Anao	1Anr	1Bsim	1Bnao	1Bnr	
Male	59	12	2	24	40	9	73
Female	110	9	4	33	75	15	123
	169	21	6	57	115	24	196
df. = 2, $k = 5.99$	$\chi^2 = 3.99$			$\chi^2 = 0.88$			
	Q1 (C)			Q1 (D)			Total
SEX	1Csim	1Cnao	1Cnr	1Dsim	1Dnao	1Dnr	
Male	25	39	9	57	12	4	73
Female	39	70	14	101	19	9	123
	64	109	23	158	31	13	196
df. = 2, $k = 5.99$	$\chi^2 = 0.23$			$\chi^2 = 1.56$			

Table 4. bivariates Distributions Q1 and SEX.

The objective of Q2 ("if a subject living in the opposite hemisphere to you, says to you that he/she had never seen any of these moons in reality, would you believe him/her?") and that of Q3 ("If a subject living your hemisphere, says to you that it had never seen some of these moons in reality, you believe it?") were the introduction of a data being able to draw the attention of the subject to conceptual aspects of the phases of the moon. It could thus oppose its own answers to those of this virtual subject, either by maintaining those brought to Q1, or not answering. We find results similar to those which we pointed for Q1. We find there also initial classification [CATn] (N = 1 to 5) exposed into 3.1.

Exploration of the trees of similarities and cohesions, and the implicative graph

Initially, we studied the data built on the 27 principal binary variables (Appendix 2) starting from the sample of the 198 individuals of which 28 lived in the northern hemisphere and 170, the southern hemisphere. We gave to a second publication, the exploitation of the 67 basic variables (Appendix 2, Appendix 3) on the sample of 170 individuals of the southern hemisphere. In fact here we studied the 27 variables which relate to the whole of the total sample.

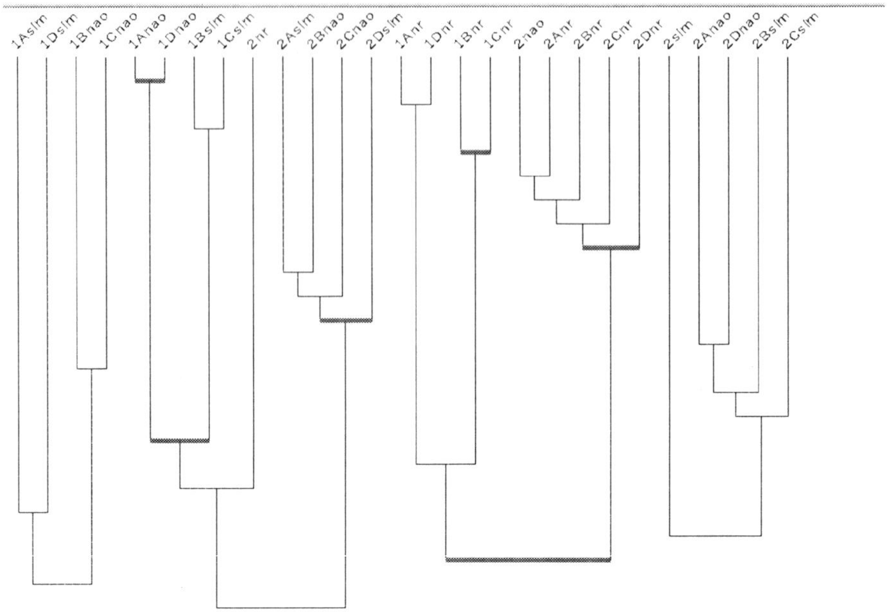

Fig. 10. Tree of similarities

By classification based on a model inspired by Lerman using probabilistic indices, we obtain a repartition of the 27 binary variables in four main classes as the tree of similarities shows (Fig. 10).

$$\begin{array}{l} \mathrm{CLS1(lev22)} = \{\mathrm{1Asim, 1Dsim, 1Bnao, 1Cnao}\} \\ \mathrm{CLS2(lev23)} = \{\mathrm{1Anao, 1Dnao, 1Bsim, 1Csim, 2nr, 2Asim} \\ \mathrm{2Bnao, 2Cnao, 2Dsim}\} \\ \mathrm{CLS3(niv21)} = \{\mathrm{1Anr, 1Dnr, 1Cnr, 2nao, 2Anr, 2Bnr, 2Cnr, 2Dnr}\} \\ \mathrm{CLS4(lev20)} = \{\mathrm{2sim, 2Anao, 2Dnao, 2Bsim, 2Csim}\} \end{array} \qquad (4)$$

This repartition reflects a control of completely coherent answer. Each class consists of methods which are logically associated

These classes result from the aggregation of binary variables explainable by the effect of the *prototypical* figures in the reading of the world by the subjects. Thus the figures A and D, on the one hand, B and C, on the other hand are strongly associated for Q1 and Q2.

The analysis of the contributions of the subjects are characterized respectively by the categorial variables "Sex", "Age", "Hemisphere" additionally though the optimal groups do not reveal outstanding effects. At most we observe a significant influence of the binary variable CHILD in the constitution of class CLS1 (lev22). That could correspond to the importance of the education with which they are confronted daily at their age and, in particular, figures that they meet in their textbooks or even in children's books.

Let us explore the implicative graph built to leave the 27 binary variables instanced on the total sample of 198 individuals. With a confidence level of 0.99, we identify 7 implicative chains (Fig. 11). Five comprise only 2 terms.

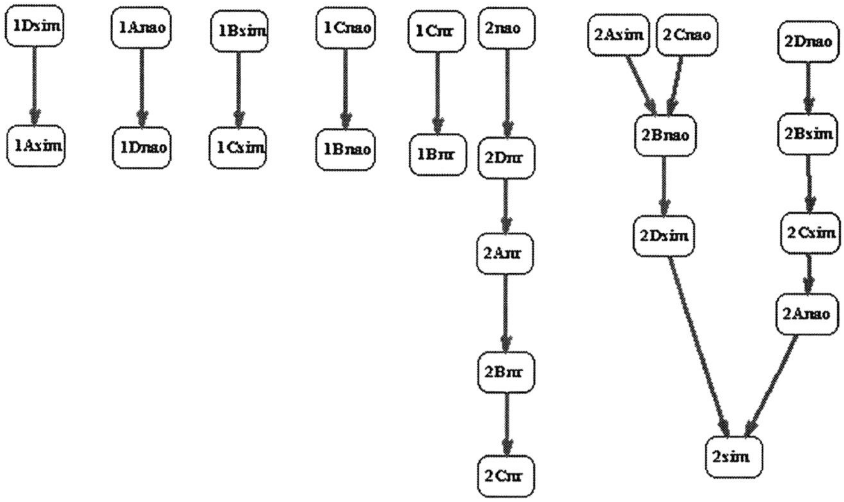

Fig. 11. Implicative Graph, [Confidence Level $1 - \alpha = 0.99$]

$$\begin{array}{l}
\text{(ch1) } [1\text{Dsim}] \Rightarrow [1\text{Asim}] \\
\text{(ch2) } [1\text{Anao}] \Rightarrow [1\text{Dnao}] \\
\text{(ch3) } [1\text{Bsim}] \Rightarrow [1\text{Csim}] \\
\text{(ch4) } [1\text{Cnao}] \Rightarrow [1\text{Bnao}] \\
\text{(ch5) } [1\text{Cnr}] \Rightarrow [1\text{Bnr}] \\
\text{(ch6) } [2\text{nao}] \Rightarrow [2\text{Dnr}] \Rightarrow [2\text{Anr}] \Rightarrow [2\text{Bnr}] \Rightarrow [2\text{Cnr}] \\
\text{(ch7.1) } [2\text{Dnao}] \Rightarrow [2\text{Bsim}] \Rightarrow [2\text{Csim}] \Rightarrow [2\text{Anao}] \Rightarrow [2\text{sim}] \\
\text{(ch7.2) } [2\text{Asim}] \Leftrightarrow [2\text{Cnao}] \Rightarrow [2\text{Bnao}] \Rightarrow [2\text{Dsim}] \Rightarrow [2\text{sim}]
\end{array} \quad (5)$$

Just as we saw we in the tree of similarities, what emerges from these chains of quasi-implications that the subjects answered (yes, no, failure to reply) by strongly associating the figures A and D, on the one hand, and the figures B and C on the other hand, as well within the framework of Q1 as Q2. The two twin variables (2Asim) and (2Cnao) also illustrate this property by the strong association of opposition between the positive designation of a prototypical figure A and that negative of the not-prototypical figure C. We observe in more than one almost perfect inclusion of {2Asim} in {2Dsim} insofar as there is one counterexample. This always confirms the dominant place of the *prototypical* figures in the mental representations and their role in the construction of the obstacles to the development of conceptualization. In our case, this relates to the acquisition of a representation of the world — here the moon — which is closer to that elaborated in learned knowledge.

Chain (Chx)							
(CH1)	(CH2)	(CH3)	(CH4)	(CH5)	(CH6)	(CH7.1)	(CH7.2)
Questions Q1					Questions Q2		
The most typical variable							
Child	Male	Male	Child	HS	Child	Female	HN
Level of risk incurred by choosing the variable like most typical							
0.149	0.117	0.278	0.0952	0.224	0.112	0.194	0.0355
The most contributive variable							
Child	Male	Male	Child	HS	Child	Female	Male
Level of risk incurred by choosing the variable like most contributive							
0.149	0.117	0.278	0.0952	0.224	0.1	0.236	0.149

Table 5. Contributions and typicality of the additional variables

By considering the weakest risk of the level, we observe that group HN is most typical of the chain (Ch7.2). It corresponds in fact to that made up by IUFM teacher trainees in Lyon, only located in the northern hemisphere and characterized by the highest academic level. The semantic characteristic of this chain is the acceptance of a different perception of the moon according to the hemispheres and the association of prototypical figures between them (A and D accepted) and, of the other, those which are not-prototypical (B and C rejected). Effects resulting from the courses on the didactics of physics are a possible interpretation of the amount of distance from these subjects.

So now we explore the cohesitive (implicative) tree, it gives six main classes from binary variables.

$$\begin{array}{l} \text{C1\_lev21}(5) = ((1\text{Anao} \Rightarrow 1\text{Dnao}) \Rightarrow (1\text{Bsim} \Leftrightarrow 1\text{Csim})) \Rightarrow 2\text{nr} \\ \text{C2\_lev20}(4) = ((1\text{Anr} \Rightarrow 1\text{Dnr}) \Rightarrow (1\text{Bnr} \Leftrightarrow 1\text{Cnr})) \\ \text{C3\_lev17}(4) = ((1\text{Bnao} \Leftrightarrow 1\text{Cnao}) \Rightarrow (1\text{Dsim} \Rightarrow 1\text{Asim})) \\ \text{C4\_lev9}(5) = (((2\text{nao} \Leftrightarrow 2\text{Anr}) \Leftrightarrow 2\text{Bnr}) \Rightarrow 2\text{Cnr}) \Leftrightarrow 2\text{Dnr} \\ \text{C5\_lev16}(4) = 2\text{Cnao} \Rightarrow ((2\text{Asim} \Leftrightarrow 2\text{Dsim}) \Rightarrow 2\text{Bnao}) \\ \text{C6\_lev14}(5) = 2\text{Dnao} \Rightarrow (2\text{Csim} \Rightarrow (2\text{Bsim} \Rightarrow (2\text{Anao} \Rightarrow 2\text{sim}))) \end{array} \quad (6)$$

In this classification, we find the classes which were formed in the approach by similarity. The formed R-rules are organized in a structure which confirms the properties of the prototypic representations and the obstacles that they induce such as we identified them through the analysis of the implicative graph.

The R-rules are conceived as an extension of the binary rules (A) $\Rightarrow$ (b) with the rules. They are affected to a d°R degree calculated according to the

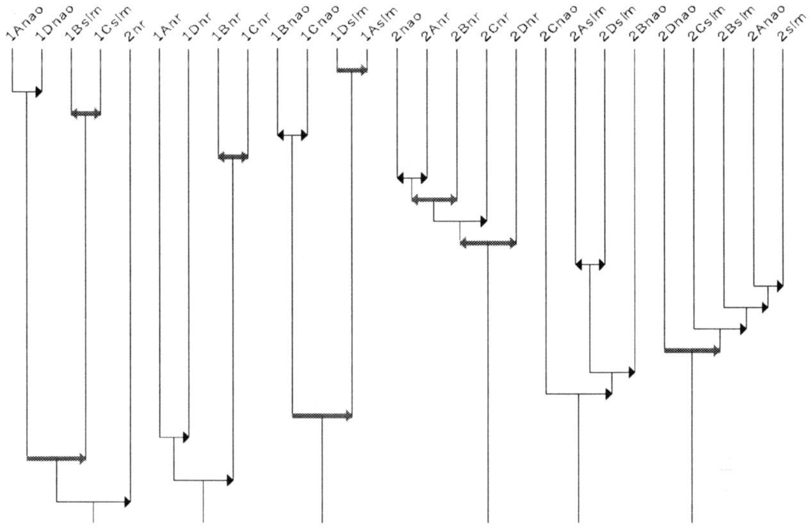

Fig. 12. Implicative Tree

following definition (Kuntz ASI 2005): a R-rule composed of a binary variable is degree d°R=0. A R-rule consists of a binary rule with a degree d°R=1. A R-rule $(R') \Rightarrow (R'')$ admits a degree d°R = d°R' + d°R" + 1.

By paying attention to the *coherence* O(C) [13] of a class C representing a R-rule to be left confrontation enters the order of the binary variables determined by the occurrences and that determined by the rules within the class. This *coherence* is measured starting from the inversions by the probability P {I> i} where i is the number of inversions observed and I the random variable "numbers of inversions"

For the class C1_lev21 (5) we obtain the order (1Anao, 1Dnao, 1Bsim, 1Csim, 2nr) within the class while the order resulting from the occurrences would give (2nr, 1Anao, 1Dnao, 1Bsim, 1Csim). We thus count 4 inversions between these two permutations. We established [13, p. 44] that the probability to have 5 or more inversions is 71/120, that is to say approximately 59.16%.

Notice that the three classes C1, C2 and C3 result from binary aggregations of variables associated with Q1, whereas the three others are associated with Q2.

When we study the typicality of each class of this division starting from the additional variables, we obtain the results in Table 6.

The C6 class incorporated at the significant level 14 consists of the variables of the chain (Ch7.1) and a R-rule of degree 4 represents. The order in the class leads to the permutation (2Dnao, 2Csim, 2Bsim, 2Anao, 2Sim)

Classify					
C1_lev21 (5)	C2_lev20 (4)	C3_lev17 (4)	C4_lev9 (5)	C5_lev16 (4)	C6_lev14 (5)
degree of R-rule					
4	3	3	4	3	4
Coherence					
$71/120 \approx$ 0.5916	$20/24 \approx$ 0.8333	$20/24 \approx$ 0.8333	$91/120 \approx$ 0.7583	$20/24 \approx$ 0.8333	$115/120 \approx$ 0.9583
The most typical variable					
MALE	HS	CHILD	CHILD	HN	FEMALE
Level of the incurred risk					
0.278	0.224	0.0334	0.1	0.0355	0.229
The most contributive variable					
FEMALE	HS	CHILD	CHILD	HN	FEMALE
Level of the incurred risk					
0.315	0.365	0.0361	0.1	0.0833	0.236

Table 6. Contributions and typicality of the additional variables

while the order resulting from the occurrences gives (2Dnao, 2Bsim, 2Csim, 2Anao, 2Sim). These two permutations reveal an inversion. From there coherence $O(C6) \approx 0.9583$. The FEMALE characteristic contributes more and remains most typical of the C6 class like chain (Ch7.1). At this level of information, we do not have a relevant interpretation of this class in relation to the contribution and the typicality of the female group.

The C5 class incorporated on level 16 corresponds to a R-rule of degree 3 and is associated with the under-chain (Ch7.2a) [2Asim] $\Leftrightarrow$ [2Cnao] $\Rightarrow$ [2Bnao] $\Rightarrow$ [2Dsim]. Its composition still reflects the association of the prototypical figures, aside, and not-prototypical, other. Taking into account the characteristics of the subjects through the additional variables enables the following interpretation: the most contributive characteristic at the same time as the most typical is the membership of the northern hemisphere. In fact because of the composition of the sample, this one merges with the group of the trainee teachers of the IUFM of Lyon. These subjects, the more well-read men of the total sample, by their answers to the Q2 question, clarify their representations, which are organized around the idea that the "other" located in the southern hemisphere cannot see the "same moon". Their training seems to lead them, through metacognitive distance, to reject the prototypical figures according to a particular condition which is to be located in the other hemisphere. This university training specific to teaching professionals could play a part in this distance as we suggested (2.6.3).

The C3 class incorporated on level 17 present the interesting results: on the one hand, it is the combined reflection of the rejection of the forms 1B and 1C and the attraction of the forms 1A and 1D; in addition, it arises that its composition is largely determined by the EchNC_Enf sub-group (Table 6).

This property occurs in the sense that we pointed to leave the analysis of the similarities. The C4 class incorporated on level 9 is also under strong influence of the conducts of answers of the subjects CHILDREN (Table 6). This directed class appears by a source rule (2nao $\Leftrightarrow$ 2Anr) which translates the fact that to answer Q2 negatively a cascade of coherent failures to reply true to fact, as a logical consequence. This property is almost tautological. However, what draws our attention is the dominant contributive share of the sub-group CHILDREN. Indeed, the refusal to believe that a virtual observer can see these moons in the opposite hemisphere evokes an obstacle of ontogenetic origin as much as an obstacle of didactic origin.

4 Conclusion

The different symbolic systems which play a role in the process of How has our methodological reasoning led to the identification of didactic or sociocultural obstacles? From the point of view of dealing with the statistical data, SIA has enabled us to establish clear links between binary variables and between class variables, of which the analysis reveals the importance of the role of didactic obstacles and of both the school and the extra-school graphic environment. From this point of view again, the resorting to implicative statistical analysis with the support of CHIC software enabled us to experience the practicalities and to pursue reflection on the issues specifically related to statistical modelling.

From the point of view of psychology we move away from the central role of the individual to take into account conceptualization. In this respect, Bruner [6] wrote: *"we were psychologists conditioned by a tradition that put the individual first. However, the symbolic systems that people use to build meaning are already installed, they are already "there" deeply ingrained in culture and language"*. In this piece of research, we observed that specific symbolic representations emphasises specific aspects of the concept and gives rise to potential didactic and/or socio-cultural obstacles. These symbolic representations are presented in temporal synchrnism with the natural language.

Therefore we spoke here of a whole comprised of linguistic and non linguistic signifiers that are engaged in the teaching-learning situations. These situations rely on the interaction between the various symbolic systems which understanding and give rise to specific problems : learning type, conceptualization level and also the nature itself of the concept as described by Vygotski in his opposition of everyday concepts and scientific concepts. Vergnaud's theory on conceptual fields taking in consideration the learning context and the characteristics of the learning method appears to be of greater interest for the interpretation of our results than the dual opposition proposed by Vygotsky. Indeed it is clear from our results that school situations and non-school situations should be differentiated since they bring up distinct focus and level of awareness.

In a school context, the focus is essentially driven toward bipolar relation [situation↔operational invariant] putting aside the whole referential situations of life experience. In this case the subject weakness consists in an inability or difficulty to recognize the situations, not exposed in the school context or even sometimes developed in this context, where the concept could be functional. For example: the learners admit that although they know the definitions they cannot apply them. They give thus the responses concerning the moon phases that are accepted in the school context, meanwhile their conceptual level remains quite weak.

On the other hand, in the non school context, the focus is mostly driven toward the bipolar relation [situation ↔ operational invariant] neglecting thus the information contained in the signifiers. In this case the subject weakness consists in the insufficiency of symbolic resources that could allow him to amplify the knowledge developed locally.

The data which came from the questionnaire corroborates with the observations made beforehand and described in this article, in the sense that where the responses seemed overly determined by school learning and by the cultural environment with the prototypical graphic representations. It must be noted however that the relationship between images and scientific learning is surely not established in the school setting. The cultural environment calls for more and more imagery resources as we have already underlined. Therefore these data bring to the fore the important role of prototypical figures in the conceptualization of the phases of the moon by well-read subjects. In most cases, it is not about symbols which represent the sensory experience, but figures which are taken as a simple recording of images which create an obstacle to a higher level of conceptualisation.

References

1. N.M. Acioly. A logica matemática no jogo do bicho: compreensão ou utilização de regras? Master's thesis, Université Fédérale de Pernambuco, Recife, Brazil, 1985.
2. N.M. Acioly. *LA JUSTE MESURE : une étude des compétences mathématiques des travailleurs de la canne à sucre du Nordeste du Brésil dans le domaine de la mesure.* PhD thesis, Université René Descartes, Paris V, 1994.
3. N.M. Acioly-Régnier. Analyse des compétences mathématiques de publics adultes peu scolarisés et/ou peu qualifiés. In *Illettrismes : quels chemins vers l'écrit?, Les actes de l'université d'été du 8 au 12 juillet 1996*, Lyon, France, 1997. Ed. Magnard.
4. G. Bachelard. *La formation de l'esprit scientifique.* Paris Lib. Vrin, 1938/1996.
5. G. Brousseau. Les obstacles épistémologiques et les problèmes en mathématiques. *Recherches en Didactique des Mathématiques*, 4(2):165–198, 1983.
6. J. Bruner. *... Car la culture donne forme à l'esprit : de la révolution cognitive à la psychologie culturell.* Paris : Editions Eshel, 1991. Original title: Acts of Meaning, Harvard University Press.

7. T.N. Carraher, D.W. Carraher, and A.D. Schliemann. Mathematics in the streets and in schools. *British Journal of Developmental Psychology*, 3:21–29, 1985.
8. J. Cashford. *The Moon Myth and image.* London: Cassell Illustrated, 2003.
9. W. Doise and G. Mugny. *Le développement social de l'intelligence.* Paris : InterÉditions, 1981.
10. G.G. Granger. *Sciences et réalité.* Paris : éditions Odile Jacob, 2001.
11. R. Gras. *Contribution à l'étude expérimentale et à l'analyse de certaines acquisitions cognitives et de certains objectifs en didactique des mathématiques.* PhD thesis, Université de Rennes 1, 1979.
12. R. Gras, S. Ag Almouloud, M. Bailleul, A. Larher, M. Polo, H. Ratsimba-Rajohn, and A. Totohasina. *L'implication statistique, nouvelle méthode exploratoire de données.* La Pensee Sauvage editions, France, 1996.
13. R. Gras, P. Kuntz, and J.-C. Régnier. Significativité des niveaux d'une hiérarchie orientée en analyse statistique implicative. *Revue des Nouvelles Technologies de l'Information*, RNTI-C-1:39–50, 2004.
14. P. Greenfield and C. Childs. Weaving, color terms and pattern representation cultural influences and cognitive development among the zinacantecos of southern mexico. *International Journal of Psychology*, 11:23–48, 1977.
15. E. Harding. *Les mystères de la femme : interprétation psychologique de l'âme féminine d'après les mythes, les légendes et les rêves.* Paris : Petite Bibliothèque Payot, 1953/2001.
16. J. Lave. Cognitive consequences of traditional apprenticeship training. *Africa in Anthropology and Educational Quarterly*, 7:177–180, 1977.
17. L. Lebart and A. Salem. *Statistique textuelle.* Paris Dunod, 1994.
18. G. Lemeignan and A. Weil-Barais. *Construire des Concepts en Physique.* Paris : Hachette education, 1993.
19. I.C. Lerman, R. Gras, and H. Rostam. Élaboration d'un indice d'implication pour les données binaires. *Revue Mathématiques et Sciences Humaines*, 74 and 75:5–35 and 5–47, 1981.
20. Mc Luhan. *Pourquoi des approches interculturelles en sciences de l'éducation.* Bruxelles : De Boeck, 1971/2002.
21. A. Luria. *Cognitive Development.* Cambridge - MA: Harvard University Press, 1976.
22. G. Mottet. Les situations-images : une approche fonctionnnelle de l'imagerie dans les apprentissages scientifiques à l'école élémentaire. work document for a paper in ASTER, n°22, 1996.
23. J. Piaget. *Psychologie et pédagogie.* Paris : éditions Denoël, 1969.
24. H. Ratsimba-Rajohn. *Contribution à l'étude de hiérarchie implicative. Application à l'analyse de la gestion didactique des phénomènes d'ostension et de contradiction.* PhD thesis, Université Rennes I, 1992.
25. E.H. Rosch. Cognitive representations of semantic categories. *Journal of Experimental Psychology*, 104:192–233, 1975.
26. A.D. Schliemann and N.M. Acioly. Mathematical knowledge developed at work: the contribution of practice versus the contribution of schooling. *Cognition and Instruction*, 6(3):185–221, 1989.
27. S. Scribner. *Thinking : reading in cognitive science*, chapter Modes of thinking and ways of speaking : culture and logic reconsidered. London : Cambridge University Press, 1977.

28. D. Toussaint. La lune est-elle menteuse ? bulletin du comité de liaison enseignants et astronomes. *Les cahiers Clairaut*, 87, 1999.
29. G. Vergnaud. Problem solving and concept development in the learning of mathematics. In *E.A.R.L.I. Second Meeting*, Tübingen, 1987.
30. G. Vergnaud. *Problems of Representation in the teaching and Learning of Mathematics*, chapter Conclusion chapter. London : Lawrence Erlbaum associates Publishers, 1987.
31. G. Vergnaud. Psychologie et didactique : quels enseignements théoriques et méthodologiques pour la recherche en psychologie. In *Colloque La Psychologie Scientifique et ses applications*, Clermont-Ferrand, 1987.
32. G. Vergnaud. Questions vives de la psychologie du développement cognitif. In *Colloque d'Aix-en-Provence*, 1987.
33. G. Vergnaud. La théorie des champs conceptuels. *Recherches en Didactique des Mathématiques*, 10(23):133–170, 1990.
34. G. Vergnaud. *Morphismes fondamentaux dans les processus de conceptualisation — Les Sciences Cognitives en débat.* Editions du CNRS, Paris, 1991.
35. L. Vygotski. *Pensée et Langage.* Paris : Messidor/Editions Sociales, 1985.

Acknowledgments

With thanks to Tim Evans, trainer, for his linguistic skills in his mother-tongue, and to Pascale Montpied, researcher at the CNRS, whose skills in English made several useful contributions. Without their help, the English version of this article would not have been possible.

Appendix 1

NB: The codes of the binary variables are to be found in the questionnaire [xxx]. The following table is reduced because of space. For Q1, Q2, Q3 and the first part of Q6, a non-response is coded by the binary variable [xxnr]

Questionnaire

SEX: ( ) M [MALE] ( ) F [FEMALE]
AGE (years) Profession:
If a teacher, subject taught, and at what level? ________
Previous training: ( ) teacher training college ( ) other. Which? ______
Number of years experience: ________
Place of residence: __________________

Q1.

Have you ever seen the moon like this in reality?

Representations of the moon

Answers							
A-Yes [1Asim]	A-No [1Anao]	B-Yes [1Bsim]	B-No [1Bnao]	C-Yes [1Csim]	C-No [1Cnao]	C-Yes [1Dsim]	C-No [1Dnao]

Fig. 13.

Q2.

If someone (from the other hemisphere) tells you that he/she has never seen any of these moons in reality would you believe that person? ( ) YES[2sim] ( ) NO [2nao]

If YES: who? ()A[2Asim]/[2Anao] ()B[2Bsim]/[2Bnao] ()C[2Csim]/[2Cnao] D[2Dsim]/[2Dnao]

Why? If NO, Why not?

Q3.

If someone (*from another country but from the same hemisphere*) tells you that he/she has never seen any of these moons in reality, would you believe that person? ( ) YES[3sim] ( ) NO [3nao]
If YES: who? ( ) A [2Asim]/[2Anao] ( ) B[2Bsim]/[2Bnao] ( ) C[2Csim]/[2Cnao] ( ) D[2Dsim]/[2Dnao]
Why? If NO, Why not?

Q4.

A primary schoolteacher explained his "technique" for teaching the phases of the moon to his pupils.

He explained that: "the moon tells lies, when it looks like a capital C it is waning (deCreasing); When it looks like a capital D it is in fact waxing (not increasing)".

What do you think of this technique? Is it:( ) adapted [4ADAP] ( ) inadapted [4INAD] ( ) efficient [4EFFI] ( ) inefficient [4INEF] Why?

Q5.

A secondary schoolteacher in France explained his "technique" for teaching the phases of the moon to his pupils. He said: "Its easy! To identify the first (premier) and last (dernier)quarter of the moon you draw a line Across its diameter. If you get a small letter p (premier) its the first quarter; if you get the small letter d its the last (dernier) quarter".

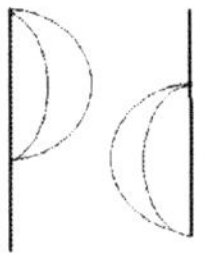

Fig. 14.

What do you think of this technique? Is it:
( ) adapted [5ADAP] ( ) inadapted [5INAD]
( ) efficient [5EFFI] ( ) inefficient [5INEF] **Why?**

Q6.

A teacher explained that the way the phases of the moon phases are perceived depends on the hemisphere. For example, it is not seem in the same way from the southern or from the northern hemisphere. Do you agree with with this point of view? ( ) YES [6sim] ( ) NO [6nao] Why? If yes: How is the moon seen from these three places? From the northern hemisphere, the southern hemisphere or the equator. [Use HS (South Hemisphere); HN (North Hemisphere) or E (Equator) to describe the corresponding representations of the moon.

Representations of the Moon				
Hemisphere	[6A_HN] [6A_E] [6A_HS]	[6B_HN] [6B_E] [6B_HS]	[6C_HN] [6C_E] [6C_HS]	[6D_HN] [6D_E] [6D_HS]

Fig. 15.

Appendix 2

	Sample 198		EchNC_Ad 119		Ech_IUFM 28		Ech_UFRPE 29		EchNC_Enf 22	
Var. bin.	Freq.	%	Freq.	%	Freq.	%	Freq.	%	Freq.	%
[1Asim]	171	86.36	100	84.03	25	89.29	26	89.66	20	90.2
[1Anao]	21	10.61	16	13.45	3	10.71	0	0.00	2	9.09
[1Anr]	6	3.03	3	2.52	0	0.00	3	10.34	0	0.00
[1Bsim]	58	29.29	43	36.13	5	17.86	4	13.79	6	27.3
[1Bnao]	116	58.59	57	47.90	23	82.14	20	68.97	16	72.7
[1Bnr]	24	12.12	19	15.97	0	0.00	5	17.24	0	0.00
[1Csim]	65	32.82	47	36.13	10	35.71	3	10.34	5	22.7
[1Cnao]	110	55.56	54	47.90	18	64.29	21	72.41	17	77.3
[1Cnr]	23	11.62	18	15.97	0	0.00	5	17.24	0	0.00
[1Dsim]	160	80.81	96	80.67	19	67.86	25	86.21	20	90.2
[1Dnao]	25	12.63	17	14.29	6	21.43	0	0.00	2	9.09
[1Dnr]	13	6.57	6	5.04	3	10.71	4	13.79	0	0.00
[2sim]	163	82.32	98	82.35	25	89.29	26	89.66	14	63.6
[2nao]	34	17.17	20	16.81	3	10.71	3	10.34	8	36.4
[2nr]	1	0.51	1	0.84	0	0.00	0	0.00	0	0.00
[2Asim]	40	20.20	25	21.01	10	35.71	3	10.34	2	9.09
[2Anao]	120	60.61	72	60.50	14	50.00	21	72.41	13	59.1
[2Anr]	38	19.19	22	18.49	4	14.29	5	17.24	7	31.8
[2Bsim]	109	55.05	69	57.98	11	39.29	18	62.07	11	50.0
[2Bnao]	48	24.24	27	22.69	11	39.29	6	20.69	4	18.2
[2Bnr]	41	20.71	23	19.33	6	21.43	5	17.24	7	31.8
[2Csim]	112	56.57	72	60.50	13	46.43	15	51.72	12	54.5
[2Cnao]	40	20.20	24	20.17	4	14.29	9	31.03	3	13.6
[2Cnr]	46	23.23	23	19.33	11	39.29	5	17.24	7	31.8
[2Dsim]	54	27.27	33	27.73	11	39.29	6	20.69	4	18.2
[2Dnao]	107	54.04	65	54.62	13	46.43	18	62.07	11	50.0
[2Dnr]	37	18.69	21	17.65	4	14.29	5	17.24	7	31.8

Appendix 3

	Sample 170		EchNC_Ad 119		Ech_UFRPE 29		EchNC_Enf 22	
Var. bin.	Freq.	%	Freq.	%	Freq.	%	Freq.	%
[3sim]	114	67.06%	74	62.18%	24	82.76%	16	72.73%
[3nao]	51	30.00%	41	34.45%	5	17.24%	5	22.73%
[3nr]	5	2.94%	4	3.36%	0	0.00%	1	4.55%
[3Asim]	31	18.24%	22	18.49%	3	10.34%	6	27.27%
[3Anao]	102	60.00%	76	63.87%	16	55.17%	10	45.45%
[3Anr]	37	21.76%	21	17.65%	10	34.48%	6	27.27%
[3Bsim]	71	41.76%	46	38.66%	16	55.17%	9	40.91%
[3Bnao]	61	35.88%	51	42.86%	3	10.34%	7	31.82%
[3Bnr]	38	22.35%	22	18.49%	10	34.48%	6	27.27%
[3Csim]	65	38.24%	46	38.66%	12	41.38%	7	31.82%
[3Cnao]	67	39.41%	51	42.86%	7	24.14%	9	40.91%
[3Cnr]	38	22.35%	22	18.49%	10	34.48%	6	27.27%
[3Dsim]	30	17.65%	22	18.49%	2	6.90%	6	27.27%
[3Dnao]	102	60.00%	75	63.03%	17	58.62%	10	45.45%
[3Dnr]	38	22.35%	22	18.49%	10	34.48%	6	27.27%
[4ADAP]	42	24.71%	28	23.53%	8	27.59%	6	27.27%
[4INAD]	55	32.35%	49	41.18%	4	13.79%	2	9.09%
[4EFFI]	64	37.65%	34	28.57%	17	58.62%	13	59.09%
[4INEF]	36	21.18%	29	24.37%	2	6.90%	5	22.73%
[5ADAP]	68	40.00%	54	45.38%	7	24.14%	7	31.82%
[5INAD]	40	23.53%	34	28.57%	4	13.79%	2	9.09%
[5EFFI]	80	47.06%	51	42.86%	18	62.07%	11	50.00%
[5INEF]	18	10.59%	11	9.24%	3	10.34%	4	18.18%
[6sim]	111	65.29%	77	64.71%	21	72.41%	13	59.09%
[6nao]	47	27.65%	32	26.89%	8	27.59%	7	31.82%
[6nr]	12	7.06%	10	8.40%	0	0.00%	2	9.09%
[6A_HN]	56	32.94%	50	42.02%	5	17.24%	1	4.55%
[6A_E]	38	22.35%	30	25.21%	6	20.69%	2	9.09%
[6A_HS]	83	48.82%	58	48.74%	17	58.62%	8	36.36%
[6B_HN]	37	21.76%	22	18.49%	10	34.48%	5	22.73%
[6B_E]	35	20.59%	27	22.69%	5	17.24%	3	13.64%
[6B_HS]	25	14.71%	21	17.65%	2	6.90%	2	9.09%
[6C_HN]	27	15.88%	17	14.29%	6	20.69%	4	18.18%
[6C_E]	41	24.12%	30	25.21%	7	24.14%	4	18.18%
[6C_HS]	31	18.24%	23	19.33%	5	17.24%	3	13.64%
[6D_HN]	62	36.47%	52	43.70%	6	20.69%	4	18.18%
[6D_E]	37	21.76%	28	23.53%	5	17.24%	4	18.18%
[6D_HS]	64	37.65%	50	42.02%	12	41.38%	2	9.09%

Appendix 4

Table Appendix 4.1

n = 198	Variables "Questions Q1 & Q2"									
GROUPS	1A	1B	1C	1D	2	2A	2B	2C	2D	
EchNCAd	10.9	19.9	20.6	11.7	8.15	9.57	6.62	8.64	5.76	χ^2
EchIUFM										
EchFRPE	0.06	0.10	0.10	0.06	0.04	0.05	0.03	0.04	0.03	$\varphi 2$
EchNCEnf										
	0.15	0.20	0.21	0.16	0.13	0.14	0.12	0.13	0.11	Tschprow
$\alpha = 0.05$	I	D	D	I	I	I	I	I	I	
	I = INDEPENDENT				D = REJECT INDEPENDENCE					

Table Appendix 4.2

	Question Q1B			Question Q1C		
Groups	*1Bsim*	*1Bnao*	*1Bnr*	*1Csim*	*1Cnao*	*1Cnr*
EchNC_Ad	(A+)	(R-)	(A+)	(A+)	(R-)	(A+)
Ech_IUFM	(R-)	(A+)	(R-)	(A+)	(A+)	(R-)
Ech_UFRPE	(R-)	(A+)	(A+)	(R-)	(A+)	(A+)
EchNC_Enf	(R-)	(A+)	(R-)	(R-)	(A+)	(R-)
	(A+)=Attraction			(R-)=Repulsion		

Table Appendix 4.3

n = 170	Variables "Questions Q1 & Q2"									
GROUPS	1A	1B	1C	1D	2	2A	2B	2C	2D	
EchNC_Ad	9.13	10.84	16.00	9.16	6.63	4.91	2.11	3.94	5.76	χ^2
Ech_UFRPE	0.05	0.06	0.09	0.05	0.04	0.03	0.01	0.02	0.03	$\varphi 2$
EchNC_Enf										
	0.16	0.18	0.22	0.16	0.14	0.12	0.08	0.11	0.13	Tschuprow
$\alpha = 0.05$	I	D	D	I	I	I	I	I	I	

Table Appendix 4.4

n = 170	Variables "Questions Q3 & Q6"						
GROUPS	3	3A	3B	3C	3D	6	
EchNC_Ad	5.46	6.55	11.5	5.35	7.01	3.00	χ^2
EchUFRPE							
EchNC_Enf	0.03	0.04	0.07	0.03	0.04	0.02	$\varphi 2$
	0.13	0.14	0.18	0.13	0.14	0.09	Tschuprow
$\alpha = 0.05$	I	I	D	I	I	I	

Part IV

Extensions to rule interestingness in data mining

Pitfalls for Categorizations of Objective Interestingness Measures for Rule Discovery

Einoshin Suzuki

Graduate School of Information Science and Electrical Engineering, Kyushu University, Japan
suzuki@i.kyushu-u.ac.jp

Summary. In this paper, we point out four pitfalls for categorizations of objective interestingness measures for rule discovery. Rule discovery, which is extensively studied in data mining, suffers from the problem of outputting a huge number of rules. An objective interestingness measure can be used to estimate the potential usefulness of a discovered rule based on the given data set thus hopefully serves as a countermeasure to circumvent this problem. Various measures have been proposed, resulting systematic attempts for categorizing such measures. We believe that such attempts are subject to four kinds of pitfalls: data bias, rule bias, expert bias, and search bias. The main objective of this paper is to issue an alert for the pitfalls which are harmful to one of the most important research topics in data mining. We also list desiderata in categorizing objective interestingness measures.

Key words: data bias, rule bias, expert bias, search bias, objective interestingness measure, rule discovery

1 Introduction

Rule discovery is one of the most extensively studied research topics in data mining as shown by the proliferation of discovery methods for finding useful or interesting (i.e. potentially useful) rules [3,5,8,29,34–38]. Such methods can be classified either objective or subjective [32]. We define that an objective method evaluates the interestingness of a rule based on the given data set while a subjective method relies on additional information typically provided by the user in the form of domain knowledge.

A subjective method often models interestingness more appropriately than an objective method but cannot be applied to domains where little or no additional information is available. A subjective method is also prone to overlooking useful knowledge due to inappropriate use of domain knowledge. An objective method is free from these problems and poses no cost to the user for preparing subjective information. In the objective approach, many

E. Suzuki: *Pitfalls for Categorizations of Objective Interestingness Measures for Rule Discovery*, Studies in Computational Intelligence (SCI) **127**, 383–395 (2008)
www.springerlink.com

researchers have proposed an objective interestingness measure which returns a real value as an estimated degree of interestingness of a discovered rule [5,8,11,12,17,23,25,26,30,34].

In general, a categorization of methods brings deep understanding on the nature of the problem and possibility for further inventions. Because there are so many objective interestingness measures, a number of papers have tried to compare, classify, and rank such measures [1,2,9,18,28,30,39]. Each of them contains interesting proposals and the authors pay attention in interpreting their experimental results properly. However, many of them seem to give false conclusions to the readers because all of the categorizations are prone to at least one of the pitfalls which we warn in this paper.

The quest for interestingness evaluation is undoubtedly important in the data mining research. For instance, some of the organizers of the ICDM conferences argued that more research for the subject is needed and put interestingness evaluation in one of the ten most important challenges in data mining (`http://www.cs.uvm.edu/~icdm/10Problems/10Problems-05.pdf`). In this paper, we mainly point out data bias, rule bias, expert bias, and search bias as pitfalls. Our intention is to foster proper research on the interestingness measures and a sound structuralism in evaluating individual research attempts for developing a rule discovery method. We also list desiderata in categorizing objective interestingness measures: respect the true objective, show experimental results in an objective manner, avoid illusion of omnipotent measures, and be a structuralist in deriving conclusions.

In the rest of this paper, we review objective interestingness measures for rule discovery and attempts for categorizing the measures in sections 2 and 3, respectively. We then point out the four pitfalls in section 4. Section 5 gives the principles to avoid such pitfalls and we try to foster astute readers who would contribute to a sound development of the data mining community. Section 6 concludes this paper.

2 Objective Interestingness Measures for Rule Discovery

2.1 Rule Discovery

The objective of rule discovery is to obtain a set Π of rules given data D and additional information α. Here α typically represents domain-specific criteria such as expected profit or domain knowledge and an element of Π represents a rule π. As we focus on the objective approach, we assume $\alpha = \emptyset$ in this paper.

The case in which D represents a table, alternatively stated "flat" data, is most extensively studied in data mining. On the other hand, the case when data D represent structured data such as time-series data and text data typically necessitates a procedure for handling such a structure (e.g. [10,22]). In order to focus on the interestingness aspect, we limit our attention to the former case. In this case, D consists of n examples $e_1, e_2, \ldots, e_n$. An example e_i is described with m attributes $a_1, a_2, \ldots, a_m$ and an attribute a_j takes one of $|a_j|$ values $v_{j,1}, v_{j,2}, \ldots, v_{j,|a_j|}$. We represent $e_k = (w_{k,1}, w_{k,2}, \ldots, w_{k,m})$, where $w_{k,j} \in \{v_{j,1}, v_{j,2}, \ldots, v_{j,|a_j|}\}$

A rule represents a local probabilistic tendency in D and can be represented as $A \rightarrow B$. Here A and B are called a premise and a conclusion, respectively, and each of them specifies a subspace of the example space. For instance, $(a_1 = v_{1,1}) \vee (a_2 = v_{2,2}) \rightarrow (a_3 = v_{3,1}) \wedge (a_4 = v_{4,4})$ is a rule. A rule can be classified into either logical or probabilistic and this paper is concerned with the latter. A probabilistic rule $A \rightarrow B$ can have a confidence $P(B|A)$ smaller than 1 i.e. $P(B|A) < 1$ while a logical rule necessitates $P(B|A) = 1$. Note that here we use each of A and B to represent a set of examples which reside in the corresponding subspaces. In the rest of this paper, we use this notation.

2.2 Example of the Objective Interestingness Measures for Rule Discovery

In computer science, research on objective interestingness measures for rule discovery goes at least back to 1960's [6, 17]. Various measures including [8] have been proposed. Since the objective of this paper is not to conduct a survey on such measures, we mainly explain the J-measure [34], which exhibits several desirable properties and is based on [6], as a representative. We leave explanation on other objective interestingness measures for rule discovery such as lift, conviction, and intensity of implication to [1, 2, 5, 8, 9, 11, 18, 25, 26, 28, 30, 39][1].

Consider a rule $Y = y \rightarrow X = x$, where the premise $Y = y$ is a single or a conjunction of "attribute = value"s and the conclusion $X = x$ is an "attribute=value"[2]. Though multiple attributes can be contained in the premise, we follow the notation of [34] and describe it $Y = y$, where Y and y correspond to a vector of attributes and a vector of attribute values, respectively. We define $f(X, Y = y)$ as the instantaneous information that the event $Y = y$ provides about X, i.e. the information that we receive about X given that $Y = y$ has occurred. It is shown in [6] that the following j-measure is the only non-negative function that satisfies $E_y[f(X; Y = y)] = I(X; Y)$, where $E_y(g)$ and $I(X; Y)$ represent the expected value of g in terms of y and the mutual information between X and Y, respectively.

[1] Intensity of implication is an excellent measure which can take the size of the data set into account. We, however, explain J-measure here because intensity of implication is explained many times in this book.

[2] These assumptions are not necessary but help understanding of a wide range of readers.

$$j(X;Y=y) = \sum_x P(x|y) \log_2 \left(\frac{P(x|y)}{P(x)} \right) \quad (1)$$

$$= P(x|y) \log_2 \left(\frac{P(x|y)}{P(x)} \right) + P(\overline{x}|y) \log_2 \left(\frac{P(\overline{x}|y)}{P(\overline{x})} \right) \quad (2)$$

(2) is derived due to the nature of rule discovery i.e. it suffices to consider the events of the conclusion $X = x$ and its negation $X = \overline{x}$. In [34], the interestingness of the rule $Y = y \rightarrow X = x$ is evaluated with the average information content of the rule under the name of the J-measure $J(X;Y = y)$.

$$J(X;Y=y) = P(y)j(X;Y=y) \quad (3)$$

Note that the J-measure represents the amount of information compressed by the rule $Y = y \rightarrow X = x$: with the rule, the code length for $P(x,y)$ examples becomes $-\log_2 P(x|y)$ from $-\log_2 P(x)$ and the code length for $P(\overline{x},y)$ examples becomes $-\log_2 P(\overline{x}|y)$ from $-\log_2 P(\overline{x})$. (3) represents the difference of the amount of information between the situations with and without the rule. From the practical viewpoint, a general rule ($P(y)$ high), an accurate rule ($P(x|y)$ high), and a rule predicting a rare event ($P(x)$ low) exhibits a high J-measure value if other parameters remain unchanged. This nature makes sense from the viewpoint of discovery of interesting rules with objective information.

It is worth noting again that a variety of interestingness measures have been proposed besides the J-measure. They include mutual information $I(X;Y)$, support $P(x,y)$, confidence $P(x|y)$, and Jaccard $P(x,y)/[P(x) + P(y) - P(x,y)]$. This situation has motivated systematic attempts to categorizing such measures.

3 Categorizations of Objective Interestingness Measures for Rule Discovery

3.1 Desiderata on Objective Interestingness Measures

In this section, we overview representative categorizations of objective interestingness measures for rule discovery [1,2,9,18,28,30,39]. It should be noted that Hilderman and Hamilton studied various objective interestingness measures extensively [13–16]. We omit them in this paper because they mainly assume that a pattern is represented by a contingency table.

Probably [30] is one of the earliest papers that discuss various objective interestingness measures for rule discovery. After a brief mention on four measures, it proposes the following desirable principles for such a measure RI.

1. $RI = 0$ if $P(x,y) = P(x)P(y)$.
2. RI monotonically increases with $P(x,y)$ when other parameters remain the same.

3. RI monotonically decreases with $P(x)$ or $P(y)$ when other parameters remain the same.

It then proposed $nP(x,y) - nP(x)P(y)$ as the simplest function that satisfies these principles. It is clear that the author was aware of the limitation of an objective interestingness measure and just proposed general principles for such measures.

[39] extends the idea of listing desiderata in [30] and overviews 21 objective interestingness measures for rule discovery. Based on discussions on examples of contingency tables of rules, they propose five additional properties to the above three principles of [30]. Each of the 21 measures is allocated either "Yes" or "No" for each of the three principles and the five additional properties. It turns out none of the 21 measures satisfy these 8 criteria and the authors conclude that there is no measure that is better than others in all application domains. We believe that the eight criteria are related with the degree of interestingness of a rule in general but they represent neither sufficient conditions nor necessary conditions of an accurate interestingness measure. An objective interestingness measure can serve as a filter to remove discovered rules which are nearly hopeless from consideration.

3.2 Cluster Analysis of Objective Interestingness Measures

[39] shows an attempt to analyze the 21 objective rule interestingness measures by ranking 10,000 synthetically generated contingency tables with the measures and investigating the correlation between pairs of measures. The results for nine kinds of ranges of support values show that many measures are highly correlated and can be grouped into a few clusters. We estimate that the authors are aware of the danger of random experiments and employed a large number of contingency tables and different ranges for support. It should be also noted that the authors did not dare to draw drastic conclusions such as "objective interestingness measures can be classified into five groups" from the experimental results. We anticipate, however, that a typical reader can misunderstand their intention and believe the experimental results as evidence of categorizations of interestingness measures.

[18] goes one step further than [39] and performs clustering instead of correlation analysis. It reports that 34 objective interestingness measures can be classified into 10 clusters based on their performance on 120,000 association rules discovered from the Mushroom data set [7]. This attempt is interesting because it provides empirical results on the categorization of interestingness measures and some of the results are on their way to be leveraged to empirical evidence based on persuasive reasons (e.g. a number of interestingness measures including confidence, Laplace, and causal confidence form a big cluster because most of them are issued from the confidence measure). Unlike [39], they seem to use one kind of parameter values thus provide one clustering result. However, the obvious danger is that even a typical reader might take

the clustering result as an empirical evidence though it depends on one data set and one kind of parameter values. We will explain these issues in section 4.

3.3 Incorporating Experts in the Analysis

[28] empirically investigates the performance of 40 objective rule interestingness measures by using a subjective evaluation of a domain expert as the ground truth. The experiments were performed using a data set on hepatitis and the domain expert was asked to classify 30 and 21 discovered rules into especially-interesting, interesting, not-understandable, and not-interesting. The performance of a measure represents how well the measure estimates the classification of the domain expert. The authors report that recall, Jaccard, kappa, CST, χ^2-M, and peculiarity demonstrated the highest performance. We believe that ranking interestingness measures based on their performance on sets of discovered rules can give a false conclusion that the ranking is valid for other rule sets. Subjective evaluation of discovered rules by domain experts can serve as a ground truth for interestingness measures but we should be careful not to overgeneralize the conclusions.

[9] investigates effectiveness of objective rule interestingness measures in estimating the subjective interestingness of a domain expert for each data set in a more systematic manner than [28]. The paper tries to neutralize the effects of data sets and rule discovery methods to overcome several shortcomings of [28]. We will explain these issues in the next section.

The methodologies of [1,2] are more drastic because they depend of classifiers which predict subjective interestingness of a domain expert from results of objective rule interestingness measures. The training data sets from which the classifiers are learned contain neither the content of the rules in question nor domain knowledge. We believe that such attempts are infeasible because the subjective interestingness of a domain expert depends on both the content of the rules and domain knowledge.

4 Four Pitfalls to Avoid

4.1 Rule Bias

The categorization depends on the rules to be evaluated. For instance, a measure called InfoChange-ADT, which requires an exception rule in evaluating a rule, can not evaluate all rules in [9]. As the authors point out, this failure was due to the method for discovering the rules: the rules were discovered by converting decision trees thus had no exception rules. The conclusions derived in [9] is specific to the method with which the rules are discovered.

As shown in section 3.2, [39] is particularly aware of this pitfall and try to avoid it by generating 10,000 contingency tables. It is a pity that the authors give no explanations how they generated these tables as we cannot assess the ratio of the generated tables for all possible tables. Anyway, it is impossible to know the rule set to be evaluated in practice. [9] selects 9 rules out of all rules to be shown to a domain expert for each data set for investigating the subjective interestingness of the domain expert[3]: 3 rules with the lowest rank, 3 rules in the middle rank, and 3 rules with the highest rank for each interestingness measure. This method serves as discriminating 3 typical situations but questions remain if it makes sense to choose only 9 rules out of hundreds or thousands of discovered rules. It seems impossible to avoid rule bias in the categorization but we will show several countermeasures for this problem in section 5.

4.2 Data Bias

In the categorization research, the rules to be evaluated are typically discovered from data sets. In such a case, the categorization is dependent on the data sets. For instance, [28] employs a single medical data set. Though the authors emphasize this fact and try to be careful in deriving general conclusions, any conclusions derived in the paper are specific to the data set. This fact was pointed out in [9], which clearly shows that the performance of interestingness measures varies in the eight data sets that they employed.

The number of possible table data sets with n binary attributes is $\Omega(2^{2^n})$ and the prior probability distribution of the table data sets is unknown. It seems impossible to avoid data bias in the categorization but we will show several countermeasures for this problem in section 5.

4.3 Expert Bias

Some of the categorization research ask domain experts to evaluate the degree of interestingness of discovered rules from their subjective viewpoints and try to obtain the ground truth of interestingness measures. We agree that subjective interestingness of domain experts is more important than objective interestingness in data mining but warn that the results depend on the domain experts. For instance, [28] asks a single medical expert to evaluate discovered rules. As the authors point out, any conclusions derived in the paper might be specific to the domain expert. We have collaborated with medical experts including the expert in related mining problems and have observed many discrepancies of opinions among the domain experts [19–21].

It is well-know that domain experts tend to have different opinions for a non-trivial technical problem, for which rule discovery is expected to be effective. For instance, [33] explains a situation in which domain experts have different opinions in classifying volcanoes on Venus and proposes a method for calculating the minimal probability that domain experts are wrong. It seems

[3] We will see this issue in section 4.3.

that resolving such discrepancies seems impossible: that is why the domain experts are conducting research.

The authors of [9] point out the weaknesses of [28] and employ eight data sets in the experiments. They clearly show that the performance of objective rule interestingness measures varies in eight data sets that they employed. Moreover, [9] tries to neutralize the effects of rule discovery methods by employing five classification algorithms. Due to the numerous settings, however, they could employ one domain expert for each data set and in each case the domain expert evaluated nine rules (i.e. three rules with the lowest rank, three rules in the middle rank, and three rules with the highest rank) for each interestingness measure. It is obvious that [9] tolerates some of the problems of [28] but could not escape from the danger of giving false conclusions to the reader. We will show several countermeasures for this problem in section 5.

4.4 Search Bias

Some of the categorization research employs a search-based procedure to categorize objective rule interestingness measures. Typically the results of a search procedure depends on the values of its parameters and/or its initial conditions. In such a case, the categorization is dependent on the search procedure.

For instance, as shown in section 3.2, [18] employs a clustering method to categorize interestingness measures. It is well-known that a typical clustering method depends on values of its parameters and its initial conditions thus their results should not be used as a ground truth. In [18], the categorization is dependent on the search procedure. As we stated, it seems that they employ one kind of parameter values and one initial condition. The authors provide no reasons for seven clusters that they obtained thus these results can be specific to the problem setting including the parameter values and the initial condition. The objective of [18] is different and the authors avoid deriving general conclusions from the results. We believe, however, some of the readers might take the results of the clustering as a truth and use it to select interestingness measures in other cases. We will show several countermeasures for this problem in section 5.

5 Desiderata for the Categorization Research

5.1 Respect the True Objective

We are aware that most of the papers that we cited as examples of the categorization research have another objective. For instance, the main objective of [18] is to develop a tool for selecting the right interestingness measure with clustering. It is highly recommended to consider the context in which the experiments were performed and not to take a categorization result of objective rule interestingness measures as a truth. In other words, we should respect the context of the paper. However, some of the readers might take a categorization result of interestingness measures as a truth. The countermeasure is to emphasize the true objective the paper and state that a categorization result represents an example under some conditions.

We have a comment on [3] which is often criticized by researchers in the "interestingness" community for its two overly simple measures i.e. support and confidence. The main objective of [3] is to develop fast algorithms for disk-resident data. As typical researchers in the database community, the authors assume that the user will select his/her interesting rules with queries thus support and confidence serve as indexes for a pre-screening. In this sense, [3] shows another example that we should respect the true objective of a paper. We are aware of this fact and respect the categorization papers as they have other objectives.

5.2 Show Experimental Results in an Objective Manner

Any papers in the categorization research should describe how much they avoided the four biases in showing experimental results in an objective manner. For instance, generating rules to be evaluated randomly is a countermeasure for the rule bias. In such a case, the authors should state to what extent they explored possible kinds of rules under what assumptions (e.g. equal prior for all rules) they generated the rules. Another typical example is to test various values for parameters to avoid search bias. For example, a typical clustering method necessitates a specification of several parameters such as the number of clusters and a threshold for terminating search. Authors who employ clustering in a similar manner to [18] should state what kinds of values they explored under what assumptions.

Data mining as well as related research fields including machine learning and artificial intelligence cope with ill-structured problems, which have no clear solution. A typical first step to such kinds of problems is to accumulate empirical evidence by analyzing various experimental results then draw general conclusions. We believe that currently no general empirical evidence on categorization of objective rule interestingness measures exist despite the numerous attempts [1, 2, 9, 18, 28, 30, 39].

The results derived in the attempts are special because they heavily depend on at least one of the data sets, discovered rules, domain experts, rule discovery methods, and analysis methods that they employed. There are attempts to tolerate the effects by employing countermeasures such as random experiments and multiple methods/experts [9,39] but their effectiveness is limited due to the huge number of possible choices. We admit that many of the papers provide reasons for experimental results but we also noticed that such reasons never consider all the four factors on which the results depend. Sadly to say, the conclusions derived in the papers such as "34 objective interestingness measures can be classified into 10 clusters" and "recall, Jaccard, kappa, CST, χ^2-M, and peculiarity demonstrated the highest performance" are not guaranteed to be valid in general thus cannot be admitted as an empirical evidence. We should be particularly cautious against such myths.

5.3 Avoid Illusion of Omnipotent Measures

The objective of machine learning is to realize software which improves its performance by a procedure called learning [27]. We have noticed that a typical researcher in data mining from the machine learning community is interested in human factors especially interestingness. At the same time, we are aware that it is (practically) impossible to realize human intelligence with software as the proliferation of weak AI shows [31]. Any scientific interest should be respected but at the same time the reality should be recognized. We should fight against the illusion that an objective interestingness measure is omnipotent, i.e. it can estimate the subjective interestingness of a discovered rule by domain experts in general.

Some of the papers [9, 18, 28, 39] try to convey this idea but at the same time they give the illusion of omnipotence by showing their results on the categorization. We believe that an objective interestingness measure can just serve as a naive filter for removing unpromising discovered rules thus should not be expected to select interesting rules automatically. Classical papers in data mining [3,30] seem to be aware of the danger and are much more conservative than recent papers.

As we stated, [3] employs two overly simple measures i.e. support and confidence, which serve as a measure for a pre-screening. This paper shows a typical situation that objective rule interestingness measures are used. The myth of omnipotent measures should be abandoned.

5.4 Be a Structuralist in Deriving Conclusions

A non-structuralist observes behaviors while a structuralist infers their true cause i.e. the structure of the behaviors. For instance, Claude Levi-Strauss is said to have discovered that complex taboos on marriage in a tribe can be clearly explained by considering a woman as a valuable exchange. A sound criticism on the experimental results is often beneficial. Unexperienced readers tend to overgeneralize experimental results even to the extent of a truth. In interpreting experimental results, discussions should include inference on the true cause.

For instance, it is easy to conclude that objective rule interestingness measures which put an emphasis on generality are effective from some experimental results in which such measures often coincide with human interests. The true reason, however, might be due to the small size of the employed data set, which resulted in many rules with very low supports. A structuralist approach is useful in avoiding the pitfalls.

6 Conclusions

Some of the criticisms might apply to us, as we showed lists of discovered patterns and sometimes asked domain experts to evaluate patterns dis covered from data sets in our papers such as [36, 38]. However, we carefully avoided generalizing the results and showed them as experimental results, and derived conclusions only after we found structural reasons. It should be also noted that those papers were devoted to proposing discovery methods for interesting rules and not for categorizing objective rule interestingness measures.

We believe that in the categorization research the authors should be more cautious because they evaluate individual efforts for developing discovery methods for interesting rules. As mentioned in section 1, categorization of methods is useful and can enhance the quality of the research community. This paper raises several requirements for a "good" categorization paper based on four biases which we point out as typical pitfalls. We repeat that each of the papers of the categorization research has its own objective as mentioned in section 5.1 and its quality should not be evaluated by the fact that it is subject to some of the biases.

Some of the researchers might request an operational procedure to circumvent the pitfalls. We believe that such an operational procedure does not exist if we employ objective interestingness measures only because interestingness is by its nature subjective. As we have argued, objective interestingness measures are useful because they serve as filtering out unpromising rules but are not omnipotent. One of the objectives of this paper is to issue an alert to beliefs on such an omnipotence. It should be noted that interesting methods which employ objective interestingness measures as a help in subjective analysis of experts exist such as [4, 24] and reports on their performance on real data is demanded.

References

1. H. Abe, S. Tsumoto, M. Ohsaki, and T. Yamaguchi. A Rule Evaluation Support Method with Learning Models Based on Objective Rule Evaluation Indexes. In *Proc. Fifth IEEE International Conference on Data Mining (ICDM)*, pages 549–552. 2005.
2. H. Abe, S. Tsumoto, M. Ohsaki, and T. Yamaguchi. Evaluating a Rule Evaluation Support Method Based on Objective Rule Evaluation Indices. In *Advances in Knowledge Discovery and Data Mining (PAKDD)*, pages 509–519. 2006.
3. R. Agrawal, H. Mannila, R. Srikant, H. Toivonen, and A. I. Verkamo. Fast Discovery of Association Rules. In *Advances in Knowledge Discovery and Data Mining*, pages 307–328. AAAI/MIT Press, Menlo Park, Calif., 1996.
4. J.-P. Barthélemy, A. Legrain, P. Lenca, and B. Vaillant. Aggregation of Valued Relations Applied to Association Rule Interestingness Measures. In *Modeling Decisions for Artificial Intelligence, LNCS 3885 (MDAI)*, pages 203–214. 2006.

5. R. J. Bayardo and R. Agrawal. Mining the Most Interesting Rules. In *Proc. Fifth ACM SIGKDD International Conference on Knowledge Discovery and Data Mining*, pages 145–154. 1999.
6. N. M. Blachman. The Amount of Information That y Gives About X. *IEEE Transactions on Information Theory*, IT-14(1):27–31, 1968.
7. C. Blake, C. J. Merz, and E. Keogh. UCI Repository of Machine Learning Databases. http://www.ics.uci.edu/~mlearn/ MLRepository.html. Univ. of Calif. Irvine, Dept. Information and CS (current May 5, 1999).
8. S. Brin, R. Motwani, and C. Silverstein. Beyond Market Baskets: Generalizing Association Rules to Correlations. In *SIGMOD 1997, Proc. ACM SIGMOD International Conference on Management of Data*, pages 265–276. 1997.
9. D. R. Carvalho, A. A. Freitas, and N. F. F. Ebecken. Evaluating the Correlation Between Objective Rule Interestingness Measures and Real Human Interest. In *Knowledge Discovery in Databases (PKDD), LNCS 3721*, pages 453–461. 2005.
10. R. Feldman and I. Dagan. Knowledge Discovery in Textual Databases (KDT). In *Proc. First International Conference on Knowledge Discovery and Data Mining (KDD)*, pages 112–117. 1995.
11. R. Gras. *Contribution à l'étude expérimentale et à l'analyse de certaines acquisitions cognitives et de certains objectifs didactiques en mathématiques.* Thèse d'Etat, Rennes 1, France, 1979.
12. R. Gras. *L' Implication Statistique.* La Pensée Sauvage, 1996. (in French).
13. R. J. Hilderman and H. J. Hamilton. Heuristic for Ranking the Interestingess of Discovered Knowledge. In *Methodologies for Knowledge Discovery and Data Mining, LNAI 1574 (PAKDD)*, pages 204–209. 1999.
14. R. J. Hilderman and H. J. Hamilton. Heuristic Measures of Interestingness. In *Principles of Data Mining and Knowledge Discovery, LNAI 1704 (PKDD)*, pages 232–241. 1999.
15. R. J. Hilderman and H. J. Hamilton. Applying Objective Interestingness Measures in Data Mining Systems. In *Principles of Data Mining and Knowledge Discovery, LNAI 1910 (PKDD)*, pages 432–439. 2000.
16. R. J. Hilderman and H. J. Hamilton. Evaluation of Interestingness Measures for Ranking Discovered Knowledge. In *Knowledge Discovery and Data Mining, LNCS 2035*, pages 247–259. 2001.
17. P. Hájek and C. M. Havel. The GUHA Method of Automatic Hypotheses Determination. *Computing*, 1:293–308, 1966.
18. X.-H. Huynh, F. Guillet, and H. Briand. Clustering Interestingness Measures with Positive Correlation. In *Proc. Seventh International Conference on Enterprise Information Systems (ICEIS)*, pages 248–253. 2005.
19. M. Jumi, M. Ohshima, N. Zhong, H. Yokoi, K. Takabayashi, and E. Suzuki. Spiral Removal of Exceptional Patients for Mining Chronic Hepatitis Data. *New Generation Computing.* (accepted for publication).
20. M. Jumi, E. Suzuki, M. Ohshima, N. Zhong, H. Yokoi, and K. Takabayashi. Spiral Discovery of a Separate Prediction Model from Chronic Hepatitis Data. In *Proc. Third International Workshop on Active Mining (AM)*, pages 1–10. 2004.
21. M. Jumi, E. Suzuki, M. Ohshima, N. Zhong, H. Yokoi, and K. Takabayashi. Multi-strategy Instance Selection in Mining Chronic Hepatitis Data. In *Foundations of Intelligent Systems, Lecture Notes in Artificial Intelligence 3488 (ISMIS-2005)*, pages 475–484. 2005.

22. E. J. Keogh and M. J. Pazzani. Scaling up Dynamic Time Warping to Massive Dataset. In *Principles of Data Mining and Knowledge Discovery, LNAI 1704 (PKDD)*, pages 1–11. 1999.
23. Willi Klösgen. Explora: A Multipattern and Multistrategy Discovery Approach. In *Advances in Knowledge Discovery and Data Mining*, pages 249–271. AAAI/MIT Press, Menlo Park, Calif., 1996.
24. P. Lenca, P. Meyer, B. Vaillant, and S. Lallich. On Selecting Interestingness Measures for Association Rules: User Oriented Description and Multiple Criteria Decision Aid. *European Journal of Operational Research.* (accepted for publication).
25. I. C. Lerman. *Classification et analyse ordinale des données.* Dunod, Paris, 1981.
26. I. C. Lerman, R. Gras, and H. Rostam. Elaboration et évaluation d'un indice d'implication pour des données binaires. *Mathématiques et Sciences Humaines*, 74:5–35, 1981.
27. T. M. Mitchell. *Machine Learning.* McGraw-Hill, Boston, 1997.
28. M. Ohsaki, S. Kitaguchi, K. Okamoto, H. Yokoi, and T. Yamaguchi. Evaluation of Rule Interestingness Measures with a Clinical Dataset on Hepatitis. In *Knowledge Discovery in Databases (PKDD)*, pages 362–373. 2004.
29. B. Padmanabhan and A. Tuzhilin. A Belief-Driven Method for Discovering Unexpected Patterns. In *Proc. Fourth Int'l Conf. Knowledge Discovery and Data Mining (KDD)*, pages 94–100. 1998.
30. G. Piatetsky-Shapiro. Discovery, Analysis, and Presentation of Strong Rules. In *Knowledge Discovery in Databases*, pages 229–248. AAAI/MIT Press, Menlo Park, Calif., 1991.
31. S. Russell and P. Norvig. *Artificial Intelligence.* Prentice Hall, Upper Saddle River, New Jersey, 1995.
32. A. Silberschatz and A. Tuzhilin. What Makes Patterns Interesting in Knowledge Discovery Systems. *IEEE Trans. Knowledge and Data Eng.*, 8(6):970–974, 1996.
33. P. Smyth. Bounds on the Mean Classification Error Rate of Multiple Experts. *Pattern Recognition Letters*, 17(12):1253–1257, 1996.
34. P. Smyth and R. M. Goodman. An Information Theoretic Approach to Rule Induction from Databases. *IEEE Trans. Knowledge and Data Eng.*, 4(4):301–316, 1992.
35. E. Suzuki. Autonomous Discovery of Reliable Exception Rules. In *Proc. Third Int'l Conf. on Knowledge Discovery and Data Mining (KDD)*, pages 259–262. 1997.
36. E. Suzuki. Undirected Discovery of Interesting Exception Rules. *International Journal of Pattern Recognition and Artificial Intelligence*, 16(8):1065–1086, 2002.
37. E. Suzuki and M. Shimura. Exceptional Knowledge Discovery in Databases Based on Information Theory. In *Proc. Second Int'l Conf. Knowledge Discovery and Data Mining (KDD)*, pages 275–278. 1996.
38. E. Suzuki and S. Tsumoto. Evaluating Hypothesis-Driven Exception-Rule Discovery with Medical Data Sets. In *Knowledge Discovery and Data Mining, LNAI 1805 (PAKDD)*, pages 208–211. 2000.
39. P.-N. Tan, V. Kumar, and J. Srivastava. Selecting the Right Interestingness Measure for Association Patterns. In *Proc. Eight ACM SIGKDD International Conference on Knowledge Discovery and Data Mining*, pages 32–41. 2002.

Inducing and Evaluating Classification Trees with Statistical Implicative Criteria

Gilbert Ritschard[1], Vincent Pisetta[2], and Djamel A. Zighed[2]

[1] Dept of Econometrics, University of Geneva, CH-1211 Geneva 4, Switzerland
gilbert.ritschard@unige.ch

[2] Laboratoire ERIC, University of Lyon 2, C.P.11 F-69676 Bron Cedex, France
{v-pisett, abdelkader.zighed}@univ-lyon2.fr

Summary. Implicative statistics criteria have proven to be valuable interestingness measures for association rules. Here we highlight their interest for classification trees. We start by showing how Gras' implication index may be defined for rules derived from an induced decision tree. This index is especially helpful when the aim is not classification itself, but characterizing the most typical conditions of a given conclusion. We show that the index looks like a standardized residual and propose as alternatives other forms of residuals borrowed from the modeling of contingency tables. We then consider two main usages of these indexes. The first is purely descriptive and concerns the a posteriori individual evaluation of the classification rules. The second usage relies upon the strength of implication for assigning the most appropriate conclusion to each leaf of the induced tree. We demonstrate the practical usefulness of this statistical implicative view on decision trees through a full scale real world application.

Key words: Classification tree, Implication strength, Class assignment, Rule relevance, Profile typicality, Targeting

1 Introduction

Implicative statistics was introduced by the French mathematician Régis Gras [7–9] as a tool for data analysis and has, since the late 90's, been exploited for deriving valuable interestingness measures for association rules of the form "If A is observed, then we are very likely to observe B too" [3,5,10,16]. The basic idea behind implicative statistics is that a statistically observed relationship is of interest only if the number of counter-examples is less than expected by chance, and that the larger the difference, the more implicative it is.

We see two major motivations for this concept of statistical implication. On the one hand, logic implication, does not admit any counter-example.

G. Ritschard et al.: *Inducing and Evaluating Classification Trees with Statistical Implicative Criteria*, Studies in Computational Intelligence (SCI) **127**, 397–419 (2008)
www.springerlink.com

Hence, it is too strong and leaves no place for dealing with the random content of statistical relationships. On the other hand, the classical confidence, which measures the chances of matching the conclusion when the condition is satisfied, is not able to tell us whether or not the conclusion is more probable than it would in case of independence from the condition. For instance, assume that the conclusion B is true for 95% of all the cases. Then, a rule with a confidence of 90% would do worse than simple chance, i.e. than deciding that B is true for all cases without taking care of the condition A. But why looking at counter-examples and not just at positive examples? Indeed, this is formally equivalent (see Section 2.2), and hence is just a matter of taste. Looking for the rarity of counter-examples makes the reasoning closer to what is done with logic rules, i.e. invalidating the rule when there are (too many) negative examples.

Though, as we will show, this concept of strength of implication is applicable in a straightforward manner to classification rules, only a little attention has been paid to this appealing idea in the framework of supervised learning. The aim of this article is to discuss the scope and limits of implicative statistics for supervised classification and especially for classification trees. One difference between classification rules and association rules is that the consequent of the former has to be chosen from an a priori set list of classes (the possible states of the response variable), while the consequent for the latter can concern any event not involved in the premise, since there is no a priori outcome variable. A second difference is that unlike the premises of association rules, those of a set of classification rules define a partition of the data set, meaning that there is one and only one rule applicable to each case. These aspects, however, do not intervene in anyway in the definition of the implication index which just requires a premise and a consequent. Hence, implication indexes are technically applicable without restrictions to classification rules. There remains, nevertheless, the question of whether they make sense in the supervised learning setting.

The implication index measures how typical the condition of the rule is for the conclusion, i.e. how much more characteristic than pure chance it is for the selected conclusion. Indeed, we are only interested in conditions under which the probability to match the conclusion is higher than the marginal proportion corresponding to pure chance. A condition with a probability lower than the marginal proportion would characterize atypical situations for the conclusion, i.e. situations in which the proportion of cases matching the conclusion is less than in the whole data set. It would thus be characteristic of the negation of the conclusion, not the conclusion itself. Looking at typical conditions for the negation of the conclusion could be useful too. Nevertheless, it does not require any special attention since it can simply be handled by looking at the implication strength of the rule in which we would have replaced the conclusion by its negation.

The information on the gain of performance over chance provided by the implication index usefully complements the knowledge provided for instance by the classical raw misclassification rate. However, we may go a step further and, by considering a so called targeting or condition typicality paradigm instead of the classification paradigm, resort to implication indexes for selecting the conclusion of a rule. Moreover, we could even imagine methods for growing trees that would optimize the implication strength of the resulting rules. Such a targeting paradigm will be adopted, for instance, by a physician who is more interested in knowing the typical profile of persons who develop a cancer than in predicting for each patient whether or not he has a cancer. Likewise, a tax-collector may be more interested in characterizing groups in which he has increased chances to find fakers than in predicting for each taxpayer whether or not he commits fraud. The most frequent class, commonly called the 'majority class' in the decision tree literature, is obviously the best choice for minimizing classification errors. However, we will see that for the targeting paradigm, the highest quality conclusion, i.e. that for which the rule has the highest implication strength, is not necessarily this majority class.

The paper is organized as follows. Section 2 shows how Gras' implication index can be applied to classification rules derived from an induced decision tree. It proposes alternatives to Gras' index inspired from residuals used in the modeling of multiway contingency tables. Section 3 discusses the use of implication strength for the individual validation of each classification rule. In Section 4 we adopt the aforementioned typical profile paradigm and consider using the implication indexes for selecting the most relevant conclusion in a leaf of a classification tree. We also briefly describe different approaches for growing trees from that typical profile standpoint. Section 5 reports experimental results that highlight the behavior of the implication strength indexes and illustrates their potential on a real world application from social sciences. Finally, we present concluding remarks in Section 6.

We start our presentation by adopting a classical classification standpoint.

2 Classification Trees and Implication Indexes

For our discussion, we consider a fictional example where we are interested in predicting the civil status (married, single, divorced/widowed) of individuals from their sex (male, female) and sector of activity (primary, secondary, tertiary). The civil status is the outcome (or response or decision or dependent) variable, while sex and activity sector are the predictors (or condition or independent variables). The data set is composed of the 273 cases described by Table 1.

2.1 Trees and Rules

Classification rules can be induced from data using classification trees in two steps. First, the tree is grown by seeking, through recursive splits of the learning data set, some optimal partition of the predictor space for predicting the

Civil status	Sex	Activity sector	Number of cases
married	male	primary	50
married	male	secondary	40
married	male	tertiary	6
married	female	primary	0
married	female	secondary	14
married	female	tertiary	10
single	male	primary	5
single	male	secondary	5
single	male	tertiary	12
single	female	primary	50
single	female	secondary	30
single	female	tertiary	18
divorced/widowed	male	primary	5
divorced/widowed	male	secondary	8
divorced/widowed	male	tertiary	10
divorced/widowed	female	primary	6
divorced/widowed	female	secondary	2
divorced/widowed	female	tertiary	2

Table 1. The illustrative data set

outcome class. Each split is done according to the values of one predictor. The process is greedy. It starts by trying all predictors to find the "best" split of the whole learning data set. Then, the process is repeated at each new node until some stopping criterion becomes true. In a second step, once the tree is grown, classification rules are derived by choosing the most relevant value, usually the majority class (the most frequent), in each leaf (terminal node) of the tree.

Figure 1 shows the tree induced with the CHAID method [11], using a 5% significance level and a minimal node size fixed at 20. The same tree is obtained with CART [4] using a minimal .02 gain value. The three numbers in each node represent the counts of individuals who are respectively 'married',

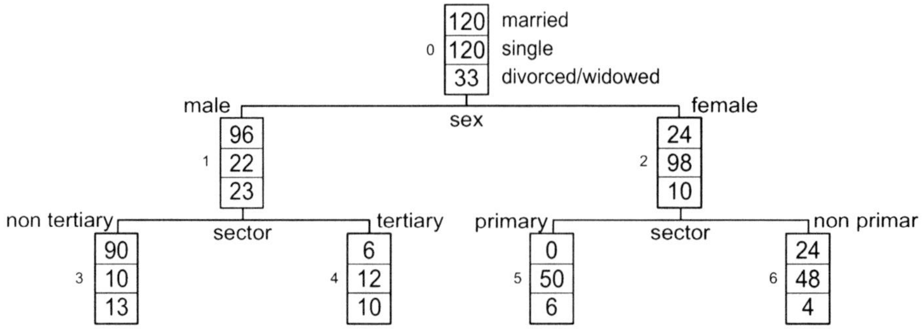

Fig. 1. Example: Induced tree for civil status (married, single, divorced/widowed)

	Man		Woman		
Civil Status	primary or secondary	tertiary	primary	secondary or tertiary	Total
Married	**90**	6	0	24	120
Single	10	**12**	**50**	**48**	120
Div./Widowed	13	10	6	4	33
Total	113	28	56	76	273

Table 2. Table associated to the induced tree

'single', and 'divorced or widowed'. The tree partitions the predictor space into groups such that the distribution of the outcome variable, the civil status, differs as much as possible from one group to the other. For our discussion, it is convenient to represent the four resulting distributions into a table that cross classifies the outcome variable with the set of profiles (the premises of the rules) defined by the branches. Table 2 is thus associated to the tree of Figure 1.

As mentioned, classification rules are usually derived from the tree by assigning the majority class of the leaf to the branch that leads to it. For example, a man working in the secondary sector belongs to leaf 3 and will be classified as married, while a man of the tertiary sector (leaf 4) will be classified as single. In Table 2, the column headings define the premises of the rules, the conclusion being given, for each column, by the row containing the greatest count. Using this approach, the four following rules are derived from the tree shown in Figure 1:

R1: Man of primary or secondary sector $\Rightarrow$ married
R2: Man of tertiary sector $\Rightarrow$ single
R3: Woman of primary sector $\Rightarrow$ single
R2: Woman of secondary or tertiary sector $\Rightarrow$ single

In contrast to association rules, classification rules have the following characteristics: i) The conclusions of the rules can only be values (classes) of the outcome variable, and ii) the premises of the rules are mutually exclusive and define a partition of the predictor space. Nonetheless, they are rules and we can then apply to them concepts such as support, confidence and, which is here our concern, implication indexes.

2.2 Counter-examples and Implication Index

The index of implication (see for instance [6] p 19) of a rule is defined from the number of *counter-examples*, i.e. of cases that match the premise but not the conclusion. In our case, for each leaf (represented by a column in Table 2), the count of counter-examples is the number of cases that are not in the majority class. Letting b denote the conclusion (row of the table) of rule j and n_{bj} the maximum in the jth column, the number of counter-examples is $n_{\bar{b}j} = n_{\cdot j} - n_{bj}$. The index of implication is a standardized form of the deviation between this number and the number of counter-examples expected when assuming that the distribution of the outcome values is independent of the premise.

Formally, the independence hypothesis H_0 states that the number $N_{\bar{b}j}$ of counter-examples of rule j results from a random draw of $n_{\cdot j}$ cases. Under H_0, letting $n_{b\cdot}/n$ be the marginal proportion of cases in the conclusion class b of rule j and setting $n_{\bar{b}\cdot} = n - n_{b\cdot}$, $N_{\bar{b}j}$ follows a binomial distribution $\mathrm{Bin}(n_{\cdot j}, n_{\bar{b}\cdot}/n)$, or, when $n_{\cdot j}$ is not fixed a priori, a Poisson distribution with parameter $n^e_{\bar{b}j} = n_{\bar{b}\cdot} n_{\cdot j}/n$ [12]. In the latter case, the parameter $n^e_{\bar{b}j}$ is both the mathematical expectation $\mathrm{E}(N_{\bar{b}j} \mid H_0)$ and the variance $\mathrm{var}(N_{\bar{b}j} \mid H_0)$ of the number of counter-examples under H_0. It is the number of cases in leaf j that would be counter-examples if they were distributed among the outcome classes according to the marginal distribution, i.e. that of the root node (right margin in Table 2).

Gras' *implication index* is the difference $n_{\bar{b}j} - n^e_{\bar{b}j}$ between the observed and expected numbers of counter-examples, standardized by the standard deviation, i.e., if we retain the Poisson model,

$$\mathrm{Imp}(j) = \frac{n_{\bar{b}j} - n^e_{\bar{b}j}}{\sqrt{n^e_{\bar{b}j}}} , \tag{1}$$

which can also be expressed in terms of the number of cases matching the rules as $\mathrm{Imp}(j) = -(n_{bj} - n^e_{bj})/\sqrt{n_{\cdot j} - n^e_{bj}}$.

Let us make the calculation of the index explicit for our example. We define for that the variable "predicted class", denoted *cpred*, which takes value 1 for each case (example) belonging to the majority class of its leaf and 0 otherwise (counter-example). By cross-classifying this variable with the premises of the rules, we get Table 3 where the first row gives the number $n_{\bar{b}j}$ of counter-examples for each rule and the second row the number n_{bj} of examples.

Likewise, Table 4 gives the expected numbers $n^e_{\bar{b}j}$ and n^e_{bj} of negative examples (counter-examples) and positive examples obtained by distributing the $n_{j\cdot}$ covered cases according to the marginal distribution. Note that these counts cannot be computed from the margins of Table 3. They are obtained by first dispatching the column total using the marginal distribution of Table 2 and then separately aggregating each resulting column according to its corresponding observed majority class (not the expected one!). This explains why Tables 3 and 4 do not have the same right margin.

From these two tables, we can easily get the implication indexes using formula (1). They are reported in the first row of Table 5. For the first rule, the index equals $\mathrm{Imp}(1) = -5.068$. This negative value indicates that the number of observed counter-examples is less than the number expected under the independence hypothesis, which stresses the relevance of the rule. For the second rule, the implication index is positive, which tells us that the rule is

Predicted class *cpred*	Man: primary or secondary	Man: tertiary	Woman: primary	Woman: secondary or tertiary	Total
0 (counter-example)	23	16	6	28	73
1 (example)	**90**	**12**	**50**	**48**	200
Total	113	28	56	76	273

Table 3. Observed numbers $n_{\bar{b}j}$ and n_{bj} of counter-examples and examples

Predicted class *cpred*	Man: primary or secondary	Man: tertiary	Woman: primary	Woman: secondary or tertiary	Total
0 (counter-example)	63.33	15.69	31.38	42.59	153
1 (example)	49.67	12.31	24.62	33.41	120
Total	113	28	56	76	273

Table 4. Expected numbers $n^e_{\bar{b}j}$ and n^e_{bj} of counter-examples and examples

less powerful than pure chance since it generates more counter-examples than would classifying without taking account of the condition.

2.3 Implication Index and Residuals

In its formulation (1), the implication index looks like a standardized residual, namely as the (signed square root of) the contribution to the Pearson Chi-square (see for example [1] p 224). The implication index is indeed related to the Chi-square that measures the divergence between Tables 3 and 4. The contributions of each cell to this Chi-square are depicted in Table 5, those of the first row being the implication indexes.

This interpretation of Gras' implication index in terms of residuals (residuals for the fitting of the counts of counter-examples by the independence model) suggests that other forms of residuals used in the framework of the modeling of the counts in multiway contingency tables could also prove useful for measuring the strength of rules. These include:

Predicted class *cpred*	Man: primary or secondary	Man: tertiary	Woman: primary	Woman: secondary or tertiary
0 (counter-example)	**-5.068**	**0.078**	**-4.531**	**-2.236**
1 (example)	5.722	-0.088	5.116	2.525

Table 5. Contributions to the Chi-square measuring divergence between Tables 3 and 4

The *deviance residual*, $res_d(j) = \text{sign}(n_{\bar{b}j} - n^e_{\bar{b}j})\sqrt{|2n_{\bar{b}j}\log(n_{\bar{b}j}/n^e_{\bar{b}j})|}$, which is the square root of the contribution (in absolute value) to the likelihood ratio Chi-square ([2] pp 136–137).

Freeman-Tukey's residual, $res_{FT}(j) = \sqrt{n_{\bar{b}j}} + \sqrt{1 + n_{\bar{b}j}} - \sqrt{4n^e_{\bar{b}j} + 1}$, which results from a variance-stabilizing transformation ([2] p 137).

Haberman's adjusted residual, $res_a(j) = (n_{\bar{b}j} - n^e_{\bar{b}j})/\sqrt{n^e_{\bar{b}j}(n_{b\cdot}/n)(1 - n_{\cdot j}/n)}$, which is the Pearson standardized residual divided by its standard error ([1] p 224).

There are thus different ways of measuring the departure from the expected number of counter-examples. It is always instructive to cross-compare values produced by such alternatives. When they are concordant, as they should be, comparison reinforces the reliability of the outcome. Divergences, on the other hand, flag situations for which we should be more cautious before drawing any conclusion from the numerical value of a given index. Section 5.1 provides some highlights on the specific behavior of each of the four alternatives considered here.

Residual		Rule R1	Rule R2	Rule R3	Rule R4
Standardized (Gras' index)	res_s	-5.068	0.078	-4.531	-2.236
Deviance	res_d	-6.826	0.788	-4.456	-4.847
Freeman-Tukey	res_{FT}	-6.253	0.138	-6.154	-2.414
Adjusted	res_a	-9.985	0.124	-7.666	-3.970

Table 6. The various residuals as alternative implication indexes

Table 6 exhibits the values of these alternative implication indexes for each of the four rules derived from the tree in Figure 1. We observe that they are concordant as expected. The standardized residual is known to have a variance that may be lower than one. This is because the counts $n_{b\cdot}$ and $n_{\cdot j}$ are sample dependent and hence themselves random. Thus $n^e_{\bar{b}j}$ is only an estimation of the Poisson parameter. Ignoring the randomness of the denominator in formula (1) leads to underestimating the strength. The deviance, adjusted and Freeman-Tukey's residuals are better suited for this situation and are known to have in practice a distribution closer to the standard normal $N(0, 1)$ than the simple standardized residual. We can see in our example that the standardized residual, i.e. Gras' implication index, tends to give lower absolute values than the three alternatives. The only exception is rule R3, for which the deviance residual provides a slightly smaller value than Gras' index. Note that R3 admits only six counter-examples.

2.4 Implication Intensity and p-value

In order to evaluate the statistical significance of the computed implication strength, it is natural to look at the p-value, i.e. at the probability $p(N_{\bar{b}j} \leq n_{\bar{b}j} \mid H_0)$. When $n^e_{\bar{b}j}$ is small, this probability can be obtained, conditionally on $n_{b\cdot}$ and $n_{\cdot j}$, with the Poisson distribution $P(n^e_{\bar{b}j})$. For large $n^e_{\bar{b}j}$, the normal distribution gives a good approximation. A correction for the continuity may be necessary, however, because the difference might be for example as large as 2.6 percent when $n^e_{\bar{b}j} = 100$. Letting $\phi(.)$ denote the standard normal distribution, we have $p(N_{\bar{b}j} \leq n_{\bar{b}j} \mid H_0) \simeq \phi\Big((n_{\bar{b}j} + 0.5 - n^e_{\bar{b}j})/\sqrt{n^e_{\bar{b}j}}\Big)$.

Residual		Rule R1	Rule R2	Rule R3	Rule R4
Standardized (Gras)	res_s	1.000	0.419	1.000	0.985
Deviance	res_d	1.000	0.099	1.000	1.000
Freeman-Tukey	res_{FT}	1.000	0.350	1.000	0.988
Adjusted	res_a	1.000	0.373	1.000	1.000

Table 7. The implication intensity and its variants (with continuity correction)

The *implication intensity* can be defined as the complement of such a p-value. Gras (see for instance [6]) defines it in terms of the normal approximation, but without the correction for continuity. We compute it as

$$\text{Intens}(j) = 1 - \phi\Big((n_{\bar{b}j} + 0.5 - n^e_{\bar{b}j})/\sqrt{n^e_{\bar{b}j}}\Big) \ . \tag{2}$$

In either case, this intensity can be interpreted as the probability of getting, under the independence hypothesis H_0, a higher number of counter-examples than the count observed for rule j. Table 7 gives these intensities for our four rules. It shows also the complement of the p-values of the deviance, adjusted and Freeman-Tukey's residuals computed with the continuity correction, i.e. by adding 0.5 to the observed counts of counter-examples. Notice that provided probabilities below 50% correspond to positive values of the indexes, i.e. bad ones, and those above 50% to negative ones. This is a direct consequence of taking the probabilities from the normal distribution, which is symmetric.

3 Individual Rule Relevance

The implication intensity and its variants are useful for validating each classification rule individually. This knowledge enriches the usual global validation of the classifier. For example, among the four rules issued from our illustrative tree, rules R1, R3 and R4 are clearly relevant, while R2, with an implication intensity below 50% should be rejected.

The question is then what shall we do with the cases covered by the conditions of irrelevant rules. Two solutions can be envisaged: i) Merging cases covered by an irrelevant rule with another rule, or ii) changing the conclusion. The possible choice of a more suitable conclusion is discussed in Section 4.1. We exclude indeed further splitting of the node, since we assume that a stopping criterion has been matched. As for the merging of rules, if we want to respect the tree structure we have indeed to merge cases of a leaf with those of a sibling leaf, which is equivalent to pruning the corresponding branch. In our example, this leads to merging rules R1 and R2 into a new rule "Man ⇒ married". Residuals for the number of counter-examples of this new rules are respectively $res_s = -3.8$, $res_d = -7.1$, $res_{FT} = -4.3$ and $res_a = -8.3$. Except for the deviance residual, they exhibit a slight deterioration as compared to the implicative strength of rule R1.

It is interesting here to compare the implicative quality with the error rate used for validating classification rules. The number of counter-examples considered is precisely the number of errors produced by the rule on the learning set. The error rate is thus the percentage of counter-examples among the cases covered by the rule, i.e. $\text{err}(j) = n_{\bar{b}j}/n_{\cdot j}$, which is also equal to $1 - n_{bj}/n_{\cdot j}$, the complement to one of the confidence. The error rate suffers that from the same drawbacks as the confidence. For instance, it does not tell us how better the rule does than a classification done independently of any condition. Furthermore, the error rate is linked with the choice of the majority class as conclusion. For our example, the error rate is respectively for our four rules 0.2, 0.57, 0.11 and 0.36. The second rule is thus also the worst from this point of view. Comparing with the error rate at the root node, which is 0.56, shows that this rate of 0.57 is very bad. Thus, for being really informative about the relevance of the rule, the error rate should be compared with the error rate of some naive baseline rule. This is exactly what the implication index does. Resorting to implication indexes, we get in addition probabilities which permits to distinguish between statistically significant and non significant relevance.

Practically, in order to detect over-fitting, error rates are computed on validation data sets or through cross validation. Indeed, the same can be done for the implication quality by computing the implication indexes and intensities in generalization.

Alternatively, we could consider, in the spirit of the BIC (Bayesian information criteria) or MDL (Minimum message length) principle, to penalize the implication index by the complexity of the condition. Since the lower the implication index of a rule j, the better it is, the index should be penalized by the length k_j of the branch that defines the condition of rule j. The general idea behind such penalization is that the simpler the condition, the lower the risk to assign a bad distribution to a case. As a first proposal we suggest the following penalized form inspired from the BIC [14] and based on the deviance residual

$$\text{Imp}_{pen}(j) = res_d(j) + \sqrt{k_j \ln(n_j)} \ .$$

For our example, the values of the penalized index are given in Table 3.

These penalized values confirm the ranking of the initial rules, which here all have the same length $k_j = 2$. In addition, the penalized index is useful for validating results of merging the two rules R1 and R2. Table 3 highlights the superiority of the merged rule "Man ⇒ married" over both rules R1 and R2. It gives a clear signal in favor of merging.

At the root node, both the residual and the number of conditions are zero. Hence, the penalized implication index is zero too. Thus, a positive penalized implication index suggests that we can hardly expect that the rule would do better in generalization than assigning randomly the cases according to the root node distribution, i.e. independently of any condition. For our example, this confirms once again the badness of rule R2.

Rule	res_d	k	Imp_{pen}
R1	-6.826	2	-3.75
R2	0.788	2	3.37
R3	-4.456	2	-1.62
R4	-4.847	2	-1.90
Man ⇒ married	-7.119	1	-4.89
Woman ⇒ single	-7.271	1	-5.06

Table 8. Implication index penalized for the rule complexity

4 Adopting a Typical Profile Paradigm

To this point, we have assumed that the conclusion of the rule was simply the majority class. This is justified when the pursued aim is classification. However, as already mentioned in the introduction, there are situations where the typical profile paradigm is better suited. Remember the example of the physician primarily interested in the characteristics of those patients who develop a cancer, and that of the tax-collector who wants to know the groups of tax payers who are at most risk of committing fraud. Social sciences, where the concern is most often to understand phenomena rather than to predict values or classes, is also a distinctive domain to which the typical profile paradigm suits well. For example, sociologists of the family may be interested in determining the profiles in terms of education, professional career, parenthood, etc. that increase chance of divorce, and in Section 5.2, we present an application where the goal is to characterize the profiles of those students who are at most risk of repeating their first year. In such situations, the majority class rule is no longer the best choice. Indeed, from this typical profile standpoint, it is more natural to search for rules with the highest possible implication strength than to minimize the misclassification rate.

Having this optimal implication strength goal in mind, we successively discuss the assignment of the most relevant conclusion to the premises defined by a given grown tree, and the use of implication strength criteria in the tree growing process.

4.1 Maximal Implication Strength versus Majority Rule

For a given grown tree, maximizing the implication strength is simply achieved by assigning to each leaf the conclusion for which the rule gets its highest implication intensity. Though ([17] pp 282–287) have already considered this way of proceeding, they do not provide a sound justification for the approach. Note also that the method has not, to the best of our knowledge, been implemented so far in any tree growing software.

			Indexes			Intensity	
Residual		married	single	div./wid.	married	single	div./wid.
Standardized	res_s	1.6	0.1	**-1.3**	0.043	0.419	**0.891**
Deviance	res_d	3.9	0.8	**-3.4**	0.000	0.099	**0.999**
Freeman-Tukey	res_{FT}	1.5	0.1	**-1.4**	0.054	0.398	**0.895**
Adjusted	res_a	2.4	0.1	**-2.0**	0.005	0.379	**0.968**

Table 9. Implication indexes and intensities of rule R2 for each possible conclusion

To illustrate the principle, we give in Table 9 the values of the alternative indexes and intensities of implication for each of the three possible conclusions that may be assigned to rule R2 of our example. The conclusion labeled "single" corresponds to the majority class. However, considering the strength of implication, the best conclusion is "divorced or widowed". All four indexes designate this conclusion as the best with an implication intensity that goes from 89.1% for Gras' index to 99.9% for the deviance residual. Indeed, to be a man working in the tertiary sector is not typical of single people since the rule would in that case generate more counter-examples than expected by chance. Concluding to "divorced or widowed" is better in that respect since the number of positive examples is in that case larger than expected by chance. Again we can notice that Gras' index seems to slightly under-estimate the implication intensity.

An important point is that unlike the majority rule, seeking the maximal implication strength favors the variability of conclusions among rules, meaning that we have more chances to create at least one rule for each value of the outcome variable. In our example, using the majority class we do not create any rule that concludes with divorced/widowed, while with the implication strength at least one rule concludes with each of the three outcome states. Indeed, we need at least as many different profiles as outcome classes if we want at least one rule concluding with each outcome state, i.e. we should have $r \leq q$ with r the number of outcome classes and q the number of rules.

By definition, if we assign the same conclusion to all rules, any negative departure from the expected number of counter-examples of a rule should be compensated for a positive departure for an other rule. Likewise, for a given rule, any negative departure from the expected number of counter examples for one of the possible conclusions should be compensated for a positive one for an other conclusion. Formally we have

$$n_{\bar{\imath}j} < n^e_{\bar{\imath}j} \;\Rightarrow \begin{cases} \text{there exists } k \neq i \text{ such that } n_{\bar{k}j} > n^e_{\bar{k}j} \text{ and} \\ \text{there exists } h \neq j \text{ such that } n_{\bar{\imath}h} > n^e_{\bar{\imath}h} \end{cases}$$

$$n_{\bar{\imath}j} > n^e_{\bar{\imath}j} \;\Rightarrow \begin{cases} \text{there exists } k \neq i \text{ such that } n_{\bar{k}j} < n^e_{\bar{k}j} \text{ and} \\ \text{there exists } h \neq j \text{ such that } n_{\bar{\imath}h} < n^e_{\bar{\imath}h} \end{cases}$$

As a consequence, all the rules cannot attain their maximal implication strength for the same conclusion, which favors indeed the diversity of the conclusions among rules. A second consequence is that at each leaf we may assign a conclusion such that the rule gets a non positive implicative index or, equivalently, an implication intensity greater or equal to 50%.

4.2 Growing Trees with Implication Strength Criteria

Let us now look at the tree growing procedure and assume that the rule conclusions are selected so as to maximize the implication strength of the rules. The question is whether there is a way to split a node so as to maximize the strength of the resulting rules. The difficulty here is that a split results indeed in more than one rule. Hence, we face a multicriteria problem, namely the maximization over sets of implication strengths.

To get simple solutions, one can think to transform the multidimensional optimization problem into a one dimensional one by focusing on some aggregated criterion. The following are three possibilities:

- A weighted average of the concerned optimal implication indexes, taking weights proportional to the number of concerned cases.
- The maximum over the strengths of the rules belonging to the set.
- The minimum over the strengths of the rules belonging to the set.

The first criterion is of interest when the goal is to achieve good strengths on average. The second one should be adopted when we look for a few rules with high implication strengths without bothering too much for the other ones, and the latter is of interest when we want the highest possible implication strength for the poorest rule.

We have not yet experimented tree growing with these criteria. It is worthwhile however to say that, from the typical profile paradigm standpoint methods such as CHAID that attempt to maximize association seem preferable to those based on entropies. Indeed, maximizing the strength of association between the resulting nodes and the outcome variable leads to distributions that departure as much as possible from that in the parent node, and hence from that of the root node corresponding to independence. We may thus expect the most significant departures from independence and hence rules with strong implication strength. Methods based on entropy measures, on the other hand, favor departures from the uniform, or equiprobable, distribution and are therefore more in line with the classification standpoint.

5 Experimental Results

We present here a series of experimental results that provide additional insights into the behavior and scope of the original implication index and the three variants we introduced. First, we study the behavior of the indexes. We then present an application, which also serves as a basis for experimental investigations regarding the effect of the continuity correction and the consequences of using maximal implication strength rules instead of the majority rule on classification accuracy, recall and precision.

5.1 Compared Behavior of the 4 Indexes

In order to gain better understanding on how the different implication indexes behave, we ran a simulation to see how they evolve when the number of counter-examples is progressively decreased from the expected number under independence to 0. At independence we expect a null implication strength, while when no counter-examples are observed we should have high implication strength.

The simulation design is as follows. We consider a dataset of size 1000 and a rule defined from a leaf containing 200 cases (20%). We suppose that a proportion p of the 1000 cases belongs to the outcome class selected as conclusion for the rule. Starting with a proportion $f = f_0$ of cases of the leaf that fall in the conclusion class, we progressively increase f in 100 constant steps until the maximum $f = 100\%$ is reached. The initial starting point corresponds to independence and the final point to a pure distribution with no counter-examples. At each step we compute, applying the continuity correction, the value of each of the 4 indexes, namely the standardized, Freeman-Tukey, adjusted and deviance residuals.

Figure 2 shows the results for $p = 10\%$, 50% and 90%. Notice the difference of scale between the three plots: The implication strengths are higher when the class of interest is infrequent in the population, i.e. when p is small. We observe that the standardized and adjusted residuals evolve linearly between independence and purity, while the increase in Freeman-Tukey's residual tends to accelerate when we approach purity. The deviance residual evolves curiously in a parabolic way. It dominates the other indexes in the neighborhood of independence, it reaches a maximum (in absolute terms) and diminishes (in absolute terms) when we approach purity. This decreasing behavior when the

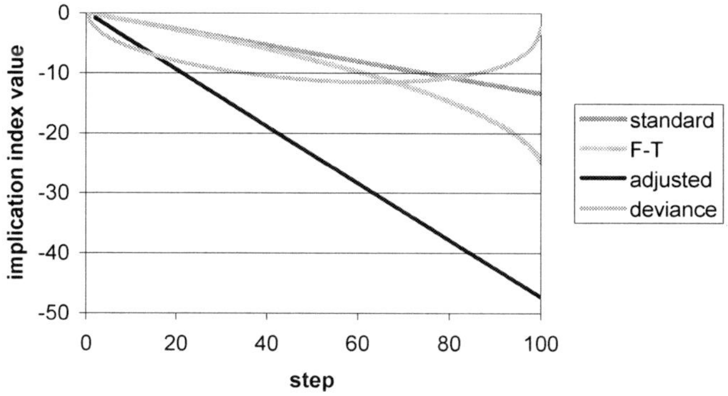

(a) 10% of cases in selected outcome class

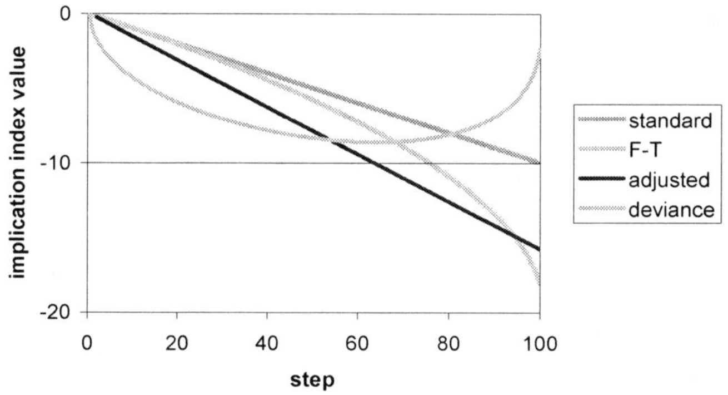

(b) 50% of cases in selected outcome class

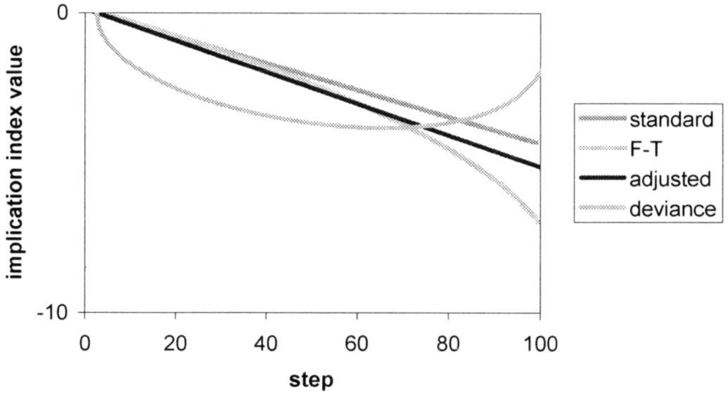

(c) 90% of cases in selected outcome class

Fig. 2. Behavior of the 4 indexes between independence (Step 0) and purity (Step 100). Values reported include the continuity correction.

number of counter-examples tends to 0 disqualifies the deviance residual as a good measure of the rule implication strength. The linear evolution of the standadized and adjusted residuals makes them our prefered measures, the latter having in addition the advantage of being the most reliably comparable with standard normal thresholds.

5.2 Application on a Student Administrative Dataset

We consider administrative data about the 762 first year students who were enrolled in fall 1998 at the Faculty of Economic and Social Sciences (ESS) of the University of Geneva [13]. The goal is to learn rules for predicting the situation (1. eliminated, 2. repeating first year, 3. passed) of each student after the first year, or more precisely to discover the typical profile of those students who are either eliminated or have to repeat their first year. For the learning data, the response variable is thus the student situation in October 1999. The predictors retained are age, first time registered at University of Geneva, chosen orientation (Social Sciences or Business and Economics), type of secondary diploma achieved (classic, latin, scientific, economics, modern, other), place where secondary diploma was obtained (Geneva, Switzerland outside Geneva, Abroad), age when secondary diploma was obtained, nationality (Geneva, Swiss except Geneva, Europe, Non Europe) and mother's living place (Geneva, Switzerland outside Geneva, Abroad).

Figure 3 shows the tree induced using CHAID with minimal node size set to 30, minimal parent node size to 50 and a maximal 5% significance for the Chi-square. Table 10 provides the details regarding the counts in the leafs. Here, our interest is not in the growing procedure, but rather in the state assigned to each leaf.

Leaf	6	7	8	9	10	11	12	13	14	Total
1 eliminated	2	17	22	56	31	16	20	18	27	209
2 repeating	1	13	15	48	10	8	16	14	5	130
3 passed	35	87	55	143	28	9	48	12	6	423
Total	38	117	92	247	69	33	84	44	38	762

Table 10. Details about the content of the leafs in Figure 3

Leaf	6	7	8	9	10	11	12	13	14
Majority class	3	3	3	3	1	1	3	1	1
Standardized residual	3	3	3	3	1	1	3	**2**	1
Freeman-Tukey residual	3	3	3	3	1	1	**2**	**2**	1
Deviance residual	3	3	3	**2**	1	1	**2**	**2**	1
Adjusted residual	3	3	3	**2**	1	1	**2**	**2**	1

Table 11. State assigned by the various criteria

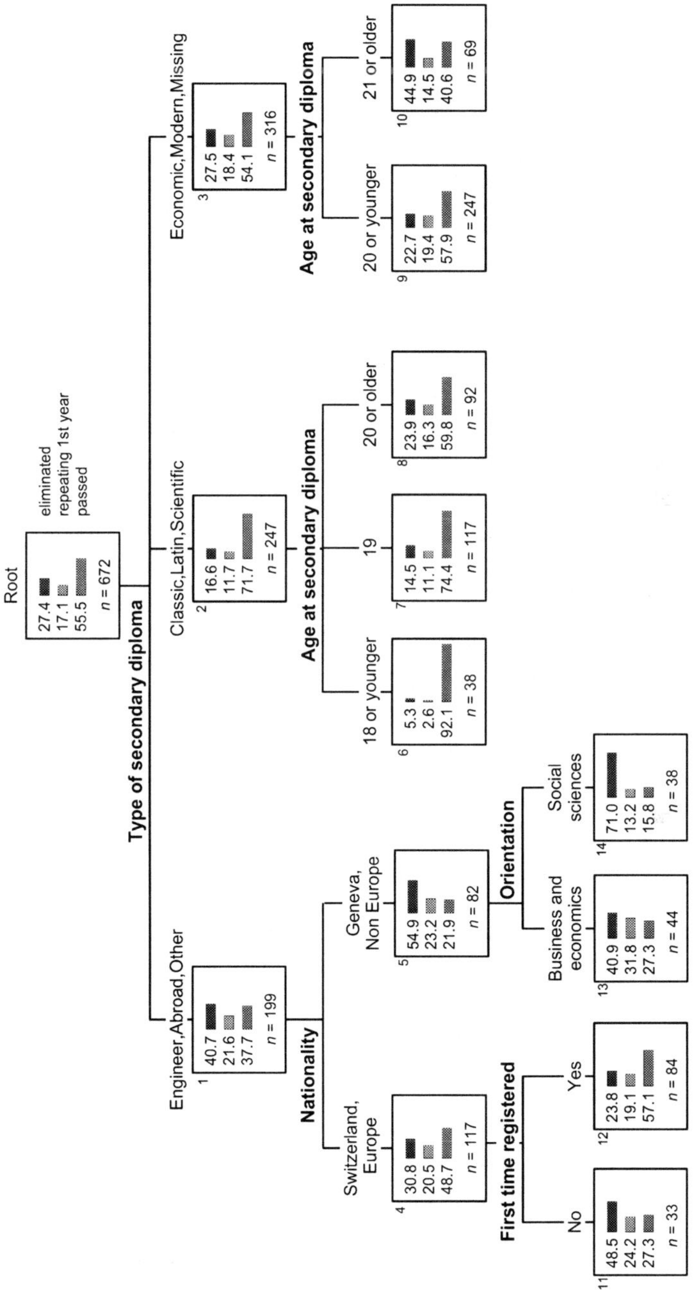

Fig. 3. CHAID induced tree for the ESS Student data. Outcome states are from top to down: eliminated, repeating 1st year, passed. Figures next to the bars are percentages.

We used successively the majority class rule and each of the four variants of implication indexes for that. Table 11 reports the results. We can see that the 5 methods agree for 6 out of the 9 leaves. The conclusion assigned to leaves number 9, 12 and 13 vary, however, among the 5 methods. All four implication indexes assign state 2, "repeating the first year", to leaf 13 where the majority class is 1, "eliminated". This tells us that belonging to this leaf, i.e having a not typical Swiss college secondary diploma obtained either in Geneva or abroad and having chosen a business and economic orientation, is a typical profile of those who repeat their first year. And this holds, indeed, despite "repeating the first year" is not the majority class of the leaf.

The deviance and adjusted residuals agree about assigning also state 2, "repeating", to leaves number 9 and 12, and the Freeman-Tukey residual agrees also with this conclusion for leaf 12. These leaves also define characteristic profiles of those who repeat their first year, even though the majority class for these profiles is "passed".

5.3 Effect of Continuity Correction

We expect continuity correction, i.e. adding .5 to the observed counts $n_{\bar{b}j}$ of counter-examples, to have only very marginal effects and to be important only in conjunction with small minimal node sizes.

For our application on the ESS student data, the continuity correction changes the conclusion only when we use the Freeman-Tukey residual for leaf 12 (with 84 cases). The conclusions remain the same for all other leaves and for all leaves when we use any of the three other residuals. Furthermore, the effect of the continuity correction vanishes when we multiply all the counts by a factor greater or equal to 1.4, which confirms our expectation.

Nevertheless, we suggest systematically introducing the error correction when computing the indexes. There are two reasons for this: First, it does not change much the index values in case of large counts and produces values best suited for comparison with standard normal thresholds in case of small counts. Secondly, it avoids possible troubles (division by zero for instance) that may occur when some observed counts are zero.

5.4 Recall and Precision

In terms of the overall error rate, selecting the majority class is no doubt the better choice. However, if we are interested in the recall rate, i.e. in the proportion of cases with a given output value c_k that are detected as having this value, we may expect the implication indexes to outperform the majority rule for infrequent classes. Indeed, highly infrequent outcome states have high chances to never be selected as conclusion by the majority rule. We may therefore expect low recall for them when we select the most frequent class as conclusion. Regarding precision, i.e. the proportion of cases classified as having a value c_k that effectively have this value, expectations are less

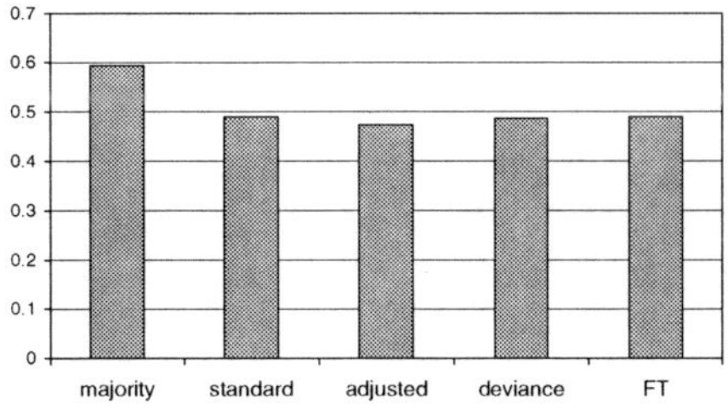

Fig. 4. Correct classification rate, 10-fold CV

clear since the relationship between the numerator and denominator seems not linked to the way of choosing the conclusion.

In order to verify these expectations on our ESS student data, we computed for the majority rule and each of the four variants of implication indexes, the 10-fold cross-validation (CV) values of the overall good classification rate, as well as of the recall and precision for each of the three outcome states. As can be shown on Figure 4 the loss in accuracy that results from using maximal implication rules lies between 12% for the adjusted residual and 10% for the standard residual.

Figure 5 exhibits the CV recall rates obtained for each of the three states. They confirm our expectations: selecting the conclusion according to implication indexes deteriorates the recall for the majority class "passed", but results

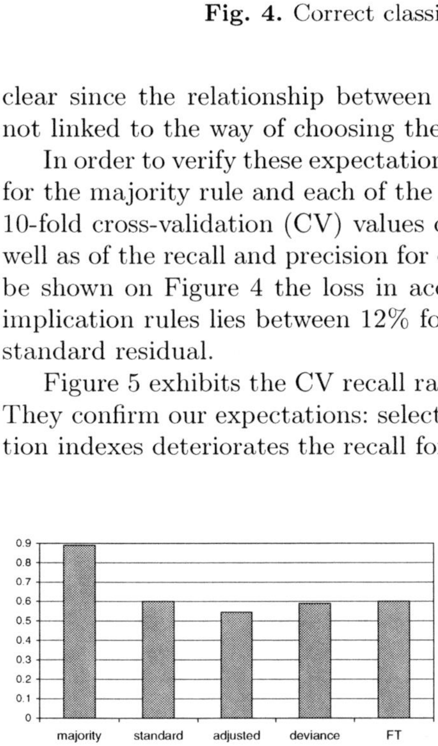

(a) passed

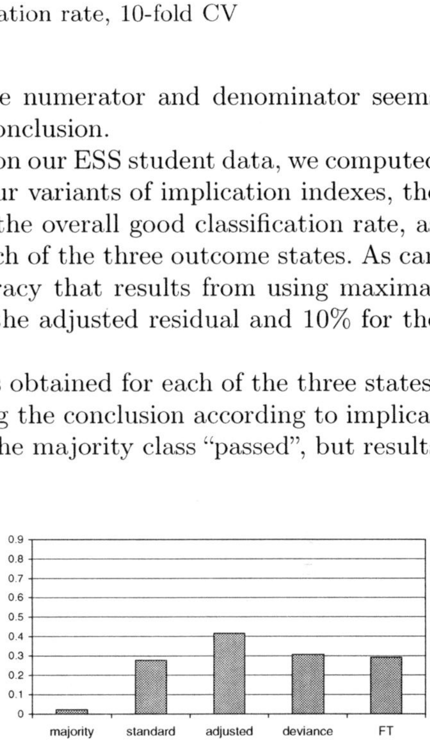

(b) repeating

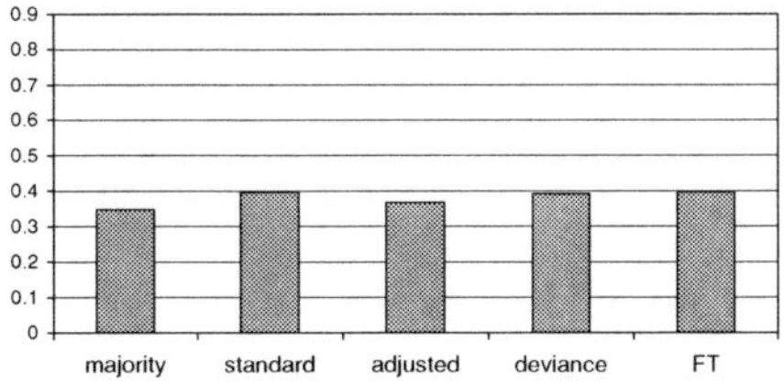

(c) eliminated

Fig. 5. Recall, 10-fold CV

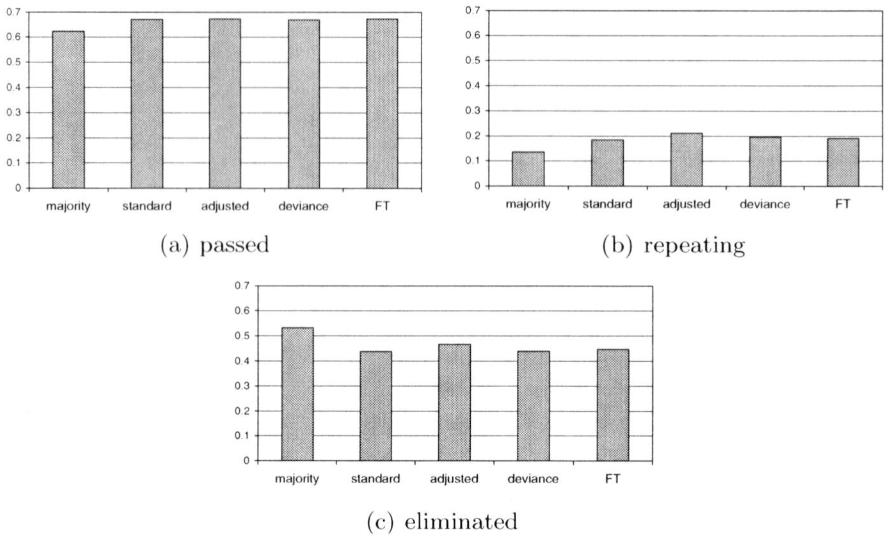

Fig. 6. Precision, 10-fold CV

in an improvement in recall for the two other classes. The improvement is especially important for the last frequent state, i.e. "repeating", for which we get recall rates ranging between 30% and 40% instead of almost 0% with the majority rule.

In Figure 6, we observe an improvement in precision for "passed" (the majority class) and "repeating" (the last frequent class) and a slight deterioration for "eliminated". This illustrates that the choice of the conclusion has apparently no predictable effect on precision. Indeed, the only thing we may notice here is that improvement concerns the two classes with a proportion of cases that is further (on either side) from the equiprobable probability $1/c$, where c is the number of outcome classes.

6 Conclusion

The aim of this article was to demonstrate the usefulness of the concept of implication strength for rules derived from induced decision trees. We have shown that Gras' implication index can be applied in a straightforward manner to classification rules and have proposed three alternatives inspired from residuals used in the statistical modeling of multiway contingency tables, namely the deviance, adjusted and Freeman-Tukey residuals. As for the scope of the implication indexes we have successively discussed their use for evaluating individual rules, for selecting the conclusion of the rule and as criteria for growing trees. We have stressed that implication indexes are a valuable complement to classical error rates as validation tools. They are especially interesting in a targeting framework where the aim is to determine the typical profile that leads to a conclusion rather than classifying individual cases. As criteria for selecting the conclusion, they may be a useful alternative to the majority rule in the case of imbalanced data. Their advantage is that in such imbalanced situation and unlike decisions based on the majority class, they favor conclusion diversity among rules as well as recall for poorly represented classes.

Four variants of implication indexes have been discussed. Which one should we use? The simulation study of their behavior has shown that the deviance residual curiously diminishes when the number of counter-examples tends to zero and should therefore be disregarded. The standard residual (Gras' index) and Haberman's adjusted residual both evolve linearly between independence and purity and thus seem to be the better choices. From the theoretical standpoint, if we want to compare the values with thresholds of the standard normal, Haberman's adjusted residual is preferable.

We have also introduced the implication intensity as the probability to get by chance more counter-examples than observed. This is indeed just a monotonic transformation of the corresponding implication index. Hence rankings based on the indexes or on the intensities will necessarily agree. Indexes seem better suited, however, to distinguishing between situations with high implication strengths. The intensities on the other hand, provide additional information about the statistical significance of the implication strength.

It is worth mentioning that, to our knowledge, implication indexes have not so far been implemented in tree growing software. Making them available is essential for popularizing them. We have begun working on implementing the maximal implication selection process and tree growing algorithms based on implication criteria into Tanagra [15] a free open source data mining software, and plan also to make these tools available in Weka.

Beside this implementation task, there are some other issues that would merit further investigation. For instance, the penalized implication index we proposed in Section 3 is not completely satisfactory. In a n-arry tree the paths to the leaves are usually shorter than in a binary tree, even if they define the same leaves. Penalization based on the length of the path as we proposed, would therefore be different for a rule derived from a binary tree than for the same rule derived from a n-arry tree. The use of implication criteria in the tree growing process needs also a deeper reflection.

Despite all which remains to be done, our hope is that this article will contribute to enlarge both the scope of induced decision trees and that of implication statistics.

References

1. Alan Agresti. *Categorical Data Analysis.* Wiley, New York, 1990.
2. Yvonne M. M. Bishop, Stephen E. Fienberg, and Paul W. Holland. *Discrete Multivariate Analysis.* MIT Press, Cambridge MA, 1975.
3. Julien Blanchard, Fabrice Guillet, Régis Gras, and Henri Briand. Using information-theoretic measures to assess association rule interestingness. In *Proceedings of the 5th IEEE International Conference on Data Mining (ICDM 2005)*, pages 66–73. IEEE Computer Society, 2005.
4. L. Breiman, J. H. Friedman, R. A. Olshen, and C. J. Stone. *Classification And Regression Trees.* Chapman and Hall, New York, 1984.
5. Henri Briand, Laurent Fleury, Régis Gras, Yann Masson, and Jacques Philippe. A statistical measure of rules strength for machine learning. In *Proceedings of the Second World Conference on the Fundamentals of Artificial Intelligence (WOCFAI 1995)*, pages 51–62, Paris, 1995. Angkor.
6. R. Gras, R. Couturier, J. Blanchard, H. Briand, P. Kuntz, and P. Peter. Quelques critères pour une mesure de qualité de règles d'association. *Revue des nouvelles technologies de l'information RNTI*, E-1:3–30, 2004.
7. R. Gras and A. Larher. L'implication statistique, une nouvelle méthode d'analyse de données. *Mathématique, Informatique et Sciences Humaines*, (120):5–31, 1992.
8. R. Gras and H. Ratsima-Rajohn. L'implication statistique, une nouvelle méthode d'analyse de données. *RAIRO Recherche Opérationnelle*, 30(3):217–232, 1996.
9. Régis Gras. *Contribution à l'étude expérimentale et à l'analyse de certaines acquisitions cognitives et de certains objectifs didactiques.* Thèse d'état, Université de Rennes 1, France, 1979.
10. Sylvie Guillaume, Fabrice Guillet, and Jacques Philippe. Improving the discovery of association rules with intensity of implication. In Jan M. Zytkow and Mohamed Quafafou, editors, *Proceedings of the Eurpoean Conference on Principles of Data Mining and Knowledge Discovery (PKDD 1998)*, volume 1510 of *Lecture Notes in Computer Science*, pages 318–327. Springer, 1998.
11. G. V. Kass. An exploratory technique for investigating large quantities of categorical data. *Applied Statistics*, 29(2):119–127, 1980.
12. I. C. Lerman, R. Gras, and H. Rostam. Elaboration d'un indice d'implication pour données binaires I. *Mathématiques et sciences humaines*, (74):5–35, 1981.
13. Claire Petroff, Anne-Marie Bettex, and Andràs Korffy. Itinéraires d'étudiants à la Faculté des sciences économiques et sociales: le premier cycle. Technical report, Université de Genève, Faculté SES, Juin 2001.
14. Adrian E. Raftery. Bayesian model selection in social research. In P. Marsden, editor, *Sociological Methodology*, pages 111–163. The American Sociological Association, Washington, DC, 1995.
15. Ricco Rakotomalala. Tanagra : un logiciel gratuit pour l'enseignement et la recherche. In Suzanne Pinson and Nicole Vincent, editors, *Extraction et Gestion des Connaissances (EGC 2005)*, volume E-3 of *Revue des nouvelles technologies de l'information RNTI*, pages 697–702. Cépaduès, 2005.
16. Einoshin Suzuki and Yves Kodratoff. Discovery of surprising exception rules based on intensity of implication. In Jan M. Zytkow and Mohamed Quafafou, editors, *Principles of Data Mining and Knowledge Discovery, Second European*

Symposium, PKDD '98, Nantes, France, September 23-26, Proceedings, pages 10–18. Springer, Berlin, 1998.
17. Djamel A. Zighed and Ricco Rakotomalala. *Graphes d'induction: apprentissage et data mining*. Hermes Science Publications, Paris, 2000.

On the behavior of the generalizations of the intensity of implication: A data-driven comparative study

Benoît Vaillant[1], Stéphane Lallich[2], and Philippe Lenca[3]

[1] IUT de Vannes, Université de Bretagne Sud, VALORIA, France
benoit.vaillant@univ-ubs.fr
[2] Université Lyon 2, Equipe de Recherche en Ingénierie des Connaissances, France
stephane.lallich@univ-lyon2.fr
[3] Institut TELECOM, TELECOM Bretagne, Lab-STICC, France
philippe.lenca@telecom-bretagne.eu

Summary. In this chapter, we present an original and synthetical overview of most of the commonly used association rule statistical interestingness measures introduced in previous works. These measures usually relate the confidence of a rule to an independency reference situation. Others relate it to indetermination, or impose a minimum confidence threshold. We propose a systematic generalization of these measures, taking into account a reference point, chosen by an expert, in order to apprehend the confidence of a rule. This generalization introduces new connections between measures. They lead to the enhancement of some measures. We then propose new parameterized possibilities. The behavior of the parameterized measures is illustrated using classical datasets, and these measures are compared to their original counter-parts. This study highlights the different properties of each of them and discusses the advantages of our proposition.

Key words: Statistical interestingness measures, intensity of implication, generalized measures.

This generalization introduces new connections between measures. They lead to the enhancement of some measures. We then propose new parameterized possibilities.

1 Introduction

In this chapter, we focus on the generalization of statistical interestingness measures. We will consider objective association rule interestingness measures, which aim at quantifying the quality of rules extracted from binary transactional datasets. Such measures are said to be objective since they only rely

B. Vaillant et al.: *On the behavior of the generalizations of the intensity of implication: A data-driven comparative study*, Studies in Computational Intelligence (SCI) **127**, 421–447 (2008)
www.springerlink.com

on frequency counts on the data in order to assess the interest of a rule, as opposed to subjective ones which are based on expressed prior knowledge. An association rule is an implication A $\rightarrow$ B, where A and B (also called *itemsets*) are conjunctions of attributes. We denote by n the total number of transactions in the database, n_a (resp. n_b, n_{ab}, $n_{a\bar{b}}$) the number of transactions matching A (resp. B, A and B, A but not B), and by p_a (resp. p_b, p_{ab}, $p_{a\bar{b}}$) the corresponding relative frequencies. Most measures are expressed as real valued functions of n, of the marginal frequencies p_a, p_b, and either p_{ab} or $p_{a\bar{b}}$, *i.e.* as functions of n, and of the *confidence* p_{ab}/p_a and marginal frequency counts of the considered rule since $p_{a\bar{b}} = p_a - p_{ab}$. Considering that the more counter-examples to a rule there are, the worse it is, we restrict our set of measures to those decreasing with $p_{a\bar{b}}$, p_a and p_b being fixed.

Support (Sup) and confidence (Conf) are the most famous of such measures, being the fundamental principles of Apriori-like algorithms [1]. These algorithms extract rules such that their Sup and Conf are above given constant thresholds, σ_s and σ_c. They are deterministic [9], and produce a large number of differing rules (see Figure 1, presenting the distribution of Sup and Conf values of 15771 automatically extracted rules on the Flag dataset), which, moreover, may not be interesting:

- an expert end-user expects from a rule that its Conf should be above a reference value, but this reference value seldom if ever equals σ_c. In this context, two main lower references are clearly identified as worthy from a user point of view. The first one is p_b, which corresponds to the independence of the itemsets A and B [33]. In this case the user will focus on rules such that the prior knowledge of A increases the knowledge of B. An alternative reference sometimes used is 0.5, as in [4]. In our opinion, the first reference is to be taken within a *targeting* strategy, and the second one when considering a *predictive* strategy. For example, let us consider an item B, corresponding to a given kind of cancer whose *a priori* probability is $p_b = 0.02$, and a rule A $\rightarrow$ B such that $p_{b|a} = 0.20$. This rule is very interesting in a targeting strategy (which aims at identifying risked groups), as the group of individuals having the characteristics of the itemset A have ten times more chance of developing the considered cancer than usual. On the other hand, the rule A $\rightarrow$ B is not interesting from a predictive point of view since an individual presenting the characteristics of the itemset A has a risk far less than 0.5 of developing a cancer. More generally, a user may be interested in taking into account a rule dependant reference value θ, $0 < \theta \leq 1$, and will consider only rules having a Conf greater than θ [25].
- what is more, the data mined is often subject to some sampling scheme. In order to take that into account, a special kind of measures has been proposed. They are called "statistical" in the sense that, unlike "descriptive" measures, their value rises with n, the relative frequencies being fixed. Let us consider the rule A $\rightarrow$ B, for which $p_a = 0.30$, $p_b = 0.50$, $p_{ab} = 0.25$ and two situations: $n = 20$ and $n = 400$. The value of the correlation coefficient

R between A and B being the same (R $= 0.22$) for each alternative, R is a descriptive measure. Another possible interestingness measure is the complement to 1 of the p-value of R, denoted by $M = 1 - P(\text{R} > \text{R}_{obs})$, where R_{obs} is the observed value of R. In this case, M rises from $M = 0.835$ for $n = 20$ up to $M = 0.999$ for $n = 400$. M is thus a statistical measure, and more precisely a probabilistic one (see section 2.1). This consideration accounts for developing an inferential approach, and retaining only rules that are significantly well evaluated by measures, in comparison to the reference chosen. Amongst the issues that arise from this approach, validating a large number of rules through the control of false rules discovery is assessed in [23].

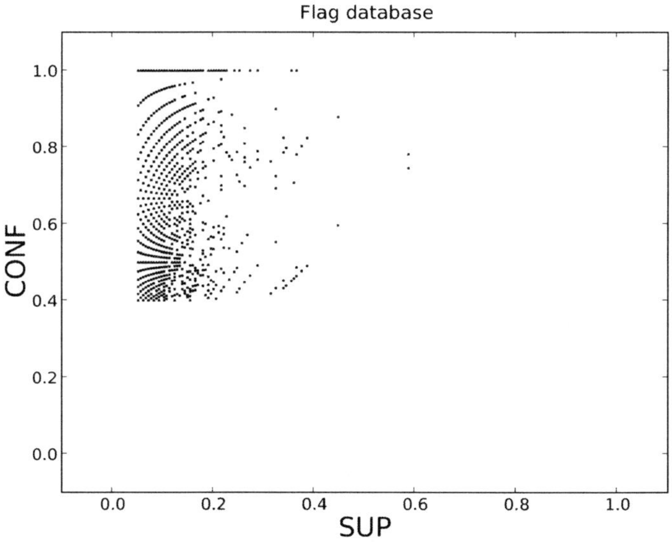

Fig. 1. SUP and CONF values of the rules extracted from the FLAG database

Figure 2 illustrates these various conflicting situations. We represent the SUP value of two families of rules. The first family, denoted by r_1, is defined such that $p_a = 0.2$ and $p_b = 0.4$. For the second family, r_2, we impose that $p_a = 0.45$ and $p_b = 0.8$. Given such characteristics, clearly, the equilibrium and independence situations appear in a different order for r_1 and r_2 when $p_{a\bar{b}}$ increases. What is more, if one tries to discard uninteresting rules using a SUP threshold, either all interesting rules from r_1 will be discarded if this σ_s is fixed considering r_2, or many uninteresting rules from r_2 will be retained if one considers r_1 as reference [39].

As previously mentioned, APRIORI-like algorithms may produce huge numbers of rules, and thus an essential step in association rule mining is

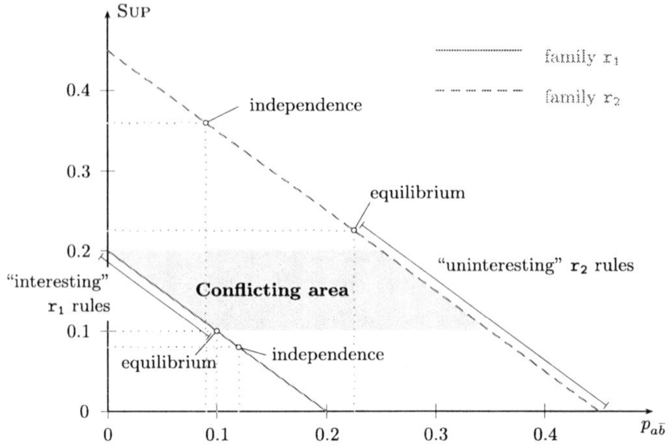

Fig. 2. Variations of SUP for two families of rules

the evaluation of their interestingness. The support and confidence framework is not satisfactory, and many new measures have been proposed. Each new measure is supposed to better highlight a user desired kind of knowledge. Various properties of interestingness measures -for various data mining tasks- have been investigated, in particular in [8, 12, 15, 16, 20, 24, 28, 32, 33, 36, 38]. In [29] we propose an extensive study of twenty well-known association rule interestingness measures based on eight user-oriented points of view.

One of these properties is related to the reference value to which the measure compares confidence, this reference value being commonly either p_b (independence), or 0.5 (indetermination). We here extend this concept to a user chosen reference.

This work is concerned with implicative statistical analysis, in which one tries to measure the strength of a rule $\mathtt{A} \rightarrow \mathtt{B}$. The basic idea behind this analysis is that the fewer counter-examples observed (i.e. in the data) there are, the more implicative the rule $\mathtt{A} \rightarrow \mathtt{B}$ is.

Since its origins –binary data, mathematical didactics situations [10, 31]– the implicative statistical analysis has achieved many developments and applications in various data ming tasks (see for example [7] and [11] for recent reviews): treatment of modal variables [2], numerical ones [19] and ordinal ones [14]; user-driven process for mining association rules [18], for classification trees [34] and [35], exception rules mining [37], classification association rules [17].

Objective association rule interestingness measures usually compare the confidence of a rule to a reference value corresponding to the independence between the antecedent and the consequent of a rule. Some of them compare the confidence of a rule to 0.5, which corresponds to an indetermination situation.

In a previous work, [25], we first suggested to parameterizing the reference value of both descriptive and statistical measures in order to compare the confidence to a reference value θ chosen by the user. The case of statistical measures, especially the intensity of implication and its generalizations has been explored in [21]. Theoretical aspects of our works were extended in [26].

In this chapter, we present our results on generalized statistical measures and we propose an original data driven comparative study of the behavior of generalized statistical measures.

This chapter is organized as follows. In section 2, we present a general synthetic overview of statistical measures making reference to independence: modeling of counter-examples distribution, construction of statistical and probabilistic measure, enhancement of the discriminating power of the statistical measures. We introduce in section 3 the statistical measures making reference to indetermination. Section 4 deals with the generalization of statistical measures. Discriminant versions of generalized measures are proposed in section 5. Finally, we conclude in section 6.

2 Statistical measures making reference to independence

2.1 Characteristics of statistical and probabilistic measures

A statistical measure evaluates how far the observed rule is from a null hypothesis H_0 corresponding to a lower reference point. From the definition of a statistical measure, which is a modeling of the kind of rules that one wishes to discover, it is then possible to define a probabilistic measure as the probability of obtaining a value of the statistical measure, at most equal to what is observed, given that the null hypothesis H_0 is true.

Classically, this null hypothesis is the hypothesis of independence between itemsets `A` and `B`, and it is tested against a one sided alternative hypothesis H_1 of positive dependence. The corresponding test can be written in terms of theoretical frequencies referring to `A` and `B` ($\pi(\cdot)$ being the theoretical frequencies):

$$H_0 : \pi_{b/a} \leq \pi_b \text{ against } H_1 : \pi_{b/a} > \pi_b$$

The modeling of the null hypothesis of independence performed in [31] can be done in three different ways, with respectively 1, 2 and 3 hazard levels.

In [25] we proposed an alternative to the first modeling, denoted $1^{'}$. The four modelings are synthesized in Table 1.

We denote by N_{ab} the random variable generating n_{ab}, and H, B and Poi refer to the hypergeometric, binomial and Poisson distributions, respectively.

- Modeling 1 (only one hazard level): margins are fixed, only the joint absolute frequencies are random, but with only one degree of freedom.
  - The modeling proposed by [31] applies to the distribution of examples, within the 4 inner possibilities of the contingency table of (A, B), n_a and n_b being fixed, following a traditional statistical process.
    Under H_0, N_{ab} here follows the hypergeometric law $H(n, n_a, p_b)$. Testing H_0 thus means testing the equality of the theoretical confidence of $\mathtt{A} \rightarrow \mathtt{B}$ and $\overline{\mathtt{A}} \rightarrow \mathtt{B}$, at fixed margins.
- Modeling 1′ (only one hazard level): still at fixed margins, an alternative approach which only takes into account the distribution of examples between AB and A$\overline{\mathtt{B}}$ is proposed in [25].
  - In this case, N_{ab} follows the binomial law $B(n_a, p_b)$. Testing H_0 then means testing the conformity of the theoretical confidence of $\mathtt{A} \rightarrow \mathtt{B}$, p_b being fixed beforehand.
- Modeling 2 (two hazard levels): modeling 2 of [31] corresponds to modeling $1'$, with n_a also randomized.
  - On a first hazard level, it is thus here assumed that N_a follows the binomial law $B(n, p_a)$.
  - On a second hazard level, conditionally to $N_a = n_a$, N_{ab} follows the binomial law $B(n_a, p_b)$. Thus N_{ab} follows the binomial law $B(n_a, p_a p_b)$.
- Modeling 3 (three hazard levels): modeling 3 of [31] once again relies on modeling $1'$, where the values of n_a, and then n are successively randomized.
  - On the first hazard level, N is assumed to follow the Poisson law $Poi(n)$.
  - On the second hazard level, it is assumed that N_a follows the binomial law $B(n, p_a)$, conditionally to $N = n$.
  - On the third hazard level, and conditionally to $N = n$ and $N_a = n_a$, it is assumed that N_{ab} follows the binomial law $B(n_a, p_b)$. In this case, N_{ab} follows the Poisson law $Poi(np_a p_b)$.

The statistical and probabilistic measures based on $N_{a\bar{b}}$ are built as follows:

- by establishing the law of N_{ab} and $N_{a\bar{b}}$ under the null hypothesis (H_0) following the chosen modeling, we can express a centered and reduced index[4] under H_0 (^{CR} notation). In order to have a decreasing quality measure with respect to $n_{a\bar{b}}$, the statistical index is defined by $\mathrm{SI}^{(i)} = -N_{a\bar{b}}^{CR}$, where i refers to the corresponding modeling.
- under standard conditions, the law of this index can be approximated by the normal distribution, leading to the definition of a probabilistic measure, defined **as the complement to 1** of the surprise of observing such an exceptional value of the index under H_0. This probabilistic index

[4] Given a random variable X, its centered and reduced expression is $x^{CR} = \frac{x-\mu}{\sqrt{v}}$, where μ is the mean of X and v its variance.

is denoted by $\mathrm{PI}^{(i)} = P(N(0,1) > n_{a\bar{b}}^{CR})$, where i again refers to one of the four modelings introduced.

The chosen modeling does not affect the expectation, but does modify the variance. [13] and [6] prefer the third modeling, which dissociates most the rules $\mathtt{A} \rightarrow \mathtt{B}$ and $\overline{\mathtt{B}} \rightarrow \overline{\mathtt{A}}$, whereas the first modeling makes no distinction between these rules. The probabilistic measure hence obtained is the intensity of implication (IntImp $= \mathrm{PI}^{(3)}$), which satisfies many properties one expects a measure should have [12, 28].

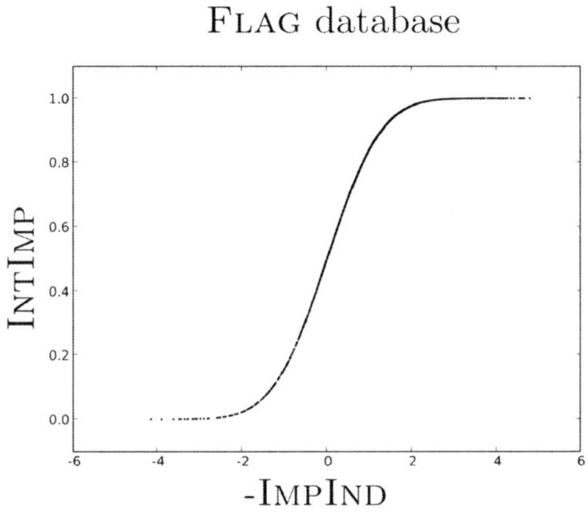

Fig. 3. Evolution of IntImp in function of -ImpInd

The statistical measure obtained with modeling 1 is $SI^{(1)} = \mathrm{R}\sqrt{n}$. The corresponding probabilistic measure $\mathrm{PI}^{(1)} = P(N(0,1) < \mathrm{R}\sqrt{n})$ is the complement to 1 of the p-value of R. It is to be noted that in the boolean case, $n\mathrm{R}^2 = \chi^2$. Hence $\mathrm{PI}^{(1)}$ is the unilateral counterpart of the complement to 1 associated with the χ^2 test of independency.

Figure 3 shows that IntImp is an anamorphosis of -ImpInd ([13], see $SI^{(3)}$ in Table 1) through the normal distribution function.

2.2 Discriminating power of statistical measures

Although it has many good properties [27, 29], one of the major drawbacks of IntImp (drawback shared by the other statistical and probabilistic measures) is the loss of discriminating power: by its definition, it will evaluate rules significantly different from independence between 0.95 and 1. If n becomes large, which is particularly true in a data mining context, the slightest divergence from an independence situation becomes highly significant, thus leading to high and homogeneous values of the measure, close to 1. It is thus difficult to select the best rules. For example, we computed the values taken by IntImp on rules extracted from three classical datasets [40]. The Breast Cancer, Contraceptive Method Choice and Housing datasets are available from the UCI repository (`http://www.ics.uci.edu/~mlearn/databases/`).

On the Breast Cancer data, containing $n = 683$ entries, 3079 rules have an IntImp value above 0.99, out of the 3095 rules generated by Apriori, with $\sigma_s = 0.10$ and $\sigma_c = 0.70$. On the Contraceptive Method Choice data, containing $n = 1473$ entries, we extracted 1035 rules having an IntImp value above 0.99, out of the 2378 rules generated (with $\sigma_s = 0.05$ and $\sigma_c = 0.60$). Finally, on the Housing data, containing $n = 506$ entries, 156 rules out of 263 were evaluated above 0.99 by IntImp, Apriori being run with $\sigma_s = 0.02$ and $\sigma_c = 0.55$.

This phenomenon of loss of discriminant power is illustrated in Figure 4, in which we represent $\text{PI}^{(1)} = P(N(0,1) < \sqrt{n}\text{R})$ for various values of n. This figure shows the loss of discriminant power of the measure as n rises, although R is not affected by such changes. For example, with $n = 323$, there are 991 rules evaluated above 0.999, 3540 when n is multiplied by 10, and 4205 when n is multiplied by 100, out of the 5402 rules.

Using the third modeling does not solve the issue as can be seen in Figure 5. In this situation almost all rules are evaluated above 0.95. On other rule sets, as presented in Figure 6, the range of values that IntImp takes is wider.

In order to counter-balance this loss of discriminating power, [30] introduce a contextual approach where ImpInd is centered and reduced on a case database $\mathcal{B}$, thus leading to the definition of the probabilistic discriminant index (a monotonically increasing transformation of IntImp contextualized on the data).

This index is defined as follows:

$$\text{PDI} = P\left[N(0,1) > \text{ImpInd}^{CR/\mathcal{B}}\right]$$

[13] propose an alternative solution by weighting IntImp through the use of an inclusion index. This index is based on the entropy of experiments B/A and $\overline{\text{A}}/\overline{\text{B}}$. We denote by $H(X) = -p_x \log_2 p_x - p_{\overline{x}} \log_2 p_{\overline{x}}$ the entropy associated with an event X. In [6] the most general form of the inclusion index is given as:

$$i(A \subset B) = \left[\left(1 - H^*(B/A)^\alpha\right)\left(1 - H^*(\overline{A}/\overline{B})^\alpha\right)\right]^{\frac{1}{2\alpha}}$$

where $H^*(X) = H(X)$ if $p_x > 0.5$, $H^*(X) = 1$ otherwise. The α parameter is chosen by the user. The value $\alpha = 2$ is advised if one wants this index to be tolerant to initial counter-examples, and we will use this value from now on.

Hence, [13] define the entropic intensity of implication as:

$$\text{EII} = \left[\text{IntImp} \cdot i(A \subset B)\right]^{\frac{1}{2}}$$

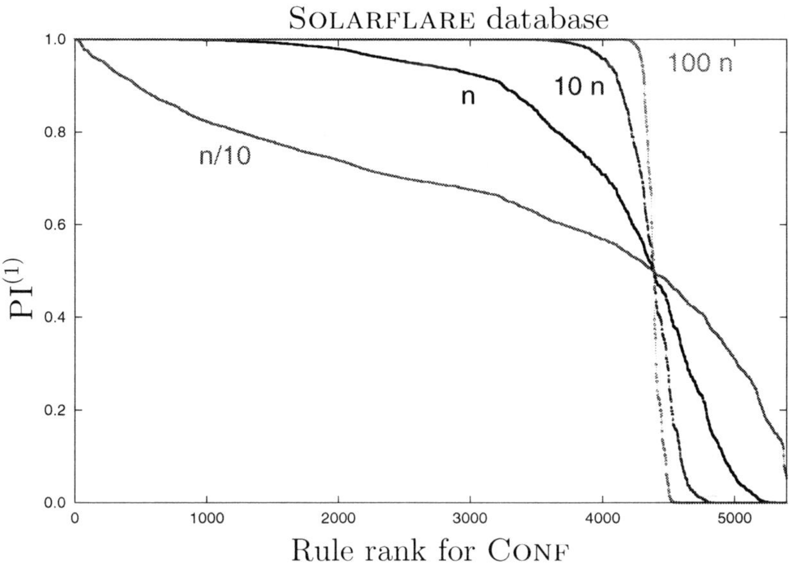

Fig. 4. PI$^{(1)}$ values of the rules, in function of their rank for CONF on the SOLARFLARE database

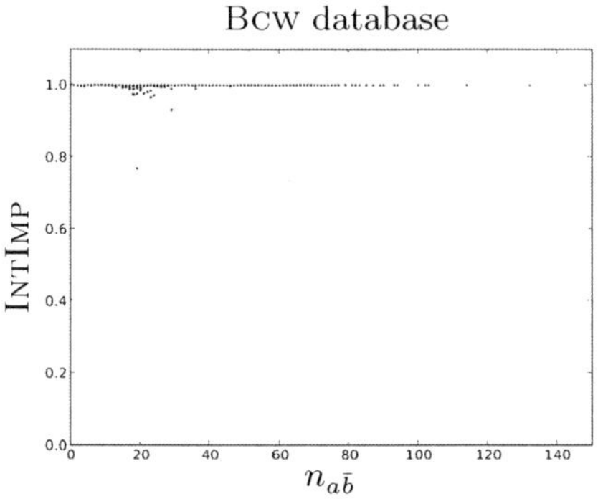

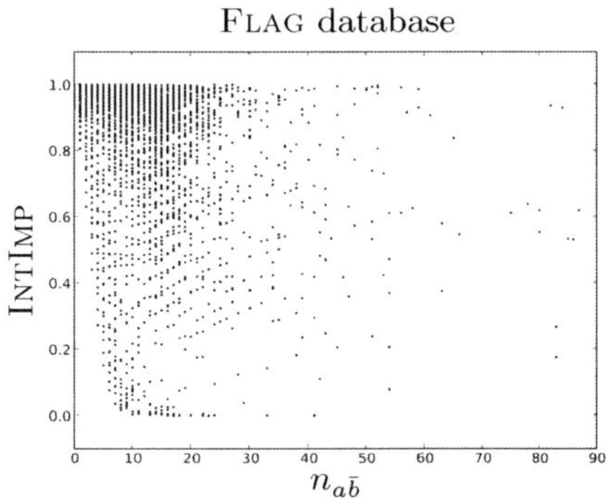

Fig. 5. Values of INTIMP in function of $n_{a\bar{b}}$ for the BCW database

Fig. 6. Values of INTIMP in function of $n_{a\bar{b}}$ for the FLAG database

The shift from $H(X)$ to $H^*(X)$ aims at discarding uninteresting situations, such as $p_{b/a} < 0.5$ or $p_{\overline{a}/\overline{b}} < 0.5$, and complies with a predictive strategy. In a targeting strategy, the value of $p_{b/a}$ should be compared to p_b, and the value of $p_{\overline{a}/\overline{b}}$ to $p_{\overline{a}}$.

The weighting of the implication of intensity by the inclusion index, although effective, is problematic. The inclusion index is a measure of the distance to indetermination based on entropy, thus being null when $p_{b/a} = 0.5$, and so is EII. However, IntImp values 0.5 at independency. Hence EII is not always null at independency: EII $= \sqrt[8]{\frac{(1-H(A)^2)(1-H(B)^2)}{16}}$ if $p_a < 0.5$ and $p_b > 0.5$, and is null otherwise.

Figures 7 to 9 show the effects of both approaches. Figure 8 shows the difference in behavior of both anamorphosis of ImpInd, namely IntImp and PDI. Since PDI is contextualized and takes into account the values of ImpInd on the data, its distribution is smoother.

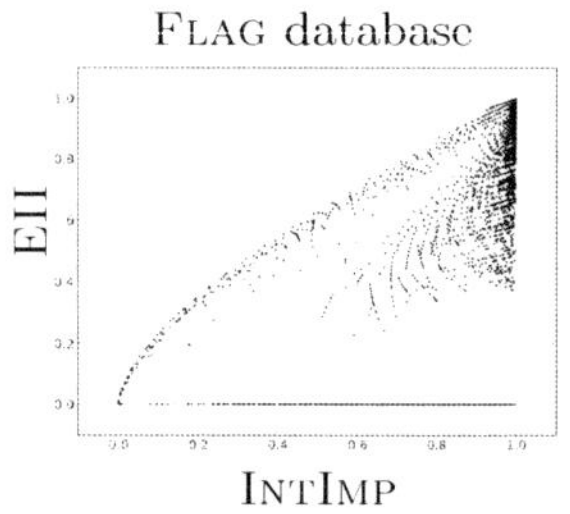

Fig. 7. Variations of EII in function of IntImp

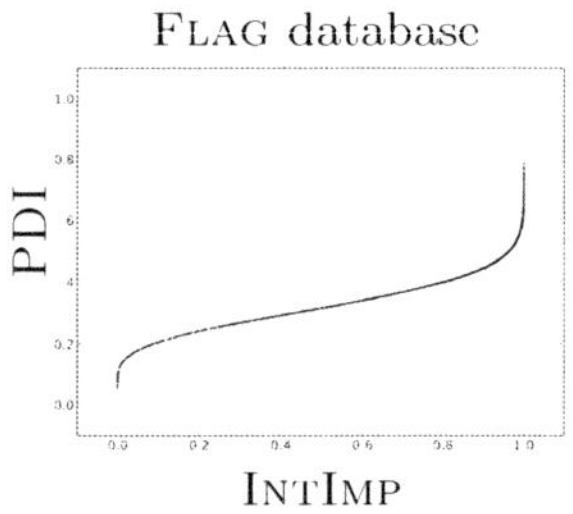

Fig. 8. Variations of PDI in function of IntImp

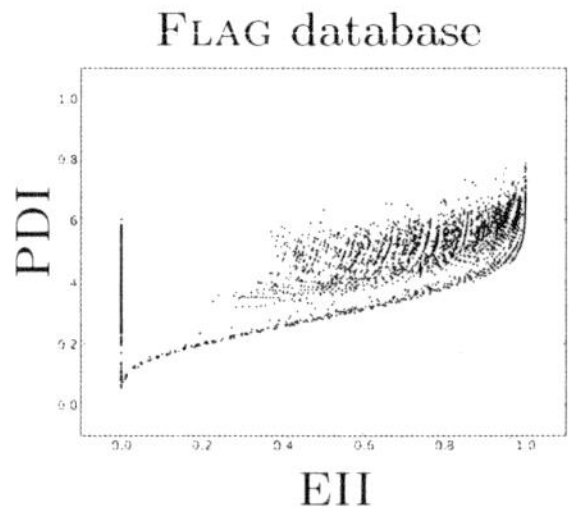

Fig. 9. Variations of PDI in function of EII

2.3 Adaptation of the entropic intensity of implication

We proposed two adaptations of EII in order to cope with the above mentioned issues: Revised EII, denoted REII and Truncated EII, denoted TEII [25,26].

Our first proposal involves replacing IntImp by IntImp* in EII where:

$$\text{IntImp}^* = \max\{2\text{IntImp} - 1; 0\}$$

This will solve the previously highlighted problems, but has the drawback of modifying the entire spectrum of values taken by EII:

$$\text{REII} = \left[\text{IntImp}^* \cdot i(A \subset B)\right]^{\frac{1}{2}}$$

Figures 10 and 11 show the joint distribution of EII and REII in function of $p_{a\overline{b}}$. Three families of rules are presented, for the first and the last ones there is no observable difference between the measures. On the contrary we see the impact of the correction added in REII on the spectrum of values of EII for the second family. In Figure 11, n is ten times smaller than in Figure 10. Here for all three families, there are observable differences.

Our second proposal only nullifies the values of EII when $p_a p_{\bar{b}} \leq p_{a\bar{b}} \leq \min\{\frac{p_a}{2}, \frac{p_{\bar{b}}}{2}\}$, without modifying its values otherwise. To achieve this, we introduce $H_t^*(X)$ an adequate truncated version of $H(X)$, and i_t a truncated version of the inclusion index i. In order to take into account both predictive and targeting strategies, a rule will have a non null evaluation by the inclusion index, and hence by TEII when the following conditions are jointly met:

- $p_{b/a} > 0.5$ (prediction) and $p_{b/a} > p_b$ (targeting); *i.e.* $p_{b/a} > \max(0.5, p_b)$
- $p_{\overline{a}/\overline{b}} > 0.5$ (prediction) and $p_{\overline{a}/\overline{b}} > p_{\overline{a}}$ (targeting); *i.e.* $p_{\overline{a}/\overline{b}} > \max(0.5, p_{\overline{a}})$

With these new conditions, TEII is null whenever the proportion of counter-examples is above $\min\left(p_a p_{\bar{b}}; \frac{p_a}{2}; \frac{p_{\bar{b}}}{2}\right)$:

$$\text{TEII} = [\text{IntImp}(A \to B) \times i_t(A \subset B)]^{\frac{1}{2}}$$

with:

- $i_t(A \subset B) = \left[(1 - H_t^*(B/A)^\alpha)\left(1 - H_t^*(\overline{A}/\overline{B})^\alpha\right)\right]^{\frac{1}{2\alpha}}$,
- $H_t^*(B/A) = H(B/A)$ if $p_{b/a} > \max(0.5, p_b)$, $H_t^*(B/A) = 1$ otherwise,
- $H_t^*(\overline{A}/\overline{B}) = H(\overline{A}/\overline{B})$ if $p_{\overline{a}/\overline{b}} > \max(0.5, p_{\overline{a}})$, $H_t^*(\overline{A}/\overline{B}) = 1$ otherwise.

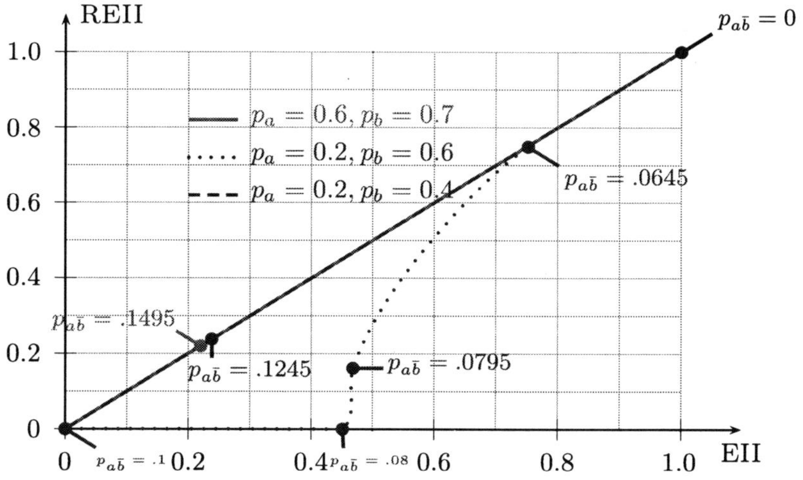

Fig. 10. Joint distributions of EII and REII in function of $p_{a\bar{b}}$, $n = 2000$

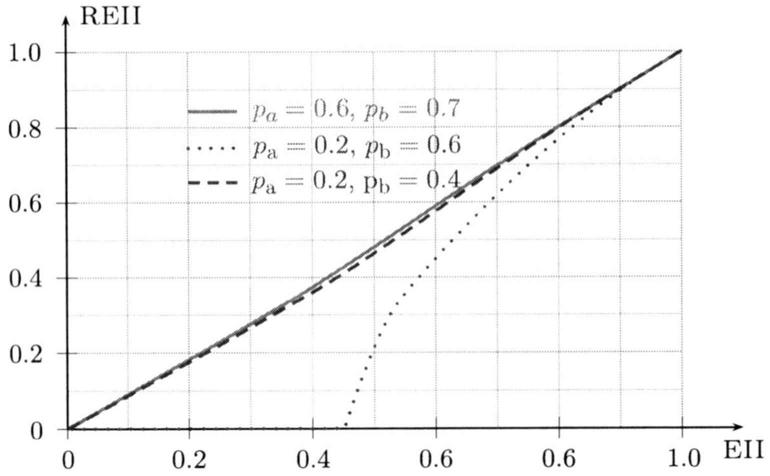

Fig. 11. Joint distributions of EII and REII in function of $p_{a\bar{b}}$, $n = 200$

3 Statistical measures making reference to indetermination

3.1 Probabilistic index

[3] propose IPEE , a probabilistic measure of deviation from indetermination (or equilibrium). The authors implicitly use modeling $1'$ since they consider $N_{a\bar{b}} \equiv B(n_a, 0.5)$ under an indetermination hypothesis, *i.e.* $N_{a\bar{b}}^{CR} = \frac{N_{a\bar{b}} - 0.5n_a}{0.5 \cdot \sqrt{n_a}}$:

$$\text{IPEE} = P\left[B(n_a, 0.5) > n_{a\bar{b}}\right] \approx P\left[N(0,1) > \frac{n_{a\bar{b}} - 0.5n_a}{0.5 \cdot \sqrt{n_a}}\right]$$

Under normal approximation, IPEE equals 0.5 at indetermination. This measure corresponds to the probabilistic index associated with modeling $1'$ (see Table 1), where p_b is replaced by 0.5. IPEE will hence inherit the weak discriminating power of this kind of measure.

As shown in Figure 12, IPEE and EII may take significantly different values for some rules.

3.2 Discriminant version

In order to enhance the discriminating power of IPEE, [5] proposed IP3E, in which IPEE is weighted by the inclusion index, following a similar method as used in EII:

$$\text{IP3E} = \left[\text{IPEE} \times 0.5(1 + i(A \subset B))\right]^{0.5}$$

There is an important difference in the construction of these entropic measures. Indeed in IP3E, the inclusion index interacts with IPEE through the

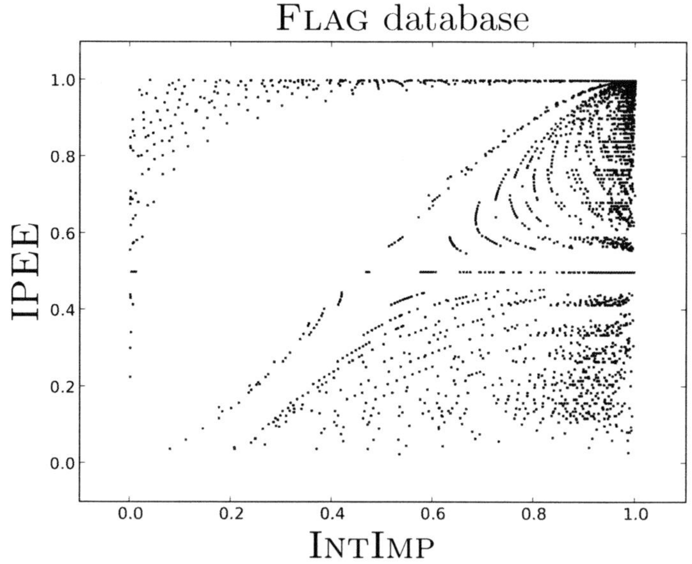

Fig. 12. Variations on IPEE in function of INTIMP

expression $0.5(1 + i(A \subset B))$. This expression takes its values between 0.5 and 1, and equals 0.5 at indetermination. Hence, in this situation, the value of IP3E is not nullified, as was the case for EII. As shown in Figure 13 the contribution of this index is of less importance in this case. This can also be seen in Figure 14 which compares IP3E to TEII.

4 Generalized statistical measures

Using the same approach as with descriptive measures [25], we generalize statistical measures and evaluate the interestingness of a rule by comparing its CONF to θ. This is done by considering in Table 1 that for each modeling under H_0, the probability of an example, conditionally to n_a, is θ:

$$N_{ab} \equiv B(n_a, \theta)$$

The results of the thus adapted modelings 1 and $1'$ are immediate, and those of modelings 2 and 3 are easily obtained through the use of the probability generating functions as detailed in [26] and recalled Table 1.

From these results, we propose a range of generalized measures (see Table 1), which are constructed in the same way as described in section 2. Generalized statistical measures are defined by $\mathrm{GSI}^{(i)}_{|\theta} = -N^{CR}_{a\bar{b}}$, while generalized probabilistic measures are defined by $\mathrm{GPI}^{(i)}_{|\theta} = P(N(0,1) > n^{CR}_{a\bar{b}})$:

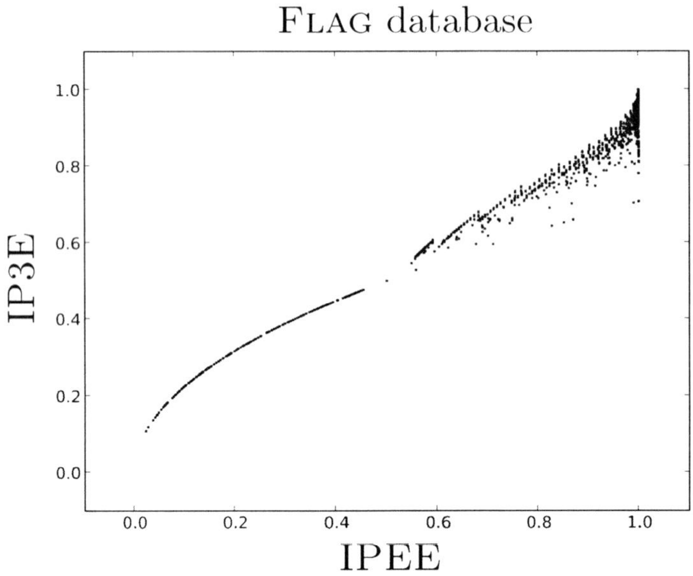

Fig. 13. Variations of IP3E in function of IPEE

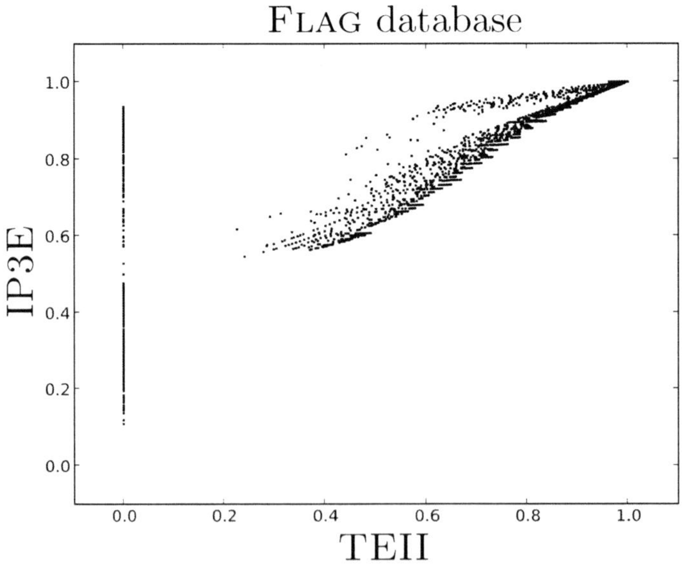

Fig. 14. Comparison of the behavior of TEII and IP3E

- by establishing the law of N_{ab} and $N_{a\bar{b}}$ under the null hypothesis (H_0) following the chosen modeling i, we can express a centered and reduced index under H_0. This statistical index is denoted by $\mathrm{GSI}^{(i)}_{|\theta}$.
- under standard conditions, the law of this index can be approximated to the normal distribution, leading to the definition of a probabilistic measure, defined **as the complement to 1** of the surprise of observing such an exceptional value of the index under H_0. This probabilistic index is denoted by $\mathrm{GPI}^{(i)}_{|\theta}$.

We will focus on two of these. The first one, $\mathrm{GPI}^{(1')}_{|\theta}$, is associated with modeling $1'$ and generalizes IPEE. For clarity reasons, it will be denoted $\mathrm{GIPE}_{|\theta}$ (we here removed the last E, since the generalized measure no longer makes reference to equilibrium). It corresponds to the chi-square goodness of fit test, assessing whether or not the B/A distribution comes from the distribution related to $(\theta;\ 1-\theta)$. The second one, $\mathrm{GPI}^{(3)}_{|\theta}$, is associated with modeling 3, and generalizes IntImp. It will thus be denoted by $\mathrm{GIntImp}_{|\theta}$.

Using $\theta = 0.9$ as lower reference value the generalized measures should focus on rules having a confidence above this threshold. Clearly, we see in Figures 15 and 16 that probabilistic indices stress the differences of evaluations near this value, discarding rules far below to a null evaluation. On the contrary rules above the reference tend to have a very good evaluation. Once more we here see the importance of the use of a discriminant version of the probabilistic measure.

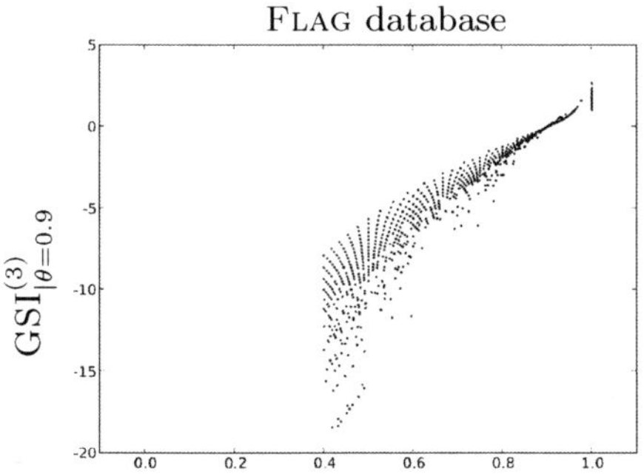

Fig. 15. Values taken by $\mathrm{GSI}^{(3)}_{|\theta=0.9}$ in function of Conf

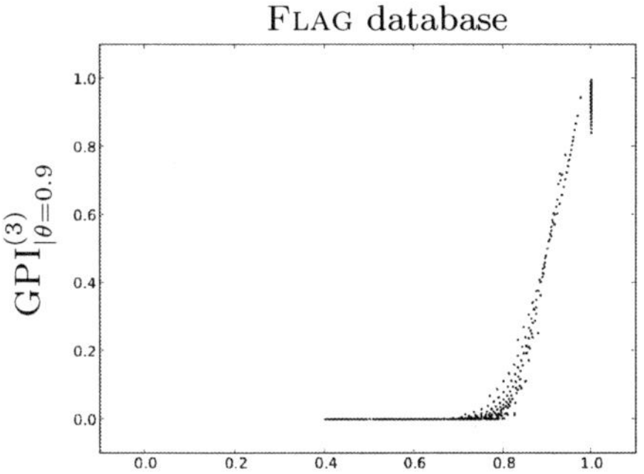

Fig. 16. Values taken by $\mathrm{GPI}^{(3)}_{|\theta=0.9}$ in function of Conf

Statistical and probabilistic indices summary

	Modelings 1 and 1′		Modeling 2	Modeling 3
Principle	1.1: n_a is fixed, N_{ab} is randomized, and both A and $\bar{A}$ are considered	1′.1: p_a is fixed, N_{ab} is randomized, and only A is considered	2.1: $N_a \equiv B(n, p_a)$ 2.2: $N_{ab} \equiv B(n_a, p_b)$ $\|N_a = n_a$	3.1: $N \equiv Poi(n)$ 3.2: $N_a \equiv B(n, p_a)$ $\|N = n$ 3.3: $N_{ab} \equiv B(n_a, p_b)$ $\|N = n, N_a = n_a$
Law N_{ab} under H_0	$H(n, n_a, p_b)$	$B(n_a, p_b)$	$B(n, n_a p_b)$	$Poi(np_a p_b)$
Law $N_{a\bar{b}}$ under H_0	$H(n, n_a, p_{\bar{b}})$	$B(n_a, p_{\bar{b}})$	$B(n, n_a p_{\bar{b}})$	$Poi(np_a p_{\bar{b}})$
Statistical index: $\mathrm{SI} = -N_{a\bar{b}}^{CR}$	$\mathrm{SI}^{(1)} = -\frac{N_{a\bar{b}} - np_a p_{\bar{b}}}{\sqrt{np_a p_{\bar{a}} p_b p_{\bar{b}}}}$	$\mathrm{SI}^{(1')} = -\frac{N_{a\bar{b}} - np_a p_{\bar{b}}}{\sqrt{np_a p_b p_{\bar{b}}}}$	$\mathrm{SI}^{(2)} = -\frac{N_{a\bar{b}} - np_a p_{\bar{b}}}{\sqrt{np_a p_{\bar{b}}(1 - p_a p_{\bar{b}})}}$	$\mathrm{SI}^{(3)} = -\textsc{ImpInd} = -\frac{N_{a\bar{b}} - np_a p_{\bar{b}}}{\sqrt{np_a p_{\bar{b}}}}$
Probabilistic index: $\mathrm{PI} = P(N(0,1) > n_{a\bar{b}}^{CR})$ $\mathrm{PI} = P(N(0,1) < \mathrm{SI})$	$\mathrm{PI}^{(1)} = P(N(0,1) < \mathrm{R}\sqrt{n})$	$\mathrm{PI}^{(1')}$	$\mathrm{PI}^{(2)}$	$\mathrm{PI}^{(3)} = \textsc{IntImp} = P(N(0,1) > \textsc{ImpInd})$

Generalized statistical and probabilistic indices summary

	Modelings 1 and 1′		Modeling 2	Modeling 3
Principle	1.1: n_a is fixed, N_{ab} is randomized, and both A and $\bar{A}$ are considered	1′.1: p_a is fixed, N_{ab} is randomized, and only A is considered	2.1: $N_a \equiv B(n, p_a)$ 2.2: $N_{ab} \equiv B(n_a, \theta)$ $\|N_a = n_a$	3.1: $N \equiv Poi(n)$ 3.2: $N_a \equiv B(n, p_a)$ $\|N = n$ 3.3: $N_{ab} \equiv B(n_a, \theta)$ $\|N = n, N_a = n_a$
Law N_{ab} under H_0	$H(n, n_a, \theta)$	$B(n_a, \theta)$	$B(n, n_a \theta)$	$Poi(np_a \theta)$
Law $N_{a\bar{b}}$ under H_0	$H(n, n_a, 1-\theta)$	$B(n_a, 1-\theta)$	$B(n, n_a(1-\theta))$	$Poi(np_a(1-\theta))$
Statistical index: $\mathrm{GSI}_{\|\theta} = -N_{a\bar{b}}^{CR}$	$\mathrm{GSI}_{\|\theta}^{(1)} = -\frac{N_{a\bar{b}} - np_a(1-\theta)}{\sqrt{np_a p_{\bar{a}} \theta(1-\theta)}}$	$\mathrm{GSI}_{\|\theta}^{(1')} = -\frac{N_{a\bar{b}} - np_a(1-\theta)}{\sqrt{np_a \theta(1-\theta)}}$	$\mathrm{GSI}_{\|\theta}^{(2)} = -\frac{N_{a\bar{b}} - np_a(1-\theta)}{\sqrt{np_a(1-\theta)(1-p_a(1-\theta))}}$	$\mathrm{GSI}_{\|\theta}^{(3)} = -\textsc{GIndImp}_{\|\theta}$ $= -\frac{N_{a\bar{b}} - np_a(1-\theta)}{\sqrt{np_a(1-\theta)}}$
Probabilistic index: $\mathrm{GPI}_{\|\theta} = P(N(0,1) > n_{a\bar{b}}^{CR})$ $\mathrm{GPI}_{\|\theta} = P(N(0,1) < \mathrm{GSI}_{\|\theta})$	$\mathrm{GPI}_{\|\theta}^{(1)}$	$\mathrm{GPI}_{\|\theta}^{(1')} = \mathrm{GIPE}_{\|\theta}$	$\mathrm{GPI}_{\|\theta}^{(2)}$	$\mathrm{GPI}_{\|\theta}^{(3)} = \textsc{GIntImp}_{\|\theta}$ $= P(N(0,1) > \textsc{GIndImp}_{\|\theta})$

Table 1. Modeling of the various statistical and probabilistic indices, and their generalized counterparts

5 Discriminant versions of the generalized statistical measures

The generalized statistical or probabilistic measures have, as the original ones do, a weak discriminating power. In order to enhance these measures, we will consider two approaches (cf. Section 2), one relying on weighting through the use of an inclusion index, like [13], the other one being contextual, like [30].

In the first approach, we propose the more general expression of the entropic generalized probabilistic index, $\mathrm{EGPI}_{|\theta}$, which is defined as the product of $\mathrm{GPI}_{|\theta}$ and an inclusion index. In order to remain coherent, we think it advisable to define a generalized inclusion index $gi_{|\theta}$, using θ as the reference value and not 0.5. This leads us to first define $\widetilde{H}_{|\theta}(X)$ a off-centered version of the entropy $H(X)$, being maximal when $p_x = \theta$, and not when $p_x = 0.5$ (see Figure 18).

5.1 A off-centered version of the entropy

In order to define this off-centered version of the entropy, we propose the following modification, so that the new index will take its maximal value 1 when $p_x = \theta$.

This index is defined as follows:

$$\widetilde{H}_{|\theta}(X) = -\widetilde{p}_x \log_2 \widetilde{p}_x - (1 - \widetilde{p}_x) \log_2 (1 - \widetilde{p}_x)$$

where $\widetilde{p}_x$ is:

$$\widetilde{p}_x = \frac{p_x}{2\theta} \text{ if } p_x \leq \theta, \ \widetilde{p}_x = \frac{p_x + 1 - 2\theta}{2(1-\theta)} \text{ otherwise (see Figure 17)}$$

We call this $\widetilde{H}_{|\theta}(X)$ index the off-centered entropic index. It is clear that $\widetilde{H}_{|\theta}(X)$ is not a strict entropy and that it must be seen as a penalization function.

The behavior of this off-centered entropy index, illustrated in Figure 18 for a B/A distribution, leads to interesting perspectives in data mining.

It could, for example, be used in a tree induction process to assess the quality of the prediction of the class variable conditionally to the predictive variables, when such a class is boolean and has a very unbalanced distribution [22].

From the definition of this new index, we build $\widetilde{H}_{|\theta}(B/A)$ and $\widetilde{H}_{|\theta}(\overline{A}/\overline{B})$ as:

- to obtain $\widetilde{H}_{|\theta}(B/A)$ from $H(B/A)$, $p_{b/a}$ is replaced by $\widetilde{p}_{b/a}$ defined as follows:

$$\widetilde{p}_{b/a} = \frac{p_{b/a}}{2\theta} \text{ if } p_{b/a} \leq \theta, \ \widetilde{p}_{b/a} = \frac{p_{b/a} + 1 - 2\theta}{2(1-\theta)} \text{ otherwise}$$

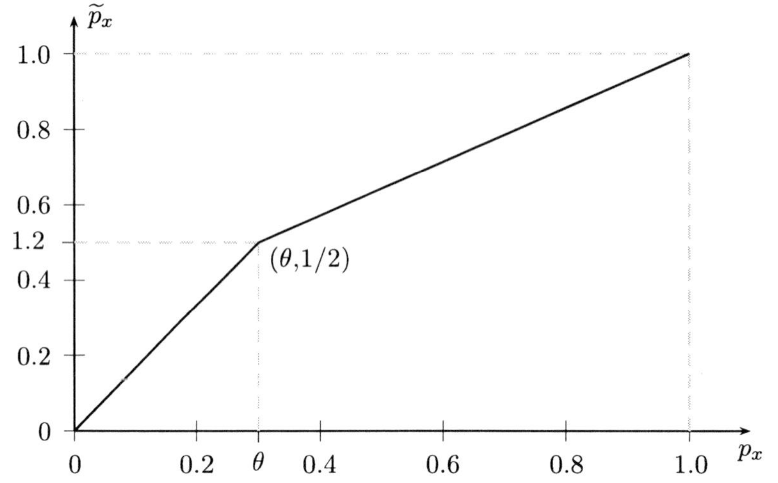

Fig. 17. Variation of $\widetilde{p}_x$ in function of p_x

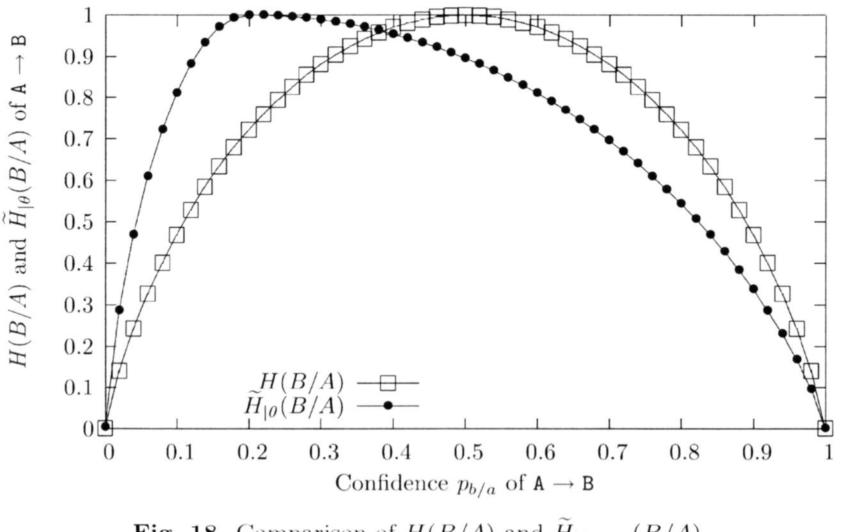

Fig. 18. Comparison of $H(B/A)$ and $\widetilde{H}_{|\theta=0.2}(B/A)$

- to obtain $\widetilde{H}_{|\theta}(\overline{A}/\overline{B})$ from $H(\overline{A}/\overline{B})$, a first possibility involves replacing $p_{\overline{a}/\overline{b}}$ by $\widetilde{p}_{\overline{a}/\overline{b}}$ defined by:

$$\widetilde{p}_{\overline{a}/\overline{b}} = \frac{p_{\overline{a}/\overline{b}}}{2\theta} \text{ if } p_{\overline{a}/\overline{b}} \leq \theta,\ \widetilde{p}_{\overline{a}/\overline{b}} = \frac{p_{\overline{a}/\overline{b}} + 1 - 2\theta}{2(1-\theta)} \text{ otherwise}$$

  This first possibility generalizes the inclusion index proposed in [13], which can be retrieved using $\theta = 0.5$.
- $\widetilde{H}_{|\theta}(\overline{A}/\overline{B})$ could also be obtained from $H(\overline{A}/\overline{B})$, by using $1 - \frac{p_a}{p_{\overline{b}}} \times (1-\theta)$ as the reference, since $p_{\overline{a}/\overline{b}} = 1 - \frac{p_a}{p_{\overline{b}}} \times (1 - p_{b/a})$. In this case, when considering independency (*i.e.* $\theta = p_b$), the reference value for $\widetilde{H}_{|\theta}(\overline{A}/\overline{B})$ is $p_{\overline{a}}$. This second possibility is the basis of a new version of the inclusion index.

5.2 Weighting approach

To define the entropic generalized probabilistic index, $\mathrm{EGPI}_{|\theta}$, the generalized index of inclusion, $gi_{|\theta}$, is first defined. To this end, $\widetilde{H}^*_{|\theta}(B/A)$ and $\widetilde{H}^*_{|\theta}(\overline{A}/\overline{B})$, are defined as:

$$\widetilde{H}^*_{|\theta}(X) = \widetilde{H}_{|\theta}(X) \text{ if } p_x > \theta,\ \widetilde{H}^*_{|\theta}(X) = 1 \text{ otherwise}$$

and $gi_{|\theta}$ as:

$$gi_{|\theta} = \left[\left(1 - \widetilde{H}^*_{|\theta}(B/A)^{\alpha}\right)\left(1 - \widetilde{H}^*_{|\theta}(\overline{A}/\overline{B})^{\alpha}\right)\right]^{\frac{1}{2\alpha}}, \text{ with } \alpha = 2.$$

From this, we deduce $\mathrm{EGPI}_{|\theta}$ which is a more discriminant version of $\mathrm{GPI}_{|\theta}$:

$$\mathrm{EGPI}_{|\theta} = \left[\mathrm{GPI}_{|\theta} \times gi_{|\theta}\right]^{\frac{1}{2}}$$

From this general expression of $\mathrm{EGPI}_{|\theta}$, we can express particular instances of generalized measures, such as $\mathrm{EGINTIMP}_{|\theta}$, the entropic generalized intensity of implication, and $\mathrm{EGIPE}_{|\theta}$, the entropic generalized probabilistic index of deviation:

- Modeling 3:

$$\mathrm{EGINTIMP}_{|\theta} = \mathrm{EGPI}^{(3)}_{|\theta} = \left[\mathrm{GPI}^{(3)}_{|\theta} \times gi_{|\theta}\right]^{\frac{1}{2}} = \left[\mathrm{G\small{INT}I\small{MP}}_{|\theta} \times gi_{|\theta}\right]^{\frac{1}{2}}$$

- Modeling 1′:

$$\mathrm{EGIPE}_{|\theta} = \mathrm{EGPI}^{(1')}_{|\theta} = \left[\mathrm{GPI}^{(1')}_{|\theta} \times gi_{|\theta}\right]^{\frac{1}{2}} = \left[\mathrm{GIPE}_{|\theta} \times gi_{|\theta}\right]^{\frac{1}{2}}$$

It must be noticed that both components of $\mathrm{EGPI}^{(3)}_{|\theta} = \mathrm{EGINTIMP}_{|\theta}$ refer to the same θ, which ensures the coherence of the measure. In particular, for $\theta = p_b$, $\mathrm{GPI}^{(3)}_{|p_b} = \mathrm{G}\textsc{IntImp}_{|p_b}$ corresponds to $\textsc{IntImp}$ and $\mathrm{EGPI}^{(3)}_{|p_b} = \mathrm{EGINTIMP}_{|p_b}$ is more coherent than EII.

In the case $\theta = 0.5$, $\mathrm{GPI}^{(1')}_{|\theta} = \mathrm{GIPEE}_{|\theta}$ corresponds to IPEE and $gi_{|\theta}$ corresponds to i. It appears that $\mathrm{EGIPE}_{|0.5}$ is slightly different from IP3E = $[\mathrm{IPEE} \times 0.5\,(i(A \subset B) + 1)]^{\frac{1}{2}}$, the entropic version of IPEE proposed by [3].

Their behavior, compared to their original counterparts, is represented in Figure 19 (for $n = 1000$, $p_a = 0.05$ and $p_b = 0.10$). They were obtained using 3 different values for θ, $\theta = p_b = 0.1$ (thus targeting independence), $\theta = 2\,p_b = 0.2$ (targeting situations in which B happens twice as often when A is true) and $\theta = 0.5$ (prediction).

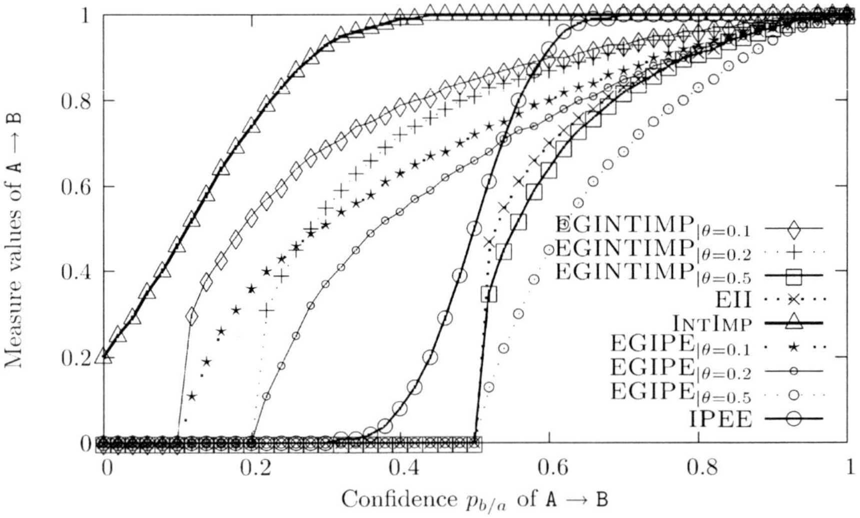

Fig. 19. Behavior of the measures, as functions of $p_{b/a}$ for $n = 1000$, $p_a = 0.05$ and $p_b = 0.10$

Figure 19 well illustrates how the θ parameter choice controls the behavior of the measures. Furthermore, we can see the effectiveness of the parametrization of the statistical or probabilistic measures, making them more discriminant.

In the specific case where the reference value is $\theta = p_b$, one could prefer the second version of the inclusion index.

Figures 20 and 21 which compare both alternatives for the third modeling to TEII put forward the best fitness of the second index.

Indeed in the first case all rules having $p_b \geq p_{\bar{a}}/p_{\bar{b}} \geq 0.5$ have a null evaluation for $\mathrm{EGPI}^{(3)}_{|\theta=p_b}$.

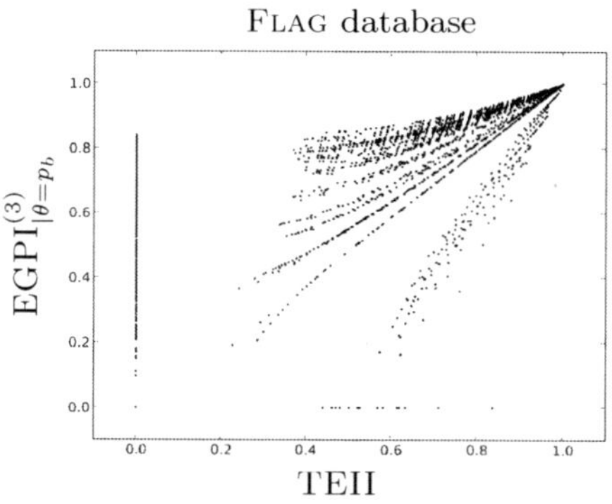

Fig. 20. Variations of $\mathrm{EGPI}^{(3)}_{|\theta=p_b}$ in function of TEII, first version of the entropic coefficient

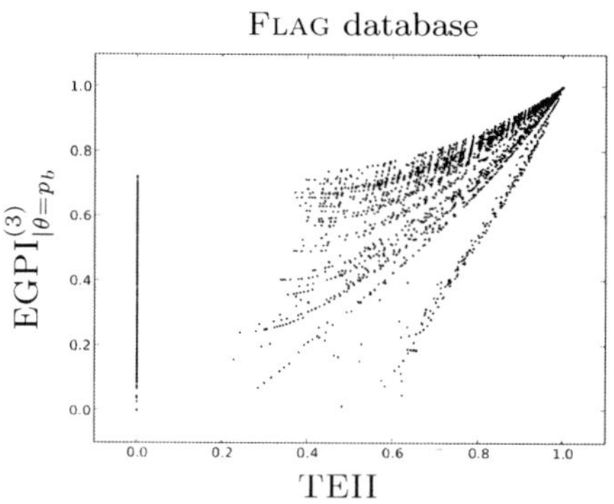

Fig. 21. Variations of $\mathrm{EGPI}^{(3)}_{|\theta=p_b}$ in function of TEII, second version of the entropic coefficient

If we consider now modeling $1'$, and compare $\mathrm{EGPI}^{1'}_{|\theta=0.5}$ to IP3E, we see that its value is null under indetermination whereas IP3E still varies. The range is therefore all the larger in our proposal (see Figure 22).

5.3 Contextual approach

In the contextual approach, $\mathrm{GSI}_{|\theta}$ is centered and reduced on a case database $\mathcal{B}$, and thus defines a contextual generalized probabilistic index, $\mathrm{CGPI}_{|\theta}$:

$$\mathrm{CGPI}_{|\theta} = P(N(0,1) > \mathrm{GSI}^{CR/\mathcal{B}}_{|\theta})$$

Depending on the modeling (1, 1′, 2 or 3), we can obtain different versions of $\mathrm{CGPI}_{|\theta}$:

- Modeling 3: $\mathrm{GSI}_{|\theta}$ corresponds to $\mathrm{GI{\scriptsize ND}I{\scriptsize MP}}_{|\theta}$. Then $\mathrm{CGPI}^{(3)}_{|\theta}$ defines a generalized probabilistic discriminant index, $\mathrm{GPDI}_{|\theta}$:

$$\mathrm{GPDI}_{|\theta} = \mathrm{CGPI}^{(3)}_{|\theta} = P(N(0,1) > \mathrm{GI{\scriptsize ND}I{\scriptsize MP}}^{CR/\mathcal{B}}_{|\theta})$$

It must be noted that $\mathrm{GPDI}_{|\theta=p_b} = \mathrm{PDI}$

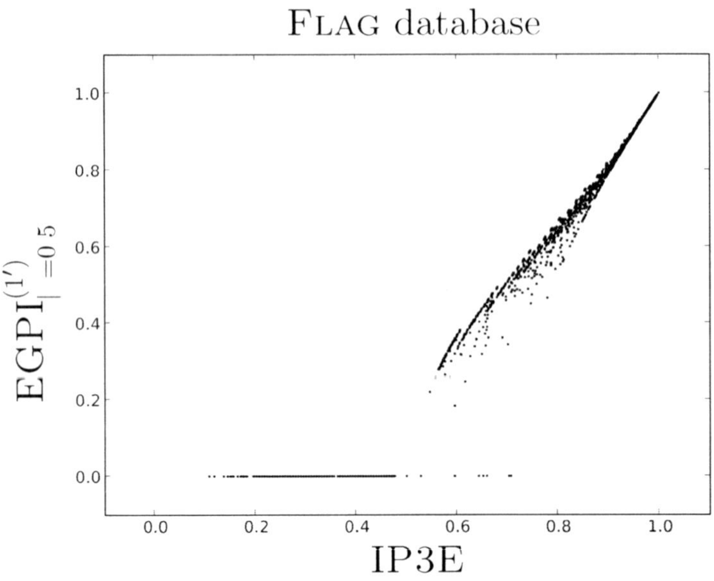

Fig. 22. Variations of $\mathrm{EGPI}^{(1')}_{|\theta=0.5}$ in function of IP3E

- Modeling 1′: $\mathrm{CGPI}^{(1')}_{|\theta} = P(N(0,1) > \mathrm{GPI}^{(1')CR/\mathcal{B}}_{|\theta})$ is a discriminant version of $\mathrm{GIPE}_{|\theta}$.
  Considering the case $\theta = 0.5$, $\mathrm{CGPI}^{(1')}_{|\theta=0.5}$ is an original contextual discriminant version of $\mathrm{GPI}^{(1')}_{|\theta=0.5} = \mathrm{IPEE}$

Figure 23 compares the contextual and entropic version of $\mathrm{GPI}^{(3)}_{|\theta=0.5}$ whereas Figure 24 compares the contextual approach using $\mathrm{GPI}^{(3)}_{|\theta}$ for the independence and indetermination situations.

6 Conclusion and perspectives

Following modeling and coherence principles, we have proposed in previous works an innovating framework from which a unified view of a large number of statistical interestingness measures can be constructed, and which clarifies some of the links between these measures.

This framework is the basis of the definition of new measures, namely the generalized intensity of implication, generalized probabilistic discriminant index, generalized probabilistic measure of deviation, and their entropic or contextual discriminant counterparts, which all compare the confidence of a

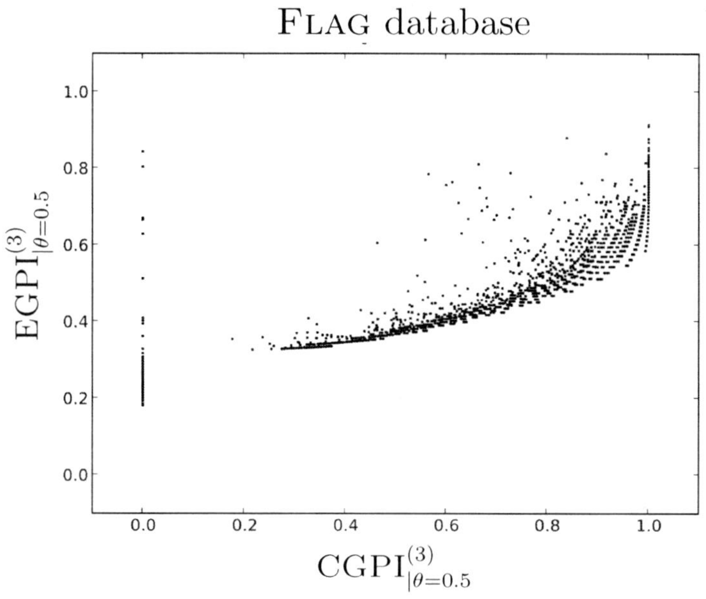

Fig. 23. Comparison of the entropic and contextual approach (modeling 3), for $\theta = 0.5$

rule to a user defined reference parameter. We extended this concept and defined an off-centered entropy. Its behavior within a supervised learning context is currently under study, and should lead to new perspectives.

Based on a sound and comprehensive framework, this chapter illustrates the use of parameterized measures within a data mining process. The behavior of the parameterized measures is illustrated using classical datasets, and these measures are compared to their original counter-parts. This study highlights the different properties of each of them. Our proposal leads to the definition of a set of measures, for which one may choose the most adapted one to user needs and data specificities.

References

1. R. Agrawal, T. Imielinski, and A.N. Swami. Mining association rules between sets of items in large databases. In P. Buneman and S. Jajodia, editors, *ACM SIGMOD International Conference on Management of Data*, pages 207–216, Washington, D.C., USA, 1993.
2. M. Bailleul and R. Gras. L'implication statistique entre variables modales. *Mathématiques, informatique et sciences humaines*, (128):41–57, 1995.
3. J. Blanchard, F. Guillet, H. Briand, and R. Gras. Assessing the interestingness of rules with a probabilistic measure of deviation from equilibrium. In J. Janssen

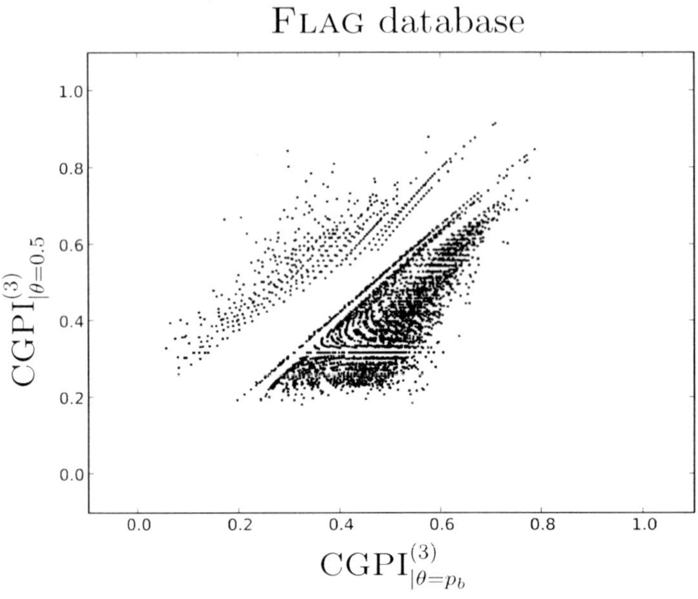

Fig. 24. Comparison of the contextual approaches (modeling 3) making reference at independence and indetermination

and P. Lenca, editors, *The XIth International Symposium on Applied Stochastic Models and Data Analysis*, pages 191–200, Brest, France, 2005.

4. J. Blanchard, F. Guillet, H. Briand, and R. Gras. IPEE : Indice probabiliste d'écart à l'équilibre pour l'évaluation de la qualité des règles. In *Atelier Qualité des Données et des Connaissances (EGC 2005)*, pages 26–34, Paris, France, 2005.
5. J. Blanchard, F. Guillet, H. Briand, and R. Gras. Une version discriminante de l'indice probabiliste d'écart à l'équilibre pour mesurer la qualité des règles. In R. Gras, F. Spagnolo, and J. David, editors, *The Third International Conference Implicative Statistic Analysis*, pages 131–137, Palermo, Italy, 2005. Supplément num. 15 de la Revue Quaderni di Ricerca in Didattica.
6. J. Blanchard, P. Kuntz, , F. Guillet, and R. Gras. Mesure de la qualité des règles d'association par l'intensité d'implication entropique. *Revue des Nouvelles Technologies de l'Information (Mesures de Qualité pour la Fouille de Données)*, (RNTI-E-1):33–43, 2004.
7. J. Blanchard, P. Kuntz, F. Guillet, and Gras R. *Statistical Data Mining and Knowledge Discovery*, chapter Implication intensity: from the basic statistical definition to the entropic version, pages 473–485. Chapman & Hall/CRC, 2003.
8. A. Freitas. On rule interestingness measures. *Knowledge-Based Systems journal*, pages 309–315, 1999.
9. A. Freitas. Understanding the crucial differences between classification and discovery of association rules - a position paper. In *ACM SIGKDD Explorations*, volume 2, pages 65–69, 2000.

10. R. Gras. *Contribution à l'étude expérimentale et à l'analyse de certaines acquisitions cognitives et de certains objectifs didactiques en mathématiques.* PhD thesis, Université de Rennes I, 1979.
11. R. Gras. Panorama du développement de l'A.S.I. à travers des situations fondatrices. In R. Gras, F. Spagnolo, and J. David, editors, *The third International Conference Implicative Statistic Analysis*, pages 9–33, Palermo, Italia, 2005. Supplément num. 15 de la Revue Quaderni di Ricerca in Didattica.
12. R. Gras, R. Couturier, J. Blanchard, H. Briand, P. Kuntz, and P. Peter. Quelques critères pour une mesure de qualité de règles d'association - un exemple : l'intensité d'implication. *Revue des Nouvelles Technologies de l'Information (Mesures de Qualité pour la Fouille de Données)*, (RNTI-E-1):3–31, 2004.
13. R. Gras, P. Kuntz, R. Couturier, and F. Guillet. Une version entropique de l'intensité d'implication pour les corpus volumineux. In H. Briand and F. Guillet, editors, *Extraction des connaissances et apprentissage (EGC 2001)*, volume 1, pages 69–80. Hermes, 2001.
14. S. Guillaume. *Traitement des données volumineuses, Mesures et algorithmes d'extraction de règles d'association et règles ordinales.* PhD thesis, Université de Nantes, 2000.
15. R.J. Hilderman and H.J Hamilton. Knowledge discovery and interestingness measures: A survey. Technical Report 99-4, Department of Computer Science, University of Regina, october 1999.
16. R.J. Hilderman and H.J. Hamilton. Measuring the interestingness of discovered knowledge: A principled approach. *Intelligent Data Analysis*, 7(4):347–382, 2003.
17. D. Janssens, G. Wets, T. Brijs, and K. Vanhoof. Adapting the CBA algorithm by means of intensity of implication. *Information Sciences*, 173(4):305–318, 2005.
18. P. Kuntz, F. Guillet, R. Lehn, and H. Briand. A user-driven process for mining association rules. In *Principles of Data Mining and Knowledge Discovery*, volume 1910 of *LNAI*, pages 483–489. Springer, 2000.
19. J.B. Lagrange. Analyse implicative d'un ensemble de variables numériques; application au traitement d'un questionnaire aux réponses modales ordonnées. *Revue de Statistique Appliquée*, XLVI(1):71–93, 1998.
20. S. Lallich. Mesure et validation en extraction des connaissances à partir des données. Habilitation à Diriger des Recherches – Université Lyon 2, 2002.
21. S. Lallich, P. Lenca, and B. Vaillant. Variations autour de l'intensité d'implication. In R. Gras, F. Spagnolo, and J. David, editors, *The Third International Conference Implicative Statistic Analysis*, pages 237–246, Palermo, Italy, 2005. Supplément num. 15 de la Revue Quaderni di Ricerca in Didattica.
22. S. Lallich, P. Lenca, and B. Vaillant. Construction of an off-centered entropy for supervised learning. In C. Skiadas, editor, *The XIIth International Symposium on Applied Stochastic Models and Data Analysis*, Chania, Crete, Greece, 2007.
23. S. Lallich, E. Prudhomme, and O. Teytaud. Contrôle du risque multiple en sélection de règles d'association significatives. In G. Hébrail, L. Lebart, and J.-M. Petit, editors, *Extraction et gestion des connaissances*, volume 1-2, pages 305–316, Clermont-Ferrand, France, 2004. Cépaduès Editions.
24. S. Lallich and O. Teytaud. Évaluation et validation de l'intérêt des règles d'association. *Revue des Nouvelles Technologies de l'Information (Mesures de Qualité pour la Fouille de Données)*, (RNTI-E-1):193–217, 2004.

25. S. Lallich, B. Vaillant, and P. Lenca. Parametrised measures for the evaluation of association rule interestingness. In J. Janssen and P. Lenca, editors, *The XIth International Symposium on Applied Stochastic Models and Data Analysis*, pages 220–229, Brest, France, 2005.
26. S. Lallich, B. Vaillant, and P. Lenca. A probabilistic framework towards the parameterization of association rule interestingness measures. *Methodology and Computing in Applied Probability*, 9(3):447–463, 2007.
27. P. Lenca, P. Meyer, B. Vaillant, and S. Lallich. On selecting interestingness measures for association rules: user oriented description and multiple criteria decision aid. *European Journal of Operational Research*, 184(2):610–626, 2008.
28. P. Lenca, P. Meyer, B. Vaillant, P. Picouet, and S. Lallich. Évaluation et analyse multicritère des mesures de qualité des règles d'association. *Revue des Nouvelles Technologies de l'Information (Mesures de Qualité pour la Fouille de Données)*, (RNTI-E-1):219–246, 2004.
29. P. Lenca, B. Vaillant, P. Meyer, and S. Lallich. *Quality Measures in Data Mining*, volume 43 of *Studies in Computational Intelligence, Guillet, F. and Hamilton, H.J., Eds.*, chapter Association rule interestingness measures: experimental and theoretical studies, pages 51–76. Springer-Verlag Berlin Heidelberg, 2007.
30. I.C. Lerman and J. Azé. Une mesure probabiliste contextuelle discriminante de qualité des règles d'association. In M.-S. Hacid, Y. Kodratoff, and D. Boulanger, editors, *Extraction et gestion des connaissances*, volume 17 of *RSTI-RIA*, pages 247–262. Lavoisier, 2003.
31. I.C. Lerman, R. Gras, and H. Rostam. Elaboration d'un indice d'implication pour les données binaires, i et ii. *Mathématiques et Sciences Humaines*, (74, 75):5–35, 5–47, 1981.
32. K. McGarry. A survey of interestingness measures for knowledge discovery. *Knowledge Engineering Review Journal*, 20(1):39–61, 2005.
33. G. Piatetsky-Shapiro. Discovery, analysis and presentation of strong rules. In G. Piatetsky-Shapiro and W.J. Frawley, editors, *Knowledge Discovery in Databases*, pages 229–248. AAAI/MIT Press, 1991.
34. G. Ritschard. De l'usage de la statistique implicative dans les arbres de classification. In R. Gras, F. Spagnolo, and J. David, editors, *The third International Conference Implicative Statistic Analysis*, pages 305–315, Palermo, Italia, 2005. Supplément num. 15 de la Revue Quaderni di Ricerca in Didattica.
35. G. Ritschard and D.A. Zighed. Implication strength of classification rules. In F. Esposito, Z.W. Ras, D. Malerba, and G. Semeraro, editors, *16th International Symposium on Methodologies for Intelligent Systems*, volume 4203 of *LNAI*, pages 463–472, Bari, Italy, 2006. Springer.
36. E. Suzuki. In pursuit of interesting patterns with undirected discovery of exception rules. In S. Arikawa and A. Shinohara, editors, *Progresses in Discovery Science*, volume 2281 of *Lecture Notes in Computer Science*, pages 504–517. Springer-Verlag, 2002.
37. E. Suzuki and Y. Kodratoff. Discovery of surprising exception rules based on intensity of implication. In J. M. Zytkow and M. Quafafou, editors, *Principles of Data Mining and Knowledge Discovery*, volume 1510 of *Lecture Notes in Artificial Intelligence*, pages 10–18, Nantes, France, September 1998. Springer-Verlag.
38. P-N. Tan, V. Kumar, and J. Srivastava. Selecting the right objective measure for association analysis. *Information Systems*, 4(29):293–313, 2004.

39. B. Vaillant. *Mesurer la qualité des règles d'associations : études formelles et expérimentales.* PhD thesis, ENST Bretagne, Université de Bretagne Sud, 2006.
40. B. Vaillant, P. Lenca, and S. Lallich. A clustering of interestingness measures. In E. Suzuki and S. Arikawa, editors, *Discovery Science*, volume 3245 of *Lecture Notes in Artificial Intelligence*, pages 290–297, Padova, Italy, 2004. Springer-Verlag.

The *TVpercent* principle for the counterexamples statistic

Ricco Rakotomalala[1] and Alain Morineau[2]

[1] Eric Laboratory – Bron, France
ricco.rakotomalala@univ-lyon2.fr
[2] Modulad – Rocquencourt, France
alain.morineau@modulad.fr

Summary. Our aim is to put into practice the principle of test value percent criterion to the counterexamples statistic, which is the basis of the well-known statistical implicative analysis approach. We show how to compute the test value in this context; what is the connection with the intensity of implication measure, on the one hand; and the index of implication, on the other hand. We evaluate the behavior of these measures on a large dataset comprising several hundred of thousands of transactions. We evaluate especially the discriminating capacity of the measures, in relation to specialized measure such as the entropic intensity of implication.

Key words: Association rule, Measure, TVpercent, Intensity of implication.

1 Introduction

Since the work of Agrawal and Srikant (1994) [1], the association rule mining has received a great deal of attention and is became one of the most popular method in the knowledge discovery community. This approach allows to produce implication rules such as "If A Then C", where A and C are sets of items or products in the analysis of market basket data. The meaning of the rule is "whenever a set of transactions contains A, than it probably contains also C".

Even if the association rule mining is very powerful, there is a pitfall which can call into question its use: the number of generated rules can be very high, it becomes difficult to distinguish the most interesting rules [13]. In this context, it is important to have a numerical indicator which makes it possible to propose the most relevant rules quickly, but also to validate them, so as to keep only the rules which show a real causation. There are many proposals of rule quality measurements these last years. Among them, we are interested in the intensity of implication measure based on the counterexamples statistic [8, 9].

In order to transform the regularity, the concomitant occurrence of the itemsets, into causation, the implication rule, we count the counterexamples

R. Rakotomalala and A. Morineau: *The TVpercent principle for the counterexamples statistic*, Studies in Computational Intelligence (SCI) **127**, 449–462 (2008)
www.springerlink.com

of the rule. The rule "If A Then C" is all the more relevant as it has few counterexamples. The intensity of implication is a measurement that is based on this idea. It was implemented in many domains. It is available on several bench test softwares [14, 23].

The intensity of implication measure relies on a classical hypothesis-testing scheme. The idea is not to test the absence or the presence of a real link between A and C, but rather to what extent we deviate from the reference situation described by the null hypothesis. In this context, we often compute the p-value of the test. It shares the property of any classical statistical index in data mining context: for the same constant proportion, the value inopportunely increases with the number of observations. In certain situations, when the number of examples is very high, the p-value cannot be computed correctly because we exceed the accuracy of the common statistical libraries. Thus, all the rules correspond to the maximum value of measurement. It is not possible to distinguish the relevant rules.

In this paper, we put into practice the test-value percent principle on the counterexamples statistic. The test-value is the expression of the p-value into the number of standard deviation of the Gaussian distribution [15, 17]. In a recent paper, we proposed a normalized version of the Test-Value named *TVpercent* criterion [18]. Applications to association rules proved it to be an interesting criterion to eliminate statistically uninteresting rules without being influenced by the number of occurrences. This criterion remains comprehensible and discriminating even if we treat a huge database. It suggests a threshold that allows to eliminate irrelevant rules. It enables to compare rules from different databases.

The organization of this paper is as follows. In section 2, we recall the computation formulas for the intensity of implication which is associated to the counterexamples statistic. In the section 3, we quickly show the TVpercent framework. Then, we extrapolate the calculation of the TVpercent to the counterexamples statistic. A new measure, v_e, is described. In section 4, we study the behavior of this measure on a large size database (340183 transactions and 468 items). We compare the TVpercent to the standard intensity of implication and the entropic intensity of implication which is designed for rules with high supports [2]. We conclude in the section 5.

2 Intensity of implication and counterexamples statistic

2.1 Computing the intensity of implication

The intensity of implication is based on the counterexamples statistic. In a rule of the type "If A then C", we check if the number of observed counterexamples is significantly infrequent. We use a statistical hypothesis testing framework. We must then define [22]: a parameter which we estimate; a formulation of the null hypothesis, the reference situation; the statistical distribution of the

Rule	A	$\bar{A}$	Total
C			
$\bar{C}$	$n_{a\bar{c}}$		$n_{\bar{c}}$
Total	n_a		n

Table 1. Contingency table — Number of transactions for a rule "If A then C"

parameter estimate under the null hypothesis; and then, an indicator which measures the deviation of the observed data from the null hypothesis.

In the Gras' approach [7], the statistical parameter is the number of counterexamples $N_{a\bar{c}}$ to a rule. Its estimation is naturally the number of observed counterexamples $n_{a\bar{c}}$ (Table 1). The null hypothesis is the independence between the antecedent and the consequent of the rule. In this situation, the probability $\pi_{a\bar{c}}$, i.e. e probability for obtaining counterexamples if the antecedent of the rule is true, is equal to $\pi_a \times \pi_{\bar{c}}$. It is the product of two marginal probabilities, they can be estimated by $\frac{n_a}{n} \times \frac{n_{\bar{c}}}{n}$. The expectation of the random variable $N_{a\bar{c}}$ under the null hypothesis is $\Lambda = n \times \pi_a \times \pi_{\bar{c}}$, estimated by $\lambda = n \times \frac{n_a}{n} \times \frac{n_{\bar{c}}}{n}$.

Although we have a hypothesis testing framework, the goal is not to accept or reject the null hypothesis, but the characterization of the deviation from this reference. In our situation, we want to characterize in what extent the observed number of counterexamples $n_{a\bar{c}}$ is less than λ. Various modeling approaches are available. We can use a hypergeometric, binomial or Poisson distribution. We mainly study here the third model with the Poisson distribution [16]. More than a simple approximation of the other distributions, this sampling scheme is interesting because it treats in a non symmetrical way the positive and negative associations, which enables to show a causation.

The p-value, the probability of obtaining a result at least as extreme or impressive as that obtained for the data, assuming the null hypothesis is true, is computed with a Poisson distribution with the parameter λ. The critical region for $N_{a\bar{c}}$is defined as the interval $[0, n_{a\bar{c}}]$. The intensity of implication, I_e, is the complement to 1 of the p-value of the test:

$$I_e = 1 - \sum_{m=0}^{n_{a\bar{c}}} \frac{\lambda^m}{m!} e^{-\lambda} \tag{1}$$

Numerical example We use an example described in the Ritschard's paper [20]. We want to characterize a rule where $n = 273$, $n_a = 76$, $n_{\bar{c}} = 153$, $n_{a\bar{c}} = 28$ (Table 2), and $\lambda = 42.59$, the intensity of implication is $I_e = 0.9884$.

2.2 Practical computation of the intensity of implication

When λ is large, about 18, it is possible to use the Gaussian approximation of the Poisson distribution, the mean and the variance are both λ. We then

Rule	A	$\bar{A}$	Total
C			
$\bar{C}$	28		153
Total	76		273

Table 2. Example of a contingency table from Ritschard [20]

compute the index of implication, which is the standardized value of the observed number of counterexamples: we use a continuity correction factor.

$$i_e = \frac{n_{a\bar{c}} + 0.5 - \lambda}{\sqrt{\lambda}} \tag{2}$$

The approximation of the intensity of implication using the Gaussian CDF (cumulative distribution function) Φ is defined as follows:

$$I_a = 1 - \Phi[i_e] \tag{3}$$

Numerical example We take again the above example (Table 2). The index of implication is $i_e = \frac{28+0.5-12.59}{\sqrt{42.59}} = -2.16$. The approximated intensity of implication is $I_a = 0.9846$. We note that the true intensity (I_e) and the approximated intensity (I_a) are similar. Very often, only the approximate formulation is referred in publications.

There is a drawback for the utilization of the intensity of implication I_a if the support of the rule n_{ac} is large: the computed value of I_a is mechanically equal to 1 because the available libraries to compute the Gaussian cumulative distribution function (CDF) are not enough accurate. For instance, if the index of implication i_e is less than -6.2, the Excel© spreadsheet gives systematically an intensity of implication equal to 1. We tested several libraries [11], none could not significantly exceed this limitation.

Using the original formulation —I_e, Poisson distribution, equation (1)— can slightly improve the results. But we deal with very small values of which are badly handled by computation libraries. In our example (Table 1), if we multiply all the values in the table by 4, counts would be equal to $n = 1092$ and $n_{a\bar{c}} = 112$. These values are not excessive. However, we find with the exact formula $I_e = 0.99999879$, and with the approximate formula $I_a = 0.999995367$. It becomes very difficult to distinguish the interesting rules. This problem is designed by the loss of the discriminating power of measurements. Elegant solutions were suggested, especially with the concept of the entropic intensity of implication where we balance the intensity of implication with an index of inclusion [10]. We present this approach below in the experimentation section.

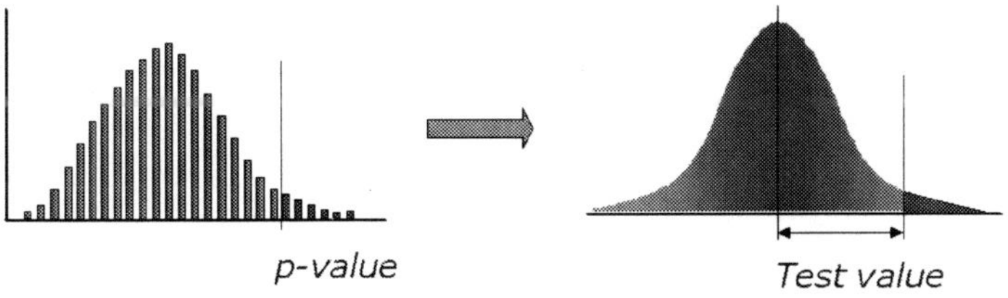

Fig. 1. From the p-value to the test value

3 TVpercent criterion based on the counter-examples statistic

3.1 Test value and normalized test value

Test value The test value also relies on a statistical framework. We try to compare the observed parameter with the theoretical parameter under the null hypothesis, which is the independence between the antecedent and the consequent of a rule.

We mainly use the p-value p to characterize the strength of the deviation between the observed number of counterexamples and the theoretical number of counterexamples under the reference situation. The p-value can take very small values, close to zero, thus not very comprehensible as soon as one deviates from the situation of reference with a large database size. In order to obtain a better adapted measure, easily interpretable, we replace it by the number of standard deviations of the standardized Gaussian distribution which should be exceeded to cover the computed p-value (Figure 1). We call the test value this criterion (equation 4).

$$TV = \Phi^{-1}(1-p) \tag{4}$$

This criterion is often used to compare proportions or conditional average for the characterization of the clusters built with clustering process [15].

Numerical example In our example (Table 2), the computed p-value for $n_{a\bar{c}} = 28$ and $\lambda = 42.59$ with the Poisson distribution is $p = 1 - I_e = 0.0116$, the corresponding test value is $TV = 2.2701$. This value is comparable to the index of implication of which the negative, v_e, can be considered as a rough approximation of the test value $v_e = -i_e = 2.16$.

TVpercent — A normalized test value In the context of the knowledge discovery, we handle a very large size databases, larger than the usual sample size in statistical inference. We must treat two kinds of problems: as we have noted it, we obtain a very small p-value, hardly handled by the algorithms of the standard libraries for CDF computation; more disturbing is a well known phenomena by the statisticians, i.e. when the sample size increases, a small deviation from the values of the parameter under the null hypothesis becomes significant, even if it corresponds to a statistical artifact.

Numerical example For instance, we assume that the number of counterexamples is equal to $n_{a\bar{c}} = 40$. The computed p-value is 0.37420 ($I_a = 1 - 0.374 = 0.62580$). The rule seems not to be relevant. When we multiply all the values of the table by 10, we obtain the $p = 0.10890$; if we multiply by 100, the p-value becomes $p = 0.00004$. It seems now that the rule is very relevant and the association is strong.

In order to avoid this pitfall, we have proposed a normalized test value [18]. The measure becomes independent of the real size of the database. We do not forget that the main goal of the measure is to rank the rules in decreasing relevance, and secondarily to suggest a cut value below which we can consider that the rule does not bring relevant information.

The main idea is to set *a priori* the size of the dataset to 100. This value corresponds to a reasonable size of the samples used when statistical inference and hypothesis testing were historically developed. The value 100 is surely an arbitrary value. But it is not more arbitrary than the usual confidence level used in statistical inference (e.g. 5%, 1%, etc.). These confidence levels are the results of the experiments of Fisher [6]. Indeed, in a not well-known process depicted by Poitevineau [19], Fisher had hesitated for setting the right value of the confidence level [5,6]. In effect, the appropriate value of the confidence level relies on the studied problem, the goal of the statistician, and the characteristics of the dataset, especially the dataset size. From this point of view, a criterion which allows to sorting the rules is surely essential. Using the same criterion in order to mechanically accept or reject a rule is doubtful.

The original process to compute the normalized test value is a Monte Carlo sampling approach. We draw randomly with replacement 100 examples from the database and we compute the p-value $p = 1 - I_e$ from equation (1) for the corresponding 2×2 contingency table (Poisson approximation). We repeat this process and then compute the average of the p-values $\bar{p}$. In the last step, the normalized test value is computed from this average $TV_{norm} = \Phi^{-1}(1 - \bar{p})$. If we use a sufficient number of repetition (e.g. 2000 samples of 100 examples with replacement), we obtain a stabilized value of the test value.

This process is known also as the bootstrap procedure, but the size of the sample is arbitrarily set to 100 here, and not equal to the dataset sample size.

This criterion makes it possible to rank the rules computed on a database. It has also the advantage, since it proceeds to the evaluation of the rules in an unique reference (100 examples), of allowing the comparison of rules computed on several similar databases, for example, on databases of different size extracted on successive dates.

Practical computation of the TVpercent

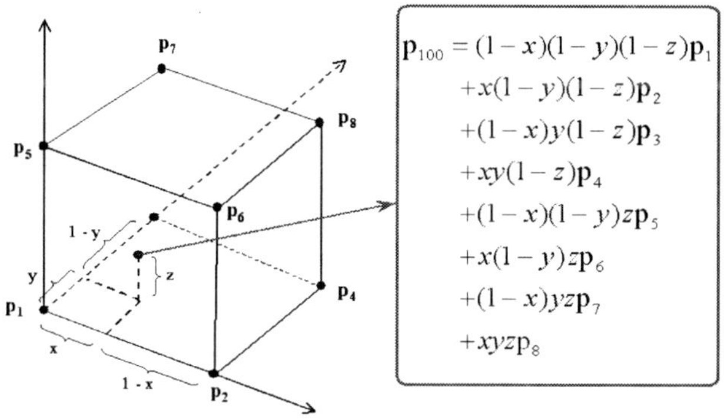

Fig. 2. Barycentric approximation of the TVpercent

Règle	A	$\bar{A}$	Total
C			
$\bar{C}$	10.25		56.04
Total	27.83		100

Table 3. Table brought back to 100 of table 2

The Monte Carlo migh be a good approach in order to obtain a stabilized value of the criterion, but it is time consuming, especially because it needs a repeated access to the database. We thus must use an approximation which is quickly computed, it must rank the rules in the same order than the original approach.

We use an interpolation procedure in order to compute the TVpercent. In the first step, we bring back the values of the contingency table to 100 (e.g. from table 2 to table 3). In the second step, we compute the p-value for each integer value which surrounds the decimal values in the table (e.g. 10.25 is surrounded by 10 and 11). Then we compute the average of 8 p-values obtained for the 8 corners of the cube (Figure 2), which correspond to the 8 contingency tables computed from the integer values. From this average p_{100} of p-values, we compute the test value which is the TVpercent, equation (5).

$$TV_{precent} = \Phi^{-1}(1 - p_{100}) \tag{5}$$

The approximation can appear naive. But it is sufficient for our goal which is to compare the relevance of the rules and to rank them [18]. We can use a more accurate procedure, but it should not to be at the expense of the computing time which is a major constraint in our context.

3.2 TVpercent criterion on the counter-examples statistic

In our first work, we use the TVpercent criterion for the co-occurrence of the antecedent and the consequent (n_{ac}). We used a hypergeometric distribution

$n_{a\bar{c}}$	n_a	$n_{\bar{c}}$	$p-value$
10	27	56	0.1127
10	27	57	0.1007
10	28	56	0.0890
10	28	57	0.0788
11	27	56	0.1769
11	27	57	0.1602
11	28	56	0.1437
11	28	57	0.1290

Table 4. The 8 configurations to evaluate for the barycentric estimation

[18]. In fact, the TVpercent principle can be extented to other parameters, such as the counterexamples statistic and the Poisson distribution.

The detail of the process is the following. We bring back the original contingency table (e.g. Table 2) to $n = 100$ (e.g. Table 3). We enumerate the 8 configurations which surrounds the decimal values. For each configuration, we compute the p-value using the Poisson distribution (e.g. Table 4). We compute the average of these p-values using the barycentric approximation (Figure 2). Then, the test value is computed from the Gaussian CDF.

Numerical example From Ritschard's example (Table 2), the 8 configurations to evaluate are displayed in the table 4. The average of the p-values is $p_{100} = 0.1410$ and the corresponding test value is $TVpercent = \Phi^{-1}(1 - 0.1410) = 1.0759$.

3.3 Test value and index of implication

When we handle large database, I_a, using the Gaussian CDF in order to get an approximation of the intensity of implication, is widely used. In our context, is the index of implication i_e is a good approximation of the test value?

For a good understanding of the problem, we must not forget that the test value $TVpercent$ is computed from the p-value p_{100} with a Gaussian CDF (symmetrical distribution), the p-value is computed from the statistical formulation with a Poisson distribution (*a priori* non symmetrical distribution). The approximation between the test value and the standardized counterexamples statistic is satisfactory if these two distributions are symmetrical. The Poisson distribution becomes approximately symmetrical when we have a high value of the λ parameter i.e. we have a large database. In our context, because we bring back n to 100, the approximation is not accurate.

Numerical example In the Table 3, $\lambda = \frac{27.83 \times 56.04}{100} = 15.6$. The negative index of implication is $v_e = -i_e = -\frac{10.25+0.5-15.6}{\sqrt{(15.6)}} = 1.2267$. The deviation from the TVpercent (1.0759) is considerable.

4 Experiments

4.1 Description of the experimentation

Database In order to evaluate the behavior of the various measures, we use a large size database (ACCIDENT) which counts accident locations [12]. The number of transactions is 340183, and the number of items is 468. Our goal is to study the behavior of the $TVpercent$ criterion in relation to state of the art measures such as intensity of implication and entropic intensity of implication.

Intensity of implication and entropic intensity of implication We want to compare the TVpercent criterion with the intensity of implication I_a measure in a large database context. It seems also interesting to compare our measure to the version of the intensity of implication dedicated to the large databases, it is the entropic intensity of implication [2, 10]. It is defined by

$$IE_a = \sqrt{I_a \times h} \tag{6}$$

where h, the index of inclusion, is equal to

$$h = \sqrt{(1 - H(\frac{n_{ac}}{n_a})) \times (1 - H(\frac{n_{\bar{a}\bar{c}}}{n_{\bar{c}}}))}$$

$H(x)$ is a modified Shannon entropy function [10]

- if $x < 0.5$, $H(x) = 1 + 0.5 \times [x \times \log_2(x) + (1 - x) \times \log_2(1 - x)]$;
- if $x \geq 0.5$, $H(x) = -0.5 \times [x \times \log_2(x) + (1 - x) \times \log_2(1 - x)]$

Computing the rules and the measures We use the Borgelt's implementation for the rule generation [3]. It is available on the web site of the author[3]. Its implementation is very efficient but it computes rules with one item only in the consequent. The parameters of the software are classical, we can choose the minimum support, the minimum confidence and the maximum length (number of items) of the rules.

The rules are then post processed in the Excel© spreadsheet. We compute the various measures that we want to evaluate in this paper (I_e, I_a, IE_a, $TVpercent$, i_e). In spite of some doubt about the accuracy of this spreadsheet, our background shows that the available very competitive specialized libraries[4] are not really more accurate. This spreadsheet is considered to be adequate in our exploratory study.

Evaluation framework Our aim is to check the concordances and the discordances between our measure ($TVpercent$) and the state of the art measures [10]. The first approach is to check if the various measures rank the rules in the same way. A scatter plot allows to check that. We can also compute

[3] `http://fuzzy.cs.uni-magdeburg.de/~borgelt/apriori.html`
[4] e.g. STATLIB library — `http://lib.stat.cmu.edu/index.php`

Indicator	Value
Number of transactions	340183
Number of items	468
Maximum length of rules	3
Support minimum	10%
Confidence minimum	75%
Number of rules	17212

Table 5. Characteristics of our experiments on the ACCIDENT dataset

a numerical indicator such as the Spearman's rank correlation coefficient [21] (correlation computed on ranks). But, if it is easy to calculate, we must use it with caution, a numerical indicator often hides a complex situation such as a non linear relation, etc. We will use it only as a rough indication.

Because in real studies, the rules are validated by experts. It is important to carefully study how the first rules are ranked by the measure. An expert, even if he is really persevering cannot check a hundred rules. In this paper, we focus on the first 20 rules. This kind of evaluation is similar to ROC curve analysis where, in certain circumstances, the AUCn criterion, focused on the first n examples, is more relevant [4].

4.2 Results and comments

The characteristics of the database and the computed rule set are described in the Table 5. The parameters of the algorithm have been chosen after several attempts. We note that the results of the various attempts are not in contradiction with the results presented here.

With a minimum support of 10%, the support of the rules runs from 34018 to 340183 transactions. The number of counterexamples $n_{a\bar{c}}$ of a rule runs from 0 (no counterexample) to 81742 (the support of the rule is 258408 in this situation). In this context, the accuracy of the computation is very important for the used measures.

The exact intensity of implication I_e: The exact formulation of the intensity of implication, equation (1), can be computed only on 2864 rules (among 17212 rules). The Excel© implementation of the Poisson CDF cannot handle some values. Even if we use some tricks to improve the accuracy, we doubt we can really improve the exact formulation on a large database.

The approximated intensity of implication I_a: The approximate formulation using the Gaussian CDF, equation (3), is more robust. It can be computed on all the rules. Another drawback appears. When the index of implication is very small (in some situations it can be equal to -209.9), the Gaussian CDF implemented in the spreadsheet is mechanically equal to 0. So the approximate intensity of implication is 1 for 9960 rules among 12712.

At the beginning, we have thought that there was a specific problem of the spreadsheet. But we found the same limitations with the specialized libraries

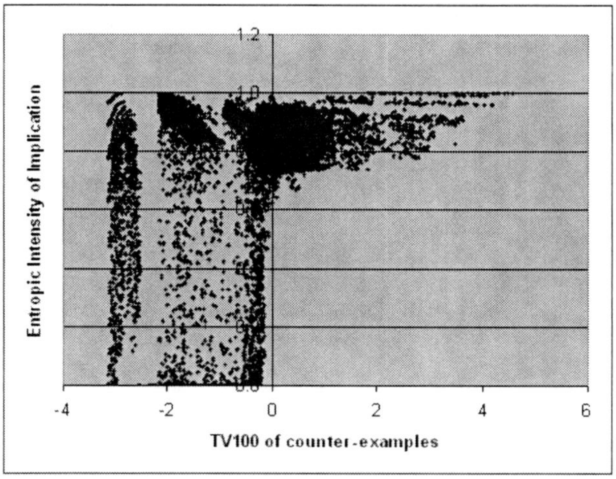

Fig. 3. Scatterplot of the TVpercent and the entropic intensity of implication

implemented in very powerful languages for numerical calculation such as FORTRAN[5]. For $i_e \leq -6.2$, the p-value cannot really be computed from the index implication with a Gaussian CDF. Using a measure which is derived from I_a for detecting the relevant rules is not à powerful approach in this context.

The entropic intensity of implication IE_a: The entropic intensity of implication, equation (6), allows to significantly overcome this drawback. The index of inclusion h takes over, to some extent, the intensity of implication when this one is saturated. So, in our experiments, the number of rules where $IE_a = 1$ is 79 (among 12712). The discriminating power of the measure is maintained. But there is some problem nevertheless. Indeed, if we rely on a manual expertise, the presentation of the first 20 rules will depend primarily on the sorting algorithm and not on the sorting criterion, this is not very satisfactory.

TVpercent for the counterexamples statistic: The computation of the TVpercent is theoretically slower than the other measures since we compute 9 CDF (8 Poisson distribution and 1 Gaussian distribution). But it is not really perceptible on our set of rules. None capacity overflow have been observed. The TVpercent has a range between -3.12 to 4.58. The rules which have equal TVpercent are those of which we observe the same values (n_a, $n_{\bar{c}}$, $n_{a\bar{c}}$). When we compare the TVpercent measure with the other measures, we note that the correlation with the entropic intensity of implication is small (0.21). There is little concordance between these measures (Figure 3).

[5] e.g. STATLIB library (http://lib.stat.cmu.edu/index.php). The available implementations are described in a book [11].

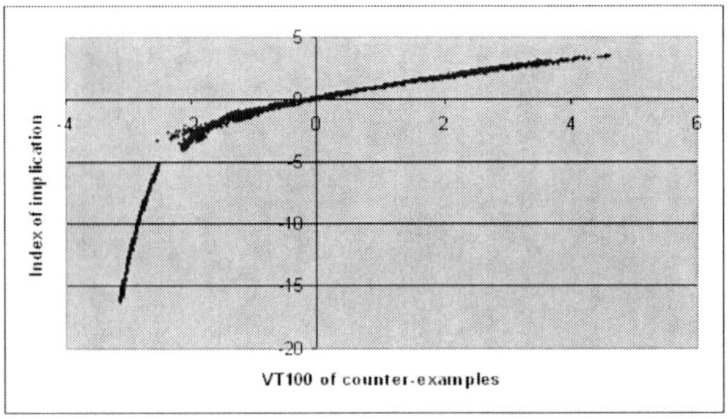

Fig. 4. Scatterplot of Index of implication and TVpercent

When we deeply studied the results, we found that we have not a symmetrical situation. On the 20 first rules according to the TVpercent, 12 have the maximal value of entropic intensity of implication $IE_a = 1$. At the opposite, there are 79 rules with $IE_a = 1$, the best rules according to the TVpercent are hidden among these rules.

TVpercent and the normalized index of implication We had already noted above the similarity between the index of implication and the TVpercent, we had also noticed that the approximation was not very precise on small dataset. What is the case when we bring back the values to $n = 100$. Nevertheless, we calculated the negative of the index of implication directly on the sample size brought back to 100. Indeed, even if the approximation of the TVpercent with the index of implication is bad, perhaps they rank the rules in a similar way?

The scatterplot shows that there is little discordance between the two measures, even if the approximation is not really accurate. The relation between these measure is clearly non linear but monotonic (Figure 4). This visual impression is corroborated with the correlation computed between the rank of the two measures (the Spearman rank correlation) which is equal to 0.999. When we focus on the 20 first rules according to the normalized index of implication, we have found the 18 best rules according to the TVpercent criterion. In a real situation where we present the rules to a human expert, these two criteria will approximately propose the same rules.

The principal difference between these two indicators is in the determination of the statistical valid rules. If we use critical values associated to usual significance levels (e.g. 2.32 for a significance level of 1%, etc.), because the TVpercent is always larger, we keep more rules with the TVpercent than the normalized index of implication. It is not really a disadvantage of the normalized index of implication. We have seen that using an arbitrary threshold value in order to keep or remove rules must be made with caution. If the user wants nevertheless to use this procedure, he must take into account this difference when he specifies the thresholds in order to avoid the elimination of interesting rules.

5 Conclusion

In this paper, we have generalized the TVpercent criterion to the counterexamples statistic. The main improvement of this new measure is the handling of the rules computed on large databases. We preserve the discrimination power of the measure i.e. the ability to rank rules without ties according to the measure. We can rank a great number of rules. In this way, it extends the field of application of the intensity of implication and constitutes an alternative to the entropic intensity of implication.

The second main result of this work is the similarity between the TVpercent and the normalized index of implication. Although the index underestimates the true value of the TVpercent, it ranks the rules in same way, and most of all, it points up the same rules if we are interested in the best rules according to these criteria.

Of course, these conclusions rely mainly on experimental evaluation. We studied the behavior of these measures on various parameters settings of the rule extraction algorithm, corresponding to the post processing of more or less large number of rules. The results described above were not called into question. However, it would be interesting to complete this study on other databases with different characteristics, for example with few transactions and a very large number of items.

References

1. R. Agrawal and R. Srikant. Fast algorithms for mining association rules. In *Proceedings of the* 20th *VLDB Conference*, pages 487–499, 1994.
2. J. Blanchard, P. Kuntz, F. Guillet, and R. Gras. *Mesures de qualité pour la fouille de données*, volume E-1, chapter Mesure de qualité des règles d'association par l'intensité d'implication entropique. Revue des Nouvelles Technologies de l'Information, 2004.
3. C. Borgelt and R. Kruse. Induction of association rules: A priori implementation. In *15th Conference on Computational Statistics*, 2002.
4. T. Fawcett. Roc graphs: Notes and practical considerations for researchers. Technical Report - HP Laboratories, 2003.
5. R.A. Fisher. *Statistical Methods, Experimental Design, and Scientific Inference*, chapter The design of experiments. Oxford University Press, 1990. 1st edition, 1935, London, Oliver and Boyd.
6. R.A. Fisher. *Statistical Methods for Research Workers.* Oxford University Press, 14 edition, 1990. 1st edition, 1925, London, Oliver and Boyd.
7. R. Gras. Contribution à l'étude expérimentale et à l'analyse de certaines acquisitions cognitives et de certains objets didactiques en mathématiques. Thèse d'Etat, 1979.

8. R. Gras. *L'implication statistique - Nouvelle méthode exploratoire des données.* La pensée sauvage, 1996.
9. R. Gras. *La fouille dans les données par la méthode d'analyse implicative*, chapter Les fondements de l'analyse statistique implicative. Ecole polytechnique de l'Université de Nantes, IRIN et IUFM Caen, 2004.
10. R. Gras, P. Kuntz, R. Couturier, and F. Guillet. Une version entropique de l'intensité d'implication pour les corpus volumineux. In *Actes de la Conférence Extraction et Gestion des Connaissances, EGC'2001*, pages 69–80. Revue Extraction et Gestion de la Connaissance, 2001.
11. P. Griffiths and I.D. Hill. *Applied Statistics Algorithms.* Ellis Horwood: Chichester, 1985.
12. K. Guerts, G. Wets, T. Brijs, and K. Vanhoof. Profiling high frequency accident locations using association rules. In *Proceedings of the 82nd Annual Transportation Research Board*, 2003.
13. F. Guillet, editor. *Mesures de la qualité des connaissances en ECD.* EGC'2004, RNTI, 2004.
14. X-H. Huynh, F. Guillet, and H. Briand. Une plate-forme exploratoire pour la qualité des règles d'association : apports pour l'analyse implicative. In *Actes des Troisièmes Rencontres de l'Analyse Statistique Implicative*, pages 339–349, 2005.
15. L. Lebart, A. Morineau, and M. Piron. *Statistique Exploratoire Multidimensionnelle.* Dunod, Paris, 1995.
16. I. Lerman, R. Gras, and H. Rostam. Elaboration et évaluation d'un indice d'implicatioon pour données binaires. *Mathématique et Sciences Humaines*, (74):5–35, 1981.
17. A. Morineau. Note sur la caractérisation statistique d'une classe et les valeurs-tests. *Bull. techn. du Centre de Statis. et d'Infor. Appl.*, (2):20–27, 1984.
18. A. Morineau and R. Rakotomalala. Critère vt100 de sélection des règles d'association. In *Actes de Extraction et Gestion de Connaissances, EGC'2006*, pages 581–592, 2006.
19. Jacques Poitevineau. L'usage des tests statistiques par les chercheurs en psychologie : aspects normatif, descriptif et prescriptif. *Mathématiques et Sciences Humaines*, 3(167):5–25, 2004.
20. G. Ritschard. De l'usage de la statistique implicative dans les arbres de classification. In *Troisième Rencontre Internationale - Analyse Statistique Implicative*, pages 305–316, 2005.
21. G. Saporta. *Probabilités, Analyse de Données et Statistique.* Technip, 1990.
22. B. Vaillant, S. Lallich, and P. Lenca. Modeling the counter-examples and association rules interestingness measures behavior. In S.F. Crone, S. Lessman, and R. Stahlbock, editors, *The 2006 International Conference on Data Mining*, pages 132–137, 2006.
23. B. Vaillant, P. Picouet, and P. Lenca. An extensible platform for rule quality measure benchmarking. In *Human Centered Processes (HCP'03), Distributed Decision Making and Man Machine Cooperation*, pages 187–191, 2003.

User-System Interaction for Redundancy-Free Knowledge Discovery in Data

Rémi Lehn, Henri Briand, and Fabrice Guillet

Laboratoire d'Informatique de Nantes Atlantique
Equipe COnnaissances & Décision
Site Ecole Polytechnique de l'Université de Nantes
La Chantrerie — BP 50609 — 44306 Nantes cedex 3
{remi.lehn, henri.briand, fabrice.guillet}@univ-nantes.fr

Summary. A classical limit of association rule at the decider's point of view is in the combinatorial nature of the association rules, resulting in numerous rules. As the overall quality of an association rule set can be considered as insight of the studied domain given to the decider by the interpretation of the rules, too many rules can make an harder interpretation then a worse quality of the overall process.

To get more readable rules and thus improve this global quality criterion, we apply techniques initially designed for redundancy reduction in functional dependencies sets to association rules. Although the two kinds of relations have different properties, this method allow very concise representations that are easily understood by the decider and can be further exploited for automatic reasoning.

In this paper, we present this method, compare it to other approaches and apply it to synthetic datasets. We end with a discussion about the information loss resulted of the simplification.

Key words: Minimal Covers, Closure, Interpretation of Association Rules, Deductive Reasonning

1 Introduction

The amount of collected data grows continuously. Decision tasks performed must take this growth into account to deal with prediction, action evaluation or validation, in the context of a large variety of application fields like management, profit optimization or analysis. The KDD (Knowledge Discovery in Databases) area scopes this range of applications in the goal of providing automated tools and adapted data representations to help an expert user in finding the evidences needed for the decision tasks. This assumes a human centered KDD process. As a human centered process involving automated procedures, it needs a targetted problem representations that are both realistic from the user's point of view and computable from a machine point of view.

R. Lehn et al.: *User-System Interaction for Redundancy-Free Knowledge Discovery in Data*, Studies in Computational Intelligence (SCI) **127**, 463–479 (2008)
www.springerlink.com

Among KDD techniques, association rules [2] allow the capture and the representation of implicative patterns that tolerate a small set of counter-examples —e.g. birds that cannot fly or sport cars that are not red. Association rules can be enhanced with statistical evaluations and filters such as the Intensity of Implication family of indices.

Association rule discovery is motivated by the exploitation of operational databases to discover a new knowledge, that was unknown before the discovery and that is potentially exploitable in a decision making process [19]. Many performant algorithms have been published to optimize the association rules search [8, 16] but they mainly focus on algorithmic optimization rather than on knowledge usability.

One of the fundamental hypothesis of association rule discovery is that the user does not specify the goal of the search. Because of the intrinsically combinatorial nature of the search and the lack of the goals, the classical use of these algorithms, chaining data selection, data formatting, frequent sets induction, rules calculation and rule presentation to the user, generally outputs quantities of rules, without order of any kind, which is in contradiction with the principle of knowledge readability and usability for a decision process. Experiments using a direct application of association rules algorithms like *A Priori*, resulted in thousands of rules. We can then seriously contest the quality of the vision of the studied domain provided by the association rules to the user if he has to explore thousands of rules. We can contest as well the quality of the induction itself if the energy that the user has to involve to interpret the association rules is nearly the same as the energy he would have to deploy to get the same domain understanding by directly browsing the database.

A classical answer to this problem is to set high thresholds on quality indices that evaluate individual rules, to eliminate the least pertinent rules as measured by these indices. But there are cases where this strategy cannot be applied: when the user doesn't know where to set thresholds corresponding to the kind of knowledge he is looking for or when he's looking for knowledge with properties other than those matched by the available indices, or when there are a lot of hidden dependencies in the data. Further global criteria have been proposed, in addition to those measured on individual rules:

- operational criteria, in precise decision making tasks [6], but, these criteria are specific and domain-dependent.
- Readability criteria, whose precise evaluation is based on a cognitive qualification of the user's perception of the represented knowledge. This criterion depends of the visualization interfaces and their adaptation to the decision making tasks. If we assume a linear, non-interactive, acquisition of the knowledge by a decision maker, the limitation of the amount of represented rules, in association with a reading convention, is an important factor of improvement of these criteria.

- The exploitation of the rules for automated tasks such as inference engines. In this case, specific properties of the rules for inference, for example the respect of logical properties, are evaluated as knowledge quality.

To meet those criteria, we propose to limit the number of association rules by not representing rules that can be inferred by the user himself in a logical reasoning. So eliminated rules are then considered as redundant as opposed to the other represented rules. This redundancy elimination considers a global criterion that is a complement to the evaluation of the quality of each individual rule.

The redundancy elimination strategy strongly depends on the theory of the represented knowledge, including a definition of the inferences that can be made by the user reading the knowledge representation. Several models have been proposed and some of these models are coupled with discovery algorithms [12,13,15,25]; other models are coupled with specific representation models such as Galois Lattices [4, 5, 10, 23, 24, 29], or with measures that allow approximating frequent itemsets [1].

Our proposed representation model is based on logical properties with the assumption of an implicative behavior of association rules. It is based on a closed itemset algorithm (see Ceglar and Roddick, 2006 [8] for a definition of this class of association rule mining algorithms), with a pure logical construction of the closure relation. This hypothesis is enforced with the assumption that a user or an automated deduction system will make logical deductions using the ruleset during the interpretation of the rules.

Efficient methods have been proposed for redundancy elimination in functional dependencies sets and functional dependencies are known to support logical properties [27]. Thus we apply one of these methods, the minimal covers, to association rules filtering. This filtering is very efficient as it gives very minimal representations, but there exists some issues where association rules do not respect the logical assumptions of our model and for which the redundancy reduction gives over-generalized rules, that, while being in respect with a logical reasoning, are in contradiction with some statistical measures of the rules. We detail in this paper these cases, and show, using synthetic examples, the information loss encountered using our method.

2 Redundant functional dependencies

Important work has been realized in the past for redundant functional dependencies elimination in relational databases, for example Ullman in the early 80's [3, 7, 9, 26, 27]; the results of this work have been directly embedded into tools that allow the automatic production of a functional dependencies representation from relational databases. *Dep-Miner* [20] is an example of such a tool.

2.1 Definitions

Functional dependency and Armstrong's axioms :

There exists a functional dependency (FD) $A \rightarrow B$ between a subset A of the attributes of a relation R and another subset of attributes B if the relation R associates one and only one set of values of attributes from B for each possible combination of values of the attributes of A [21,27]. For example, the relation R defined as:

R		
a	b	c
a1	b1	c1
a2	b1	c2
a3	b2	c3
a3	b2	c4

holds a functional dependency $a \rightarrow b$ because each different value of a is associated with one and only one value of b. This relation holds a functional dependency $c \rightarrow ab$ too.

The three main axioms for calculus over functional dependencies systems are the Armstrong's axiom [28]:

$$\text{Reflexivity}: \models A \cup B \rightarrow A. \quad (1)$$
$$\text{Augmentation}: A \rightarrow B \models A \cup C \rightarrow B \cup C. \quad (2)$$
$$\text{Transitivity}: A \rightarrow B, B \rightarrow C \models A \rightarrow C. \quad (3)$$

A quite direct application of the *augmentation* axiom to association rules was proposed by Toivonen et al., 1995 [25].

Closure on a FD set :

The closure on a FD set, $F = (A, \cup, \rightarrow)$, is defined on an attribute set A as the set F^+ of FD that can be written from F by the repeated application of Armstrong's axioms (definition from Ullman, 1989 [28]). Two FD sets F_1 and F_2 are equivalent if $F_1^+ = F_2^+$. The closure of a subset $X \subset A$ of attributes over a FD set $F = (A, \cup, \rightarrow)$ is the set

$$X_F^+ = \cup_i \{Y_i \mid (X \rightarrow Y_i) \in F^+\} \quad (4)$$

For example, the closure on the FD set $F = \{a \rightarrow b, ab \rightarrow c\}$ is the set $F^+ = \{a \rightarrow a, ab \rightarrow a, ac \rightarrow a, abc \rightarrow a, b \rightarrow b, ab \rightarrow b, bc \rightarrow b, abc \rightarrow b, c \rightarrow c, ac \rightarrow c, bc \rightarrow c, abc \rightarrow c, ab \rightarrow ab, abc \rightarrow ab, ac \rightarrow ac, abc \rightarrow ac, bc \rightarrow bc, abc \rightarrow abc, a \rightarrow b, a \rightarrow ab, ac \rightarrow bc, ac \rightarrow abc, ab \rightarrow c, ab \rightarrow ac, ab \rightarrow bc, a \rightarrow c, a \rightarrow ac, ab \rightarrow abc\}$. Note that, among others, $a \rightarrow c$ is member of the closure.

2.2 Functional dependency decomposition

FD can be rewritten using union and decomposition as defined by the following equivalence:

$$\{A \rightarrow B\} \equiv \{A \rightarrow \{b\} \mid b \in B\}. \tag{5}$$

Functional dependencies decomposition consists in a FD set rewriting as the right hand side of this equivalence (5). There are two advantages in this rewriting: eliminating the redundancy dues to union and decomposition and simplifying the processing of the FDs [7, 28]. At the decider's point of view, the writing of the FDs as the left hand side of the equivalence (5) gives a more concise representation of the FDs and then has to be considered after the redundant FD elimination.

3 Minimal covers

The minimal covers is the minimal FD set, $\hat{F}$ computed from FDs F such as $\hat{F}^+ = F^+$ and $\hat{F}$ is minimal. $\hat{F}$ is minimal if it does not contain neither *redundant FDs*, nor *superfluous attributes*. A FD is said to be redundant if it can be written using the Armstrong's axioms on the FD system with the exception of this FD: $X \rightarrow Y$ of F is redundant if $F \subset (F \setminus \{X \rightarrow Y\})^+$. This condition is satisfied if $(X \rightarrow Y) \in (F \setminus \{X \rightarrow Y\})^+$. By using the definition of the closure of an attribute set over a FD set (4), a FD can be qualified as redundant if $Y \subset X^+_{(F \setminus \{X \rightarrow Y\})}$. Ullman [28] shows that if $Y \subset X^+_F$, then $(X \rightarrow Y) \in F^+$.

An attribute x of the left hand side of a FD $X \rightarrow Y$ is superfluous if the FD $(X \setminus x) \rightarrow Y$ can be computed using the Armstrong's axioms on the FD system, i.e.
$F \subset ((F \setminus \{X \rightarrow Y\}) \cup \{(X \setminus x) \rightarrow Y\})^+$. This condition is satisfied if $((X \setminus x) \rightarrow Y) \in F^+$ or $Y \subset (X \setminus x)^+_F$.

For example, the minimal covers of $F = \{a \rightarrow b, ab \rightarrow c, ac \rightarrow d, a \rightarrow c\}$ is $\hat{F} = \{a \rightarrow b, ab \rightarrow c, a \rightarrow d\}$, because $\{a \rightarrow b, ab \rightarrow c\}$ allows to infer $a \rightarrow c$, then c is superfluous in $ac \rightarrow d$ and $a \rightarrow c$ is redundant.

3.1 Proposed algorithms

The direct application of the previous definitions allows the computation of the minimal covers. It is however non-deterministic at the state of the choice of a FD to evaluate and the choice between eliminating a redundant FD or a superfluous attribute [3]. The closure computation is a NP-hard problem, as the rewritings of formulae only using the reflexivity axiom, producing $n \times 2^{n-1}$ FDs given n attributes, solves the boolean satisfiability problem (SAT).

The closure computation can be avoided by providing a boolean function that is true if an attribute subset is included in the closure of another attribute subset over a FD set. Furthermore, the decomposition of the FDs giving FD whose right-hand sides have only one attribute (5) allows to consider only whether a single attribute belongs to the closure.

Determination of an attribute belonging to the closure of an attribute set over a FD set :

The reflexivity axiom (1) can be rewritten into

$$\begin{aligned} &\models\!\!> (A \cup A) \rightarrow A \\ (A \cup A) \rightarrow A &\models\!\!> A \rightarrow A; \end{aligned} \tag{6}$$

therefore, $y \in X_F^+$ *if* $y \in X$; because x is in the right hand side of a FD belonging to F^+ for which the left hand side is included into $\{x\}$.

Armstrong's augmentation axiom (2) allows the rewriting of a FD $(A \rightarrow B) \in F$ into

$$\begin{aligned} &A \rightarrow B &&\models\!\!> (A \cup A) \rightarrow (A \cup B) \\ &and\ (A \cup A) \rightarrow (A \cup B) &&\models\!\!> A \rightarrow (A \cup B); \end{aligned} \tag{7}$$

the addition of the FD $((A \cup B) \rightarrow C) \in F$, the transitivity axiom (3) allows to write

$$\left.\begin{aligned} A \rightarrow (A \cup B)\ \ (7) \\ (A \cup B) \rightarrow C \quad\ \ \end{aligned}\right\} \models\!\!> A \rightarrow C. \tag{8}$$

The same rewritings can be achieved by the application of Armstrong's axioms on FDs whose left hand side is included in the attribute set $A \cup B$. The only demonstration of $B \subset A_F^+$ is enough to determine that if $(A \cup B \rightarrow C) \in F$, then $C \subset (A \cup B)_F^+$. Furthermore, there aren't any other rewriting starting from $A \rightarrow B$ giving FDs with right hand side that contain only subsets of $A \cup B$, then

$$y \in X_F^+ \ if\ \exists A \rightarrow B \in F \mid A \subset X \ and\ y \in (X \cup A)_{(F \setminus \{A \rightarrow B\})}. \tag{9}$$

The determination of the belonging of an attribute to the closure of an attribute set over a FDs set can be written as

$$y \in X_F^+ \ if \left\{ \begin{aligned} &\quad y \in X \\ &or\ \ \exists A \rightarrow B \in F \mid A \subset X \ and\ y \in (X \cup A)^+_{(F \setminus \{A \rightarrow B\})}. \end{aligned} \right. \tag{10}$$

This recursive definition can be easily translated into an iterative form (terminal recursion) giving the algorithm described by algorithm 1. The proposed algorithms are based on a greedy algorithm (algorithmé2) : for each step of this algorithm (lines 17 to 25) one or more rules are evaluated to update the partial closure X_i (lines 22 to 23) or the end condition is reached.

As rules are only evaluated once, the maximum number of iterations of the closure main loop (lines 17 to 25) is the number of rules in the ruleset. Thus this algorithm performs linearly with the number of rules of the computed ruleset.

Data : A ruleset F.
Result : The minimal covers $\hat{F}$ of F.

1 Let $F_k = \emptyset$;
2 **foreach** $(X \to Y) \in F$ **do**
3   Let $X_k = \emptyset$;
4   **foreach** $x \in X$ **do**
5   **if** $Y \not\subset (X \setminus x)^+_F$ **then**
6             $X_k = X_k \cup \{x\}$;
7   **if** $X_k \neq \emptyset$ **then**
8         $F_k = F_k \cup \{X_k \to Y\}$;
9 $\hat{F} = F_k$;
10 **foreach** $(X \to Y) \in F_k$ **do**
11   **if** $Y \subset X^+_{(\hat{F} \setminus \{X \to Y\})}$ **then**
12         $\hat{F} = (\hat{F} \setminus \{X \to Y\})$;

Algorithm 1: Minimal covers.

3.2 Examples

Figures 1, 2 and 3 show examples of the application of the proposed algorithms. Line numbers in these examples refer to line numbers in algorithms 1 and 2.

4 Related work

4.1 Propositional logic

Armstrong's axioms are theorems of the propositional logic [28]. It can then be proven that every computable expression using Armstrong's axioms are true formulae for the propositional logic. Kaufman gave the proof that the theory of FD redundancy is valid for logical implications [17], therefore, that a FD system $F = (A, \cup, \to)$ shares its properties with a world in propositional logic $w = (A, \wedge, \to)$ where A is a set of propositions, $\wedge$ is the logical conjunction and $\to$ is the logical implication.

Data :
- F : a ruleset.
- X : a set of attributes.
- y : an attribute.

Result : a boolean : *true* if $y \in X_F^+$, *false* else.

Let $F_i = F$;
Let $X_i = X$;
Let $F_k = \emptyset$;
closed = false;
while *not closed and* $y \notin X_i$ **do**
    closed = true;
    $F_k = \emptyset$;
    foreach $A \to B \in F_i$ **do**
        if $A \subset X_i$ **then**
            $X_i = X_i \cup B$;
            closed = false;
        else
            $F_k = F_k \cup \{A \to B\}$;
    $F_i = F_k$;
if $y \in X_i$ **then**
    $y \in X_F^+$!;
else
    $y \notin X_F^+$!;

Algorithm 2: Determination if an attribute y belongs to the closure of an attribute set X over a ruleset F.

This properties have been applied to association rules in several approaches [22, 25], using monotony properties of conjonctions of items for redundancy elimination.

4.2 Conceptual lattices

Minimal covers and conceptual lattices share the use of inclusions between extensions of the represented attribute combinations to limit the amount of represented knowledge.

Association rules and Galois lattices :

Logical implications involved in the computation of the minimal covers are captured by pseudo-intent (non-closed descriptions) are implicit and are not represented on the Galois lattice but can be inferred by the user [23] (example shown by figure 4). Galois lattices can represent the other, non-logical, association rules. Furthermore, it has been proven that the representation of a limited subset of association rules using a Galois lattice is sufficient for the

1	superfluous attributes:	$F = \{ a \rightarrow b, a \wedge c \rightarrow b \}$
3	superfluous attributes:	$(X \rightarrow Y) = (a \rightarrow b)$
3	superfluous attributes:	$(X \rightarrow Y) = (a \wedge c \rightarrow b)$
5	superfluous attributes:	$(X \setminus x) = (\{a, c\} \setminus \{a\})$
	closure:	$b \in X_F^+$?
	closure:	$F = \{a \rightarrow b, a \wedge c \rightarrow b\}$, $X = c$
15	closure:	$F_k = \emptyset$
20	closure:	$(A \rightarrow B) = (a \rightarrow b)$
24	closure:	$A \not\subset X_i$
	closure:	$F_k = \{a \rightarrow b\}$
	closure:	$X_i = \{c\}$
20	closure:	$(A \rightarrow B) = (a \wedge c \rightarrow b)$
24	closure:	$A \not\subset X_i$
	closure:	$F_k = \{a \rightarrow b, a \wedge c \rightarrow b\}$
	closure:	$X_i = \{c\}$
28	closure:	$b \notin X_F^+$!
5	superfluous attributes:	$a \notin (\{c\})_F^+$
5	superfluous attributes:	$(X \setminus x) = (\{a, c\} \setminus \{c\})$
	closure:	$b \in X_F^+$?
	closure:	$F = \{a \rightarrow b, a \wedge c \rightarrow b\}$, $X = a$
15	closure:	$F_k = \emptyset$
20	closure:	$(A \rightarrow B) = (a \rightarrow b)$
22	closure:	$A \subset X_i$
	closure:	$F_k = \{\}$
	closure:	$X_i = \{a, b\}$
20	closure:	$(A \rightarrow B) = (a \wedge c \rightarrow b)$
24	closure:	$A \not\subset X_i$
	closure:	$F_k = \{a \wedge c \rightarrow b\}$
	closure:	$X_i = \{a, b\}$
25	closure:	$F_i = F_k$
	closure:	$F_i = \{a \wedge c \rightarrow b\}$
27	closure:	$b \in X_i$!
	closure:	$b \in X_F^+$!
5	superfluous attributes:	$c \in (\{a\})_F^+$
5	superfluous attributes:	c is superfluous.
8	superfluous attributes:	$(X_k \rightarrow Y) = (a \rightarrow b)$

Fig. 1. Example of superfluous attributes elimination over the set of rules $\{a \rightarrow b, a \wedge c \rightarrow b\}$.

inference of the whole association rule system [10, 23, 29] with the following reading conventions:

1. the support of a non-closed description (pseudo-intent) which is not represented on the Galois lattice is equal to the support of the closed set in which it is included; in the example given figure 4, the support of a (non closed) is the same as the support of $a \wedge b \wedge c$ (closed). [10, 23];

9 redundant FD: $\hat{F} = F_k = \{\, a \to b, a \to c, b \to c \,\}$
11 redundant FD: $(X \to Y) = (a \to b)$
11 redundant FD: $\hat{F} \setminus (X \to Y) = \{\, a \to c, b \to c \,\}$
closure: $b \in X_F^+$?
closure: $F = \{b \to c, a \to c\}$, $X = a$
15 closure: $F_k = \emptyset$
20 closure: $(A \to B) = (b \to c)$
24 closure: $A \not\subset X_i$
closure: $F_k = \{b \to c\}$
closure: $X_i = \{a\}$
20 closure: $(A \to B) = (a \to c)$
22 closure: $A \subset X_i$
closure: $F_k = \{b \to c\}$
closure: $X_i = \{a, c\}$
25 closure: $F_i = F_k$
closure: $F_i = \{b \to c\}$
15 closure: $F_k = \emptyset$
20 closure: $(A \to B) = (b \to c)$
24 closure: $A \not\subset X_i$
closure: $F_k = \{b \to c\}$
closure: $X_i = \{a, c\}$
28 closure: $b \notin X_F^+$!
11 redundant FD: $b \notin (\{a\})^+_{(\hat{F} \setminus (X \to Y))}$!
12 redundant FD: $\hat{F} = \{\, a \to b, a \to c, b \to c \,\}$
11 redundant FD: $(X \to Y) = (b \to c)$
11 redundant FD: $\hat{F} \setminus (X \to Y) = \{\, a \to b, a \to c \,\}$
closure: $c \in X_F^+$?
...
(in a similar way, it can be proven that:)
11 redundant FD: $c \notin (\{b\})^+_{(\hat{F} \setminus (X \to Y))}$!
12 redundant FD: $\hat{F} = \{\, a \to b, a \to c, b \to c \,\}$

Fig. 2. Example of redundant rule elimination over the ruleset $\{a \to b, b \to c, a \to c\}$: examination of the rules $a \to b$ and $a \to c$.

2. every description on the frontier of the search[1] is closed [23];
3. every association rules and theirs confidences can be inferred using the represented intents of the inferred pseudo-intents [23].

Galois lattices however need the computation and the representation of the whole lattice -a non represented description means a non closed description- where the representation of the minimal covers allows correct inferences. In the previous example (figure 4), if we only represent $b \to b \wedge c$ and $c \to b \wedge c$ it is impossible to infer $a \to b \wedge c$ as the information about the closure of a is not represented. Another limit to Galois lattices is when there are no logical

[1] The *frontier* of the search is the set of the most specific intents.

11 redundant FD: $(X \to Y) = (a \to c)$
11 redundant FD: $\hat{F} \setminus (X \to Y) = \{ a \to b, b \to c \}$
closure: $c \in X_F^+$?
closure: $F = \{a \to b, b \to c\}$, $X = a$
15 closure: $F_k = \emptyset$
20 closure: $(A \to B) = (a \to b)$
22 closure: $A \subset X_i$
closure: $F_k = \{\}$
closure: $X_i = \{a, b\}$
20 closure: $(A \to B) = (b \to c)$
22 closure: $A \subset X_i$
closure: $F_k = \{\}$
closure: $X_i = \{a, b, c\}$
25 closure: $F_i = F_k$
closure: $F_i = \{\}$
27 closure: $c \in X_i$!
closure: $c \in X_F^+$!
11 redundant FD: $c \in (\{a\})^+_{\hat{F} \setminus (X \to Y)}$!
redundant FD: $a \to c$ is redundant !
12 redundant FD: $\hat{F} = \{ a \to b, b \to c \}$

Fig. 3. Example of redundant rule elimination over the ruleset $\{a \to b, b \to c, a \to c\}$: examination of the rule $b \to c$.

implication. In this case, every discovered description is closed, then there are no pseudo-intent and then the Galois lattice is equivalent with the description inclusion lattice.

Algorithms are proposed to compute frequent closed itemsets [29], as a replacement for the original frequent itemsets step in the classical association rule discovery process.

δ-free sets :

Boulicaut et al. [5] proposed a new notion, the δ-free descriptions, that meet this latter problem. It is an enhancement of the closure definition, taking into account quasi-inclusions[2]. A description is said to be δ-free if there aren't any rule between subsets of these description, that is invalidated by at most δ examples. The δ factor is supposed to be small. The set of δ-free frequent descriptions allows to approximate the set of the descriptions, with an error rate defined by support and confidence thresholds. It reduces the number of represented descriptions while reducing the time required for their computation.

[2] that can be seen as extents statistical implications.

❶ ***Relation :***

a	b	c
X	X	X
	X	X
		X
	X	

❹ ***Galois lattice :***

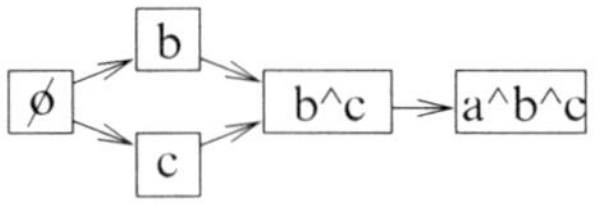

❸ ***minimal covers :*** a−>b^c

❷ ***33 logical implications :***

	a	b	c	a^b	a^c	b^c	a^b^c ⇦ *left hand sides*
a	X			X	X		X
b	X	X		X	X	X	X
c	X		X	X	X	X	X
a^b	X			X	X		X
a^c	X			X	X		X
b^c	X			X	X	X	X
a^b^c	X			X	X		X

⇧ *right hand sides*

❸The minimal covers of the initial 33 rules ❷ can be used to infer the logical rules ❷.

❹The conceptual lattice of the relation can be used to infer the whole set of rules, including the minimal covers. In this example, to infer the minimal covers, we use the following reasoning: an attribute a appears in the closed set $a \wedge b \wedge c$, but it does not appear on any other represented closed set. It means that descriptions a, $a \wedge b$ and $a \wedge c$, which are described by a are described by b and c too and then that $a \rightarrow b \wedge c$.

Fig. 4. Comparison between a minimal covers and a Galois lattice.

5 Experimental results on the extension of the minimal covers to association rules

Armstrong axioms are not valid for every kind of implicative systems where statistical implications are considered. The figure 5 is an illustration of the limits of the transitivity axiom for statistical implications.

There are the same limits for the augmentation axiom.

5.1 Further propositions to circumvent these limits

The use of logical propositions for association rules are interesting in three categories of applications:

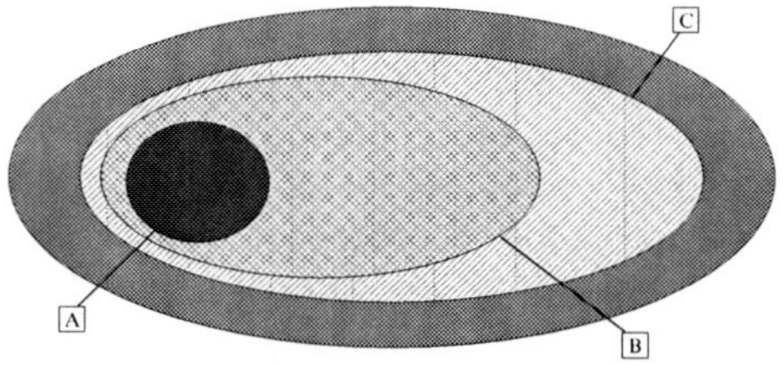

Note : A (B, ...) here represents the set of objects for which a is true (b, ..., respectively)

❶ Both statistical implications $a \rightarrow b$ and $b \rightarrow c$ are observable, as well is observed the statistical implication $a \rightarrow c$.
$\Rightarrow$ valid transitivity !

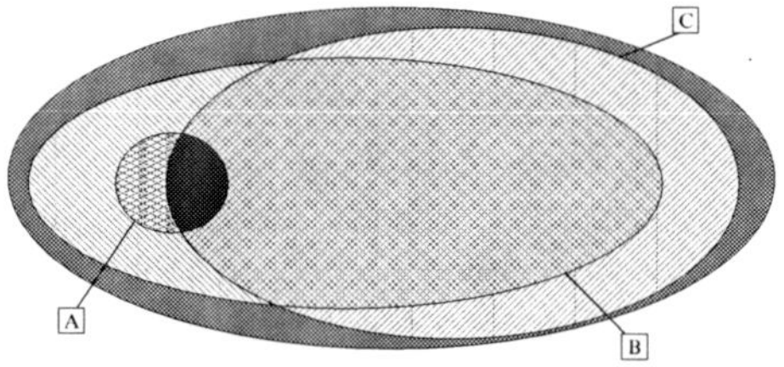

❷ $a \rightarrow b$ and $b \rightarrow c$ are observable, but no relation exists between a and c
$\Rightarrow$ no transitivity !

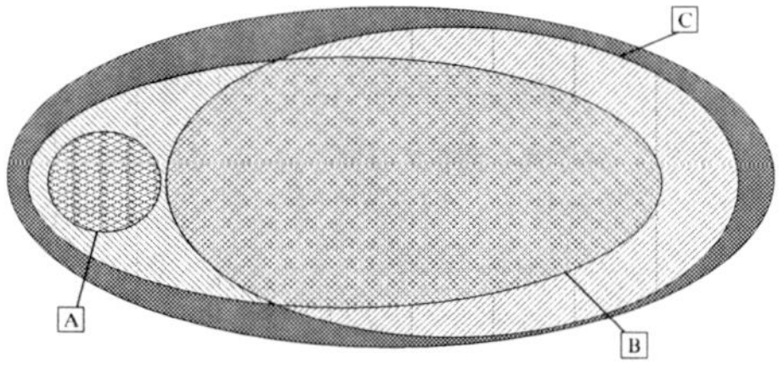

❸ $a \rightarrow b$ and $b \rightarrow c$ are observable, but the statistical implication $a \rightarrow \neg c$ is observable
$\Rightarrow$ anti-transitivity !

Fig. 5. Limits of the axiom of transitivity for statistical implications.

1. the hypothesis that association rules behave like logical propositions can be investigated with the validation of an expert. This is required to use the rules as a knowledge base for inference engines of an expert system.
2. Setting thresholds on quality indices of the rules (e.g. a low support and an high confidence) aims at getting a behavior of association rules that is near logical propositions. There isn't unfortunately any formula giving the correct thresholds that will ensure a complete logical behavior, except the trivial case of a confidence threshold of 1.
3. The redundancy elimination can be considered as a part of an interactive process, providing the user a way to infirm or confirm his hypothesis during his reasoning [18]. This interaction can be useful in finding exceptions to the logical behavior of rules or sets of rules.

The first two points are confirmed by experiments whose results are presented by figure 6 and table 7[3]. These represents experiments on 100 rulesets, varying *min_conf*[4] in $\{0.8, 0.9, 1\}$ and φ[5] in $\{0.8, 0.9, 1\}$. For each ruleset, we check the minimal covers and the closure with the same quality criteria to determine the ratio of valid inferences. Here, we assume that the user is able to infer the closure himself from either the original ruleset or its minimal covers. These experiments show that the higher the *min_conf* threshold is set, the better the minimal covers and inferences on the basis on the minimal covers are correct. A similar but weaker behavior is observed with the intensity of implication.

6 Conclusion

We applied to association rules a redundancy elimination method for functional dependency systems. This method eliminates redundant rules, that can be inferred by a user in a logical reasoning. Properties of functional dependencies are an example of rewriting rules that are both useful for propositional logic and useful for human interpretation for data model design. We showed that this method can strongly reduce the quantity of rules exposed to the user, at the price of some approximation: there is no practical way for the user to compute the quality indices associated with each individual inferred rule, and the user can be disoriented to infer wrong rules, that does not exist in the initial rule set, according to the quality criteria set during the rule discovery.

However the conciseness of the achieved representation, the resort on simple and usual logical properties, and the existence of efficient, polynomial, algorithms for the minimal covers computation make this method useful for a

[3] The program used to present these results can be downloaded at the following URL : `http://www.fc.univ-nantes.fr/~remi/felix/min-covers`.

[4] threshold on confidence.

[5] Here, the original definition of intensity of implication is used [14].

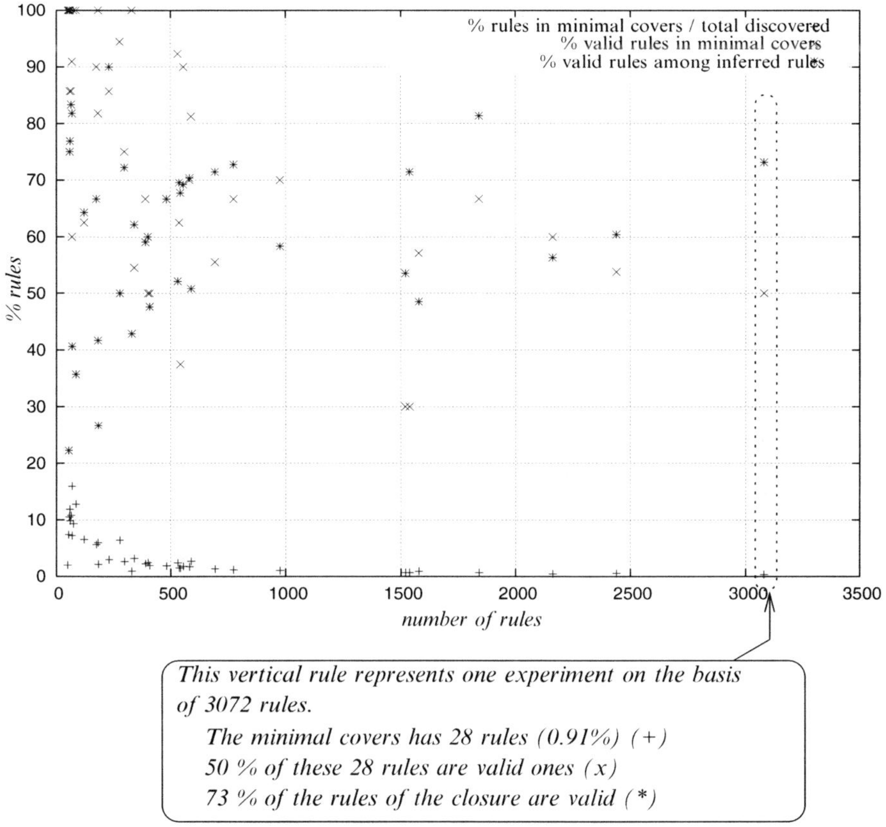

Fig. 6. Sample experimental results

first acquisition of rules by a user, before starting an interactive mining among the rules.

This method is also an alternative for the quality evaluation of a set of rules, considering the global quality of a rule set used for reasoning and decision making, in addition to the sum of the qualities of individual rules.

References

1. F.-N. Afrati, A. Gionis, and H. Mannila. Approximating a collection of frequent sets. In *Proceedings of the Tenth ACM SIGKDD International Conference on Knowledge Discovery and Data Mining*, pages 12–19. ACM, 2004.
2. R. Agrawal, H. Mannila, R. Srikant, H. Toivonen, and A. Inkeri Verkamo. Fast discovery of association rules. In Fayyad et al. [11], pages 307–328.
3. J. Atkins. A note on minimal covers. *SIGMOD RECORD*, 17(4):16–21, December 1988.

$\min\_conf$	percentage of successful experiments, in term of confidence of rules inferred from the minimal covers.
0.8	66.1%
0.9	70.5%
1	98.1%
$\min\_\varphi \in \{0.9, 0.95, 1\}$	

$\min\_\varphi$	percentage of successful experiments, in term of intensity of implication of rules inferred from the minimal covers.
0.9	72.4%
0.95	73.3%
1	76%
$\min\_conf \in \{0.8, 0.9, 1\}$	

Note : here φ represents R. Gras' intensity of implication [21].

Fig. 7. Experimental results : confirmation of the inferred rules using tests with quality measures.

4. Y. Bastide, N. Pasquier, R. Taouil, G. Stumme, and L. Lakhal. Mining minimal non-redundant association rules using frequent closed itemsets. In *Proceedings of the First International Conference on Computational Logic - CL2000, LNCS 1861*, pages 972–987. Springer, 2000.
5. J.-F. Boulicaut, A. Bykowski, and C. Rigotti. Approximation of frequency queries by means of free-sets. In *Proceedings of Principles of Data Mining and Knowledge Discovery*, Lecture Notes in Computer Science, pages 75–85. Springer-Verlag, 2000.
6. J.R. Brachman and T. Anand. The process of knowledge discovery in databases: a human-centered approach. In Fayyad et al. [11], pages 37–58.
7. H. Briand, J.B. Crampes, Y. Hebrail, D. Herin Aime, J. Kouloumdjian, and R. Sabatier. *Les systèmes d'information.* éditions DUNOD, 1986.
8. Aaron Ceglar and John F. Roddick. Association mining. *ACM Comput. Surv.*, 38(2):5, 2006.
9. C. Delobel and M. Adiba. *Bases de données et systèmes relationnels.* DUNOD Informatique, 1982.
10. L. Dumitriu, C. Tudorie, E. Pecheanu, and A. Istrate. A new algorithm for finding association rules. In *Proceedings of Data Mining*, volume 2, pages 195–202. Wessex Institute of Technology, WIT Press, 2000.
11. U.M. Fayyad, G. Piatetsky-Sapiro, and P. Smyth, editors. *Advances in Knowledge Discovery and Data Mining.* AAAI Press, 1996.
12. L. Fleury. Adaptation d'une méthode de recherche de la couverture minimale d'un ensemble de dépendances fonctionnelles pour l'élimination des redondances dans un système de règles. INFORSID, Aix en Provence, 1994.
13. J.-G. Ganascia. *AGAPE et CHARADE : deux techniques d'apprentissage symbolique appliquées à la construction de bases de connaissances.* PhD thesis, Université de Paris Sud, 1987.
14. R. Gras, S. Almouloud, M. Bailleuil, A. Lahrer, M. Polo, H. Ratsimba-Rajohn, and A. Totohasina. *L'implication statistique : nouvelle méthode exploratoire de données.* Edition de la pensée sauvage, 1996.

15. J.-L. Guigues and V. Duquennes. Familles minimales d'implications informatives résultant d'un tableau de données binaires. In *Mathématiques et sciences humaines*, number 95, pages 5–18. 1986.
16. J. Hipp, U. Güntzer, and G. Nakhaeizadeh. Mining association rules: deriving a superior algorithm by analysing today's approaches. In *Proceedings of the 4th European Conference on Principles of Data Mining and Knowledge Discovery*, volume 1910 of *Lecture Notes in Computer Science*, pages 159–168. Springer Verlag, 2000.
17. A. Kaufmann. *Nouvelle logique pour l'intelligence artificielle.* Mathématiques appliquées. Editions Hermes, 1987.
18. P. Kuntz, R. Lehn, and H. Briand. A user-driven process for mining association rules. In *Proceedings of Principles of Data Mining and Knowledge Discovery*, volume 1910 of *Lecture Notes in Computer Science*, pages 483–489. Springer-Verlag, 2000.
19. R. Lehn, F. Guillet, P. Kuntz, H. Briand, and J. Philippé. Felix: An interactive rule mining interface in a kdd process. In *Proceedings of the 10th Mini-Euro Conference, Human Centered Processes, HCP'99*, pages 169–174. Ecole Nationale Supérieure des Télécommunications de Bretagne, 1999.
20. S. Lopes, J.-M. Petit, and L. Lakhal. Efficient discovery of functional dependancies and armstrong relations. Rapport de recherche LIMOS, Université Blaise Pascal, Clermont-Ferrand II, 1999.
21. H. Mannila and K.-J. Räihä. *The Design of Relational Databases.* Addison-Wesley, 1992.
22. B. Padmanabhan and A. Tuzhilin. On characterization and discovery of minimal unexpected patterns in rule discovery. In *IEEE Transactions on Knowledge and Data Engineering*, volume 2, pages 202–216. IEEE Press, 2006.
23. N. Pasquier. *Data Mining : Algorithmes d'Extraction et de Réduction des Règles d'Association dans les Bases de Données.* PhD thesis, Université de Clermont-Ferrand II, 2000.
24. N. Pasquier, Y. Bastide, R. Taouil, and L. Lakhal. Efficient mining of association rules using closed itemset latices. *Information Systems*, 24(1):25–46, 1999.
25. H. Toivonen, M. Klemettinen, P. Ronkainen, K. Hatonen, and H. Mannila. Pruning and grouping of discovered association rules, 1995.
26. J.D. Ullman. *Principles of Database Systems.* Computer Science Press, 1982.
27. J.D. Ullman. *Principles of Database and Knowledge–base Systems*, volume 1. Computer Science Press, 1989.
28. J.D. Ullman. *Reasoning about functional dependencies*, chapter 7.3, pages 382–392. Volume 1 of [27], 1989.
29. Mohammed Javeed Zaki. Generating non-redundant association rules. In *Knowledge Discovery and Data Mining (KDD'00)*, pages 34–43. ACM Press, 2000.

Fuzzy Knowledge Discovery Based on Statistical Implication Indexes

Maurice Bernadet

LINA / Ecole Polytechnique de l'Université de Nantes,
Rue Christian Pauc, La Chantrerie,
BP 60601, 44306 Nantes Cedex 3, France
maurice.bernadet@univ-nantes.fr

Summary. We describe one application of statistical implication indexes to fuzzy knowledge discovery. After recalling principles of fuzzy logics, we explain how we have adapted statistical indexes to fuzzy knowledge: the support, the confidence and a less common index, the intensity of implication. These indexes highlight statistical links between conjunctions of fuzzy attributes and fuzzy conclusions, but do not evaluate the associated fuzzy rules, which depend of the chosen fuzzy operators. Since fuzzy operators are numerous, we evaluate their sets by applying the generalized modus ponens on the database and by comparing its results to the effective conclusions. We give a summary of the results on several databases, and we present the sets of fuzzy operators that appear to be the best. Studying methods to aggregate fuzzy rules, we show that in order to keep classical reduction schemes, fuzzy operators must be chosen differently. However, one of these possible operator sets is also one of the best for processing the generalized modus ponens.

Key words: Statistical implication, fuzzy knowledge discovery, fuzzy implication, fuzzy operators.

1 Introduction

Fuzzy logics are extensions of classical logics, which allow intermediate truth-values between *True* and *False* [31]. They may express knowledge in a more natural way than classical Boolean logics, allowing graduated attributes as in the sentence "*X is rather high*" (for "*X is high*" is rather true) and then assigning, for instance, a truth value of 0.8 to the proposition *X is high*. Fuzzy logics offer many logical operators, which permits a good translation of various kinds of knowledge. In the domain of knowledge discovery [13], considering that crisp intervals on continuous attributes are difficult to interpret and that strict thresholds are often too abrupt, one may think that fuzzy logics could improve the expressiveness of extracted knowledge.

M. Bernardet: *Fuzzy Knowledge Discovery Based on Statistical Implication Indexes*, Studies in Computational Intelligence (SCI) **127**, 481–506 (2008)
www.springerlink.com

Some methods, specific to fuzzy knowledge discovery, have already been defined, allowing to extract fuzzy rules; these methods are often holistic: they may use neural networks [10, 17, 18] or genetic algorithms [8, 23, 26]. However, we have preferred to adapt classical knowledge discovery methods to fuzzy knowledge because classical methods are more analytical, hence allowing a follow-up of the mechanisms during their progress. The methods we have defined extract interesting fuzzy rules by computing statistical indexes to evaluate statistical implications between fuzzy premises and fuzzy conclusions. Once statistical implications have been highlighted, they are used to build fuzzy rules based on fuzzy operators, which must be chosen according to the considered application.

We here summarize a significant part of our work on this topic, examining it in the perspective of its relation to statistical implication analysis. We have added recent results about fuzzy rule reduction, which shows one important difference between fuzzy rules and statistical rules: while generally statistical rules may not be aggregated, a proper choice of fuzzy operators allows the use of classical rule reduction schemes.

The first part describes the initial process of our fuzzy knowledge discovery methods, which requires a definition of fuzzy partitions to convert numerical or symbolic attributes into fuzzy ones. The second part justifies the classical indexes we have chosen to evaluate the statistical implications; these indexes are the classic support and confidence of a rule, associated with a less common index, the intensity of implication. We then explain how we have adapted these indexes to fuzzy attributes and we describe one algorithm we use to explore the set of possible rules.

Such exploration algorithms, based on statistical indexes, highlight statistical links between conjunctions of fuzzy attributes and fuzzy conclusions, but they do not evaluate the associated fuzzy rules, which depend on the chosen fuzzy operators. Since fuzzy logics offer numerous possibilities to define logical operators, the next paragraph describes the methods we use to evaluate sets of fuzzy operators. These methods are based on another statistical index, which evaluate fuzzy rules, given one set of fuzzy operators; we call this index the GMP-pertinence (GMP stands for generalized modus ponens). The next two parts describe the results given by this method on a simplified example and summarize the results on real databases found in the UCI repository. We then review the best sets of fuzzy operators found with this method. Subsequently, we have studied sets of fuzzy operators which allow easy reduction of fuzzy rules. We have highlighted that most of these sets are not the most GMP-pertinent; one must distinguish two kinds of applications for which the chosen fuzzy operator sets should be different: on the one hand, there are knowledge-based systems using the GMP for decision aids, and on the other hand, there are analytical systems to help experts to explore knowledge extracted from databases. However, associating Gödel-Brouwer's implication with standard Zadeh's minimum and maximum appears as a good compromise.

2 The "Fuzzification" Process

Translation of classical attributes, numeric or symbolic, into fuzzy ones is called "fuzzification". To carry out this translation we need at first to define *pseudo fuzzy partitions*, allowing a classical attribute to belong to several fuzzy classes (generally at most two), with degrees less than 1 and which usually sum up to 1. Once these pseudo fuzzy partitions have been defined, one may translate the values of each classical attribute into its degrees of membership to the different modalities of the corresponding fuzzy attribute.

2.1 Definition of Fuzzy Partitions

Let us first recall that fuzzy logics evaluate the truth-value of a fuzzy proposition *"X is A"*, as the degree to which X belongs to the fuzzy set A: if $\mu_A(X)$ is the membership (or characteristic) function of the fuzzy set A, one may write *Truth("X is A")* $= \mu_A(X)$,

Fuzzy sets allow the definition of fuzzy C-partitions or "pseudo partitions" in which each value of a continuous attribute may be classified into several fuzzy classes, with a total membership of 1. These fuzzy pseudo partitions allow the conversion of continuous attributes into fuzzy ones, then giving the truth-value of fuzzy propositions. For a continuous attribute CA, varying from $minCA$ to $maxCA$, one can define a fuzzy pseudo partition in several ways [7, 22].

The simplest method divides the interval $[minCA,\ maxCA]$ in n sub-intervals, with a small percentage of coverage between two adjacent ones, giving each sub-interval a symbolic name related to their position. For instance, one may divide the interval $[minCA,\ maxCA]$ in 5 sub-intervals with an overlap of 20%, then giving 5 fuzzy modalities for this attribute, such as *strong negative, rather negative, medium, rather positive* and *strong positive* (Fig. 1).

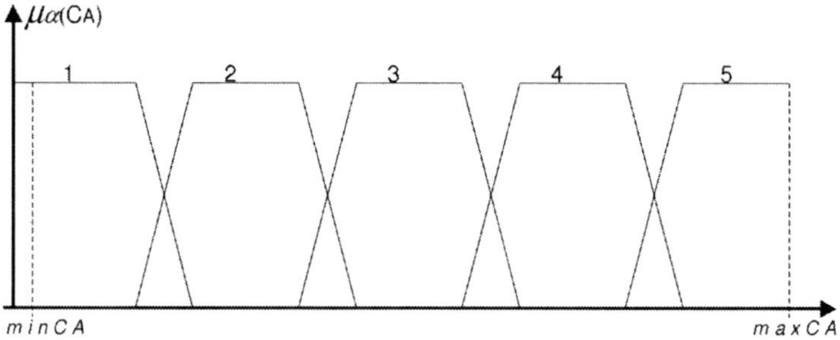

Fig. 1. A fuzzy C-partition (α =1, strong negative; α =2, rather negative; $\alpha = 3$, medium; $\alpha = 4$, rather positive; α =5, strong positive).

The fuzzy classes may also be defined by experts; otherwise one may choose 3 or 5 classes as standard options. Different numbers of classes may also be used, but too high a number of classes might too heavily slow down the knowledge discovery process. It is often interesting to try several numbers of classes because it is difficult a priori choosing the number of classes that will give a good partition is difficult.

Another kind of method extracts the number of classes and defines the fuzzy C-classes from the database. These methods consider values of the attributes giving the same conclusion and, whenever possible, cluster these values into the same fuzzy sets, with a membership value equal to the rate of samples giving this conclusion. These methods often use histograms of attribute values for each possible conclusion. Moreover, it is possible to develop a more satisfactory method, by generalizing optimal discretization methods such as those studied in [33] to fuzzy logics; we have recently defined such a method, based on clustering, which gives more satisfying results than the previous ones, but with the drawback of needing more computing time [5].

2.2 Fuzzification of a Database

Once the fuzzy classes have been defined for each attribute, one may convert the related values of each item by mapping these values to the membership values of each fuzzy class associated to the considered classical attribute (Fig. 2).

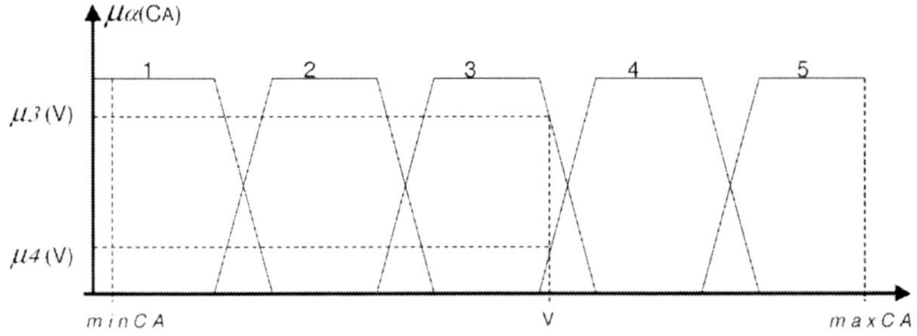

Fig. 2. Mapping from the value V of the continuous attribute CA into membership values of fuzzy attributes (here, only $\mu 3$ and $\mu 4$ are non zero).

3 Choosing Statistical Implication Indexes

Several indexes may be used to evaluate classical rules [1], and we have chosen three of them [3, 6]: the confidence, the support and a less known index, the intensity of implication. Let us consider two propositions a and b associated respectively with A and B, the sets of elements which verify them.

- *The confidence* of a rule such as *"if a then b"* expresses the conditional probability of b to be true when a is true; calling $n_{a \wedge b}$, the number of items verifying *"a and b"* and n_a the number of items verifying a, the confidence may be evaluated by

$$Confidence(a \Rightarrow b) = \frac{n_{a \wedge b}}{n_a} \tag{1}$$

- *The support* of a rule *"if a then b"* may be defined as the rate of occurrences of items verifying *"a and b"*, related to all items of the database; calling $n_{a \wedge b}$ the number of items verifying *"a and b"* and n_E the total number of items in the database, the support may be evaluated by

$$Support(a \Rightarrow b) = \frac{n_{a \wedge b}}{n_E} \tag{2}$$

- *The intensity of implication* is an index expressing the quality of a rule. This index, defined by R. Gras and A. Larher [15], is based on simple probability concepts: since the cardinalities of two subsets A and B of a reference set E are determined by the objects of the database belonging to A and B, we consider two random subsets X and Y having respectively the same cardinalities as A and B. The implication $a \Rightarrow b$ is characterized by the relation $A \subset B$ and its counter-examples are associated to the subset $A \cap \overline{B}$. So, we compare the cardinality of $A \cap \overline{B}$ (given by the database) to the random variable given by the cardinality of $A \cap \overline{B}$, supposing that there is no statistical link between X and Y (Fig. 3).

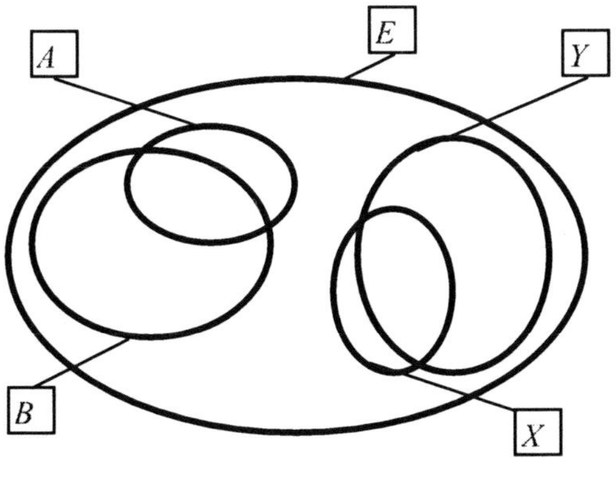

Fig. 3. X and Y vary randomly in E

If the cardinality of $A \cap \overline{B}$ is unusually small compared to the expected value of the distribution of the cardinalities of $X \cap \overline{Y}$, we consider $a \Rightarrow b$ as a quasi-implication. The intensity of $a \Rightarrow b$ is therefore the difference between

1 and the probability for the random variable "cardinality of $X \cap \overline{Y}$" to be smaller than the cardinality of $A \cap \overline{B}$. Intuitively, the intensity of implication measures the degree of statistical astonishment at the size of $A \cap \overline{B}$, considering the sizes of A, B and E, and assuming there is no a priori link between A and B.

One may note that the confidence does not answer the question: "What probability is there to have an implicative link between propositions a and b?". A conditional probability rather answers the question "What is the probability of proposition b when proposition a is true? " So, these two measures are complementary: in a learning approach, the intensity of implication allows the withdrawal of little pertinent implications, while conditional probabilities give for each rule an inference mechanism for uncertain reasoning in an expert system. Therefore, the quality of one implication is better when the number of its counter-examples is smaller than their expected number, that is to say when the quantity $P(Card(X \cap \overline{Y}) \leq Card(A \cap \overline{B}))$ is small.

Consequently, it is the observation of the *"smallness of* $Card(A \cap \overline{B})$ *compared to* $Card(X \cap \overline{Y})$*"* which is taken as a basic evaluator of the interest of the quasi-implication $a \Rightarrow b$. The intensity of implication is then defined by the function

$$\varphi(a, \overline{b}) = 1 - P\left(Card(X \cap \overline{Y}) \leq Card(A \cap \overline{B})\right) \tag{3}$$

Let us call $n = Card(E)$, $n_A = Card(A)$, $n_{\overline{A}} = Card(\overline{A})$, $n_B = Card(B)$, $n_{\overline{B}} = Card(\overline{B})$, $n_{A \cap B} = Card(A \cap B)$, $n_{A \cap \overline{B}} = Card(A \cap \overline{B})$; the random variable $Card(X \cap \overline{Y})$ obeys a hypergeometric distribution:

$$P\left(Card(X \cap \overline{Y}) = k\right) = \frac{C_{n_A}^{k} \cdot C_{n-n_A}^{n-n_B-k}}{C_n^{n-n_B}} = \frac{C_{n_A}^{k} \cdot C_{n_{\overline{A}}}^{n_{\overline{B}}-k}}{C_n^{n_{\overline{B}}}} \tag{4}$$

and

$$P\left(Card(X \cap \overline{Y}) \leq Card(A \cap \overline{B})\right) = \sum_{\substack{i = 0 \\ i \geq n_A - n_B}}^{Card(A \cap \overline{B})} \frac{C_{n_A}^{i} \cdot C_{n_{\overline{A}}}^{n_{\overline{B}}-i}}{C_n^{n_{\overline{B}}}} \tag{5}$$

A detailed comparison of the intensity of implication with other statistical indexes has been made in [9] and [12]. Let us summarize these studies:

- the value of the intensity of implication increases with the size of the learning set, while other indexes remain constant,
- the intensity of implication reflects the human way of withdrawing a previous opinion: some new counter-examples for a strong implication do not change much this index, but progressively doubts appear, and finally a few more counter-examples cause the withdrawal of this opinion;

- for similar reasons, the use of intensity of implication is well adapted to noisy data since a small number of counter-examples do not necessarily invalidate the implication;
- finally, the intensity of implication prohibits the generation of rules such as $a \Rightarrow b$ when the proposition b is true for nearly all examples of the learning set: in that case it is not surprising that the set of examples for which a is true is nearly included in the set of examples for which b is true.

4 Generalization of Statistical Indexes to Fuzzy Knowledge

We then have generalized these three indexes to fuzzy knowledge, to search statistical implicative links between fuzzy attributes. To allow fuzzy inferences, each discovered link must be associated to the corresponding fuzzy rule. However, results from [27] and [14], show that statistical implications give more satisfying semantical results than two main fuzzy implications. That is due to the fact that fuzzy implications generalize classical implications such as $p \Rightarrow q$; such implications are true when p is false, whatever the value q. In the same situation, statistical analysis reasonably judges that there is no link between p and q. Nevertheless, since we not only wanted to study the semantical aspects of implications, but also wanted to study inference mechanisms on extracted rules, we have studied fuzzy implications to use with the main inference mechanism of fuzzy logics, the Generalized Modus Ponens.

To handle fuzzy knowledge, machine learning indexes may exclusively be based on the theory of fuzzy sets, like those described in [22] or in [2]; but these indexes may as well be a generalization of indexes developed in classical logic, like those described in [25, 28] or [32]. We have adopted this second method and we have generalized the three classical indexes we have retained to fuzzy knowledge; for this purpose we have applied a definition given by Zadeh [30] for the probability of a fuzzy event.

Considering a set of objects E, with n domains of reference $D_1, D_2, \ldots, D_n$, on which we define fuzzy attributes for objects in E, the set of classical propositions D becomes the set of all fuzzy propositions that can be expressed on objects of E. The number of elements of E that satisfy a proposition p associated to the fuzzy set $\widetilde{P}$ with the characteristic function μ_P, may be evaluated by the crisp cardinality of the fuzzy set $\widetilde{P}$ on E: $Card(\widetilde{P}) = \sum_{x \in E} \mu_P(x)$.

This notion of cardinality has recently been criticized: when a strong proportion of items has a low membership in a fuzzy set, comparisons using this cardinality can lead to absurd results [11]. This problem should not occur within our approach, because the fuzzy subsets we use constitute a pseudo-partition, built by consultation of an expert or by a clustering method, and because these modes of construction produce fuzzy sets with a kernel (the set of elements with a membership of 1) covering a large part of the support (the set of elements with a membership greater than 0). However, to prevent a bad partitioning, we verify that the support of each fuzzy set is not larger than a threshold σ of the percentage of its kernel. If this is not true, we take into account only the degrees of membership superior or equal to 0.5 (also called the alpha cut to level 0.5) and we use as the cardinality of a fuzzy set $\widetilde{P}$ on a referential E, a formula inspired of [24] and [29]: $Card(\widetilde{P}) = \sum_{x \in E,\ \mu_P(x) \geq 0.5} \mu_P(x)$.

More often, however, we use Zadeh's crisp cardinalities of fuzzy sets. The confidence of a rule, its support and its intensity of implication are then expressed by the same formulas as above, by replacing cardinalities of crisp sets by cardinalities of fuzzy sets. Thus, if one calls $\top$ (t-norm) the fuzzy "and" operator with a fuzzy complement $\mu\overline{A}(x) = 1 - \mu A(x)$, one can write:

$$nA = Card(A) = \sum_{x \in E} \mu A(x), \tag{6}$$

$$n\overline{A} = Card(\overline{A}) = \sum_{x \in E} \mu\overline{A}(x) = \sum_{x \in E} (1 - \mu A(x)), \tag{7}$$

$$n\overline{B} = Card(\overline{B}) = \sum_{x \in E} \mu\overline{B}(x) = \sum_{x \in E} (1 - \mu B(x)), \tag{8}$$

$$nA \cap B = Card(A \cap B) = \sum_{x \in E} \mu A \cap B(x) = \sum_{x \in E} \top(\mu A(x),\ \mu B(x))), \tag{9}$$

$$nA \cap \overline{B} = Card(A \cap \overline{B}) = \sum_{x \in E} \mu A \cap \overline{B}(x) = \sum_{x \in E} \top(\mu A(x),\ (1 - \mu B(x))) \tag{10}$$

The confidence of the rule may still be expressed by

$$Confidence(X\ is\ A \Rightarrow Y\ is\ B) = \frac{n_{A \cap B}}{n_A}, \tag{11}$$

Its support by

$$Support(X\ is\ A \Rightarrow Y\ is\ B) = \frac{n_{A \cap B}}{n_E}, \tag{12}$$

And its intensity of implication by

$$\varphi(a, \overline{b}) = 1 - P\left(Card(X \cap \overline{Y}) \leq Card(A \cap \overline{B})\right) \tag{13}$$

with

$$P\left(Card(X \cap \overline{Y}) \leq Card(A \cap \overline{B})\right) = \sum_{\substack{i = 0 \\ i \geq n_a - n_b}}^{Card(A \cap \overline{B})} \frac{C_{n_a}^{i} \cdot C_{n_{\overline{a}}}^{n_{\overline{b}} - i}}{C_n^{n_{\overline{b}}}} \tag{14}$$

Once the indexes to evaluate rules have been chosen, a fuzzy knowledge extraction algorithm may be used, with the same principles as a classical knowledge extraction algorithm. The algorithm we use is an exploratory search in the tree of possible rules.

5 A Knowledge Extraction Algorithm

This algorithm builds all the rules that can be constructed from a set of propositions, computes their confidence, their support and their intensity of implication and keeps the rules for which these indexes are above three respective thresholds. To limit the number of rules studied and the exploration depth, we also restrict to a maximum the number of propositions in the premises of a rule. So, we use 4 thresholds α, β, γ, δ and one rule is kept if its confidence is greater than α, its support greater than β, its intensity of implication over γ and if its premises have at most δ propositions. The thresholds α, β, γ, δ are chosen by the users in accordance to the number of rules that they wish to obtain.

Rules are structured in a tree; the root of this tree (level 0) is the "rule with the empty premises", level 1 has rules using only 1 proposition in their premises, ..., level i has rules using i propositions in their premises, and so on. The algorithm uses a depth first strategy; this search is not carried deeper when the current rule has not the minimal support γ, or when the size of the current rule is above δ.

Let us call

- c, the confidence of a rule, which must be greater than the threshold α;
- s, the support of a rule, which must be over the threshold β;
- i, the intensity of implication, which must be greater than the threshold γ;
- l, the length of the rule (the number of propositions in its premises, which must not be more than the threshold δ);
- $E = e_1, e_2, \ldots, e_n$, the learning set;
- $P = p_1, p_2, \ldots, p_n$, the set of propositions describing examples in E;
- C, the set of propositions associated to the conclusions;
- $D = a_1, a_2, \ldots, a_m$, the set of attributes in the possible propositions of the premises;
- $F_{decision}$, the fuzzy partition associated to the attribute of the classifying decision;
- $nF_{decision}$, the cardinality of this partition;
- R, the set of rules produced.

Our algorithm, described below, uses two scanning procedures:
-"`Forward`" adds, when possible, a fuzzy proposition not used yet at this level to the premises of the rule,
-"`Backward`" removes the last fuzzy proposition of the premises (the one on the further right), and, if possible, replaces it by the following one. When this proposition has no following one (because it uses the last modality of the last attribute), the new most to the right proposition is removed and replaced, if possible, and so on. When there are no more propositions in the premises, the tree of premises has been completely explored and the algorithm ends.

```
Algorithm: Knowledge Extraction
Begin
  R = ∅;
  For all values v_i ∈ F_decision do
   Let T, the tree of rules with the conclusion {a_decision = v_i};
   Let B, the set of observations in E with {a_decision = v_i} true;
   CurrentNode = Forward(T, R);
   While the tree T has not been totally searched do
     Let r : Premise ⇒ p_i ∈ C, the rule associated to CurrentNode;
     Let A, the set of examples in E in which Premise is true;
     Compute c, s, i from the cardinalities of E, A and B;
     Let l, the length of Premise;
     If(c ≥ α) and (s ≥ β) and (i ≥ γ) and (l ≤ δ)
        Then R = R ∪ { Premisse ⇒ p_i };
      End If
      If (s < β) or (l > δ) or (CurrentNode terminal)
        Then CurrentNode = Backward(T, CurrentNode);
        Else CurrentNode = Forward(T, CurrentNode);
      End If
    End While;
  End For;
End.
```

For instance, let us consider three attributes $\{a,\ b,\ c\}$, each with three modalities: $\{L{=}low,\ M{=}medium,\ H{=}high\}$; the rule tree will be explored by successively considering premises of rules accordingly to Table 1.

6 From Statistical Implications to Fuzzy Rules

The above algorithm highlights statistical links between conjunctions of fuzzy attributes and possible fuzzy values for the attribute in conclusion. However, to apply the modus ponens of fuzzy logics, called generalized modus ponens (GMP), one must consider fuzzy rules.

6.1 Differences between Fuzzy Rules and Statistical Implications

A fuzzy rule may be considered as the generalization of a classic logical rule; it associates in its premises a conjunction of fuzzy propositions with, as conclusion, a fuzzy proposition. The considered propositions correspond to the fuzzy attributes of a same item of a database. If *X1, X2, ..., Xn* and *Y* are fuzzy attributes, having associated respective modalities *A1, A2, ..., An* and *B*, a fuzzy rule has the form: *If "X1 is A1" and "X2 is A2" ... and "Xn is An" then "Y is B"*, or, more formally *"X1 is A1"* $\wedge$ *"X2 is A2"* ... $\wedge$ *"Xn is An"* $\Rightarrow$ *"Y is B"*, where *A1, A2, ..., An* and *B* are fuzzy subsets instead of classical subsets.

(1)	(2)	(3)
$a = L$	$a = M$	$a = H$
$a = L \wedge b = L$	$a = M \wedge b = L$	$a = H \wedge b = L$
$a = L \wedge b = L \wedge c = L$	$a = M \wedge b = L \wedge c = L$	$a = H \wedge b = L \wedge c = L$
$a = L \wedge b = L \wedge c = M$	$a = M \wedge b = L \wedge c = M$	$a = H \wedge b = L \wedge c = M$
$a = L \wedge b = L \wedge c = H$	$a = M \wedge b = L \wedge c = H$	$a = H \wedge b = L \wedge c = H$
$a = L \wedge b = M$	$a = M \wedge b = M$	$a = H \wedge b = M$
$a = L \wedge b = M \wedge c = L$	$a = M \wedge b = M \wedge c = L$	$a = H \wedge b = M \wedge c = L$
$a = L \wedge b = M \wedge c = M$	$a = M \wedge b = M \wedge c = M$	$a = H \wedge b = M \wedge c = M$
$a = L \wedge b = M \wedge c = H$	$a = M \wedge b = M \wedge c = H$	$a = H \wedge b = M \wedge c = H$
$a = L \wedge b = H$	$a = M \wedge b = H$	$a = H \wedge b = H$
$a = L \wedge b = H \wedge c = L$	$a = M \wedge b = H \wedge c = L$	$a = H \wedge b = H \wedge c = L$
$a = L \wedge b = H \wedge c = M$	$a = M \wedge b = H \wedge c = M$	$a = H \wedge b = H \wedge c = M$
$a = L \wedge b = H \wedge c = H$	$a = M \wedge b = H \wedge c = H$	$a = H \wedge b = H \wedge c = H$
(4)	**(5)**	**(6)**
$b = L$	$b = M$	$b = H$
$b = L \wedge c = L$	$b = M \wedge c = L$	$b = H \wedge c = L$
$b = L \wedge c = M$	$b = M \wedge c = M$	$b = H \wedge c = M$
$b = L \wedge c = H$	$b = M \wedge c = H$	$b = H \wedge c = H$
(7)	**(8)**	**(9)**
$c = L$	$c = M$	$c = H$

Table 1. The sequence of premises tested by our algorithm

For example, one can write rules such as

"If the pressure is high then the weather is cold" or
"If temperature is low and humidity is high then saturation is near".

A statistical implication expresses that, when premises are true, it is very likely its conclusion is true. In a different way, a fuzzy rule allows the evaluation of the fuzzy set associated with its conclusion, given the truth degree of its premises. The conclusion is then a set of possible values for one fuzzy attribute, some values being partially appropriate. The fuzzy rule aggregates its possible counter-examples by extending the range of the possible values of the attribute in conclusion and by giving them a small degree of membership to the associated fuzzy set.

To completely define a fuzzy rule, one needs to specify the fuzzy operators to which it refers. Generally, only one conjunction and one implication are sufficient, but one often needs one complement and one disjunction as well. Unfortunately, the number of possible fuzzy operators is infinite, but generally one may restrict them to some standard sets.

6.2 Main Sets of Fuzzy Operators

With the indexes we have chosen only a basic fuzzy conjunction ("and" operator) is needed, but if the extracted rules are to be processed by a knowledge based system, a fuzzy implication is also necessary, often with a fuzzy disjunction ("or") and a fuzzy complement ("not"). Fuzzy logics offer a great choice of logical operators [21]. Let us summarize the main fuzzy operators.

When there is no ambiguity, we will simplify our notation $\mu(a)$ into a, representing both the proposition a and its truth value.

•***A fuzzy complement*** (fuzzy negation) is generally realized with the standard complement $C(a) = 1 - a$; we have only considered this possibility.

•***A fuzzy conjunction*** (fuzzy "and") must be defined by a t-norm (triangular norm) $\top$, which is a function from $[0, 1] \times [0, 1]$ into $[0, 1]$ characterized by:

$$\begin{array}{l} \top(0,0) = \top(0,1) = \top(1,0) = 0, \qquad \top(1,1) = 1, \\ \top(a,b) = \top(b,a) \text{ (commutativity)}, \\ \top(a,\top(b,c)) = \top(\top(a,b),c) \text{ (associativity)}, \\ \forall a,a',b,b' \quad a \leq a' \text{ and } b \leq b' \Rightarrow \top(a,b) \leq \top(a',b') \text{ (monotony)}. \end{array}$$

The axioms on the first three lines keep the properties of the classical conjunction for classical sets. Other axioms are often added, such as the continuity of $\top(x, y)$ and/or its under-idempotence: $\top(a, a) \leq a$. One frequently uses:

. Zadeh's minimum: $\top(a, b) = \min(a, b)$,
. the probabilistic intersection: $\top(a, b) = a \times b$ or
. Lukasiewicz's bounded (or bold) difference: $\top(a, b) = \max(0, a+b-1)$.

•***A fuzzy disjunction*** (fuzzy "or") must be defined by a t-conorm (triangular conorm) $\perp$, which is a function from $[0, 1] \times [0, 1]$ into $[0, 1]$, characterized by:

$$\begin{array}{l} \perp(1,1) = \perp(0,1) = \perp(1,0) = 1, \qquad \perp(0,0) = 0, \\ \perp(a,b) = \perp(b,a) \text{ (commutativity)}, \\ \perp(a,\perp(b,c)) = \perp(\perp(a,b),c) \text{ (associativity)}, \\ \forall a,a',b,b' \quad a \leq a' \text{ and } b \leq b' \Rightarrow \perp(a,b) \leq \perp(a',b') \text{ (monotony)}. \end{array}$$

The first three lines keep the properties of the classical disjunction. Some other axioms are often added, such as the continuity of $\perp(x, y)$ and/or its over-idempotence: $\perp(a, a) \geq a$. One frequently uses:

. Zadeh's standard union (or Gödels's t-norm): $\perp(a, b) = \max(a, b)$,
. the probabilistic union: $\perp(a, b) = a + b - a \times b$ or
. the bounded sum (or Lukasiewicz's): $\perp(a, b) = \min(1, a + b)$.

The conjunction and the disjunction should be bound by de Morgan's laws:

$$C(\top(a,b)) = \perp(C(a),\ C(b)) \quad \text{and} \quad C(\perp(a,b)) = \top(C(a), C(b)).$$

In this case, the t-norm $\top$ and the t-conorm $\perp$ are said to be dual relatively to the fuzzy complement C; amongst t-norms and t-conorms dual relatively to the standard fuzzy complement $C(a) = 1 - a$, one may choose

. the minimum and maximum:

$$\top(a,b) = \min(a,b), \quad \perp(a,b) = \max(a,b),$$

. the probabilistic product and sum:

$\top(a,b) = a \times b, \quad \bot(a,b) = a + b - a \times b$ or

. the bounded difference and sum:

$\top(a,b) = \max(0, a+b-1), \quad \bot(a,b) = \min(1, a+b).$

•***A fuzzy implication*** is a fonction I from $[0,1] \times [0,1]$ into $[0,1]$ which defines for all truth values of two fuzzy propositions a and b, the truth value $\mathrm{I}(a,b)$ of "*if a then b*". This function I may be defined in different ways in fuzzy logics, which are equivalent in classical logics.

1) In classical logics one may define I by:

$$\mathrm{I}(a,b) = \neg\, a \vee b, \tag{15}$$

which becomes in fuzzy logics:

$$\mathrm{I}(a,b) = \bot(C(a), b) \tag{16}$$

2) Classical logics allow also to define I by:

$$\mathrm{I}(a,b) = \max\{x \in \{0,1\} \mid (a \wedge x) \leq b\}, \tag{17}$$

which becomes in fuzzy logics:

$$\mathrm{I}(a,b) = \sup\{x \in [0,1] \mid \top(a,x) \leq b\} \tag{18}$$

3) Formula (15) may also be written

$$\mathrm{I}(a,b) = \neg\, a \vee (a \wedge b), \tag{19}$$

or

$$\mathrm{I}(a,b) = (\neg\, a \wedge \neg\, b) \vee b; \tag{20}$$

these formulas become in fuzzy logics:

$$\mathrm{I}(a,b) = \bot(C(a), \top(a,b)), \tag{21}$$

or

$$\mathrm{I}(a,b) = \bot(\top(C(a), C(b)), b), \tag{22}$$

where $\top, \bot$ and C must satisfy de Morgan's law.

Definitions (15), (17) (19) and (20) are equivalent, which is not true for definitions (16), (18), (21) and (22), which consider fuzzy truth values instead of classical truth values 0 or 1; these 4 last formulas allow definitions of several classes of fuzzy implications.

- ***S-implications*** are defined using formula (16), which specifies a fuzzy implication I given a t-conorm $\bot$: $\mathrm{I}(a,b) = \bot(C(a), b)$. One may then define:
  - for the maximum (dual intersection: the minimum),
    Kleene-Dienes' implication: $\mathrm{I}(a,b) = \max(1-a, b)$;
  - for the probabilistic union (dual intersection: the probabilistic product),
    Reichenbach's implication: $\mathrm{I}(a,b) = 1 - a + a \times b$;
  - for the bounded sum (dual intersection: the bounded difference),
    Lukasiewicz's implication: $\mathrm{I}(a,b) = \min(1, 1-a+b)$.

- ***R-implications*** are defined by formula (21) which specifies one implication given one t-norm $\top$: $\mathrm{I}(a,b) = \sup\{x \in [0,1] \mid \top(a,x) \leq b\}$. This allows to define:

- for the minimum, Gödel's implication:
  $I(a, b) = 1 \quad \text{if } a \leq b \quad \text{oror} \quad b \text{ if } a > b.$
- for the probabilistic product, Goguen's implication:
  $I(a, b) = 1 \quad \text{if } a \leq b \quad \text{oror} \quad b/a \text{ if } a > b.$
- for the bounded difference, Lukasiewicz's implication:
  $I(a, b) = \min(1,\ 1 - a + b).$

- ***QL-implications*** use relation (22) with one t-norm $\top$ and its dual t-conorm $\perp$. This class of implications did not prove interesting within our studies, so we will not go into it.

Our algorithms also need one ***aggregation operator***, which may be defined in numerous ways. However, since we want an averaging evaluation of the implication and since we need a mechanism to allow exclusion of abnormal records, we have chosen the arithmetic mean, which allows the use of standard deviations:

$$Aggregation(\mu_1(x, y),\ \ldots,\ \mu_n(x, y)) \ = \ {}^1\!/_n \sum_{i=1}^{n} \mu_i(x, y). \qquad (23)$$

7 Fuzzy Operators to be Used with Generalized Modus Ponens

When the extracted rules are to be used in a knowledge-based system by application of the generalized modus ponens (GMP), the choice of fuzzy operators must be coherent with the operators chosen then. Let us recall that the GMP is the following inference scheme:

If X is A then Y is B
X is A'

Y is B'

For one implication $\mu_{a \Rightarrow b}(\mu_a(x), \mu_b(y)) = I(\mu_a(x),\ \mu_b(y))$ and one t-norm $\top$, one can write:

$$\mu_{b'}(y) = \sup_{x \in A'} \{\top(\mu_{a'}(x),\ I(\mu_a(x),\ \mu_b(y)\ )\} \qquad (24)$$

7.1 Our Method

To choose a set of fuzzy operators, we have defined a specific index which evaluates fuzzy rules given this set of operators [4]. This index, called the *GMP-pertinence*, is determined by applying the GMP on the considered rule for a random sample of a database. If the distance between the truth-values of the inferred conclusions and the observation is below a threshold chosen by the operator, the example is added to the set of records for which the test is correct, otherwise it is added to the set of records for which the test fails. We define the *GMP-pertinence* of the rule given a set of fuzzy operators as its rate of good examples: $GMPpert = \sum_x \sum_y n_+ / \sum_x \sum_y (n_+ + n_-)$.

A comparative study of fuzzy implication operators [20] has shown that the GMP gives good results by combining the minimum and bold intersections with Lukasiewicz's, Kleene-Dienes' and Gödel-Brouwer's implications. We have added to these operators the probabilistic conjunction and Goguen's implication because of their probabilistic nature. So, we have limited our trials to combinations of one of these t-norms with one of these implications. To illustrate our method, we first present a simplified example, then we summarize the results obtained on more consequent databases.

7.2 One Simple Example

Let us consider two attributes "Size" and "Shoe size" in a set of individuals who can be grouped in two fuzzy classes: the small persons with small size feet, and the tall persons with large size feet. To these classes correspond 2 rules:

"small size" ⇒ *"small shoe size"*
"tall size" ⇒ *"large shoe size"*

First Benchmark
In a first benchmark, the items that do not belong to any class and which may be considered as noisy data make up 5% of all data. We have studied the GMP-pertinences for the 4 possible rules associating one fuzzy modality of the first attribute (*size*) to one modality of the second attribute (*shoe size*). Figure 4 describes the data set, in which points outside the ellipses represents noisy data.

A) For the rule *"size=small"* ⇒ *"shoe size=small"*

The GMP-pertinences of the fuzzy operators calculated with various t-norms are then rather close: 0.866 for bold intersection, 0.850 for probabilistic intersection and 0.844 for Zadeh's intersection. This rule is interesting because such levels of confidence indicate that about 85% of the persons of small size have a small shoe size. Table 7.2 shows the GMP-pertinence for sets of operators combining each considered implication with every t-norm. Results are rather close whichever t-norm or implication is used.

B) For the rule *"size=tall"* ⇒ *"shoe size=small"*
Confidence levels calculated now with different t-norms are 0.148 for bold intersection, 0.164 for probabilistic intersection and 0.170 for Zadeh's intersection. This rule is not interesting since it is verified for only about 16% of tall persons.

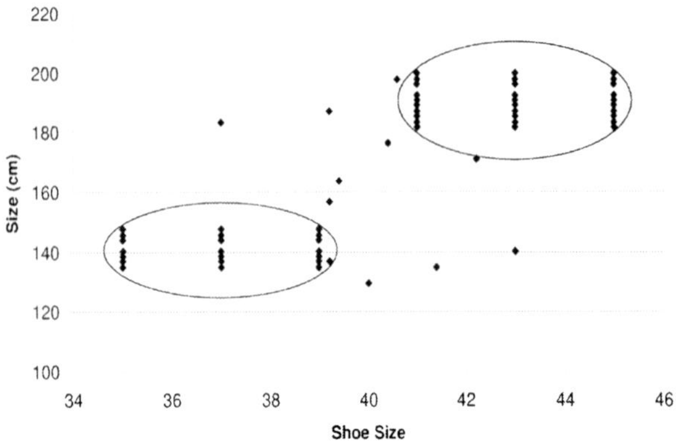

Fig. 4. Data distribution in the first benchmark

Implication	bold conjunction	Zadeh's conjunction	probabilistic conjunction
Gödel-Brouwer	0,895923	**0,900154**	0,897859
Goguen	0,899513	0,899546	**0,900154**
Kleene-Dienes	0,887846	0,897010	0,890855
Lukasiewicz	**0,900154**	0,896241	0,897908

Table 2. GMP-pertinence of the fuzzy operators for the rule *"size=small"* ⇒ *"shoe size=small"*

C) For the rule *"size=tall"* ⇒ *"shoe size=large"*

Confidence levels with different t-norms are 0.780 for bold intersection, 0.760 for probabilistic intersection and 0.754 for Zadeh's. This rule is interesting with a conditional probability about 76.5%. Table 7.2 below shows rather significant differences (1 to 5%) on the quality of the operators for each t-norm. With bold t-norm, three implications, Lukasiewicz's, Goguen's and Gödel-Brouwer's give good results. With the probabilistic t-norm, implications of Goguen, Gödel-Brouwer and Lukasiewicz give good and rather close results, since their quality is above 92%.

Implication	bold conjunction	Zadeh's conjunction	probabilistic conjunction
Gödel-Brouwer	0,921295	**0,926686**	0,923805
Goguen	0,92604	0,926031	**0,926686**
Kleene-Dienes	0,876628	0,919381	0,889767
Lukasiewicz	**0,926686**	0,918095	0,921885

Table 3. GMP-pertinence of the fuzzy operators for the rule *"size=tall"* ⇒ *"shoe size=large"*

Second Benchmark
In a second benchmark (Fig. 5) we have increased the rate of noisy data, which accounts now for 37% of all data.

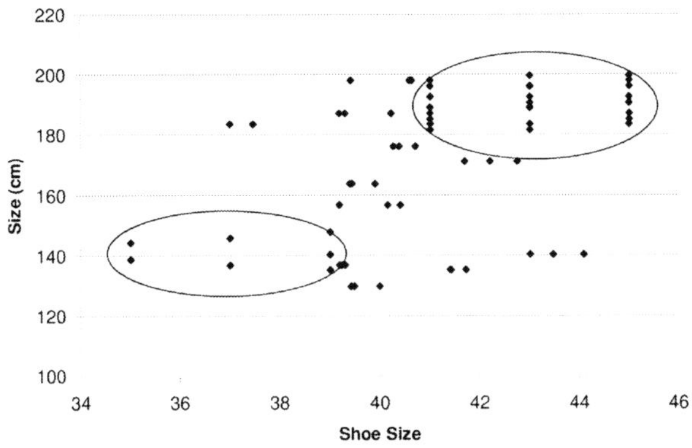

Fig. 5. Data distribution in the 2nd benchmark

A) For the rule *"size=small"* ⇒ *"shoe size=small"*

Confidence levels calculated with various t-norms are rather near: 0.523 for bold intersection, 0.510 for probabilistic intersection and 0.516 for Zadeh's intersection. This sample is critical with a confidence about 52%, which justifies its interest although results of this rule may be erroneous. This interest is confirmed by a support near 28%.

Table 7.2 shows that qualities of operators are rather low (less than 85%). Again, some implications may be grouped accordingly with the t-norms, particularly the implications of Gödel-Brouwer and Goguen on the one hand, and those of Lukasiewicz and Goguen on the other hand.

Implication	bold conjunction	Zadeh's conjunction	probabilistic conjunction
Gödel-Brouwer	0,800913	**0,848043**	0,825563
Goguen	0,834291	0,840908	**0,848043**
Kleene-Dienes	0,766130	0,797044	0,778592
Lukasiewicz	**0,848043**	0,781826	0,808287

Table 4. GMP-pertinence of the fuzzy operators for the rule *"size=small"* ⇒ *"shoe size=small"*

B) For the rule *"size=tall"* ⇒ *"shoe size=large"*

Confidence levels calculated with different t-norms are rather close: 0.726 for bold intersection, 0.700 for probabilistic intersection and 0.688 for Zadeh's. So, this rule is interesting with a conditional probability about 70%.

Implication	bold conjunction	Zadeh's conjunction	probabilistic conjunction
Gödel-Brouwer	0,915288	**0,936121**	0,925429
Goguen	0,930192	0,930027	**0,936121**
Kleene-Dienes	0,872485	0,906273	0,882153
Lukasiewicz	**0,936121**	0,897394	0,91496

Table 5. GMP-pertinence of the fuzzy operators for the rule *"size=tall"* ⇒ *"shoe size=large"*

Table 7.2 shows rather significant differences (1 to 5%) on the evaluation of the operators for each t-norm. One can remark that with bold t-norm, three implications, Lukasiewicz's, Goguen's and Gödel-Brouwer's, give good results. With the probabilistic t-norm, implications of Goguen, Gödel-Brouwer and Lukasiewicz give good and rather close results: their GMP-pertinence is above 92%.

Results of Table 7.2 are similar to those of 7.2, with a better GMP-pertinence. One can again associate the same operators as with table 3, but with much higher scores, which confirms these associations.

7.3 Results on real databases

We have tried out our algorithms on several databases found in the UCI repository [19], in particular "Wisconsin Breast Cancer Database", which consists of 699 items with 10 attributes and two classes, "Wine Recognition Database" with 178 items, 13 attributes and three classes and "Ionosphere Database" with 351 items, 39 attributes and 2 classes. The results are similar to those highlighted by our previous example, but the differences on GMP-pertinences of operators are less strong than in our second benchmark, for which the proportion of noisy data had been deliberately strengthened. Let us consider a few examples extracted from our results.

For "Wisconsin Breast Cancer Database", with 3 fuzzy partitions on each attribute, a minimal confidence of 0.8, a minimal intensity of implication of 0.9, a support of 5% and at most 3 propositions in the premises, we get 331 rules; if one pushes the search to 6 propositions, we obtain 814 rules. Going to 9 propositions brings few supplementary rules with a total of 870.

Considering the evolution of the rule numbers according to the maximum number of premises and the number of classes (Table 7.3), one remarks that the number of rules decreases when the number of classes increases, until 8 classes.

	at most 3 premises	at most 6 premises	at most 9 premises
2 classes	407 rules	908 rules	954 rules
3 classes	331 rules	814 rules	870 rules
4 classes	250 rules	632 rules	678 rules
5 classes	267 rules	464 rules	466 rules
6 classes	253 rules	471 rules	474 rules
7 classes	255 rules	432 rules	434 rules
8 classes	242 rules	410 rules	412 rules
9 classes	395 rules	672 rules	674 rules

Table 6. Evolution of the number of rules with the number of classes for "Wisconsin Breast Cancer Database"

The profusion of rules with small numbers of classes is offset by the imprecision of the rules: the average confidence of rules with 2 classes is much weaker than that obtained with more classes. With 2 classes and at most 6 premises, only 35 rules (7%) have a confidence of 1, while with 9 classes and 6 premises, 388 rules (57%) have the same confidence of 1. Increasing the number of premises beyond 6 brings little improvement because addition of attributes to rules only specializes the rules with a confidence under 1.

For example, the rule *If Clump Thickness is "very small" then Class is "benign"*, appears with a confidence of 0.964, a support of 28% and a GMP-pertinence of 91.1%, 90.3% or 89.6% depending on the fuzzy operators. Specialization of this rule by adding supplementary attributes increases confidence by reducing the support, until it reaches a confidence of 1; the rule is then: *If Clump Thickness is "very small" and Single Epithelial Cell Size is "very small" and Bare Nucleii="very little" then Class is "benign"*. Its support is then 26% with a GMP-pertinence of 1.

Results on other databases are similar, but, due to the highest number of attributes, the numbers of generated rules are much larger. Using the same thresholds and 3 classes by attribute, one obtains for "Wine Data Base" 1092 rules of at most 3 premises and 13470 of at most 6 premises. With "Ionosphere Data Base" one extracts 13824 rules with at most 3 premises. A more severe choice of thresholds is then needed to reduce these high numbers of rules.

Let us consider another example, from the results on "Wine Data Base". The rule *If Magnesium is "very little" then Wine is "type2"* appears with a confidence of 0.894, and when specializing it by adding one attribute, we get 3 new rules with a confidence of 1, such as the rule *If Magnesium is "very little" and Flavanoids is "little" then Wine is "type2"*. Adding one more attribute gives 9 more rules with a full confidence of 1. Comparisons between qualities of operators using GMP-pertinences confirm again our conclusions on the choice of operators, the association between Lukasiewicz' implication and Lukasiewicz' bounded sum and difference appearing slightly better.

8 Synthesis of these studies

We have remarked on all examples that, whatever the rule, for a given t-norm, the same implications always have the best GMP-pertinence in the evaluation of the GMP. For the bold t-norm, one can group Goguen's implication and Lukasiewicz's, these two implications being the best. For Zadeh's t-norm, implications of Gödel-Brouwer and Goguen are the best. For probabilistic t-norm, one can group implications of Gödel-Brouwer and Goguen, as the best ones. To summarize these results, the sets of operators that appear then as the best with the GMP associate:

- Lukasiewicz's t-norm and Lukasiewicz's implication,
  $\top(a, b) = \max(0, a + b - 1)$
  $\mathrm{I}(a, b) = \min(1, 1 - a + b)$.
- Gödel's t-norm (Zadeh's minimum) and Gödel-Brouwer's implication,
  $\top(a, b) = \min(a, b)$
  $\mathrm{I}(a, b) = 1 \quad \text{if } a \leq b \quad \text{or} \quad b \quad \text{if } a > b$.
- Probabilistic t-norm and Goguen's implication.
  $\top(a, b) = a \times b$
  $\mathrm{I}(a, b) = 1 \quad \text{if } a \leq b \quad \text{or} \quad b/a \quad \text{if } a > b$.

Thus, the sets of operators that appear experimentally the best to use with the GMP, are those that associate a t-norm $\top$ with the R-implication I that it defines:

$$\mathrm{I}(a, b) = \sup\{x \in [0, 1] \mid \top(a, x) \leq b\} \tag{25}$$

This result is justified by theory, in particular by [16], which shows that the best implication to apply with the GMP for a given t-norm is the residue of this t-norm (the definition of an R-implication is indeed the residue of the associated t-norm). The proof lies on the facts that for a given truth value a of the premises, a given truth value function $\mathrm{I}(a, b)$ of the implication between a and the conclusions b, the GMP should be non decreasing between its arguments, since the truer the antecedent and the truer the implication, the truer the consequent should be. Moreover, since the neutral element of the GMP should be 0, and its unit 1, the GMP should be realized with a t-norm. In order to have the most powerful rules, one has to choose $\mathrm{I}(a, b)$ as large as possible, giving then definition (25).

9 Operators for Fuzzy Rule Reduction

When rules are not extracted to build knowledge based systems, but to give human experts a synthetic view of the database, the number of rules extracted is often too high to be studied, and methods to reduce sets of fuzzy rules are welcome. For this kind of applications of knowledge discovery, we have studied methods allowing aggregation of rules.

9.1 A First Scheme of Rule Reduction

In classical logics, one may write $(a \Rightarrow c) \wedge (b \Rightarrow c) \vdash (a \vee b) \Rightarrow c$, and if one wants to process similarly with fuzzy rules without having to reevaluate the inferred fuzzy rule $(a \vee b) \Rightarrow c$, it should be rigorously correct to write

$$\forall (x,y) \in [0,1] \times [0,1]: \; \mu_{(a \Rightarrow c) \wedge (b \Rightarrow c)}(x,y) \;=\; \mu_{((a \vee b) \Rightarrow c}(x,y),$$

or, with a t-norm $\top$ and its complementary t-co-norm $\perp$:

$$\forall (x,y) \in [0,1] \times [0,1]: \; \mu_{a \Rightarrow c}(x,y) \top \mu_{b \Rightarrow c}(x,y) \;=\; \mu_{(a \vee b) \Rightarrow c}(x,y)$$

If we note $\alpha = \mu_a(x,y)$, $\beta = \mu_b(x,y)$, $\gamma = \mu_c(x,y)$ and $\mathrm{I}(\alpha, \beta) = \mu_{a \Rightarrow c}(x,y)$, we had to find which, if any, of the operator sets allows to write the condensed form:

$$\mathrm{I}(\alpha, \gamma) \top \mathrm{I}(\beta, \gamma) \equiv \mathrm{I}(\alpha \perp \beta, \gamma) \quad \text{or}$$
$$\top(\mathrm{I}(\alpha, \gamma), \mathrm{I}(\beta, \gamma)) \equiv \mathrm{I}(\perp(\alpha, \beta), \gamma)) \tag{26}$$

We have then considered four fuzzy implications:
- Kleene-Dienes' implication: $\mathrm{I}_{KD}(a,b) = \max(1-a, b)$;
- Gödel-Brouwer's implication: $\mathrm{I}_{GB}(a,b) = 1$ if $a \leq b$ or b if $a > b$;
- Goguen's implication: $\mathrm{I}_{Gog}(a,b) = 1$ if $a \leq b$ or b/a if $a > b$;
- Lukasiewicz's implication: $\mathrm{I}_L(a,b) = \min(1, 1-a+b)$.

with 3 dual pairs of t-norm and t-conorm:
- Zadeh's minimum and maximum:
  $\top_Z(a,b) = \min(a,b)$, $\perp_Z(a,b) = \max(a,b)$,
- the probabilistic product and sum:
  $\top_{\mathrm{Pr}}(a,b) = a \times b$, $\perp_{\mathrm{Pr}}(a,b) = a + b - a \times b$,
- Lukasiewicz's difference and sum:
  $\top_L(a,b) = \max(0, a+b-1)$, $\perp_L(a,b) = \min(1, a+b)$.

We have then proved that with Zadeh's t-norm and t-conorm, all the implications we considered verify relation (26), but that none of these implications combined with the probabilistic t-norm/t-conorm or with Lukasiewicz's bounded sum and difference verify it.

Proof
A) With Zadeh's $\top_Z(a,b) = \min(a,b)$ and $\perp_Z(a,b) = \max(a,b)$, one may take into account the fact that fuzzy implications are monotonously decreasing with their first argument: the truth-value of any fuzzy implications must increase as the truth of its antecedent decreases. So, for any fuzzy implication I

- when $\alpha \leq \beta$

$$\mathrm{I}(\alpha,\gamma) \geq \mathrm{I}(\beta,\gamma), \text{ so } \top_Z(\mathrm{I}(\alpha,\gamma), \mathrm{I}(\beta,\gamma)) = \min(\mathrm{I}(\alpha,\gamma), (\beta,\gamma)) = \mathrm{I}(\beta,\gamma),$$
$$\text{while } \mathrm{I}(\perp_Z(\alpha,\beta),\gamma) = \mathrm{I}(\max(\alpha,\beta),\gamma) = \mathrm{I}(\beta,\gamma).$$

Therefore, in this case, $\top_Z(\mathrm{I}(\alpha,\gamma), \mathrm{I}(\beta,\gamma)) = \mathrm{I}(\beta,\gamma) = \mathrm{I}(\perp_Z(\alpha,\beta),\gamma)$, and by symmetry on α and β, this result is always true:

$$\top_Z(\mathrm{I}(\alpha,\gamma), \mathrm{I}(\beta,\gamma)) \equiv \mathrm{I}(\perp_Z(\alpha,\beta),\gamma).$$

B) With the probabilistic t-norm and t-conorm, let us consider counter-examples:

- for Kleene-Dienes' implication, with $\alpha = 0.5$, $\beta = 0.5$, $\gamma = 0.6$,
  $\top_{\text{Pr}}(I_{KD}(\alpha,\gamma), \text{I}_{KD}(\beta,\gamma)) = 0.3$, but $\text{I}_{KD}(\bot_{\text{Pr}}(\alpha,\beta),\gamma) = 0.6$;
- for Lukasiewicz's implication, with $\alpha = 0.5$, $\beta = 0.7$, $\gamma = 0.5$,
  $\top_{\text{Pr}}(\text{I}_L(\alpha,\gamma), \text{I}_L(\beta,\gamma)) = 0.8$, but $\text{I}_L(\bot_{\text{Pr}}(\alpha,\beta),\gamma) = 0.65$;
- for Gödel-Brouwer's implication with $\alpha = 0.5$, $\beta = 0.5$, $\gamma = 0.6$,
  $\top_{\text{Pr}}(\text{I}_{GB}(\alpha,\gamma), \text{I}_{GB}(\beta,\gamma)) = 1$, but $\text{I}_{GB}(\bot_{\text{Pr}}(\alpha,\beta),\gamma) = 0.6$;
- for Goguen's implication with α= 0.5, β=0.5, γ= 0.6
  $\top_{\text{Pr}}(\text{I}_{Gog}(\alpha,\gamma), \text{I}_{Gog}(\beta,\gamma)) = 1$, but $\text{I}_{Gog}(\bot_{\text{Pr}}(\alpha,\beta),\gamma) = 0.8$.

C) With Lukasiewicz's difference and sum, let us consider counter-examples:

- for Kleene-Dienes' implication, with $\alpha = 0.5$, $\beta = 0.5$, $\gamma = 0.6$,
  $\top_L(\text{I}_{KD}(\alpha,\gamma), \text{I}_{KD}(\beta,\gamma)) = 0.1$, but $\text{I}_{KD}(\bot_L(\alpha,\beta),\gamma) = 0.6$;
- for Lukasiewicz's implication, with $\alpha = 0.5$, β =0.7, $\gamma = 0.5$,
  $\top_L(\text{I}_L(\alpha,\gamma), \text{I}_L(\beta,\gamma)) = 0.8$, but $\text{I}_L(\bot_L(\alpha,\beta),\gamma) = 0.5$;
- for Gödel-Brouwer's implication and $\alpha = 0.5$, $\beta = 0.5$, $\gamma = 0.6$,
  $\top_L(\text{I}_{GB}(\alpha,\gamma), \text{I}_{GB}(\beta,\gamma)) = 1$, but $\text{I}_{GB}(\bot_L(\alpha,\beta),\gamma) = 0.6$;
- for Goguen's implication and $\alpha = 0.5$, $\beta = 0.5$, $\gamma = 0.6$,
  $\top_L(\text{I}_{Gog}(\alpha,\gamma), \text{I}_{Gog}(\beta,\gamma)) = 1$, but $\text{I}_{Gog}(\bot_L(\alpha,\beta),\gamma) = 0.6$.

So, whatever the implication, among the considered pairs of dual t-norms and t-conorms, only Zadeh's minimum and maximum verifies relation (26), which allows easy fuzzy rule reduction.

9.2 A Second Scheme of Rule Reduction

Similarly, to keep the classical reduction scheme $(a \Rightarrow b) \wedge (a \Rightarrow c) \vdash a \Rightarrow (b \wedge c)$ or, with the same condensed notations as above

$$\text{I}(\alpha,\beta)\top\text{I}(\alpha,\gamma)) \equiv \text{I}(\alpha,\beta\top\gamma) \quad \text{or}$$
$$\top(\text{I}(\alpha,\beta),\text{I}(\alpha,\gamma)) \equiv \text{I}(\alpha,\top(\beta,\gamma)), \tag{27}$$

we have proved that with Zadeh's t-norm and t-conorm, all the considered implications verify relation (27), but that none of these implications combined with probabilistic t-norm or with Lukasiewicz's t-norm verify it.

Proof

A) With Zadeh's $\top_Z(a,b) = \min(a,b)$ and $\bot_Z(a,b) = \max(a,b)$, one may take into account the fact that fuzzy implications are monotonously increasing with their second argument: the truth-value of any fuzzy implication must increase when the truth of its conclusion increases. So, for any fuzzy implication I

- when $\beta \leq \gamma$
  $\text{I}(\alpha,\beta) \leq \text{I}(\alpha,\gamma)$, so $\top_Z(\text{I}(\alpha\beta),(\alpha,\gamma)) = \min(\text{I}(\alpha,\beta),\text{I}(\alpha,\gamma)) = \text{I}(\alpha,\beta)$,
  while $\text{I}(\alpha,\top_Z(\beta,\gamma)) = \text{I}(\alpha,\min(\beta,\gamma)) = \text{I}(\alpha,\beta)$.

Therefore, in this case, $\top_Z(\mathrm{I}(\alpha,\beta),\mathrm{I}(\alpha,\gamma)) = \mathrm{I}(\alpha,\beta) = \mathrm{I}(\alpha,\top_Z(\beta,\gamma))$, and by symmetry on β and γ, this result is always true:

$$\top_Z(\mathrm{I}(\alpha,\beta),\mathrm{I}(\alpha,\gamma)) \equiv \mathrm{I}(\alpha,\top_Z(\beta,\gamma)).$$

B) With probabilistic t-norm and t-conorm, let us consider counter-examples:

- for Kleene-Dienes' implication and $\alpha = 0.4$, $\beta = 0.8$, $\gamma = 0.5$, $\top_{\mathrm{Pr}}(\mathrm{I}_{KD}(\alpha,b),\mathrm{I}_{KD}(\alpha,\gamma)) = 0.48$, but $\mathrm{I}_{KD}(\alpha,\top_{\mathrm{Pr}}(\beta,\gamma)) = 0.6$;
- for Lukasiewicz's implication and $\alpha = 0.6$, $\beta = 0.3$, $\gamma = 0.5$, $\top_{\mathrm{Pr}}(\mathrm{I}_L(\alpha,b),\mathrm{I}_L(\alpha,\gamma)) = 0.63$, but $\mathrm{I}_L(\alpha,\top_{\mathrm{Pr}}(\beta,\gamma) = 0.4$;
- for Gödel-Brouwer's implication and $\alpha = 0.6$, $\beta = 0.7$, $\gamma = 0.8$, $\top_{\mathrm{Pr}}(\mathrm{I}_{GB}(\alpha,\beta),\mathrm{I}_{GB}(\alpha,\gamma)) = 1$, but $\mathrm{I}_{GB}(\alpha,\top_Z(\beta,\gamma) = 0.56$;
- for Goguen's implication and $\alpha = 0.6$, $\beta = 0.6$, $\gamma = 0.6$, $\top_{\mathrm{Pr}}(\mathrm{I}_{Gog}(\alpha,\beta),\mathrm{I}_{Gog}(\alpha,\gamma) = 1$, but $\mathrm{I}_{Gog}(\alpha,\top_{\mathrm{Pr}}(\beta,\gamma) = 0.6$.

C) With Lukasiewicz's t-norm and t-conorm, let us consider counter-examples:

- for Kleene-Dienes' implication and $\alpha = 0.4$, $\beta = 0.6$, $\gamma = 0.7$, $\top_L(\mathrm{I}_{KD}(\alpha,\beta),\mathrm{I}_{KD}(\alpha,\gamma) = 0.3$, but $\mathrm{I}_{KD}(\alpha,\top_L(\beta,\gamma)) = 0.6$;
- for Lukasiewicz's implication and $\alpha = 0.6$, $\beta = 0.3$, $\gamma = 0.6$, $\top_L(\mathrm{I}_L(\alpha,\beta),\mathrm{I}_L(\alpha,\gamma)) = 0.7$, but $\mathrm{I}_L(\alpha,\top_L(\beta,\gamma) = 0.5$;
- for Gödel-Brouwer's implication and $\alpha = 0.5$, $\beta = 0.6$, $\gamma = 0.6$, $\top_L(\mathrm{I}_{GB}(\alpha,\beta),\mathrm{I}_{GB}(\alpha,\gamma)) = 1$, but $\mathrm{I}_{GB}(\alpha,\top_L(\beta,\gamma) = 0.2$;
- for Goguen's implication and $\alpha = 0.5$, $\beta = 0.6$, $\gamma = 0.6$: $\top_L(\mathrm{I}_{Gog}(\alpha\beta),\mathrm{I}_{Gog}(\alpha\gamma)) = 1$, but $\mathrm{I}_{Gog}(\alpha,\top_L(\beta,\gamma) = 0.4$.

So, whatever the implication, among the considered t-norms only Zadeh's minimum verifies relation (27), which allows easy fuzzy rule reduction.

One may remark here that among the sets of fuzzy operators that appear as the best with the GMP, we have found Gödel's t-norm (Zadeh's minimum) and Gödel-Brouwer's implication. Therefore, this set of fuzzy operators may be considered as the most interesting when one wants to extract rules for a knowledge based system and also to reduce the extracted rules within a same application. These results also illustrate that fuzzy rules cannot generally be treated as classical rules.

10 Conclusion

We have described a generalization of statistical implication indexes to fuzzy knowledge discovery. The first operation needed to compute these indexes is the choice of fuzzy partitions to convert numerical or symbolic attributes into fuzzy ones. We have justified our choice of three classical statistic indexes: the support, the confidence and the less common, but powerful, intensity of implication. We have then explained how we have adapted these indexes to fuzzy attributes by replacing cardinalities of crisp sets by cardinalities of fuzzy sets, then we have described one algorithm to explore the set of possible rules.

The indexes we use to extract fuzzy rules highlight statistical links between conjunctions of fuzzy attributes and fuzzy conclusions, but they do not evaluate the associated fuzzy rules, which depend on the chosen fuzzy operators. Since fuzzy operators are numerous, we have evaluated sets of standard operators by applying the generalized modus ponens (GMP) on the database items and by comparing its results to the effective conclusions. After a simplified example illustrating these mechanisms, we have given a summary of our results on more consequent databases.

We have then observed that the sets of fuzzy operators which give the best results with the GMP, associate a t-norm and the related R-implication:

- Lukasiewicz's t-norm (bold t-norm) and Lukasiewicz's implication,
- Gödel's t-norm (Zadeh's minimum) and Brouwer-Gödel's implication,
- probabilistic t-norm and Goguen's implication.

To allow an easy reduction of the number of rules proposed to human experts, we have studied methods to cluster fuzzy rules. We have then shown that among the considered sets of operators, using Zadeh's minimum as t-norm and maximum as t-conorm is the best choice, independently of the implication. Since Gödel-Brouwer's implication gives the best results with the GMP when one uses Zadeh's minimum and maximum, this implication seems a rather good compromise. Associating Lukasiewicz's implication with Lukasiewicz's t-norm and t-conorm shows better GMP-pertinence, but does not allow classical schemes of rule reduction.

Finally, we must remark that the increase of computing complexity induced by the use of fuzzy logics is relatively small for rule extraction, since instead of increasing by one (integer number) the counters of good and bad examples, fuzzy logics add membership degrees (real numbers). The operations of fuzzification, the choice of fuzzy operators and the reductions of rules are more complex, but the advantages of using fuzzy logics may compensate for this, because intervals on continuous attributes are expressed by more expressive fuzzy labels and since abrupt threshold are avoided.

References

1. R. Agrawal, T. Imielinski, and A. N. Swami. Mining association rules between sets of items in large databases. In Peter Buneman and Sushil Jajodia, editors, *Proceedings of the 1993 ACM SIGMOD International Conference on Management of Data*, pages 207–216, Washington, D.C., 1993.
2. J. Aguilar-Martin and R. Lopez De Mantaras. The process of classification and learning the meaning of linguistic descriptors of concepts. In M. M. Gupta and E. Sanchez, editors, *Approximate reasoning in decision analysis*, pages 165–175. North Holland, 1982.
3. M. Bernadet. Basis of a fuzzy knowledge discovery system. In *Conf. PKDD'2000 - LNAI 1910*, pages 24–33. Springer-Verlag, 2000.

4. M. Bernadet. A comparison of operators in fuzzy knowledge discovery. In *Conf. IPMU'2004*, volume 2, pages 731–738, July 2004.
5. M. Bernadet. A study of data partitioning methods for fuzzy knowledge discovery. In *Conf. IPMU'2006*, pages 1396–1402, July 2006.
6. M. Bernadet, G. Rose, and H. Briand. Fiable and fuzzy fiable: two learning mechanisms based on a probabilistic evaluation of implications. In *Conf. IPMU'96 (Information Processing and Management of Uncertainty in Knowledge-Based Systems)*, volume 2, pages 911–916, July 1996.
7. J. C. Bezdek and J. D. Harrisand. Fuzzy partitions and relations: An axiomatic basis for clustering. *Fuzzy Sets and Systems*, 1:111–127, 1978.
8. A. Bonarini. Evolutionary learning of fuzzy rules: competition and cooperation. In *Fuzzy Modeling: Paradigms and Practice*, pages 265–284. Kluwer Academic Press, 1996.
9. H. Briand, L. Fleury, R. Gras, Y. Masson Y. J., and Philippe. A statistical measure of rules strength for machine learning. In *WOCFAI 1995*, pages 51–62, 1995.
10. J. J. Buckley and K. Hayashi. Neural networks for fuzzy systems. *Fuzzy Sets and Systems*, pages 265–276, 1995.
11. M. Delgado, D. Sánchez, and M. A. Vila. Fuzzy cardinality based evaluation of quantified sentences. *Int. Journal of Approximate Reasoning*, 23:23–66, 2000.
12. L. Fleury and Y. Masson. Intensity of implication: a measurement in machine learning. In *IEA/AIE'95*, pages 621–629, June 1995.
13. W. J. Frawley, G. Piatetsky-Shapiro, and C. J. Matheus. Knowledge discovery in databases: An overview. *AI Magazine*, 13:57–70, 1992.
14. R. Gras, R. Couturier, F. Guillet, and F. Spagnolo. Extraction de règles en incertain par la méthode implicative, extraction des connaissances : Etat et perspectives. In *RNTI-E-5*, pages 385–389. Cepaduès Editions, 2006.
15. R. Gras and A. Larher. L'implication statistique, une nouvelle méthode d'analyse de données. *Mathématiques, Informatique et Sciences Humaines*, 120:5–31, 1992.
16. P. Hajek. *Metamathematics of Fuzzy Logic.* Kluwer Academic, 1998.
17. S. K. Halgamuge and M. Glesner. Neural networks in designing fuzzy systems for real world applications. *Fuzzy Sets and Systems*, 65:1–12, 1994.
18. A. Heinz. Adaptive fuzzy neural trees. In *IDA-95 Symposium*, pages 70–74, August 1995.
19. S. Hettich and S. D. Bay. *The UCI KDD Archive.* Univ. of California, Irvine, Department of Information and Computer Science, 1999. [http://kdd.ics.uci.edu].
20. E. E. Kerre. A comparative study of the behavior of some popular fuzzy implications on the generalized modus ponens. In *Fuzzy Logic for the management of uncertainty*, pages 281–295. John Wiley & Sons, Inc., 1992.
21. G. J. Klir and Bo Yuan. *Fuzzy Sets and Fuzzy Logics - Theory and Applications.* PrenticeHall. Englewood Cliffs, 1995.
22. L. Lesmo, L. Saitta, and P. Torasso. Fuzzy production rules: a learning methodology. In *Advances in Fuzzy Sets, Possibility Theory and Applications*, pages 181–198. Plenum Press, 1983.
23. T. Murata, H. Ishibuchi, H. Nakashima, and M. Gen. Fuzzy partition and input selection by genetic algorithms for designing fuzzy rule-based classification systems. In *LNCS 1447*, pages 407–416. Springer-Verlag, 1998.
24. A. Ralescu. Cardinality, quantifiers and the aggregation of fuzzy criteria. *Fuzzy Sets and Systems*, 69:355–365, 1995.

25. J. Rives. FID3: Fuzzy induction decision tree. In *ISUMA'90*, pages 457–462, December 1990.
26. J. A. Roubos, M. Setnes, and J. Abonyi. Learning fuzzy classification rules from data. In *Developments in Soft Computing*, pages 108–115. Springer-Verlag, 2001.
27. F. Spagnolo and R. Gras. A new approach in Zadeh classification: fuzzy implication through statistic implication. In *NAFIPS-IEEE 3rd Conference of the North American Fuzzy Information Processing Society*, pages 425–429, June 2004.
28. R. Weber. Fuzzy-ID3: a class of methods for automatic knowledge acquisition. In *2nd International Conference on Fuzzy Logic and Neural Networks*, pages 265–268, July 1992.
29. M. Wygralak. Questions of cardinality of finite fuzzy sets. *Fuzzy Sets and Systems*, 102:185–210, 1999.
30. L. A. Zadeh. Probability measures of fuzzy events. *Journal of Mathematical Analysis and Applications*, 23:421–427, 1968.
31. L. A. Zadeh. Fuzzy logic and its application to approximate reasoning. *Information Processing*, 74:591–594, 1974.
32. J. Zeidler and M. Schlosser. Continuous valued attributes in fuzzy decision trees. In *Conf. IPMU'96 (Information Processing and Management of Uncertainty in Knowledge-Based Systems)*, pages 395–400, 1996.
33. D. A. Zighed, S. Rabaseda, R. Rakotomalala, and F. Feschet. Discretization methods in supervised learning. In *Encyclopedia of Computer Science and Technology*, volume 40, pages 35–50. Marcel Dekker, 1999.

About the editors

Régis Gras is an Emeritus professor at Polytech'Nantes (Polytechnic graduate School of Nantes University, France) and he is a member of the "KnOwledge and Decision" team (KOD) in the Nantes-Atlantic Laboratory of Computer Sciences (LINA, CNRS UMR 6241), since 1998. He received a PhD degree of mathematics in 1979 from the University of Rennes, France. He used to be the Chair of the French International Commission on Teaching of Mathematics (1995–1998), then a member of the Committee on Teaching of Mathematics of the European Mathematics Society (1997–2003). He has designed a set of methods gathered in his "Statistical Implicative Analysis" original approach. Since, he continues to develop and extend this approach to data mining issues. He served as PC Chairs of the 4 editions of SIA conference (France 2001, Sao Paulo Brazil 2003, Palermo Italy 2005, and Castellon Spain 2007). He is the author of 4 books, and a co-editor of 3 books of chapters.

Einoshin Suzuki received his Bachelor, Master, and Doctor of Engineering Degrees all from the University of Tokyo in 1988, 1990, and 1993, respectively. He joined Tokyo Institute of Technology as an assistant professor (1993–1996) then Yokohama National University as a lecturer professor (1996–1997) and was promoted to an associate professor (1997–2006). Since April 2006, he is with Kyushu University as a full professor. He has obtained the Best Paper Award of the Japanese Society for Artificial Intelligence twice. He has served as the Honorary Chair of EGC-07, the PC Chair of DS-04, and the PC Vice Chair of ICDM-04 and is serving as the Steering Committee Chair of the International Conference on Discovery Science and a PC Co-Chair of PAKDD-08.

Fabrice Guillet is an associate professor in Computer Science at Polytech'Nantes, and he is a member of the "KnOwledge and Decision" team (KOD) in the Nantes-Atlantic Laboratory of Computer Sciences (LINA, CNRS UMR 6241) since 1997. He received a PhD degree in Computer Sciences in 1995 from the Ecole Nationale Superieure des Telecommunications de Bretagne. He is a founder of the "Knowledge Extraction and Management"

French-speaking association of research (EGC[1]), and he is also involved in the steering committee of the annual EGC French-speaking conference since 2001. His research interests include knowledge quality and knowledge visualization in the frameworks of Data Mining and Knowledge Management. He has recently co-edited with H. Hamilton a refereed book of chapters entitled "Quality Measures in Data Mining" published by Springer in 2007.

Filippo Spagnolo received his Bachelor degree in 1972 (Università di Palermo, Italy), and PhD (Didactics of Mathematics) in 1995 (University of Bordeaux, France). Researcher in Groups of research in University of Palermo in "Matematiche Complementari" (Mathematics Education, History of Mathematics and Fundamenta of Mathematics), 1979–2001. Associate Professor in "Matematiche Complementari", since 2004. Editorial Board of many international reviews: Mediterranean Journal for Research in Mathematics Education, Canadian Journal of Science Mathematics and Technology Education, Acta Didactica Universitatis Comenianae (Mathematics). Editorial in Chief of review "Quaderni di Ricerca in Didattica (Mathematics)", G.R.I.M. since 1990, Palermo, Italy. Co-ordinator of PhD "Storia e Didattica della Matematica, Storia e Didattica della Fisica, Storia e Didattica della Chimica", University of Palermo with a consortium of 4 Universities of Italy and 4 Universities in Europe.

About the manuscript coordinator

Bruno PINAUD received his engineer diploma in computer science from Polytech'Nantes (Polytechnic graduate School of Nantes University, France) in 2001 and his PhD in 2006. He is currently an adjunct assistant Professor at Polytech'Nantes and an associate member of the "KnOwledge and Decision" team (KOD) in the Nantes-Atlantic Laboratory of Computer Sciences (LINA CNRS 2729). His current main research activities are about graph drawing with metaheuristics and some applications in data-mining and knowledge visualization.

[1] http://www.polytech.univ-nantes.fr/associationEGC

Index

Printed in the United States
125098LV00002B/217-243/P

9 783540 789826